CARBON FIBERS

CARBON FIBERS
THIRD EDITION, REVISED AND EXPANDED

EDITED BY
JEAN-BAPTISTE DONNET
TONG KUAN WANG
JIMMY C. M. PENG
*Ecole Nationale Supérierure de Chimie
and Université de Haute-Alsace
Mulhouse, France*

SERGE REBOUILLAT
*DuPont de Nemours International S. A.
Geneva, Switzerland*

Marcel Dekker, Inc.   New York • Basel • Hong Kong

Library of Congress Cataloging-in-Publication Data

Carbon fibers/edited by Jean-Baptiste Donnet ... [et al.]. —3rd ed., rev. and expanded.
  p. cm.
  Includes bibliographical references and index.
  ISBN 0-8247-0172-0
  1. Carbon fibers. II. Donnet, Jean-Baptiste
  TA418.9.F5C36   1998
  620.1'97—dc21   98–9327
                                                                  CIP

This book is printed on acid-free paper.

Headquarters
Marcel Dekker, Inc.
270 Madison Avenue, New York, NY 10016
tel: 212-696-9000; fax: 212-685-4540

Eastern Hemisphere Distribution
Marcel Dekker AG
Hutgasse 4, Postfach 812, CH-4001 Basel, Switzerland
tel: 44-61-261-8482; fax: 44-61-261-8896

World Wide Web
http://www.dekker.com

The publisher offers discounts on this book when ordered in bulk quantities. For more information, write to Special Sales/Professional Marketing at the headquarters address above.

Copyright © 1998 by Marcel Dekker, Inc. All Rights Reserved.

Neither this book nor any part may be reproduced or transmitted in any form or by any means, electronic or mechanical, including photocopying, microfilming, and recording, or by any information storage and retrieval system, without permission in writing from the publisher.

Current printing (last digit):
10 9 8 7 6 5 4 3 2 1

PRINTED IN THE UNITED STATES OF AMERICA

PREFACE

Carbon fiber, commercially available since the 1960s, has attracted worldwide interest as a high-performance material. With its light weight and excellent engineering properties, in the form of high modulus and high strength fiber materials it was initially developed for aircraft and aerospace use.

Despite initially high expectations and the remarkable scientific and technical fertility of the domain where we have seen the introduction of cellulosic, PAN, pitch and mesophase pitch fibers, and more recently vapor grown fibers, each of them initiating an impressive development, the commercial success was not immediate and it experienced successive up and downs.

The 1970s prediction of a world market rapidly growing to 10,000 tons/yr was not fulfilled until 1995. These commercial difficulties resulted in an ever changing scene, with constant spin-offs of industrial companies and first class chemical groups going into production, development, and marketing, then leaving it, sometimes temporarily, to newcomers. The market is now expanding and, despite the reduction of the military market in recent years, the success of the civil aerospace, recreation and other applications have helped sustain a world market of 12,500 tons/yr in 1997, with forecasts to reach 50,000 tons/yr in the next five to ten years.

In the past eight years an enormous amount of new information on carbon fibers has been produced, expanding the existing literature from the time of the first edition (1984) and the revised second edition (1989) of *Carbon Fibers*, which has been out of print for a few years. Answering many demands, we finally decided to launch a new edition!

This revised and expanded third edition is prepared and published as a completely new version: each chapter of this book has been written by specialists in their respective fields. These contributors are internationally acknowledged, and come from all around the world (Belgium, China, France, India, Korea, Russia, Switzerland, Taiwan, and the U.S.A.). They offer the reader authoritative information on today's carbon fiber science and technology. In addition, encouraged by a positive reader response to previous editions, the general organization of this edition has followed suit, with additions and changes reflecting the rapid development of carbon fiber science and technology in the past eight years. For example, many scientific techniques which have been implemented throughout the world and used as tools to characterize carbon fiber properties will be introduced and explored in this volume.

Carbon Fibers, Third Edition, like the previous editions, provides a comprehensive up-to-date review of carbon fiber science and technology. The materials offered by specialists in this new book are profoundly strengthened as compared to those in the previous editions and it shall certainly be beneficial to the carbon community. We hope, this book will provide complete information on the manufacturing, structural, surface, mechanical, electronic, thermal, and magnetic properties of carbon fibers and their applications in the industry. Over 1360 papers and patents are quoted and analyzed in this *reference book*.

I express my sincere gratitude to all the authors involved in this book who made this new edition possible. We are thankful to our colleagues in the carbon fiber field for the very positive response to the previous editions. We hope this new book will evoke an even greater response and be a valuable guide in the field for the beginning of the twenty-first century.

Last but not least I have to express my deepest gratitude to my three co-editors, whose enthusiastic cooperation and very important work were for me the best support during this long and sometimes difficult venture.

Jean-Baptiste Donnet

Contents

Preface   *iii*
Contributors   *vii*

1. **Manufacture of Carbon Fibers**   1
   O. P. Bahl, Zengmin Shen, J. Gerard Lavin, and
   Roger A. Ross

2. **Structure and Texture of Carbon Fibers**   85
   Agnès Oberlin, Sylvie Bonnamy, and Khalid Lafdi

3. **Surface Treatment of Carbon Fibers**   161
   Jimmy C. M. Peng, Jean-Baptiste Donnet, Tong Kuan Wang,
   and Serge Rebouillat

4. **Surface Properties of Carbon Fibers**   231
   Tong Kuan Wang, Jean-Baptiste Donnet, Jimmy C. M. Peng,
   and Serge Rebouillat

5. **Mechanical Properties of Carbon Fibers**   311
   L. H. Peebles, Yuri G. Yanovsky, Anatoly G. Sirota,
   Valery V. Bogdanov, and P. M. Levit

6. **Electrical and Thermal Transport in Carbon Fibers** 371
   Jean-Pierre Issi and B. Nysten

7. **Carbon Fiber Applications** 463
   Serge Rebouillat, Jimmy C. M. Peng, Jean-Baptiste Donnet, and Seung-Kou Ryu

   Author Index 543
   Subject Index 567

CONTRIBUTORS

O. P. Bahl, Ph.D. Deputy Director, National Physical Laboratory, New Delhi, India

Valery V. Bogdanov, Ph.D. Engineering-Cybernetic Faculty, St. Petersburg State Technical University, St. Petersburg, Russia

Sylvie Bonnamy, Ph.D. Centre de Recherche sur la Matière Divisée, CNRS-Université, France

Jean-Baptiste Donnet, Ph.D. Ecole Nationale Supérieure de Chimie and Université de Haute-Alsace, Mulhouse, France

Jean-Pierre Issi, Ph.D. Unité de Physico-Chimie et de Physique des Matériaux, Université Catholique de Louvain, Louvain-la-Neuve, Belgium

Khalid Lafdi, Ph.D. Materials Center Technology Center, Southern Illinois University, Carbondale, Illinois

J. Gerard Lavin, Ph.D. Experimental Station, DuPont Central Research & Development, Wilmington, Delaware

P. M. Levit, Ph.D. Engineering-Cybernetic Faculty, St. Petersburg State Technical University, St. Petersburg, Russia

B. Nysten[*] Unité de Physico-Chimie et de Physique des Matériaux, Université Catholique de Louvain, Louvain-la-Neuve, Belgium

Agnès Oberlin, Ph.D. Mas Andrieu, Saint Martin de Londres, France

L. H. Peebles, Ph.D. Annandale, Virginia

Jimmy C. M. Peng, Ph.D. Ecole Nationale Supérieure de Chimie and Uiversité de Haute-Alsace, Mulhouse, France

Serge Rebouillat, Ph.D. European Technical Centre, DuPont de Nemours International S. A., Geneva, Switzerland

Roger A. Ross, Ph.D. Fiber Engineering Technology Center, DuPont Nylon, Chattanooga, Tennessee

Seung-Kou Ryu, Ph.D. Department of Chemical Engineering, Chungnam National University, Taejon, Korea

Zengmin Shen, Ph.D. Institute of Carbon Fibers, Beijing University of Chemical Technology, Beijing, People's Republic of China

Anatoly G. Sirota, Ph.D. Engineering-Cybernetic Faculty, St. Petersburg State Technical Institute, St. Petersburg, Russia

Tong Kuan Wang, Ph.D. Ecole Nationale Supérieure de Chimie and Université de Haute-Alsace, Mulhouse, France

Yuri G. Yanovsky, Ph.D. Institute of Applied Mechanics, Russian Academy of Sciences, Moscow, Russia

---

[*] *Current affiliation*: Unité de Chimie et de Physique des Hauts Polymères.

1
MANUFACTURE OF CARBON FIBERS

O. P. Bahl
National Physical Laboratory, New Delhi, India

Zengmin Shen
Beijing University of Chemical Technology, Beijing, People's Republic of China

J. Gerard Lavin
DuPont Central Research & Development, Wilmington, Delaware

Roger A. Ross
DuPont Nylon, Chattanooga, Tennessee

1.1 HISTORY OF CARBON FIBERS

Carbon fibers are fibrous carbon materials with carbon content more than 90%. They are transformed from organic matter by 1000–1500°C heat treatment, which are the substance with imperfect graphite crystalline structure arranged along the fiber axis [1,2]. There are two ways to produce carbon fibers: one is from organic precursor fibers, the other is from gas growth [2]. The first commercially produced carbon filament was made from a cellulosic precursor for its application as incandescent lamp filament in 1879 [3,4]. Cotton threads and, later bamboo fibers were formed into filaments of desired size and shape and then baked to a substantially all carbon replica of the original filament. Systematic work regarding carbonization of rayon and PAN yarns and fabrics was initially investigated by Union Carbide Corporation during World War II as a possible substitute material for control grids in the vacuum tube power amplifiers. Two processes for manufacturing high-strength and high modulus carbon fibers from organic precursor fibers (rayon and PAN) were invented almost simultaneously in 1959 and 1961. Four years latter in 1963 a high modulus carbon fiber made from pitch was invented [5,6]. According to the chemical abstracts the total number of carbon fiber related articles and patents published up to 1994 is around 24,000.

Since the final carbon fibers contain almost 100% carbon, any fibrous material with carbon back-bone could be used as precursor which would yield

Since the final carbon fibers contain almost 100% carbon, any fibrous material with carbon back-bone could be used as precursor which would yield carbonaceous residue upon heat treatment. Many naturally occurring materials such as cotton, jute, linen, ramie, sisal and man made polymers like polyester, polyamides, polyvinyl chloride, polyvinyl alcohol phenolic resins, syndiotactic 1,2 polybutadiene, polythene, etc. have been tried as possible precursors for carbon fiber manufacture.

Ultimately, three precursors namely viscose rayon (regenerated cellulose) pitch and polyacrylonitrile have received major attention for making carbon fibers with attractive mechanical properties.

1.2 POLYACRYLONITRILE-BASED CARBON FIBERS

Polyacrylonitrile (PAN) is the only precursor which has found maximum attention, as of today, for developing high performance carbon fibers. Almost seventy to eighty percent of commercially available carbon fibers are basically derived from PAN polymer [(CH$_2$CHCN)$_n$]. It scores over other two predominant precursors namely viscose rayon and pitch, mainly because of the following reasons [7].

1. Its structure, shown in Figure 1, permits faster rate of pyrolysis without much disturbance to its basic structure and to the preferred orientation of the molecular chains along the fiber axis present in the original fiber.
2. It decomposes before melting.
3. Higher degree of preferred orientation of the molecular chains is possible during spinning wherein PAN can be stretched to as high as 800%. Further improvement in the orientation is also possible during thermal stabilization when it becomes plastic at around 180°C and through various post spinning modifications.
4. It results in high carbon yield (50–55%) when pyrolyzed to 1000°C and above.

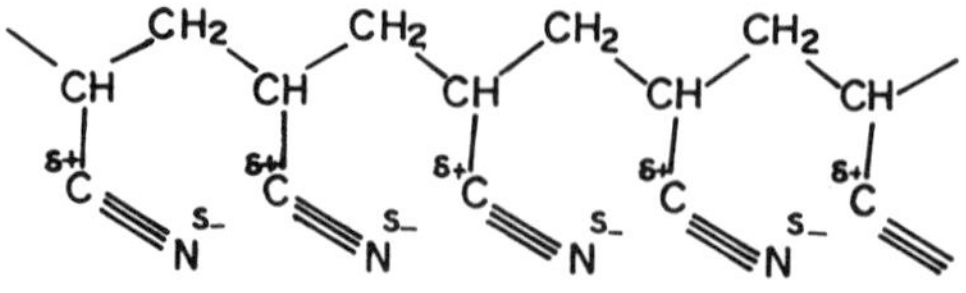

Figure 1 Structural formula of polyacrylonitrile (PAN).

1.2.1 Background

Thermal stability of polyacrylonitrile fibers was recognized [8], soon after its development by DuPont company in the 1940s. At the time, cellulose fibers and PAN (homopolymer) fibers were the only fibrous substances which did not melt below their charring temperatures and were therefore capable of retaining their fiber identity. Winter, of Union Carbide's National Carbon Co., research Laboratories, investigated DuPont's early PAN fibers as precursor for carbon fibers near the end of World War II [8]. Some of these fibers had been thermally stabilized (early versions of "black Orlon") by Coxe of DuPont [9].

The first public report that acrylic fibers could be rendered fireproof by heat treating them in air or inert atmosphere at 200°C was published by Houtz in 1950 [10]. The material was popularly called "black Orlon." Several companies in the U.S. patented fireproofed fabrics based on this process, including DuPont, John-Manville, and Carborundum [11]. The first extensive study of carbonization and graphitization of PAN fibers was reported in 1961 by Shindo [12] in Japan. Shindo recognized the importance of an oxidative heat treatment step prior to carbonization and thus improving the carbon yield from PAN. He demonstrated good tensile strength and modulus of 0.75 GPa and 112 GPa, about three times those available from rayon precursor carbon fibers of that time.

These early fibers, however, were neither high-strength nor high-modulus by today's standards. One problem was that most acrylic fibers of that time were really not suitable for conversion to carbon fibers. They either contained wrong comonomer, possessed unsuitable structure, or contained too many impurities from the point of carbon fibers. In England, a suitable fiber did exist and was used as the basis for experiments by Watt and co-workers at the Royal Aircraft Establishment (RAE) [13]. These workers discovered that the application of tension during the initial oxidation step at 220°C was important for maintaining or even enhancing molecular alignment during carbon fiber processing to 2500°C [14]. This high modulus carbon fiber was subsequently designated Type I. The RAE workers also discovered that an optimum final heat treatment temperature of only 1000–1500°C would yield a higher strength carbon fibers designated as Type II. The tensile strength of Type II, high–strength carbon fibers has steadily risen from 1.4 GPa in late sixties to 2.7 GPa in the early 1970s to 4 GPa in 1980s and to 7 GPa in the 90s.

1.2.2 PAN to Carbon Fibers

Figure 2 describes some of the main steps involved in the production of various grades of carbon fibers using acrylonitrile as starting material. First step in the processing is polymerization. Next is of spinning to obtain the polyacrylonitrile fibers. For developing carbon fibers from PAN fibers, first step is thermal stabilization. It is achieved by heating PAN fibers in the oxidizing atmosphere, usually air, at 200–300°C under application of load. This stabilized fiber is then treated in inert atmosphere to 1500°C to get the high strength carbon fibers known as "type II" carbon fibers. If the heat treatment is carried further to beyond 2500°C, the carbon fibers obtained are of high modulus and known as type I carbon fibers.

1.2.3 Precursor Preparation

It has been shown [10] that the mechanical properties of resulting carbon fibers depend, to a large extent, on the mechanical properties of PAN precursor fibers. The quality and properties of precursor fiber decide the ultimate properties of carbon fibers. A suitable precursor should have low diameter, maximum possible orientation of molecular chains along the fiber axis, maximum crystalline content, low activation energy for cyclization etc. (described in detailed in Section 2.1.9.). Therefore, development of special grade of PAN fibers holds key to the whole carbon fiber industry.

1.2.4 Effect of Comonomers

As shown in Figure 1 PAN has highly polar nitrile groups, and as a result, strong dipole–dipole forces exist between adjacent chains. These dipolar forces are responsible for creating hindrances in molecular alignment during spinning. Further, homopolymer PAN as a precursor results in a poorer quality carbon fibers than does a copolymer of PAN (with a 2–5% percentage of comonomer)

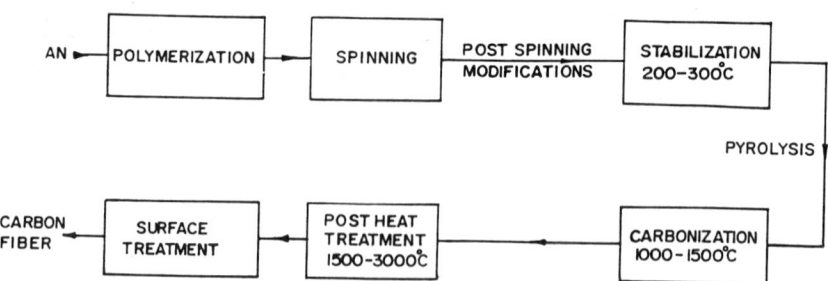

Figure 2 Process steps of carbon fiber production from PAN.

[15]. Various comonomers used are: vinyl comonomers having acidic groups such as acrylic acid [16], methacrylic acid [17], and itaconic acid [18]; esters such as methacrylate [19]; amides such as acrylamide [10]; and ammonium salts of vinyl compounds such as quaternary ammonium salts of amino ethyl–methyl propenoate [21]. Chemical structures of these comonomers are shown in Table 1. Combination of two or more comonomers has also been tried; for example, methacrylic acid/methacrylate and itaconic acid/methacrylate [22–24].

The role of comonomers in the stabilization reactions of the precursor fiber is fairly understood for the case of acidic comonomers which help in lowering the temperature of initiation of the cyclization reaction [25] in addition to other advantages. Cyclization, which is one of the critical steps in the processing of carbon fibers, takes place around 180°C [26], depending on the nature of the comonomer used. The cyclization initiates through a radical mechanism in the case of PAN homopolymer and through ionic mechanism in the presence of acidic comonomers. Heat is evolved at rapid rate with cyclization in the case of PAN homopolymer [27]. Dissipation of this heat is a serious problem because of local steep rise in the temperature. Such excessive heating may cause chain scission thus resulting in poor quality carbon fibers.

For this reason, a slower propagation step is preferred to a faster one. The use of comonomer is thus recommended because of its role in lowering the initiation temperature and slowing down the propagation step in PAN. Among all the comonomers, itaconic acid seems to be superior for making acrylic precursor for carbon fibers due to the occurrence of two carboxylic groups [27] in its structure.

In addition to the nature of the comonomer, its proportion also causes a significant variation in the processes of stabilization and carbonization and the

Table 1 Various Comonomers for Acrylic Precursors

No.	Comonomers	Chemical structure
1	Acrylic acid	$CH_2 = CHCOOH$
2	Methacrylic acid	$CH_2 = C(CH_3)COOH$
3	Methacrylate	$CH_2 = CHCOOCH_3$
4	Acrylamide	$CH_2 = CHCOONH_2$
5	Itaconic acid	$CH_2 = (COOH)CH_2 COOH$

final carbon fiber properties [15]. An increase in the comonomer content reduces the optimum time required for thermoxidative stabilization and improves mechanical properties of the carbon fibers. However, it reduces the yield of carbon fibers [15]. The general understanding based on the results reported by various workers [28–30] is that about 2–5 mol% is the most appropriate content for acidic comonomers. A higher comonomer content causes increase in the rate of stabilization (or more specifically, of its propagation step) which causes reduction in chain length and eventually results in poor quality carbon fibers.

1.2.5 Polymerization

The best quality PAN precursor for developing the high performance carbon fibers can only be obtained by appropriate selection of polymerization conditions like control of molecular weight, molecular weight distribution, molecular defects, and initiation temperature of the cyclization reaction for the precursor fibers. The acrylic precursor is produced by polymerizing acrylonitrile and the appropriate comonomer by one of the following polymerization methods:

1. Solution polymerization [31–35].
2. Solvent-water suspension polymerization [36–38].

1.2.5.1 Solution polymerization
Solution polymerization has been successfully used for the preparation of copolymer precursor material [32–35]. Significant features of solution polymerization are the preparation of a precursor with a lower temperature of stabilization and fewer molecular defects. In addition to these, the solution can be used as a dope directly after removing the unreactive acrylontrile monomer. However, a draw back associated with this method is inherent in the use of organic solvents, which generally lower the molecular weight of the resulting PAN polymer.

1.2.5.2 Solvent-water suspension polymerization
The solvent–water suspension method appears to be a better method for the production of an acrylic precursor because it produces a polymer of higher molecular weight (as compared to solution polymer) with fewer molecular defects and a lower stabilization temperature (or exothermic peak temperature) [39].
 Solvent-water suspension polymerization is initiated by azo initiators and the product is obtained as a slurry [36–38].

1.2.6 Molecular Defects in Acrylonitrile Copolymers

Molecular defects which remain in the form of enaminonitrile and ß-ketonitrile groups influence the initiation temperature for the oligomerization reaction of PAN precursor along with the solvent fragments present in the polymer chain [40]. These defects are generated during polymerization as shown in Figures 3 and 4.

The amount of these defects that remain are indirectly related to the molecular weight of the polymer and the monomer concentration in the reaction mixture [41].

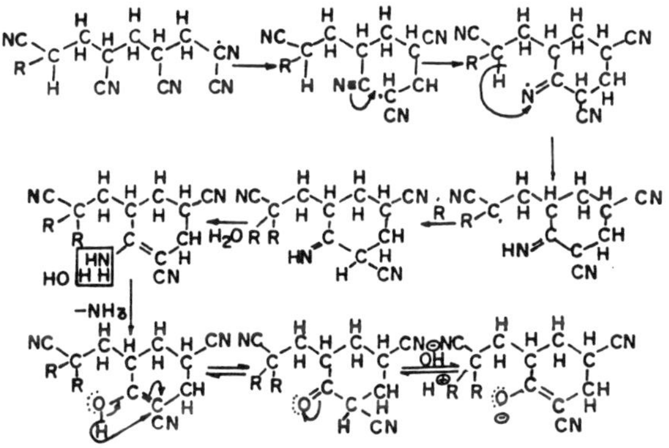

Figure 3 Mechanism of generation of structural defects in acrylonitrile polymer during polymerization (from ref. 41).

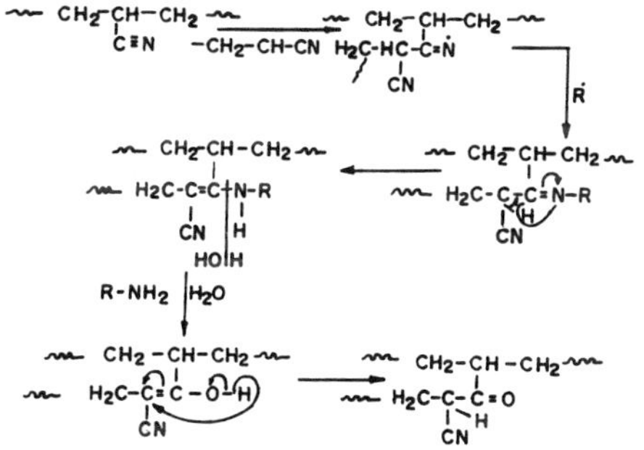

Figure 4 Mechanism of generation of structural defects in acrylonitrile polymer through branching (from ref. 41).

1.2.7 Spinning of Fibers

Spinning of acrylic fibers is generally done by

1. Wet spinning [42–45].
2. Dry-jet wet spinning [46–49].
3. Dry spinning [50–52].
4. Melt spinning [53–55].

The most widely used technique for the spinning of special acrylic fibers (SAF) is wet spinning which is currently being replaced by dry-jet wet spinning since the fibers produced by this technique possess better mechanical properties. A brief illustration of the influence of spinning conditions on the properties of the precursor fiber are shown as follows:

1.2.7.1 Wet spinning
Acrylic precursor fibers in their gel state are obtained by spinning the polymer solution into a coagulation bath having a higher percentage of solvent at low temperature [56]. Wide variations in the properties of wet spun fibers are possible by varying the spinning parameters [42–45]. The denier or diameter of the fiber depends upon the spinneret's orifice diameter, the throughput rate, and the take up velocity, while the stretchability (elongation-at-break) depends upon the stretching of the fiber in the spinline. The strength of the wet spun fibers mainly depends upon the conditions of coagulation bath and the stretching parameters.

The stretching of gelled fibers is performed to obtain better orientation and mechanical properties [57–58]. The orientation of the molecular chains in the gelled state can easily be improved by stretching as in this state the trapped solvent decreases the cohesive forces among the nitrile groups of the polymer chains [59]. The stretching of gel fibers also results in the unfolding of polymer chains and in the formation of an oriented network morphology [60]. The treatment of the acrylic fibers with ultrasonic waves in a coagulation bath for an improvement in strength and a reduction in tenacity variation along the length of the fiber has also been reported [61]. Japan Exlan Co., Ltd. has produced acrylic fibers of increased toughness and tenacity 1.03 GPa [42] by stretching the fibers during cooling, while the fibers not stretched during cooling were reported to have lower tenacity of the order of 0.66 GPa.

1.2.7.2 Dry-jet wet spinning
This technique is similar to wet spinning except that the fibers are extruded a few millimeters (~10 mm) above the coagulation bath [52], so as to achieve molecular orientation before the coagulation. The dry-jet wet spinning for the

production of acrylic precursor fibers is becoming more popular because of its inherent advantages over wet spinning, prominent among them is a higher spinning speed by which fibers of very fine linear density (< 1 tex) can be produced [63]. The dope can be spun at a higher temperature and therefore solutions of higher concentrations may be used for spinning. The production of filaments of controlled non circular cross-section is claimed by this technique [63]. The filaments spun by dry-jet wet spinning technique are stronger and more extensible than filaments spun by an immersed jet.

Toray Industries Inc. [49] claimed improvement in the properties of acrylic precursor fibers spun by dry-jet wet spinning by treating the fibers with ultrasonic waves in a coagulation bath. These precursors are reported to produce carbon fibers of higher tenacity compared to carbon fibers prepared by precursors not treated with ultrasonic waves in coagulation bath. Acrylic fibers of 1.08 GPa tenacity and 17.19 GPa Young's modulus were prepared by Toyobo Co., Ltd. [48].

1.2.7.3 Dry spinning

This technique involves extrusion of a spinnable dope into a hot chamber having circulation of hot gases to evaporate the solvent from the extruded fibers. The spinning speed in this technique is much higher than in wet spinning (kept at about 1000 m/min). However, the number of filaments extruded per spinneret is less than for the wet spinning process.

Japanese patent owned by Mitsubishi Rayon Co., Ltd. [51] described the manufacture of acrylic fibers showing 30% solids content were dry spun and stretched in boiling water, relaxed at 150°C, and finally stretched to a stretch ratio of 1.5 at 120°C adopted a dry spinning method to produce acrylic fibers of strength 0.15 GPa were produced by the scientists at Bayer [50].

1.2.7.4 Melt spinning

Melt spinning even though difficult as acrylic polymers tend to degrade well below their melting point but achieved successfully [53–55] by mixing the polymer with water and water-soluble polyethylene glycol (PEG). These additives decrease the melting point of the acrylic polymer through a plasticizing effect, and thus avoid the complication of degradation reactions [64–65]. Grove et al. [66] investigated the structure and mechanical properties of a melt spun water plasticized PAN-based precursor. The morphology of these fibers is reported to be similar to that of wet and dry spun acrylic precursors. However, these fibers possess surface defects and internal voids which lead to poor quality of the resulting carbon fibers. It is believed that the properties of the carbon fibers resulting from these precursors may be improved by removing impurities from the plasticized melt and thereby reducing the surface flaws and defects.

The acrylic precursor fibers produced by wet spinning and dry spinning have circular and dog-bone-shaped cross-sections depending on the rate of coagulation. However, melt spinning of acrylic precursor produces fibers of trilobal, multilobal, or any other type of cross-section. The noncircular or multilobal carbon fiber can be employed for making composites [67] as these fibers have better strength and modulus than circular fibers and additionally exhibit a greater surface area which provides better adhesion with matrix.

Among the acrylic fibers prepared by various spinning techniques, those prepared by dry-jet wet spinning show superior mechanical properties to those prepared by dry spinning. A tow of a large number of filaments (50,000) can be produced by wet spinning, while dry and dry-jet wet spinning routes are not suitable for extruding very large number of filaments from a single spinneret [63]. On the other hand, dry and dry-jet wet spinning techniques are advantageous in obtaining different cross-sections of the fibers.

1.2.8 Stretching Parameters

Stretching of fibers generally improves the orientations of the molecular chains in the direction of the fiber axis. The stretching of PAN precursor fibers can be made more effective by reducing the dipole-dipole interactions among the nitrile groups. These dipolar interactions obstruct the molecular chains from becoming oriented during stretching and by the use of suitable solvent these interactions can be reduced, for example a 50% increase in the extent of orientation was observed when the precursor fibers were stretched to 70% [68] in the presence of a solvent (DMF).

Stretching done in the gel state (i.e., the state prior to coagulation of the extruded dope) is known to produce better orientation than stretching of the fully coagulated fiber [69]. For this reason, the gel fiber is made to pass through several baths containing varying compositions of coagulation mixture in order to slow down the rate of coagulation and thus produce a higher degree of orientation. Heating by infrared radiation before stretching of the fiber is another way to improve the degree of orientation [70]. The temperature for initiation of the cyclization reaction and the energy of activation decrease with increasing stretch ratio, as shown in Figures 5 and 6.

1.2.9 Importance of Spinning in Clean Room Conditions

Moreton and Watt [71] observed that spinning of polyacrylonitrile under clean room conditions produces better precursor fibers. They spun acrylic fibers in a clean room equipped with five filter units. Four of the units provided laminar flow of clean air directed horizontally, while the fifth central unit provided a downward flow and was fitted underneath with an extraction duct to remove the vapors from the fibers at the hot stretching stage. Extracted air was replaced through an air conditioning and filtration plant which drew air from outside the building. To prevent dirt from getting into the clean room, entry was made via an outer room. The motors, gear boxes, and drive wheels were positioned at the front of the spinning apparatus so that any dust from the moving parts would be carried away from the fiber by the flow of filtered air.

The strength of the carbon fibers obtained from precursors spun under clean room conditions was found to continuously increase with the heat treatment temperatures, whereas control fibers had a maximum strength at around 1000°C as shown in Figure 7. The strength of carbon fibers obtained from precursor fibers spun under clean room conditions had a strength improvement of 84% after being heated up to 1400°C and 82% increase upon heat treatment up to 2500°C over the fiber treated under normal conditions.

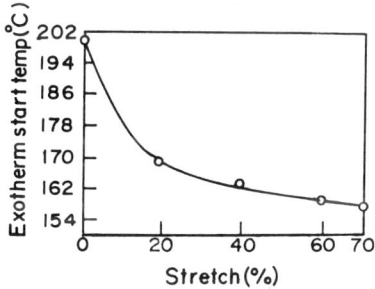

Figure 5 Temperature for start of exotherm (Ti) as a function of stretch ratio (from ref. 68).

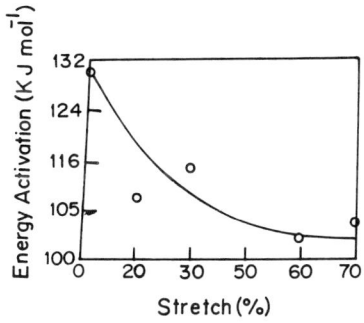

Figure 6 Energy of activation for cyclization reaction as a function of stretch ratio (from ref. 68).

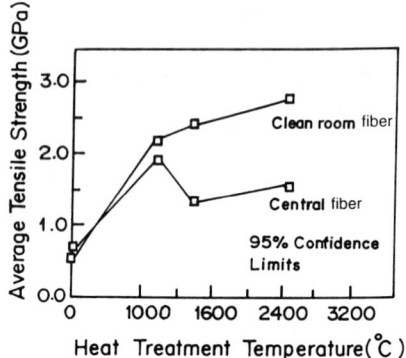

Figure 7 Effect of heat treatment temperature on fiber strength (from ref. 71).

1.2.10 PAN Precursor Characteristics

A PAN precursor fiber for developing carbon fibers needs to have:

* High molecular weight (weight average ~2.6×10^5) [72–73].
* An appropriate molecular weight distribution (MWD).
* Minimum molecular defects.
* Average angle of orientation of molecular chains should be minimum possible.
* Low comonomer content (2–5%).
* A fine denier (0.7–1.2 denier) [75–77]. A fine denier of the precursor fiber is helpful for the dissipation of heat during conversion of precursor to carbon fibers since the heat evolved during the exothermic oligomerization reaction may lead to low carbon yield.
* High strength and high modulus [78].
* A broad exothermic peak due to nitrile group oligomerization during heating, and it should preferably start at a lower temperature [79].
* High carbon yield [80].

Chari et al. [78] have reported a direct relationship between Young's modulus of PAN fiber and Young's modulus of resultant carbon fibers thus

Manufacture of Carbon Fibers

establishing the importance of having good quality PAN precursor for developing high performance carbon fibers.

1.2.11 Shelf Life of PAN Fibers

In a study done by Bahl et al. [81] it was observed that there is a change in the properties of the PAN precursor fiber with time which will eventually affect the properties of the resulting carbon fibers. The tensile strength and Young's modulus of precursor fiber increased from 0.59 GPa and 4.8 GPa in 1983 to 0.62 and 8.3 GPa in 1988. This was also reflected in the angle of orientation z which decreased from 18° to 15°C and crystallite size increased from 43 to 56 during the same period. Molecular chains in the amorphous phase of the PAN structure are under constant motion including rotations around the C–C bonds. These rotations may give rise to more ordered isotactic/syndiotactic chain segments thereby bringing them into the folds of an already existing crystalline phase. The overall crystallite size of the PAN should therefore increase with time. This also affects the thermal behavior of PAN fiber where the activation energy of cyclization is found to decrease from 144 KJ/mole (in 1983) to about 127 KJ/mole (in 1988).

1.2.12 Post Spinning Modifications[1]

A systematic study on the physical, mechanical and thermal characteristics of different PAN precursors has led to the basic understanding [74] that these characteristics vary markedly for different precursors and the mechanical properties of the resulting carbon fibers also bear a direct correlation with characteristics of the starting precursor [82–89]. Studies [72–80] show that precursor fiber should have characteristics as are mentioned in the previous section. There are however, some limitations during spinning, which do not permit to produce a precursor with all the desired characteristics combined together [87]. For example it would be necessary to have higher comonomer contents to reduce the dipolar forces for achieving higher degree of stretch ratios, but on the contrary, from the carbon fiber point of view, comonomer contents should be minimum. This therefore puts a limit on the ultimate stretching that can be imparted during spinning. In order to overcome these drawbacks, post spinning modifications on PAN have been attempted which have resulted in certain improvements in the quality of PAN precursor vis-à-vis the resulting carbon fiber properties.

These modifications can be put into three broad categories:

  1. Modification through coating with certain resins [88].

2. Chemical modification using certain reagents like Lewis acids, organic bases, inorganic compounds etc. which act as catalysts to enhance the rate of cyclization.
3. Post spinning stretching using different plasticizers like DMF, steam, nitrogen etc. to improve the structure of PAN precursor.

1.2.12.1 Modification through coatings

These are the lubricant, antistatic agents, and emulsifiers which are basically used as spin finish on the precursor fibers. These reagents are required to reduce the entangling, fusion, and metal to fiber adhesion of acrylic precursor fibers during thermooxidative stabilization and carbonization process. Lubricant reduces the metal to fiber adhesion and improves interfiber cohesion. Antistatic agents, which are polar compounds, help in dissipation of static charges built up during the processing of tow of the fibers. Emulsifiers improve the stability of the finish. In addition to this, coating of fibers with these chemicals improves the properties of the resulting carbon fibers [88]. Interfiber cohesion is important for precursor fibers because a tow of a large number of filaments is used for carbonization. The importance of a spin finish is further emphasized because it improves the unwinding stability of the carbon fibers if it is applied to the precursor fiber after stabilization [89]. Some commonly used lubricants are silicone oil [89–94], fatty acid derivatives [95,96] and guar gum [97].

1.2.12.2 Chemical modifications

This modification involves impregnation of PAN precursor fiber with, solutions of cobalt salts [98], a salt of iron [99], hydrogen peroxide [100], CuCl [101], hydrazine [102], hydroxyl amine [103], HCl [104], guanidine carbonate [105] tin dibutyldimethoxide [106], tin hydrochloride [107], potassium dichromate [108], etc. These treatments reduce the time required for stabilization of PAN fibers by lowering the activation energy of stabilization reaction [109–111].

PAN fibers when impregnated with Lewis acids not only improve the kinetics of cyclization of PAN fibers but also help in filling the defect centers by metallic ions which leads therefore to carbon fibers with improved properties [111–112]. Impregnation of PAN with strong bases shows substantial reduction in cyclization exotherm.

Post spinning modification using $KMnO_4$ has been studied in detail recently [110,111,124]. It is observed that with $KMnO_4$ treatment of PAN fibers, the tensile strength of ultimate carbon fibers (HTT 1000°C) goes up to 4.2 GPa as against 2.1 GPa for the carbon fibers from unmodified fibers [114].

The reasons for this improvement are attributed to following:

Manufacture of Carbon Fibers

1. The plasticizing effect of $KMnO_4$ during oxidation imparts better orientation to molecular chains.
2. Catalytic effect of $KMnO_4$ converts PAN into a ladder polymer in short duration without overoxidizing the backbone (Figure 8).
3. Treatment of $KMnO_4$ reduces the activation energy of cyclization from 120 kJ/mol to 90 kJ/mole (Table 2) [114].
4. Diffusion of MnO_4^- in the fiber promotes cyclization from within the core thus avoiding the core/sheath formation during stabilization process [7]. The uniformity of stabilization across the fiber is essential to avoid stress concentration in the resultant carbon fiber structure.

Salient points about some of the chemical modifications are described briefly. The increase in tensile strength and modulus of resultant carbon fibers up to 15–40% and 10–20% respectively were observed at HTT 1300°C by treating the PAN fibers with cobaltous chloride before stabilization [115]. The treatment increases the crystallite size, the crystallinity, and orientation in the PAN fiber because formation of ladder polymer during stabilization is slow in the modified PAN fibers and results in carbon fibers having better modulus. Carbon fibers from PAN fibers treated with cuprous chloride are found to have much higher mechanical properties than those obtained from unmodified fibers [116].

Irradiation of acrylic precursor fibers with gamma rays has also been reported [117]. A dose of 20 Mrd accelerates the dehydrogenation reaction which imparts higher stability to the polymer back bone chain and improves the mechanical properties of the stabilized fibers. However, higher doses of irradiation lead to higher weight losses during subsequent carbonization. The thermooxidative stabilization of acrylic precursor fiber is promoted by infrared radiation heating [118]. Lowering of the energy of activation for stabilization has been observed after modification by $N_2H_4 \times H_2O$ [102]. Catalytic surface treatment of PAN fibers by dibutyltindimethoxide reduces the time required for stabilization by 2 hrs [106].

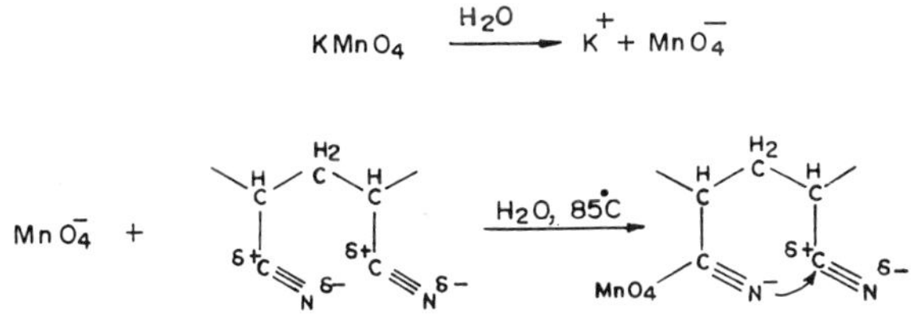

Figure 8 Proposed reaction mechanism of $KMnO_4$ treatment of PAN fibers (from ref. 110, 113, 114).

1.2.12.3 Post spinning stretching using different plasticizers.

As mentioned earlier, highly polar nitrile groups in PAN result in strong dipole dipole interactions among nitrile groups. This makes the stretching of PAN fiber difficult. Therefore, plasticizer is to be added for stretching PAN fibers. Generally, heat acts as a plasticizer during stabilization which decreases the interaction among nitrile groups and after application of load, the required stretching in fiber is achieved.

The post spinning stretching before stabilization using certain chemicals is also used. This is mainly done to improve the structure of PAN fibers and specially the angle of orientation of molecular chains with respect to fiber axis which leads to the carbon fibers with better mechanical properties. Stretching in super saturated steam [119,120], hot solutions of CuCl [116] and nitrogen [121] led to improved physical characteristics of PAN precursor and ultimately, to better carbon fiber properties.

Recent trend, however has been towards thinning of the precursor fiber by using certain plasticizers which ultimately lead to carbon fibers with diameters of 4.5 to 5 μm only. Such carbon fibers should contain less defects per unit volume and hence should possess very superior mechanical properties, in conformity to Weibull's weakest link principal [122]. A process for making low diameter carbon fibers is patented by Peeper et al. [123], in which fibers are stretched during stabilization in presence of carboxylic acid (other than formic acid) more than 300%. Benzoic acid [124] and DMF [68] were also tried as plasticizers. Due to large size, benzoic acid could not diffuse to the core of the fiber. So during stretching the molecular chains on the surface and those inside the core suffered a differential degree of stretch, leaving behind residual stresses in the fiber structure. To overcome this problem, acetic acid was tried (Figure 9) [125]. Reduction in diameter and improvement in tensile strength were observed with acetic acid modification of PAN fibers.

Table 2 DSC Studies of the Unmodified and Modified Samples [109]

Sample	S.T (°C)	E.A. KJ/mol	Heat flow J/g	P.T. °C
P	195	119	−1229	268
K-3	185	111	−1262	263
K-5	185	102	−1263	265
K-10	180	90	−1266	268
K-20	160	90	−917	267
K-30	160	109	−349	212

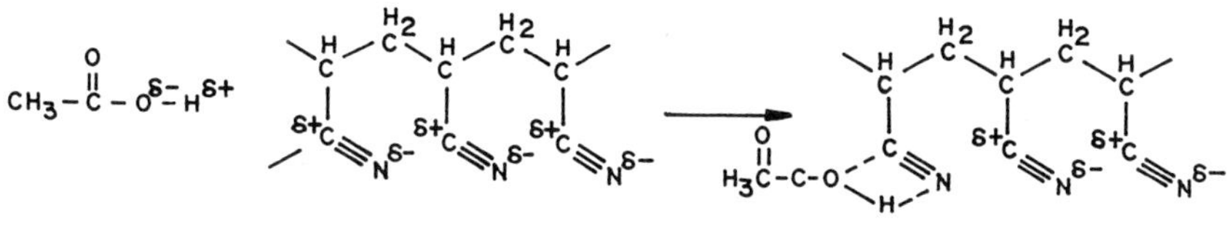

Figure 9 Proposed reaction mechanism of acetic acid treatment of PAN fibers (from ref. 114,125).

1.2.12.4 Bimodification
A new technique named as Bimodification has recently [125] been developed which takes the advantage of both catalyst and plasticizer for the post spinning modification. $KMnO_4$ (catalyst) and acetic acid (plasticizer) were used for modification. It was observed that the carbon fibers prepared after bimodification of precursor fiber possess much better mechanical properties than the carbon fibers prepared after the modification by $KMnO_4$ or acetic acid of precursor fiber individually.

1.2.13 Stabilization of the Precursor Fiber

This is the second step in the processing of PAN fibers into carbon fibers. This process involves heat treatment at lower temperature under tension in an oxidative atmosphere and produces changes in chemical structure of the acrylic precursor such that the product becomes thermally stable to subsequent high temperature treatments [80,126]. The stabilization is considered to be the most decisive step because it largely governs the final structure of the fiber and hence its ultimate mechanical properties. Stabilization is carried out by controlled heating of the precursor fiber in an oxidizing atmosphere e.g., air in the temperature range of 180°C to 300°C. The heating rate is usually 1–2°C/min. The main reactions taking place are cyclization, dehydrogenation, and oxidation. Dehydrogenation reactions precede cyclization, but they continue during and even after cyclization [127]. These reactions are responsible for the change in color of the precursor fiber to yellow, brown, and ultimately black. Berlin et al. [128] and Fester [129] believe it to be due to the development of polyene structure during heat treatment of PAN, while others are of the opinion that the black color appears due to the formation of a condensed ring structure

[130–132] (Figure 10) containing carbon-nitrogen double bond. This structure is often known as ladder structure. Most of the reactions are exothermic. The heat of dehydrogenation and cyclization was calculated to be –242.67 KJ/mol and –58 KJ/mole, respectively.

Several groups [133–135] have carried out Fourier transform IR studies of the initial cyclization reactions occurring during the degradation of PAN copolymers at temperature below the onset of the major exothermic reaction i.e., < 200°C and under reduced pressure of 5×10^{-2} torr. The degradation of PAN was found to involve initiation by an anion X^-, which is presumably derived from an impurity or a degradation product (Figure 11). The initiation was followed by cyclization to yield an imine structure, followed by oxidation to give the final pyridone structure. Cyclization involves oligomerization of nitrile groups. The triple bond of a nitrile group changes to a double bond, and the nitrogen of this nitrile group forms a bond with the carbon of the succeeding nitrile group of the chain. This is normally believed to occur around 180°C [136]. Cyclization initiates at several points of the PAN molecule during heat treatment, and it grows until it terminates reaching another conjugate or cyclized unit. The cyclization reaction depends upon the initiation centers already present in the polymer. Another opinion is that it is initiated by the attack of oxygen [132]. The cyclization reaction is initiated by a radical mechanism in the homopolymer and by an ionic mechanism in the copolymers (as discussed in Section 2.1.3) but the structure of the stabilized product does not differ very much [137].

Cyclization is not continuous throughout the chain. So, there are some uncyclized units left at random in the molecular chain [138]. These uncyclized units act as sites for chain scission and hence source of the evolved gases that account for the weight loss. The weight loss depends upon the average length of the cyclized sequence and the volatility of the fragments. Out of the various volatile products evolved, HCN is released during chain scission from the units which do not undergo cyclization and NH_3 is produced from the terminal imine structure [139,140]. Preheating of PAN fibers in air causes dehydrogenation reaction, resulting in formation of double bonds in the chain backbone which impart greater stability to the chain. Thus it is recommended that precursor fiber be heated slowly to get well stabilized fibers. The stabilization process is greatly influenced by the variables of tension on the fiber, the heat treatment temperature, the treatment medium, and the prestabilization treatment, salient features which are discussed in the following sections.

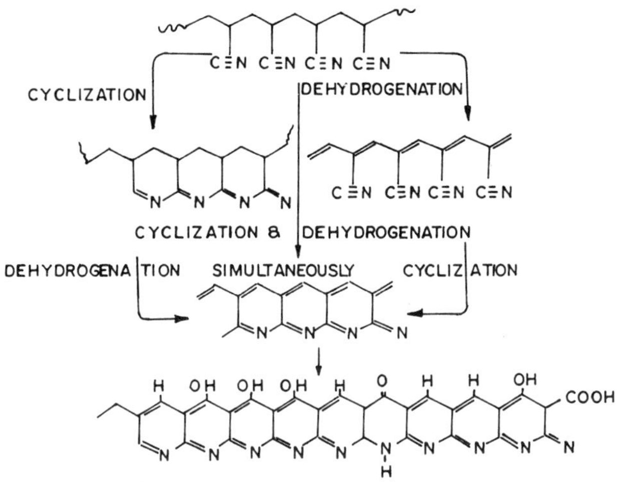

Figure 10 Sequence of reactions during thermooxidative stabilization of PAN precursor.

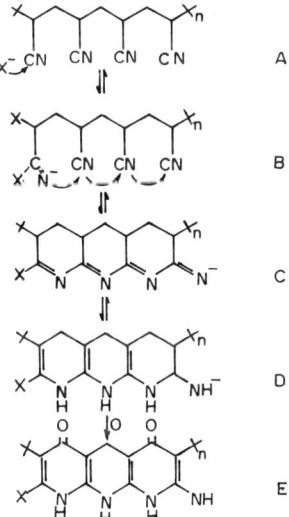

Figure 11 Decomposition of PAN at 200°C under reduced pressure (from ref. 134).

1.2.13.1 Effect of tension
During stabilization, if no load is applied, shrinkage is observed in the fiber which has been studied by various workers [141–144]. It is believed that there are two types of shrinkage viz. physical and chemical as shown in Figure 12. Physical shrinkage, which is attributed to the entropy recovery of a drawn and quenched material [145] is explained on the basis of the helical form of the PAN molecule with intermolecular repulsive forces between the neighboring nitriles group dipoles. Stretching of the fibers during their manufacture develops strains in the molecules which relax during heat treatment and, consequently, the fiber shrinks.

Chemical shrinkage on the other hand, occurs as a result of chemical reactions during the stabilization process, which leads to the formation of cyclized ladder polymer [146,147]. The rate and magnitude of this shrinkage depends upon various factors such as the surrounding atmosphere, the temperature of treatment, the applied load, the stretch given to the precursor fiber, the heating rate, etc. [145]. This shrinkage may be considered to be a measure of cyclization reactions [148].

Physical shrinkage plays an important role in cyclization and the subsequent reactions which occur at 180°C and above. A regular positioning of nitrile groups in the molecular chains is required for these reactions [146,147]. This conformation is adversely affected during heat treatment when the molecular chains relax and physical shrinkage appears. The molecular relaxation can be prevented by applying tension to the fiber during heat treatment. The application of a very high tension during stabilization may cause deterioration of the mechanical properties of the fiber because it may disturb the preferred orientation required for the cyclization reaction causing rupture of bonds [149].

1.2.13.2 Effect of heating
Mass transfer, heat transfer, and shrinkage are the three main phenomena taking place during the stabilization process. Mass transfer involves diffusion of oxygen into the precursor fibers and evolution of volatile products such as HCN, NH_3, H_2, etc. [150]. Heat transfer takes place due to the occurrence of various exothermic chemical reactions. PAN is a bad conductor of heat and the heat generated during stabilization can cause damage to the fibers. It is therefore imperative to use lower heating rates etc. during stabilization with a view to avoid overheating of PAN fibers, such as chain scission.

The effect of heating rate reported [151] for acrylic precursors showed a maximum tensile strength of the resulting carbon fibers for a heating rate of 1–3°C/min. The maximum strength is higher for a final heat treatment temperature (HTT) of 260 to 280°C rather than 290°C, as shown in Figure 13. The heating rate up to initiation temperature should not be higher than 5°C/min.

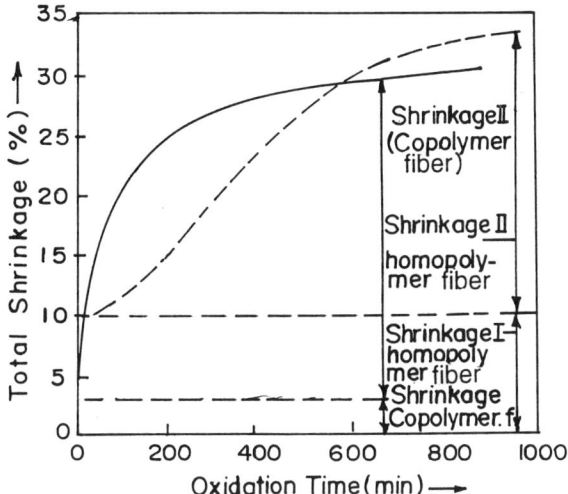

Figure 12 Shrinkage versus oxidation time of polyacrylonitrile fibers at 215°C in the presence of O_2 for copolymer and homopolymer fibers (from ref. 148).

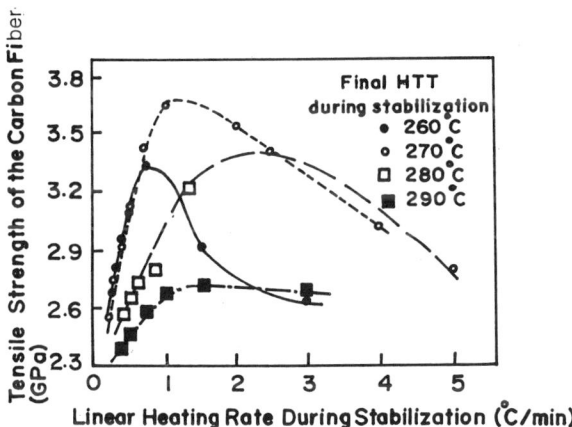

Figure 13 The effect of heating rate and final HTT during stabilization on the strength of carbon fiber (from ref. 151).

1.2.13.3 Effect of medium

The stabilization of precursor fiber has been studied in various media, via air [152,153], oxygen [154], sulfur dioxide, hydrogen chloride gas, nitrous oxide, a mixture of bromine and oxygen and a mixture of hydrogen chloride and oxygen [154]. Precursor fibers stabilized in an oxidizing medium produce better carbon fibers compared to precursors stabilized in an inert atmosphere [155,156]. Kinetic data calculated from DTA exotherms [157] show a higher energy of activation of stabilization reactions for the case of heat treatment in an oxidizing atmosphere than in an inert atmosphere. The rate of disappearance of –CN groups was found to be higher in air than in an inert atmosphere [158] which suggests that oxygen helps in the reaction of the nitrile groups, leading to the formation of a ladder polymer. Precursor fibers also show greater chemical shrinkage [148] in oxidative than inert atmosphere (Figure 14).

Fitzer et al. [157] have demonstrated that oxygen acts in two opposite ways during stabilization. On the one hand, it initiates the formation of active centers for cyclization, while on the other hand it retards the reactions by increasing the activation energy. Even so, stabilization in an oxidizing medium is preferred because it results in the formation of some oxygen–containing groups (such as –OH, C = O, –COOH) in the backbone of the ladder polymer. These groups subsequently help in generating cross links during carbonization [159]. As already stated, stabilization is also carried out in certain media other than oxygen, such as sulfur dioxide and hydrogen chloride [154–156]. These acidic media increase the reaction rate and show a greater degree of stabilization.

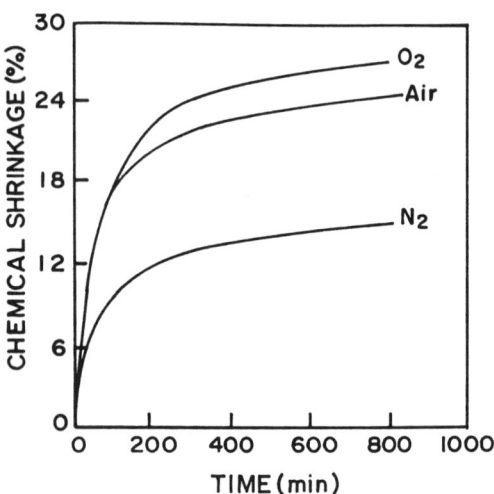

Figure 14 Chemical shrinkage of PAN fibers as a function of heating time at 215°C in the presence of oxygen, air and nitrogen (from ref. 148).

1.2.13.4 Structure of stabilized fibers

Oxidation of PAN fibers has been the subject of numerous investigations, and consequently several different structures have been proposed for the stabilized PAN material. Houtz [10] was the first to put forward a structure for Orlon black fibers obtained by heating them in air (Figure 15, I). The structure was later supported by other investigators [160,161]. However, his structure contains no oxygen and cannot account for the evolution of CO and CO_2 during the carbonization of the oxidized fiber up to 1000°C [162]. Standage and Matkowsky [163], on the basis of IR studies and elemental analysis of oxidized PAN, proposed a different structure with oxygen forming bridging ether links. (Figure 15, II). Watt [164] has proposed the formation of a ladder polymer (Figure 15, III) during heating of PAN in air, in which oxygen is present only as carbonyl groups. This structure has been proposed on the basis of mechanical properties, elemental analysis, IR studies etc. This structure was supported by Bahl and Manocha [159]. Kirby et al. [165] and Friedlander et al. [132] after a study of the model compound, proposed a nitrone-type ladder structure with each nitrogen atom donating lone pair of electrons to an oxygen (Figure 15, IV).

However, all these structures have been criticized [166] on the basis of stereochemistry. The carbon atoms in these ladder type polymer chains are not in one plane, thus leading to a cyclized and not aromatic character. Besides, these structures possess additional polar groups, suggesting that the oxidized PAN fiber should show better mechanical properties than the cyclized PAN fiber. In addition, the non aromatic character of these structures would suggest a lower thermal stability of preoxidized PAN fibers, although they actually have greater thermal stability.

Goodhew et al. [167] suggested structures A and B (Figure 16) as the most likely structures for oxidized and inert heat treated PAN fibers. These structures are adequate to explain the occurrences in oxidative and inert heat treatments. The principal structural difference between the two treatments is that while oxidation results in fully aromatic heterocycles carrying hydroxyl and carbonyl groups, the treatment in inert atmosphere produces hydrogenated heterocylces with carbon–nitrogen conjugation. Thus the oxidative treatments produce more thermally stable structures. It may, however, be pointed out that even this structure can not account for the evolution of CO_2 during carbonization by heating in inert atmosphere up to 1000°C, as reported by several workers [162,167]. Clearly there is still great uncertainty regarding the reactions which occur during the pyrolysis and oxidation of PAN in air.

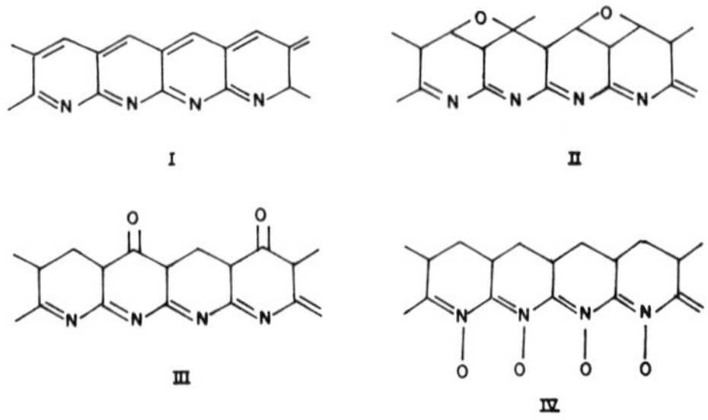

Figure 15 Structural models proposed by different authors for the oxidized PAN (from refs. I (10), II (163), III (159), IV (165)).

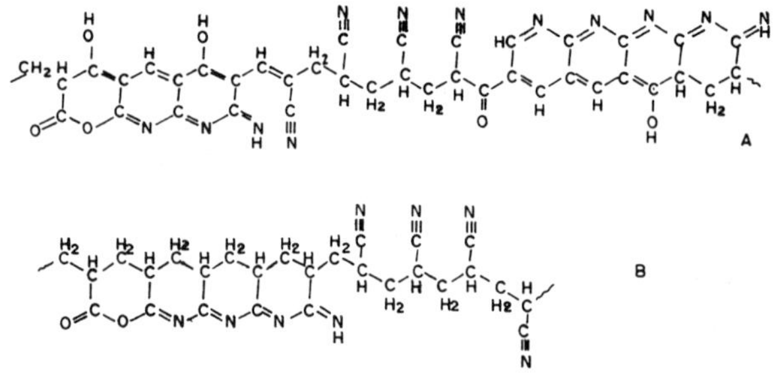

Figure 16 Structure proposed by Godhow, (A) stabilized in air, (B) stabilized in inert atmosphere i.e. N_2 (from ref. 167).

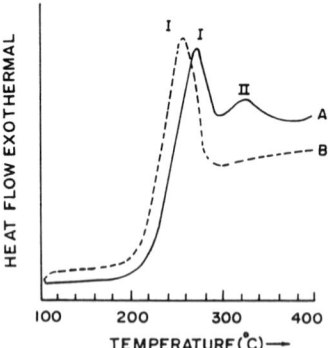

Figure 17 DSC curves of PAN fibers, A: in presence of air, B: in presence of nitrogen, heating rate 5°C/min (from ref. 114,168,170).

1.2.13.5 *A New process of stabilization*

Several authors [121,127] have taken recourse to DSC curves for carrying out proper thermal stabilization of PAN fibers before carbonization, as it gives vital information regarding the total heat flow, energy of activation, cyclization reaction starting temperature as well as the final temperature of oxidation etc. A DSC curve of PAN precursor (Figure 17) shows that a sharp exotherm starts at about 200°C, completes at 300°C with a maxima centered around 265°C. This maxima corresponds to cyclization and thermal stabilization of PAN fibers. However, another small and relatively broad exotherm maxima is observed which starts at 300°C, completes at 400°C, and with a maxima centered at 350°C [114,168,169].

It has been observed by Mittal et al. [114,170,169] that thermal stabilization of PAN fibers is not at all completed at less than 300°C, normally used for stabilization. Fibers stabilized up to temperatures 300–400°C (maxima II) are better stabilized. It has been found that additional aromatization and intermolecular crosslinking reactions (Figures 18 and 19) in the fiber take place during heating in the temperature range 300–400°C. This leads to formation of highly ordered and compact structural units in the stabilized fiber (Figure 20 [114,170]). The molecular chains in the fiber undergo realignment, reorientation and reinforcement along the fiber axis, to cause pockets of discrete repeat units of finite size maintaining a distance of 6.8 Å. The repeat units grow in number and size as the stabilization temperature increases.

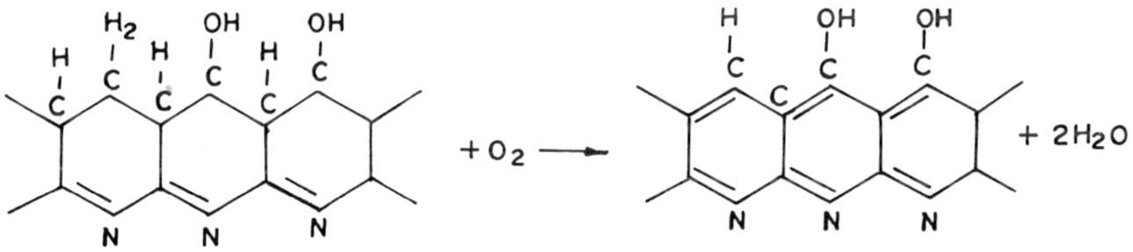

Figure 18 Aromatization reaction during stabilization of PAN fibers in temperature range 300–400°C in presence of oxygen (from ref. 114,170).

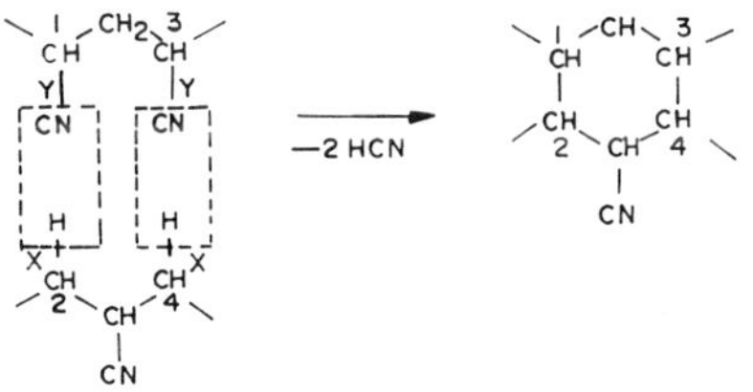

Figure 19 Intermolecular cross-linking reaction during stabilization of PAN fibers in temperature of 300–400°C in presence of oxygen (from ref. 114,170).

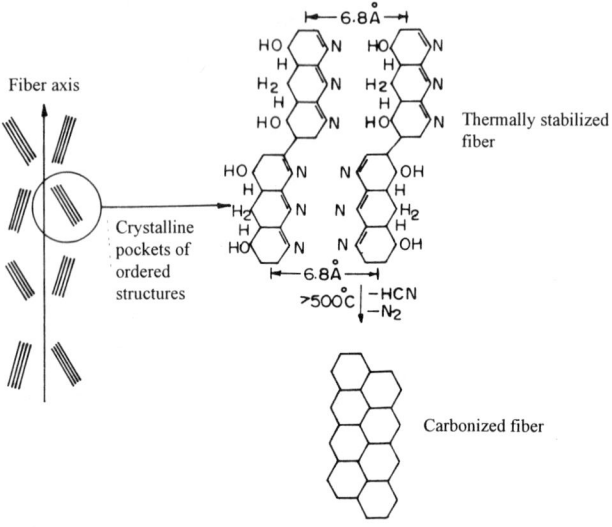

Figure 20 Structure of thermally stabilized PAN fiber (from ref. 114,170).

1.2.14 Carbonization Process

1.2.14.1 Heat treatment temperature up to 1500°C
Stabilized precursor fibers are converted into carbon fibers in the process of carbonization, which involves heat treatment in an inert atmosphere [7] under low or very little tension up to about 1500°C [80]. In this process, all elements other than carbon are eliminated in the form of appropriate byproducts [171] and a graphite like structure is formed.

In the carbonization process, the heating rates in the two zones are crucial. The first zone, i.e., up to abut 600°C, requires a low heating rate (less than 5°C/min) so as to make the mass transfer slow. A faster mass transfer at higher heating rates may cause surface irregularities in the form of pores due to the diffusion of involved gases, etc. [151]. This zone is very crucial because it involves most of the chemical reactions and the evolution of volatile products. The evolution of water is due to the crosslinking of ladder polymer chains through the oxygen containing groups as shown in Figure 21. In the second zone i.e., between 600°C and 1500°C, higher heating rates can be used because of the reduced possibility of damage due to exothermic reactions or the evolution of byproducts since such reactions have already been completed by

Manufacture of Carbon Fibers

600°C. This heating zone involves the evolution of N_2, HCN, and H_2. Nitrogen and hydrogen cyanide are evolved due to intermolecular crosslinking of the polymer chains, and hydrogen evolves as a result of dehydrogenation. During crosslinking, the carbon atoms of one cyclized sequence fit in to the spaces left by the nitrogen of the adjacent sequence [172]. This helps in the growth of a graphite like structure in the lateral direction, as shown in Figure 22. An inert atmosphere is necessary during the carbonization step in order to avoid oxidation at such high temperatures. For this purpose nitrogen, argon, and other non oxidizing media, such as HCl [173], BBr_3 [174], and ZnO [175], have been used.

During the early stages of carbonization in the temperature range 400–500°C, the hydroxyl groups present in the oxidized PAN fibers start crosslinking condensation reactions which help in reorganization and coalescence of the cyclized sections. This crosslinking probably fixes the structure of the polymer, while the remaining linear segments either become cyclized or undergo chain scission evolving the gaseous products. These cyclized structures undergo dehydrogenation [176] and begin to link up in the lateral direction, producing a graphite like structure consisting of three hexagons in the lateral direction and bounded by nitrogen atoms [168]. Watt [172] suggested that the lateral molecular growth occurs by condensation reactions in which the carbon atoms in one cyclized sequence fit into spaces left by the nitrogen atoms in an adjacent cyclized sequence.

Oxidized PAN fibers are capable of producing high performance carbon fibers in greater yield because they have preferentially an aromatic character which prevents the backbone carbon chain from extensive splitting. In the case of PAN fibers stabilized in inert atmosphere, the aromatization does not readily occur at low temperatures, and therefore it has to be induced at higher temperatures which causes considerable splitting of the main carbon chain and results in a low yield and poor mechanical properties. The carbon fiber yield can further be enhanced by carbonizing oxidized PAN fibers in an atmosphere of HCl vapors [173] or by spinning the PAN fiber in clean room [71]. Carbonization in HCl vapors decreases the amount of HCN by eliminating nitrogen as ammonia. It also has a marked dehydrating action and eliminates oxygen as water vapor. Consequently, the yield of carbon increases. Spinning of the PAN fiber in a clean atmosphere results in a decrease in the average number of surface flaws, giving rise to carbon fiber with better mechanical properties. Tests conducted on the precursor fibers collected at different stages of heat treatment during the carbonization process showed that the modulus and the strength of the fibers increase rapidly up to about 1500°C. A decrease of the strength of the carbon fibers has been reported for heat treatment temperatures above 1500°C [177–179]. However, the modulus keeps increasing at a slower rate [179].

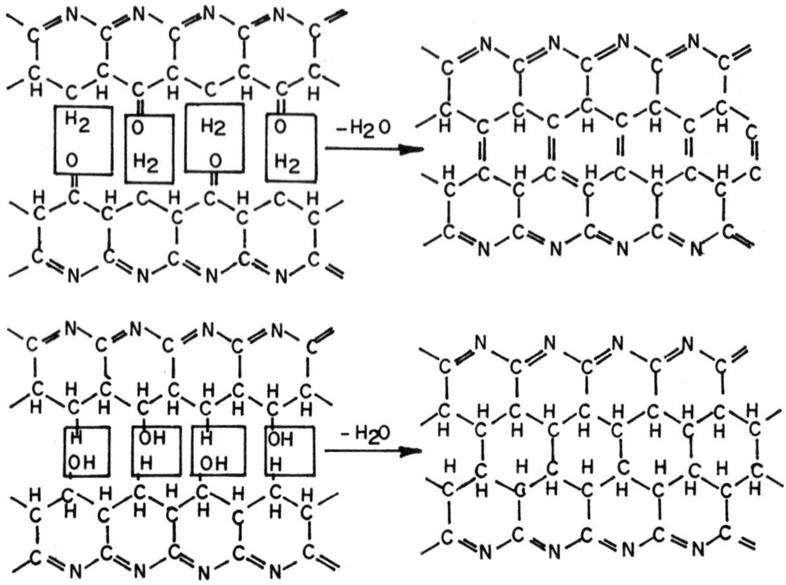

Figure 21 Intermolecular cross-linking of stabilized fibers through oxygen-containing groups (from ref. 148).

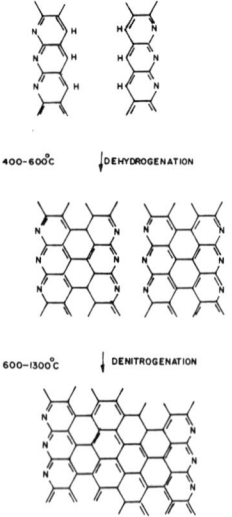

Figure 22 Schematic representation of carbon fiber preparation form PAN (from ref. 172).

Manufacture of Carbon Fibers

1.2.14.2 Heat treatment temperature up to 3000°C

Following the carbonization process, the fibers have a structure with small crystallites. Perfecting the structure by increasing the size and alignment of the crystallites is achieved by heat treating the fibers to temperatures beyond 1500°C. The process involves heating the carbon fibers under tension in an inert atmosphere, usually at about 2000–2500°C and sometimes up to 3000°C [7]. Nitrogen cannot be used as the inert medium because above 2000°C it becomes active and forms cyanogen by reacting with the carbon fibers [82]. Improvements in the crystal perfection are also achieved by passing an electric current through the carbon fibers when yarn is at 1800–3000°C [180]. Certain metal oxides such as chromium oxide, manganese dioxide, vanadium oxide, and molybdenum oxide, are used as catalysts [181] for promoting the growth of crystallites during heat treatment of carbon fibers.

Bromination of the carbon fibers produces a remarkable improvement in strength [182]. The addition of boron vapors also increases the stiffness of the fibers [183]. It is believed that boron not only increases the crystallinity but also acts as a solid solution hardening element in the graphite, thus preventing shear in the crystallites and maintaining the modulus as well as the strength. Young's modulus of the carbon fibers has been found to be directly related to the final heat treatment temperature [179].

1.2.14.3 Carbonization behavior of high temperature stabilized fibers

The high temperature stabilized (300–400°C) fibers can be carbonized at a very fast heating rate of the order to 200–300°C/min. Further, the mechanical properties of the resulting carbon fibers show significant improvement over the one prepared by stabilization in the temperature range (200–300°C) [114]. Not only this, the evolution of tarry products during carbonization is much less comparatively as a result of additional dehydrogenation achieved during the stabilization between 300–400°C. So, this process of the stabilization of PAN fiber up to 300–400°C helps in considerably improving the economics of the carbon fiber production. Further, carbon fibers (HTT 1000°C) prepared from stabilized fibers using this novel technique, possess Young's modulus of around 240 GPa and tensile strength of around 2.1 GPa. As one would have expected, Young's modulus value is extremely high [114]. The tensile strength and Young's modulus further increase to 3.7 GPa and 300 GPa respectively, after the heat treatment to 1700°C as shown in Figures 23 and 24 [184].

These high temperature stabilized fibers can be directly graphitized to 2500°C at a rate of 100–200°C/min and still having the strength of about 2 GPa and modulus of 425 GPa which is contrary to the general belief that the rate of carbonization should be slow and in steps to produce the good quality carbon fibers [7]. Some of the commercially available carbon fibers along with their mechanical properties are listed in Table 3.

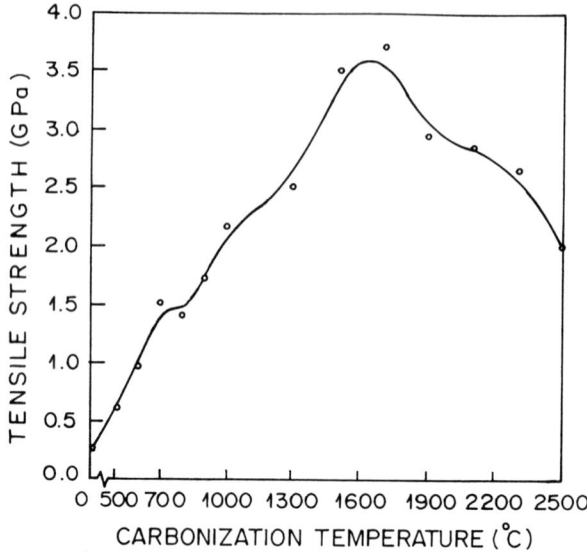

Figure 23 Tensile strength behavior of carbon fibers (stabilized at 350°C) carbonized to different heat treatment temperatures (HTT) (from ref. 184).

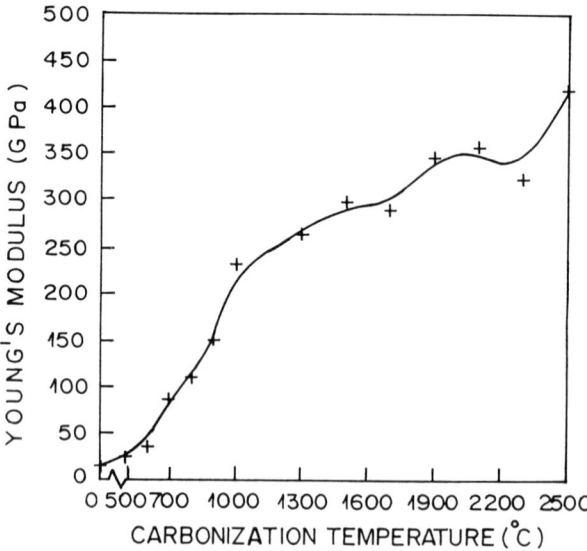

Figure 24 Young's Modulus behavior of carbon fibers (stabilized at 350°C) carbonized to different heat treatment temperatures (HTT) (from ref. 184).

Table 3 Mechanical Properties of Some Commercially Available PAN Based Carbon Fibers

Manufacturer	Product name	T.S (MPa)	Y.M (GPa)	Strain to failure %
Hercules Inc. (U.S.A.)	AS-4	4000	235	1.6
	IM-6	4880	296	1.73
	IM-7	5300	276	1.81
Torey Indust. (Japan)	T300	3530	230	1.5
	T800H	5490	294	1.9
	T1000G	6370	294	2.1
	T1000	7060	294	2.4
	M46J	4210	436	1.0
	M40	2740	392	0.6
	M55J	3920	540	0.7
	M60J	3920	588	0.7
Amoco Corp. (U.S.A.)	Thornel T600	4160	241	1.72
	Thornel T700	3720	248	1.83
Toho Beslon (Japan)	HTA-7	3840	234	1.64
	ST111	4400	240	1.80
Mitsubishi Rayon (Japan)	Purofil T1	3330	245	1.40
	Purofil M1	2550	353	0.7

1.3 RAYON–BASED CARBON FIBERS

Another important raw material for producing carbon fibers which degrades without melting is cellulose. Natural cellulose fibers such as those of cotton and ramie are not preferred for the production of carbon fibers because these are discontinuous in nature and exhibit a low degree of orientation in addition to having impurity materials such as lignin. Accordingly, they are considered inferior to synthetic cellulose fibers. The most commonly employed cellulosic fibers for producing carbon fibers are textile grade rayon, regenerated cellulose, and cuprammonium rayon, although polynosic rayon [185] and viscose rayon fibers [186] have also been used as precursors for making carbon fibers.

1.3.1 Background

As mentioned earlier, the first carbon fibers produced commercially in 1879 were from cellulosic rayon precursor [3,4]. Cotton threads as well as bamboo fibers were formed into filaments and then pyrolyzed to a substantially all carbon filament. Carbonization of rayon yarns and fabrics was briefly investigated by Union Carbide Corporation as a possible substitute material for control grids in vacuum tube power amplifiers [187]. However, the first commercial venture into multifilament carbon fibers was setup in 1957 to manufacture carbon fibers in various forms from cotton and rayon precursors [187]. These materials were developed for use as high temperature thermal insulation as particle filters for hot, corrosive gases or liquids and as activated carbon fibers.

Union Carbide began commercial production of continuously processed various grades of carbon fibers produced from rayon in 1963. Rayon based carbon fibers were used successfully as heat shield and later for exit cone applications as well. In 1964 the company announced the first true high modulus carbon fibers from rayon [187]. The fibers were prepared by hot stretching rayon based carbon yarn at temperatures exceeding 2800°C [188]. A series of increasingly high modulus fibers were soon produced commercially, beginning with 'Thornel'25 in late 1965 and continuing until 1978. The production of these carbon fibers has been closed down because mainly of the high cost of production.

These high modulus rayon based fibers were used extensively in the development of polymer matrix composites for space structures. In recent years, the less costly high modulus PAN based carbon fibers have substituted for the rayon based fibers. However, because of the properties of low density, high purity (content of alkaline and alkaline–earth metal is less than 50–100 ppm), low heat conductivity and high strain, rayon–based carbon fibers are now still used in aerospace industry, especially used as missile's thermal protection material. In addition, they are also used in medical and other industries and there is still production in the United States and Russia [189–191].

1.3.2 Processing

Following are the three main stages involved in the development of carbon fibers from rayon.

     a. Low temperature decomposition (< 400°C).
     b. Carbonization (< 1500°C).
     c. Graphitization (> 2500°C).

Manufacture of Carbon Fibers

The yield of carbon fibers obtained by pyrolysis of rayon is usually in the range of 10 and 30% and depends upon the nature of the precursor fibers and the processing parameters. The molecular weight of the cellulosic precursor plays a vital role as it determines the number of end groups available for initiation of the degradation process. A slow rate of heating results in a higher yield but it is less economical, and for that reason, pyrolysis is carried out in a suitable reactive atmosphere [192,193]. This changes the course of the decomposition process so that it occurs at a lower temperature but at a much faster rate and results in a greater yield of the carbon fiber [194]. Since cellulosic materials suffer disruption of each ring to ring linkage and extensive weight loss, the graphitization is usually carried out under stress so that the atoms can be juxtaposed to fit into the graphite crystal.

1.3.3 Low Temperature Degradation

1.3.3.1 Inert atmosphere
Adsorbed water is first of all removed from rayon by heating it at 100°C in inert atmosphere. Using a slow heating rate, temperature is raised to 400°C and the fiber is ready for more rapid carbonization. Removal of tarry products is effectively accomplished if the heat treatment is carried out in an oil bath [195]. This technique has the additional advantage of providing a better heat transfer medium. The repeat unit of cellulose in rayon (Figure 25) is seen to contains hydrogen and oxygen which can result in elimination of five molecules of water giving a weight loss of 55.5%. The actual weight loss is of the order of 70–90%. This excessive weight loss has been attributed to chain splitting reactions involving CO_2, CO, alcohols, ketones, and a number of other carbon containing compounds [196,197]. In addition, the intermediate dehydrated stage produces levoglucosan when the pyrolysis is carried out in an inert atmosphere. This levoglucosan decomposes to give volatile carbon products during later stages of the pyrolysis, resulting in a further loss of carbon.

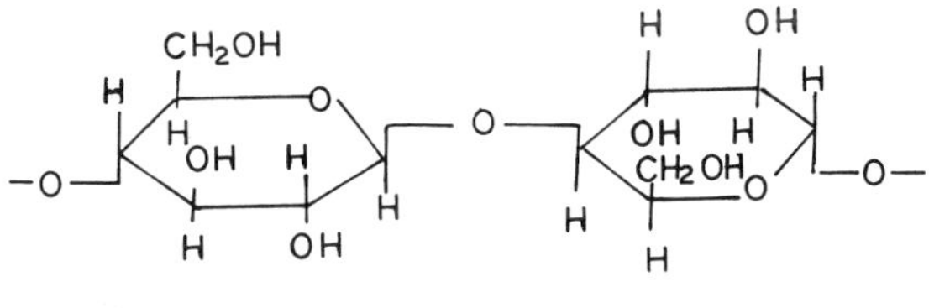

Figure 25 Repeat unit of cellulose in rayon.

Bacon and Tang [198] have studied the pyrolysis of rayon in a comprehensive manner. Although the pyrolysis in the low temperature region involves four different stages (briefly described below), the major pyrolytic decomposition takes place between 210 and 320°C with an ultimate weight loss of 70%–90%. The rate of heating has to be kept low to avoid excessive tar formation and tar deposition onto the fibers, which results in brittleness of the final carbon fibers.

Stage I (25–150°C) Absorbed water is eliminated during this temperature range, resulting in weight loss of around 12%. In addition, a small increase in the degree of lateral order has been observed [199] during this stage.

Stage II (150–240°C) Splitting up of the structural water occurs in stage II (150–240°C) from hydrogen and hydroxyl fragments present, preferable in equatorial positions. Formation of $C = O$ and $C = C$ bonds has been confirmed from the infrared analysis indicating that dehydration process is essentially intramolecular.

Stage III (240–400°C) Thermal degradation starts during this stage. Thermal degradation results from thermal cleavage of C–O and C–C linkages occurs (shown in Figure 26) thus leading to the formation of large amounts of tar, H_2O, CO, and CO_2.

Stage IV (> 400°C) Each cellulose unit breaks down into a residue containing four carbon atoms which then repolymerize into a carbon polymer ultimately producing a graphite like structure above 400°C (stage IV) by condensation reactions involving the removal of hydrogen. The possible scheme of reactions [198] for the various stages of the polymerization process is reproduced in Figure 26. A yield of a four carbon residue from each cellulose ring unit corresponds to an ultimate weight loss of 70.4%, which is actually obtained under certain conditions.

1.3.3.2 Degradation of rayon in reactive atmosphere
The carbon fiber yield and the processing rates are quite low when the pyrolysis is carried out under inert atmosphere. These can be markedly enhanced by carrying out the low temperature pyrolysis in the presence of a reactive atmosphere, such as air or oxygen [200], chlorine [201], and HCl vapors [202]. The presence of a reactive atmosphere inhibits the formation of tars [203] and promotes dehydration of cellulose. The dehydration reactions begin to occur much earlier than they do when heat treatment is performed in an inert atmosphere, and the major part of the pyrolysis takes place over a wider range

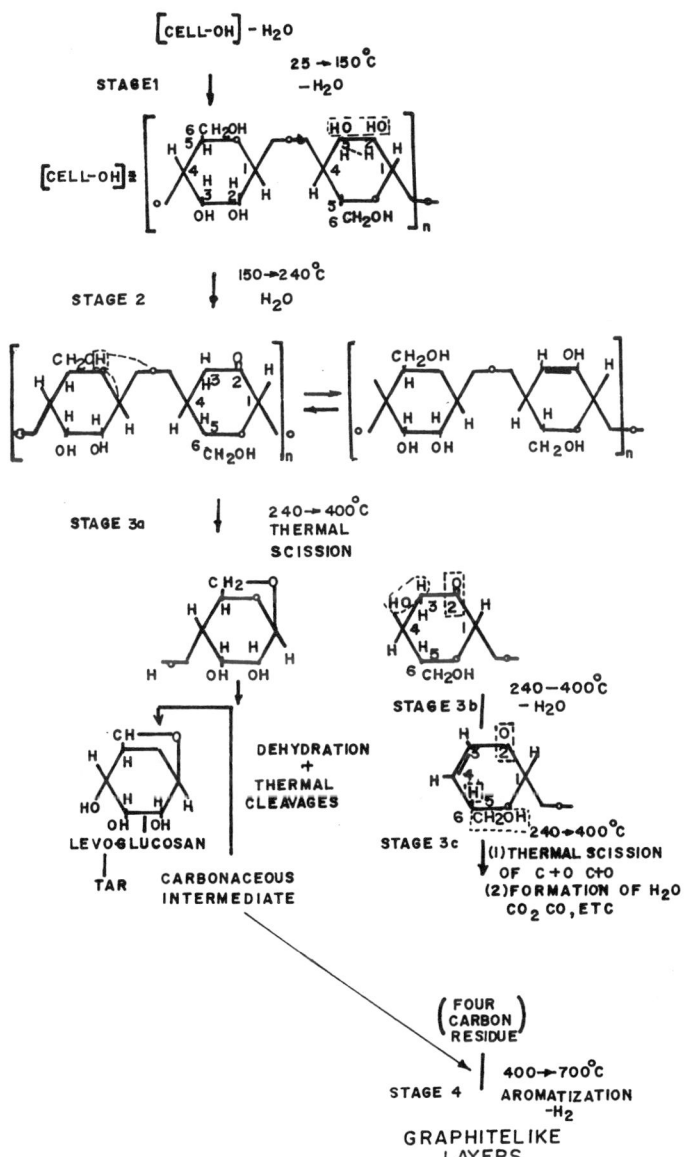

Figure 26 Reactions involved in the conversion of cellulose into carbon fibers (from ref. 195).

of temperature. It is evident (Figure 24) that the formation of levoglucosan, which later produces tar, occurs through primary hydroxyl groups. Pyrolysis in the presence of oxygen [204,205] results in the oxidation of C–6 methylol groups into carboxylic groups and thus prevents the formation of levoglucosan. In addition, the carboxylic acid groups, because of hydrogen bonding, can give rise to a highly cross-linked intermediate, resulting in a better carbon yield. The dehydration starts much earlier but continues through a wider temperature range.

According to Shindo [192], pyrolysis of rayon fibers in HCl vapor starts as early as 110°C and the weight loss occurs in a stepwise manner mainly through the elimination of water. The ultimate yield of the fiber corresponds to a yield of five carbon atoms from each cellulose ring unit as depicted in Figure 27. The processing time and the carbon yield vary with the nature of the reactive atmosphere. The time of processing has been found to vary between 20 minutes and 10 hours, and the carbon yield for rayon carbonized up to 1000°C varied from 23% in air or oxygen atmospheres [200] to 27% for a chlorine atmosphere [201], and to 35 % in HCl vapor [192].

Structural and mechanical property changes during the heat treatment of rayon in HCl [192] and in oxygen [188] atmospheres are shown in Figures 28 and 29. Tensile strength falls very rapidly during the initial stage of decomposition, which seems to occur exclusively in the amorphous regions. The tensile strength reaches a minimum near the point at which the destruction of the crystalline cellulose is complete and then begins its remarkable recovery. Contrary to the initial behavior of the tensile strength, the fiber modulus increases during the initial pyrolysis. Thereafter, tensile strength and modulus both fall together, but the modulus continues to fall long after the tensile strength has begun its recovery. Finally, the modulus begins to increase markedly as the graphitic layer structure forms and the fiber becomes a brittle carbon material.

1.3.3.3 Degradation of rayon in the presence of flame retardants
The efficiency of a reactive atmosphere in increasing processing rate and improving carbon yield, is controlled by the rate of diffusion of the reactive gaseous species into the interior of the fiber. Change in the rate of diffusion by altering the temperature condition leads to the degradation of fiber structure. This diffusion limitation was overcome by impregnating rayon fibers with certain reactive chemicals before carrying out the heat treatments.

A variety of chemicals viz. nitrogenous salts of strong acids [206], acids or acidic salts [207], metal halides [208], and various derivatives of phosphoric acid [209] were used to impregnate rayon fibers prior to heat treatment. These chemicals being flame retardant in nature alter the course of carbonization process by catalyzing the reaction of elimination of hydroxyl groups [210–212].

Manufacture of Carbon Fibers

This, thus, promotes the dehydration of the cellulose structure at low temperatures (150 to 200°C) and stabilize the molecule against excessive formation of volatile tars, the substances responsible for the flammability of cellulose [211].

The pyrolysis of rayon in the presence of flame retardants produces dehydrocellulose exclusive of any levoglucosan. Thus flame retardants also act as carbonization promoters since they speed up the reaction rates, spread the reactions out over wider temperature range and improve carbonization yields.

The tensioning or stretching is found to be ineffective in case of rayon during the low temperature heat treatment [188]. The reason of this is degradation of polymeric structure due to thermal and hydrolytic scission of the ether linkages containing cellulose units. Consequently no amount of tension or stretching is capable of maintaining or increasing molecular preferred orientation which otherwise is a possible mean of increasing the preferred orientation of final carbon fibers.

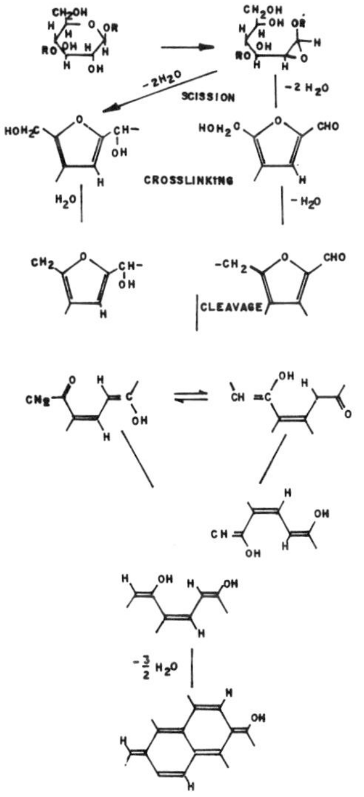

Figure 27 Conversion of cellulose into carbon in the presence of hydrogen chloride (from ref. 189).

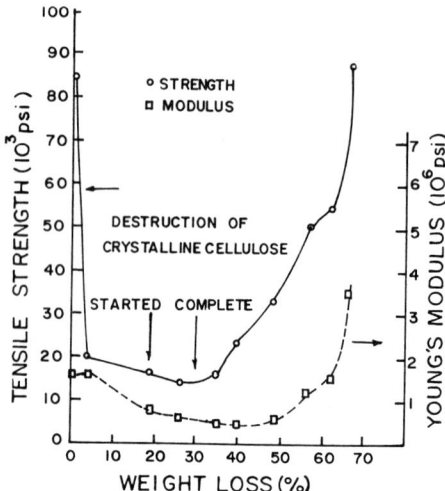

Figure 28 Tensile strength and modulus of rayon heat treated under HCl vapor at various stages of heat treatment and carbonization to 1000°C. Heating rate was 60°C (from ref. 189).

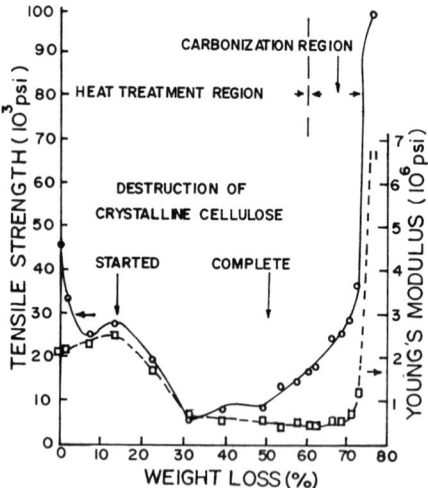

Figure 29 Tensile strength and modulus of rayon fiber heat treated under oxygen, at various stages of heat treatment and carbonization to 1300°C. Heating schedule: 18 min under oxygen to 40 weight loss: 12 min under nitrogen to 60% weight loss: 16 min to 73% weight loss (from ref. 188).

1.3.4 Carbonization

Carbonization of the properly heat treated rayon can be carried out under inert atmosphere in the temperatures usually ranging from 1000 to 1500°C in a time considerably less than 1 min. Investigators have also used various reactive atmospheres to carbonize the rayon fiber, for example, hydrochloric acid [192], ammonia, and alkylamines [188,213]. The use of tension during the carbonization of heat treated rayon yarn increases the degree of preferred orientation and hence an improvement in fiber mechanical properties is observed by Bacon [188]. Schalamon and Bacon in their patent [214] revealed the importance of stretching which states that a polymerized structure begins to form during carbonization, and stretching orients this structure to some degree. It also stated that stretching during carbonization raised the Young's modulus of carbon fiber (HTT 900°C) from 42 GPa to 68 GPa.

1.3.5 Graphitization

Rayon based carbon fibers even after heat treatment to 1500°C do not exhibit the properties desired for advanced applications. It therefore becomes an experimental necessity to increase the mechanical properties of these carbon fibers so as to make them of practical use. It is well known that all carbons become plastic at a temperature of around 2000°C and as a result development of graphitic structure takes place rapidly in the carbon. Accordingly to Ruland [215] and also as described in Section 1.3 viscose rayon fibers degrade in such a way that the original orientation texture of the fiber is almost complete at around 240°C. During further heat treatment to final carbonization temperature between 900–1000°C, there is a limited improvement in the orientation of the carbon fibers. Carbon fibers at this stage will possess a good number of cross-links between fibrils. Additionally, viscose derived carbon fibers are hard carbons and show over resistance to the improvement in the orientation until these are subjected to the graphitization temperature of 2000°C and above.

Thus, graphitization which is usually carried out above 2800°C [188] for a fraction of second under stress is an effective way to obtain a high modulus carbon fiber. The Young's modulus of the graphite fibers depends on the effective stretch during graphitization and on the graphitization temperature. Young's modulus increases with increase in effective stretch and graphitization temperature as illustrated in Figures 30 and 31. The samples that stretched over 100% during heating for an additional 45 to 59 minutes to about 2800°C has been found to have an increased tensile strength. The mechanical properties of some of commercially available rayon based carbon fibers are listed in Table 4.

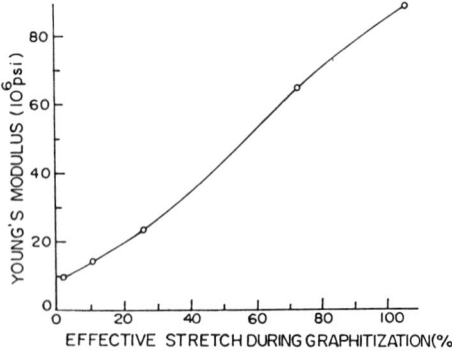

Figure 30 Effect of stretching during graphitization to 2800°C on graphite fiber Young's Modulus (from ref. 188).

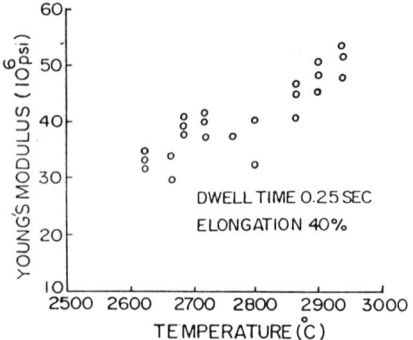

Figure 31 Effect of graphitization temperature on Young's modulus [from ref. D. W. Gibson, a poster ACS Polymer reprints, 9: 1376 (1968)].

Table 4 Mechanical Properties of Some Commercially Available Rayon Based Carbon Fibers

Manufacturer	Product name	T.S (MPa)	Y.M (GPa)
	Thornol 50	2070	345
Amoco Corp.	Thornol 75	2520	517
	Thornol 40	1720	276
Hitco	HMG 50	.70	345
(U.S.A.)	HMG 40	1720	276
	HMG 25	1030	172

1.4 PITCH BASED CARBON FIBERS

Pitch-based carbon fibers provide properties not readily obtainable with PAN-based fibers. Whereas the PAN-based fiber has excellent strength at a modulus of 200 GPa, strength decreases as the modulus is increased, and the upper limit of modulus is about 650 GPa. Pitch-based fibers are capable of modulus levels up to the theoretical modulus of graphite, 1000 GPa, and have much better thermal and electrical conductivity properties than PAN-based fibers. Pitch and PAN-based fibers should be considered complementary, each filling a different set of commercial needs.

The raw materials for pitch are abundant and cheap. They are by-products of petroleum and coal-chemical industry (such as petroleum refined residue, coal tar, and pitch, the residue of solvent refined coal and petroleum) and some pure aromatic hydrocarbon (such as naphthalene, anthracene). Isotropic and anisotropic pitch with proper softening point and good spinnability can be purified by the methods of thermal modification [216–220], solvent modification [221–224], hydrogenation [225–229] and catalytic modification [230–232] etc. Then, followed melt spinning, stabilization, carbonization, graphitization, surface treatment and sizing, general performance carbon fiber (GPCF) and high performance carbon fiber (HPCF) can be manufactured, respectively. A manufacturing process of carbon fibers from pitch is presented in Figure 32, and the properties of GPCF and HPCF are shown in Table 5.

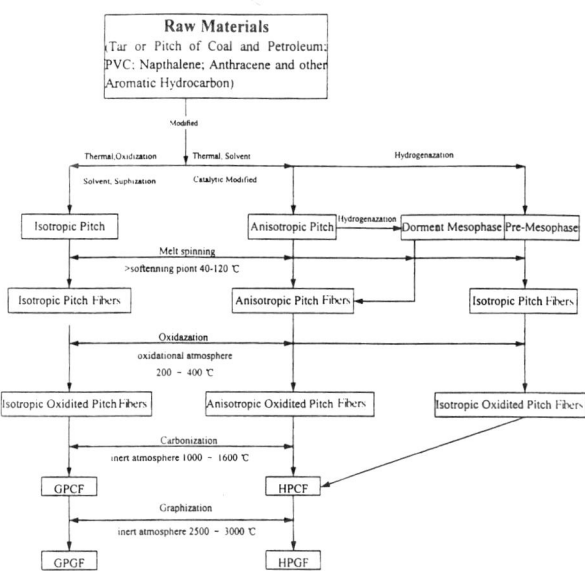

Figure 32 Schematic process of manufacture for pitch-based carbon fibers (from ref. 260).

1.4.1 Background

There are two distinct types of pitch-based carbon fiber available commercially. There is the "General Purpose" fiber, which has a tensile strength of up to 700 MPa and modulus of 30 to 60 GPa, and the "High Performance" fiber, which has tensile strength up to 4 GPa and modulus from 180 to 900 GPa. There is also a version of the "High Performance" fiber optimized for thermal conductivity, which, at 1000 $Wm^{-1}K^{-1}$ is 2.5 times more conductive than copper. The pitch making and spinning technologies used to make the two types of fiber are quite different.

General performance carbon fiber (GPCF) was developed by Otani in 1963 [216,217]. These fibers are sometimes referred to as "isotropic" carbon fibers, since they are made from isotropic pitch. These pitches are prepared from high boiling fractions of petroleum feedstocks, usually heavy slurry oils produced in catalytic cracking of crude oil. Based on Otani's work, in 1973 Kureha chemical industry company using naphtha and by-product tar of ethylene and acetylene through the high temperature water steam cracks as raw materials developed carbon fiber factory with production capacity of 120 tons per year [218–220]. This kind of carbon fiber can be used as insulator materials of furnace, electrode boards of fuel cell and concrete filler. With the expanding of the market, the output of carbon fiber was increasing continuously. Some other companies also developed this kind of material. For example, in 1982 Ahsland company of USA built a production line with capacity of 100 tons/year and enlarged it to 160–180 tons/year in 1985 [233]. This line was sold to Anshan East Asia Carbon Fibers Inc. of China in 1993, and has already been put into production in 1994.

GPCF is cheap, and it has the typical advantages of carbon materials, such as light weight, electrical conductivity, chemical inertia, heat resistance, abrasion resistance and so on. It can replace some of the asbestos products (such as break disc, seal materials, abrasion resistance and insulator materials). GPCF can also be used as shield cover, static electricity resistance brush in electrical conductive plastic, fire–resistant clothes in textile industry, as well as to produce ACF for the protection clothes and respirator in military industry [238].

Taylor and Brooks developed high performance carbon/graphite fibers (HPC/GF) from the finding of carbonaceous mesophase, which is a scientific basis of pitch based carbon fiber, and consequently its industrial technology was developed by L. S. Singer of Union Carbide company (UCC) in 1970 [239–241]. The key idea was to convert the pitch raw material to a mesophase, or liquid crystal pitch, which was known to be capable of orienting through flow and shear [242]. UCC commercialized HPCF in 1976 by building a production line with capacity of 230 tons per year. From 1979 to 1982, Thornel

P–100, –120 (tensile strength 2.2 GPa, tensile modulus 724 and 827 GPa respectively) were put into market.

Mesophase pitch fibers can be converted to carbon fibers with 80–90% yields by weight and very easily developed to high modulus fibers merely by heat treatment. These fibers are much more graphitic than PAN or rayon based fibers and thus exhibit very high values of thermal conductivity, electrical conductivity, low value of thermal expansion etc. For example, the thermal conductivity of a mesophase pitch–based carbon fibers, Thornel P–140, is twice that of copper, and its specific thermal conductivity (thermal conductivity / density) is eight times that of copper. All of which make them particularly suitable for many of the critical aerospace applications [243].

In the 1980s, in order to lower the spinning temperature, improve the spinnability of mesophase, and reduce the cost of carbon fibers, lots of researchers developed new types of mesophase, especially in Japan and USA. So the patents of neo-mesophase [222–224], pre-mesophase [225–227], dormant-mesophase [228, 229] and soluble-mesophase [230–232], which are the precursors for the manufacture of high performance carbon and graphite fibers were rapidly appeared. In 1987, Mitsubishi company built a factory with capacity of 500 tons pitch-based HPCF [233]. The highest tensile strength was 3.0 GPa, and modulus was 734 GPa (Table 5).

Table 5 Properties and Production of Commercially Available and Developing Pitch Based Carbon Fibers

Sort	Manufacturer	Tensile strength (GPa)	Modulus (GPa)	Capability (T/Y)
GP	Asland	686	41	140
	Donac	686	34	300
	Kureha	590–790	30–33	900
	Nittobo	657–980	40–49	in development
	Nippon	784–980	39–49	in development
HP	Amoco	1300–2400	170–960	140
	Donoc	1800–3000	140–600	in development
	Kureha	1800–4000	150–400	in development
	Mitsubishi	1800–3000	176–735	500
	Chem	3230–3300	392–686	50(building)
	Nihon Sekiyu	2450–2940	392–588	in development
	Nippon Steel	1470–3000	147–784	12/24(building)
	Petoca	5000	>500	in development
	Teijia	2940	490–686	12(pilot)
	Ton Nenryo			
ACF	Donac	78245	Specific area 1000–2000M^2/g	in development

1.4.2 Pitch

Pitch is a mixture of fused ring aromatic compounds (possessing 3–8 rings) with alkyl side-chains and heterocyclic compounds, which has a very complex structure and chemical composition. Its molecular weight is between 300 to 400 [244]; carbon content more than 80% and softening point below 120°C. Its normal state is black and shiny solid. The properties of pitch depend on its chemical composition and molecular weight distribution. Although there are many kinds of industrial pitches, not all of them are suitable for manufacturing carbon fibers, so it is necessary to make some adjustment and do some modification pretreatment.

1.4.2.1 Fined pitch
Raw tar and pitch, especially the by-product of coal-chemical industry, generally contain some solid impurity and primary quinolin insoluble (QI). When heated, they easily form mesophase. However, many impurity and free carbon are adsorbed to hinder the sphere growth and coalescence, which makes it difficult to form continuous large anisotropic domain with good rheology properties and spinnability; as well as to form the homogeneous isotropic pitch. The fiber containing this type of fine particles develops crack and pores, leading to poor quality carbon fiber. So whether preparing isotropic pitch or mesophase pitch, the solid impurity and primary QI need to be removed to get fined pitch mainly composed of hydrocarbon [245].

Generally, coal tar pitch is refined by solvent refinement or thermal filtration method, but petroleum pitch is by vacuum distillation or wiped film evaporator method. Two methods which have been adopted in the industrial process are introduced in the followings. One is solvent oil static deposited method developed by Mitsubishi Chemical company [246]. In this method, the light distillation fraction of petroleum and coal tar (180~300°C) were used as solvent oil into pitch. It was static for 1~3 hours at 150~190°C, then the pure pitch with QI < 0.1% and yield of 70% was obtained. The other one is called "CHERRY–T" method developed by Osaka Gas [247]. In this method, coal tar was introduced into the primary autoclave, after thermal condensation for 5 hours at 0.9 MPa and 410°C, the F pitch (content primary QI and poor stability part) was removed, the 203~360°C distillation fraction was then introduced into the secondary autoclave. After thermal condensation for 10 hours at 0.9 MPa and 450°C, the pitch with softening point 65~110°C and QI free was obtained, which is called C–pitch. This kind of pitch is an excellent precursor for carbon fibers. In recent year, the super critical fluid extraction (SCFE) method has been developed [248,249] which has both the function of distillation and extraction—a new technology for purifying raw pitch. Thus, the qualified precursors after purification, using various methods mentioned above, for carbon fibers with low ash content, low heteroatom, low metal ion and high aromaticity can be obtained.

1.4.2.2 Preparation of isotropic pitch
Pitch is a kind of thermal-plastic material. When heated to a certain temperature higher than its softening point, it will soften and melt. In the process of manufacturing pitch-based carbon fibers, pitch precursors need to be undertaken oxidation process. Otherwise, in the higher temperature of carbonization process, it melts or fuses and can not keep its fiber form but convert into coke. After oxidation process, the large planar structure of pitch fused ring hydrocarbon converts into thermal stable oxygen-bridge structure. Generally, the oxidation temperature is about 200~400°C, and the softening point of pitch is 230~280°C. In most cases, during the thermal condensation process to increase the softening point, the anisotropic pitch easily appears. The spinnability of the isotropic pitch is poor because of the existence of anisotropic pitch. An important factor to fabricate GPCF is to increase the softening point of pitch and avoid the appearing of mesophase at the same time, thus the isotropic pitch with high softening point and spinnability could be achieved.

There are several methods available for preparing isotropic pitch: reduce pressure thermal condensation [250], wiped film evaporator [245], air blow oxidation [251], sulphidation [252], additive method [253], and PVC method [217,254,255]. The common ground of these methods is that, the light fraction from the purified pitch is removed during the thermal condensation (> 350°C), i.e., when temperature is increased, dehydrogenation, cross-linking, condensation etc. occur, and release H_2, H_2O, H_2S, and low molecular weight compounds. Therefore, the molecular weight increases, H/C ratio decreases, softening point increases. For example, ethylene tar has obvious changes after reduced pressure condensation (380°C, 2 hrs.): the average molecular weight from 556 to 1027; H/C from 0.95 to 0.64; softening point from 40 to 265°C [250]. The wiped film evaporator method and blow air oxidation method have been developed and applied in industry for increasing softening point and avoiding the forming of mesophase. The key technology of the first method is increasing the evaporate rate to avoid the mesophase appearing. The later, through control of oxygen pressure in air-blow and thermal treatment temperature to arrest the formation of mesophase, the pitch with high softening point (250~300°C) and good spinnability can be achieved.

As stated above, different purified pitches are chosen as raw materials and modified by various methods to get isotropic pitch. For general performance carbon fiber (GPCF), isotropic pitches usually need to be qualified having [250]:

1. Softening point: 230~280°C.

2. H/C: 0.55~0.70.
3. QI: less than 5%.
4. Mw/Mn: about 1.13.
5. Nonthixotropic character.

1.4.2.3 Preparation of anisotropic (mesophase) pitch
Heat treatment of pitches above 350°C for a prolonged time results in the succession of dehydrogenative condensation reactions, amongst the molecules, thus forming large molecules which further aggregate into liquid crystalline phase with nematic order. This phase, known as mesophase (Greek for changing phase), has higher surface tension than the low molecular weight isotropic liquid phase from which it grows. It forms small liquid droplets which grow in size, coalesce into larger spheres, and eventually become extended anisotropic regions [258]. The amount of mesophase continues to increase, and when it reaches about 50%, a phase inversion occurs and the anisotropic phase becomes continuous and surrounds the droplets of the remaining isotropic material. The mesophase polymerizes on further increase in heat treatment temperature and results in the solid formation which is well-known as coke. This solidification is one of the limiting factors in the production of carbon fibers which must be avoided.

In a word, mesophase is a kind of mesogen state of transition from isotropic to anisotropic semi-coke. The formation of mesophase is closely related to molecular structure and weight of mother pitch and the carbonization condition. It needs certain size and a smooth degree of aromatic layer. From the time Brooks and Taylor discovered the nematic character of mesophase in 1965, mesophase was considered to be a kind of irreversible chemotropic liquid crystalline at the early time [259,260]. It was later proved that the nematic liquid crystalline of laminar aromatic molecular has multiplicity state. When studying the pyrolysis of naphthalene, Lewis discovered the thermotropic mesophase [261] in 1978, later lyotropic mesophase was found [262,263]. Both mesophases have one common character that the formation and conversion of mesophase is reversible. The phase transform of thermotropic mesophase with change of temperature is due to the high thermal stability, low reactivity and low molecular weight of the mesophase molecules. Lyotropic mesophase convert with the mesophase molecular concentration and the temperature of the system or both of them, that is, the mesophase deposited at saturated concentration in mother pitch and dissolved again when adding mother pitch. Riggs and Diefendorf [262] also discovered the lyotropic character when they extracted mesophase with solvent, based on the theory of neo-mesophase. The difference between carbonaceous liquid crystalline and general liquid crystalline is that the former is disc-like nematic liquid crystalline composed of laminar aromatic molecules.

Manufacture of Carbon Fibers

The formation of oriented large laminar aromatic molecules inside the mesophase sphere is affected by the types and composition of raw materials and the preparing condition, leading to different arrangement forms [264,265]. Aided by polarized microscope, four different types of arrangement were observed. As illustrated in the Figure 33, (a) is typical B–T type; (b) is oblate laminar; made from naphthalene and anthracene, $AlCl_3$ as catalyst, heat treated 30 mins at 400 magnetic field [266]. In addition, it can also be formed when pitches contain several percent of carbon black or graphite powder [267]; (c) is radial type, obtained from pitch heat treatment at 300°C for long time [268]; (d) is concentric circle type, from decacylene heat treated at 540~550°C [269]. From the view of molecular composition of inside carbonaceous mesophase sphere, no matter what kind of mesophase, the laminar is parallel stack by large fused ring aromatic molecules. A new structure model [270] shown in the Figure 34 indicates that aromatic ring carrying naphthenes or short alkyl chain, less loose orderly arranged by fused ring aromatic hydrocarbon of different molecular weight form disc nematic liquid crystal. It has good rhology character at high temperature. With the increase of temperature, mesophase spheres coalescence and break up, leading to form bulk mesophase, in which the misalignment transform termed «disclination» was discovered [271]. The orientation of the discotic layers and the presence of disclination influence the structure and properties of the final carbon materials.

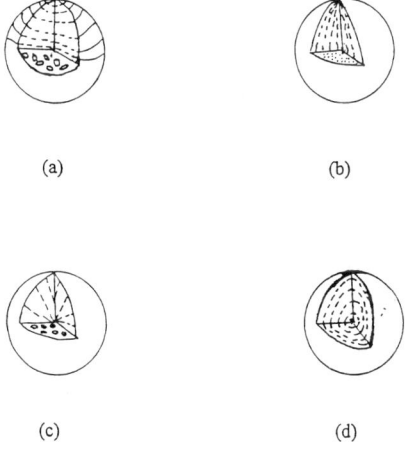

Figure 33 Schematic orientation arrangement of molecule in mesophase sphere (a) typical B–T type; (b) oblate laminar; (c) radial type; (d) concentric circle type (from ref. 264).

The methods of preparing mesophase pitch include thermal modification, solvent modification, hydrogenation and catalyst modification. For high performance carbon fibers (HPCF), mesophase pitch needs to have:

1. Low ash content, low metallic ion and hetero-atom.
2. 100% anisotropic content.
3. Softening point in the range of 230 to 280°C.
4. Low viscosity at the spinning temperature, and little change of viscosity with temperature.
5. Good spinnability and thermal stability.
6. High oxidization activity and high carbon yield.

Therefore, raw material must be chosen carefully, purified and modified properly. There are lots of patents of preparing mesophase pitch, some of them are briefly described as followings:

1. Thermal modification: It includes one stage atmospheric pressure method [272–275], two stage pressure-vacuum method [276, 277], and reflux–air blowing [278] method etc. One stage atmospheric pressure method was developed by UCC. This method is to remove the light fraction which hinders the formation of mesophase from the autoclave to prevent the excess condensation. In the patent of UCC, a process with vigorous stirring and large nitrogen blowing was described to prepare mesophase pitch with low viscosity (no greater than 200 poises at 380°C), low molecular weight (average molecular weight below 1000), appropriate softening point (SP, no higher than 350°C), 100% anisotropic content (AC) and excellent spinnability. Whereas, pressure–vacuum method and reflux–air blowing method give the mesophase pitch with AC: 100%; SP: 305–325°C and AC: 95%, SP: 315°C, respectively. However, the weakness of these methods is the resulted pitch with high softening point and poor spinnability. In order to lower the softening point, the aromatic hydrocarbon with naphthenes and alkyl side chain is chosen as the raw material. 100% anisotropic content with good spinability and non–thixotropic pitch [275] could be possibly reached through one stage atmospheric pressure with nitrogen blowing method, resulting in mesophase with lower softening point (280°C), and low viscosity (a viscosity no greater than 20 poises at 370°C).

2. Solvent modification: Based on the "micella model" and solubility parameter theory of Diefendorf and Riggs [279], it was developed into a method to produce neo–mesophase by EXXON company [280]. They used a solvent extraction process to remove mesophase forming molecules from the isotropic pitch. This approach was extended by Greenwood to obtain a "heart cut" pitch [234]. First, the slurry oil would be heat soaked for a moderate period of time, say 10 or 12 hours. Then the product would be dissolved in a solvent,

such as benzene or toluene, and the insoluble fraction (high molecular weight material) rejected. Next, a poor solvent was added to the solution, causing precipitation of the higher molecular weight material remaining. The lowest molecular weight material remained in solution, and was rejected. The precipitate was then dried and prepared for spinning. Such pitches typically had negligible quinoline insoluble content. This was a particularly effective way of making very high strength and modulus fibers [235].

3. Hydrogenation: Mesophase pitch turns into partly hydrogenated fused ring aromatic hydrocarbon through hydrogenated reaction. Due to the change of carbon atom orbit and the molecular structure, the π-electron conjugation system is reduced and the steric hindrance increases. The interaction between molecules is weakened, resulting in the formation of pitch with higher molecular weight, lower melt temperature and lower viscosity. Yamada and Honda et al. [281–283] discovered the pre-mesophase and dormant-mesophase by treating 1,2,3,4, tetrahydric-quinoline and pitch through hydrogen transfer reaction to form partly naphthenated pitch molecules. After rapid thermal treatment, the pre-mesophase is appeared, which, after carbonization, shows an anisotropic character. Otani [284] converted anisotropic pitch into isotropic pitch through hydrogenation, then obtained dormant-mesophase by heat treatment. It has potential orientation, exhibiting anisotropic character under the shear force in the spinning process (see Figure 32).

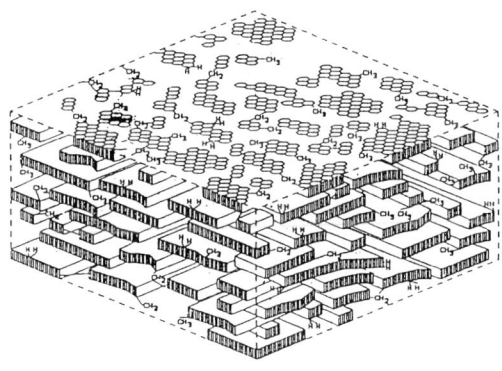

Figure 34 Structure model of soluble mesophase pitch (from ref. 270).

4. Catalytic modification: Based on the non-dehydrogenation cation polymerization theory [285–287], Mochida et al. used naphthalene, anthracene and other aromatic hydrocarbon as raw materials, with aid of Lewis acid (e.g., $AlCl_3$, HF/BF_3), at low temperature through catalytic polymerization and obtained the mesogen with naphthalenes and alkyl chains. After thermal condensation, soluble mesophase with AC 100% is reached, which has low softening point, low viscosity, high heat stability, good oxidation reaction character, and good spinnability [288–290]. The processes are illustrated in Figure 35. Through co-carbonization reaction to prepare mesophase pitch, the alkyl and naphthenic group can be adjusted by the amount and type of co-carbonization agents. Therefore, the spinnability of mesophase pitch can be adjusted through controlling the molecular structure [291]. For example, Yoon et al.. [292] used methyl-naphthalene and naphthalene as starting materials and HF/BF_3 as catalyst, and got the mesophase through co-carbonization reaction with softening point of 235°C, AC: 100%. The graphitized fiber from this pitch further heat treated at 2500°C for 2 mins shows high values of tensile strength (4.75 GPa) and Young's modulus (975 GPa). Especially, it shows an large value of elongation of 0.5% [292]. There are also many researchers using different types of catalyst such as the super acid of ZrO_2/SO_3^{2-} [293] and the iron system catalysts (ferrocene, iron benzoate, and naphthoate) [294,295] to prepare mesophase pitches.

Mitsubishi Gas Chemical (MGC) plans to build an installation with capability of 1000 tons/year by catalytic modification [296]. Compared with other method, it has advantage of high yield and good properties, low cost, etc., and may have a splendid future.

1.4.3 Melt Spinning

Pitch is spun by centrifugation [297], spurt [298], and vortex [299] into isotropic pitch fiber as shown in Figure 36. Pitch is feed into the screw extrude with gas-off (single or double screw). After melt evenly mixed, pitch is sent to spinning component (filter layer, distribution board and spinneret) by gauge pump, and melt pitch exits spinneret by pressure. Anisotropic or isotropic continuous pitch fibers can be reached after drawn, solidified, collected, oiled, and winded processes [300,301].

Generally, both isotropic and anisotropic pitches are pseudo-plastic fluid with character of shear thinning [275,302]. Their viscosity decreases with the increase of shear rate, as seen in Figure 37. The flow activation energy of pitch is very high (100–200 KJ/mol), about 2 to 4 times higher than that of polyester (54 KJ/mol), so pitch is very sensitive to the temperature. As a result, it is necessary for spinnable pitch to have good viscosity-temperature properties (shown in Figure 38). The relationship of shear rate and shear stress with different temperature is shown in Figure 39 [303]. When temperature is below T2, hysteresis loop appears, which indicates the thixotropic character of pitch. Only until temperature is above T2 does pitch show plastic flow which is suitable for spinning. Research results show that the structure and properties of carbon (graphite) fibers have close relationship with those of pitch fiber. For example, the thinner the diameter of pitch fiber, the higher the tensile strength of carbon fiber; the performance of carbon fiber from pitch fiber with onion structure is better than radial structure. In order to get such fibers, it is necessary to modify pitch precursor, change spinneret structure and adjust the spinning technology.

From the view point of mass and force balance along the fiber, during the pitch melt spinning process, the spun fibers should be drawn and wound up quickly to get fiber of 8–14 μm diameter. Because the weak pitch fiber (with tensile strength of 10–50 MPa) is easy to break, thus wind–up plays an important role in producing good pitch fiber. There are two methods of winding: one is direct winding [304] seen in Figure 40. In this process, mesophase exits from spinneret, followed by high speed drawing, air blowing, collected, and wound with a speed of 10–1000 m/min. The other method is to collect the fibers in a conveyer belt or container on the conveyer belt, then oxidize the pitch fiber as illustrated in Figure 41 [305]. In this method, anisotropic pitch with softening point of 272°C is used, and spun at 305°C; fiber was drawn at the speed of 250 m/min, and conveyer belt moved at 2 cm/min. The length of furnace is about 6 m with an inlet temperature 150°C and outlet temperature 260°C. After stabilization, five bundle fibers are plied into 1K to go through carbonization furnace at 2 m/min. Continuous carbon fibers with length of 6 km can be reached.

Through controlling the spinning technical parameters (such as increasing the drawing speed), fiber diameter can be reduced to 8 μm. Melt spinning temperature [306] and spinneret structure [307,308] also directly affect the quality of the pitch-based carbon fibers. For example, at higher temperature, onion-radial transverse structure is obtained; but at low temperature, radial structure was instead (seen in Figure 42). The tensile strength of onion-radial and radial structures is 2.88 GPa and 2.78 GPa, respectively; with tensile modulus of 584.3 and 433.9 GPa [309]. By utilizing filament formation geometry to establish preferred flow profiles and spin conditions that complement them, structure can be manipulated and controlled. Example geometries, when coupled with appropriate feedstocks and operating conditions, leading to structure control and resultant product responses, are shown in Figure 43. Fiber cross-sectional structure, as observed by SEM, is schematically represented while product categorizations of physical and thermal properties are noted. The typical fiber structures illustrated here have been labeled by several researchers as "pacman" radial, wavy radial and severe "pacman." Other structures such as random, onion-skin and "PanAm" have also been produced and categorized. The fibers with "pacman" cross-sections have longitudinal splits which may adversely affect physical properties. Downstream processing, within limits, appears to have minimal influence in changing the general "structure" established in the filament formation step. Subsequent heat treatment densifies the initial structure, i.e., increases the packing to increase tensile and thermal properties and modulus.

Catalytic Modified Method:

A) Alcl₃ Catalytic Modified Method [284]:

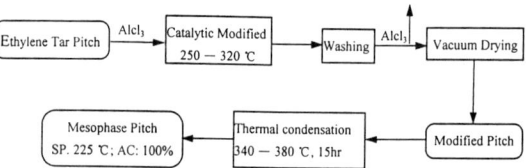

B) HF/BF₃ Catalytic One Stage Method[285,286]:

1)
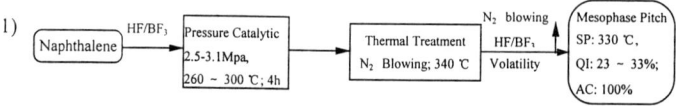

2)
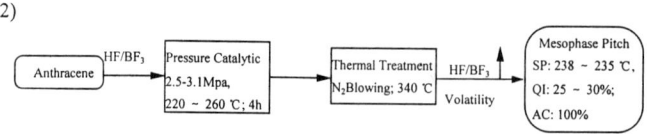

C) HF/BF₃ Catalytic Two Stage Method[285]:

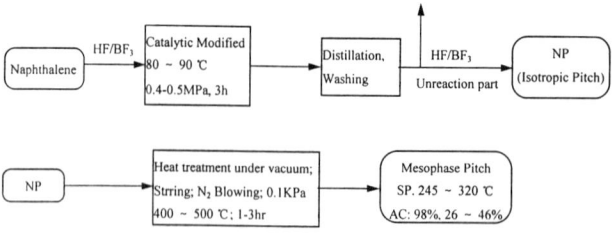

Figure 35 Schematic process of catalytic modification methods for preparing mesophase pitch (from ref. 288–290).

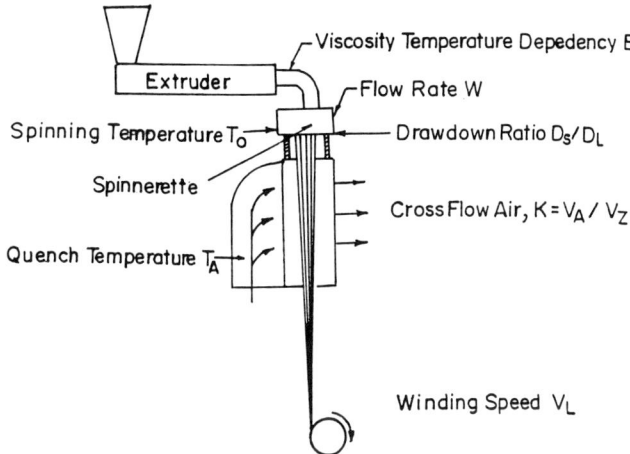

Figure 36 Commercial melt spinning equipment for obtaining the multifilament mesophase yarn (from ref. 301).

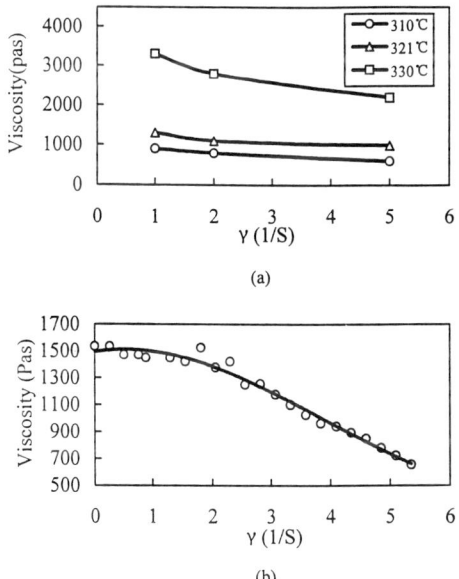

Figure 37 Effect of shear rate on viscosity of pitch (from ref. 275,302).

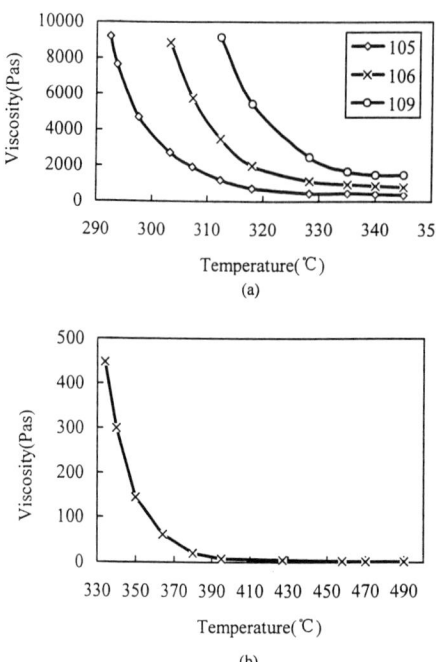

Figure 38 Effect of temperature on viscosity of pitch (from ref. 275,302).

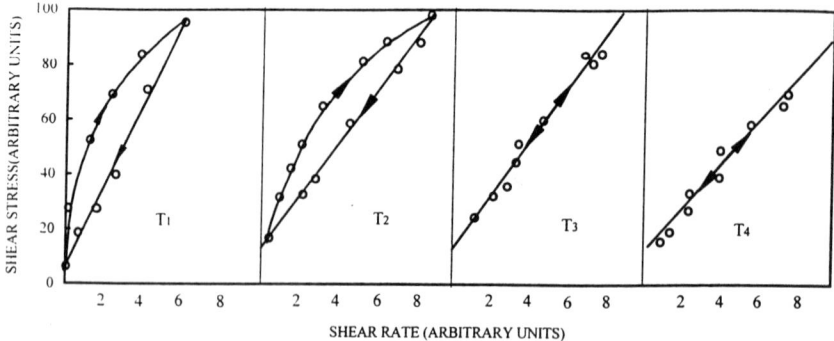

Figure 39 Relationship of shear stress and shear rate at different temperature (from ref. 303).

Manufacture of Carbon Fibers 55

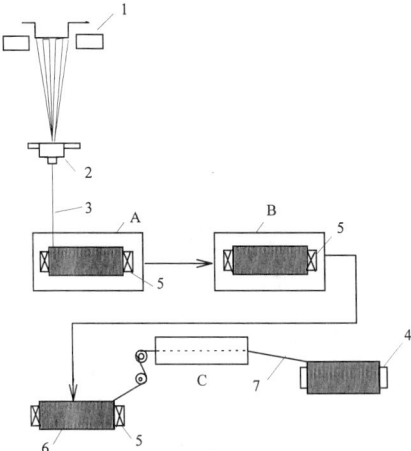

Figure 40 Schematic process of directly winding pitch fiber, oxidation, carbonization and graphitization. (1: spinning apparatus, 2: drawing gas, 3: fiber bundle, 4: bobbin, 5: heat-resisting bobbin, 6: pre-carbonized fiber, 7: carbon/graphite fiber, A: stabilization, B: pre-carbonization, C: carbonization/ graphitization) (from ref. 304).

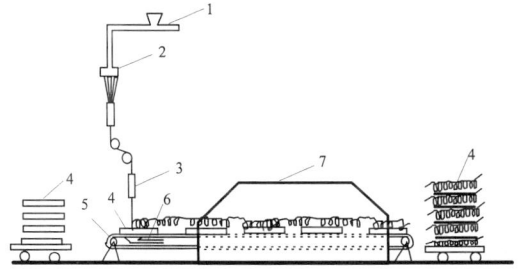

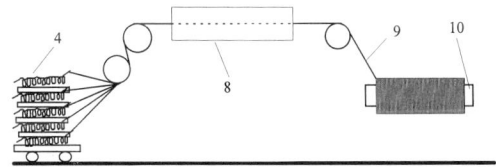

Figure 41 Schematic process of collecting pitch fiber in container, oxidation, carbonization and graphitization. (1: extruder; 2: spinneret; 3: fiber guide; 4: container; 5: conveyor belt; 6: suction stallion, 7: stabilization furnace, 8: carbonization/graphitiation, 9: carbon / graphite fiber, 10: bobbin) (from ref. 305).

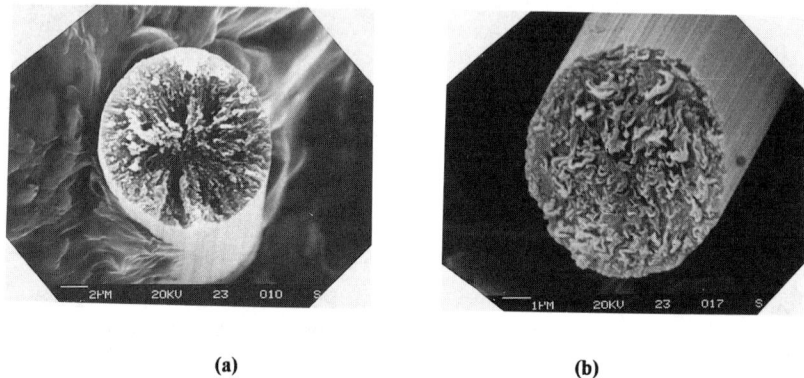

(a)   (b)

Figure 42 SEMs of mesophase pitch based carbon fiber, a: at low temperature (340), b: at high temperature (375) (from ref. 309).

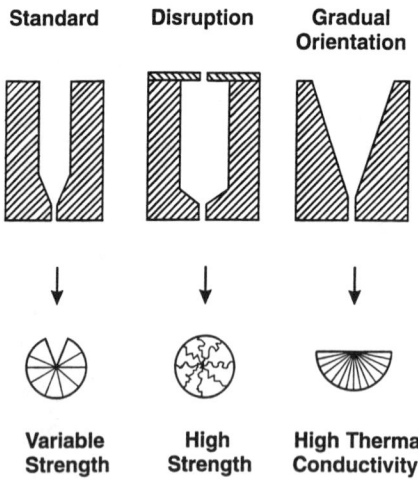

Figure 43 Control of pitch-based carbon fiber "structure/texture" and resultant properties.

1.4.4 Stabilization

Pitch is a kind of thermal plastic materials and can not be carbonized in nitrogen atmosphere without stabilization. This is because pitch fibers can not keep their fiber form but soften and melt at high temperature. Before carbonization, oxidation is necessary to change the thermoplastic character into thermosetting character. Oxidation proceeds in both gas phase and liquid phase. Gas may be oxygen, ozone, air, mixture of oxygen and nitrogen (40/60) [313], NO_2, SO_2, SO_3, and other oxidative atmosphere. Liquid phase may be nitric acid, sulfuric acid, potassium permanganate, and hydrogen peroxide. oxidative liquids. There are a number of alternate possibilities to accomplish stabilization, but the most common process is air oxidation in order to reduce the cost and lower the pollution, and the processes are schematically illustrated in Figures 40 and 41.

The main technological parameters of oxidation process include temperature, time, concentration of oxidant and the stress on the fibers. Oxidation degree is a function of concentration, temperature and time. The properties of carbon fibers are directly affected by the degree of oxidation. If oxidation degree is low, fibers melt during carbonization; if oxidation degree is too high, the properties of carbon fibers become poorer [314–316]. According to different structures and compositions, the best oxidation time of isotropic pitch fiber (more than 3 hours) is longer than that of anisotropic pitch fiber (less than one hour) at the same amount of air condition. The best condition for the oxygen content of isotropic pitch fibers is about 20%, while anisotropic about 8–10%. The final oxidation temperature of isotropic fiber is 325–340°C, while anisotropic is 300–310°C. That is because when the condensation degree of isotropic is lower, softening point also lower, and the hydrogen content is higher. Therefore, the formation of infused oxygen bridge structure of isotropic pitch through oxidation, dehydrogenation, cross–linking and cyclization reactions during thermal oxidation process needs a longer time and forms a more disordered structure than that of anisotropic pitch. There are different views on the effect of stress imposed on fiber during oxidation process. Some researchers considered that pitch has large planar molecular structure, so the stress on fibers during oxidation process was unnecessary [317–319], but others [320–322] proposed that stress did help the microcrystalline of fibers preferred orientation, so tensile strength and modulus increased. Stress can avoid the laceration caused by tensile stress during melt spinning and shrinkage resulting from the change of chemical structure during oxidation process [315], thus the drawing can improve carbon fibers properties. The properties of carbon fibers decrease while the stress is a little stronger than the interacting force of pitch molecules—in the excess stress case, pitch fiber will break. To improve carbon fiber properties, it is very important to select proper stress according to thermal stress-strain curve.

The study on isotropic and anisotropic pitch fibers oxidation in air by DSC and TG [315,323] shows that the interaction of oxygen with pitch is a weight increased and exothermic reaction. The element analysis result indicates that during oxidation process, H/C ratio decreases, while O/C ratio increases. FT–IR results [315,323,324] (see Figure 44) show that the peak intensity of $-CH_3$ and $-CH_2$ Ar–H decreases with the depth of oxidation. Results from the above experiment may be drawn as the following: during the stabilization process in the air, dehydrogenation, cross-linking, cyclization, etc. take place and give off CO, CO_2, H_2O, and low molecular weight hydrocarbon. More stable oxygen-bridge structures such as acid anhydride, carbonyl group, aromatic ether and phenolic hydroxyl group are formed in the process. The thermoplastic character of pitch has been changed into thermosetting after stabilization process thus pitch fibers could keep the fiber form during the next carbonization step. The reaction model scheme is shown in Figure 45.

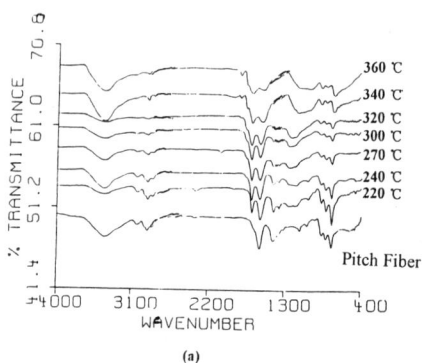

(a)

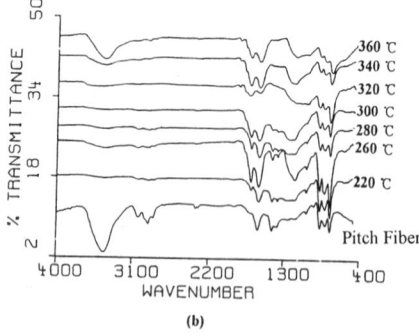

(b)

Figure 44 FT–IR spectrum of pitch fiber oxidation process, a: isotropic pitch fiber; b: anisotropic pitch fiber (from ref. 323).

Manufacture of Carbon Fibers

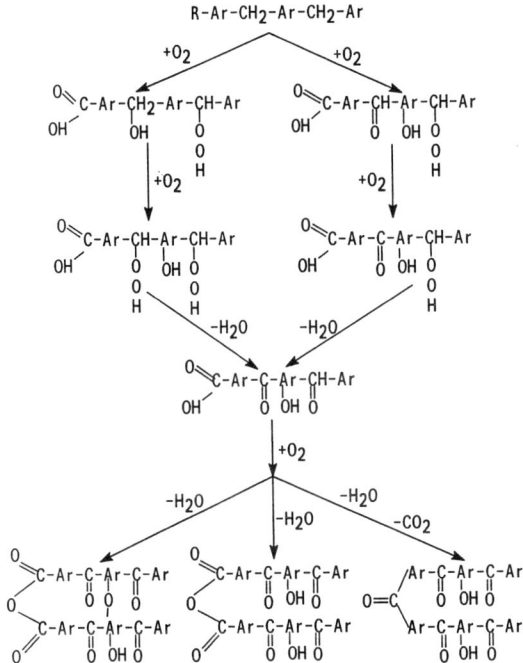

Figure 45 Reaction model of pitch fiber oxidation process (from ref. 314, 309).

1.4.5 Carbonization

The purpose of carbonization (< 2000°C) is to eliminate the heteroatoms, forming turbostatic graphite like structure with carbon content more than 96%, and to improve mechanical, electrical and thermal properties of the final product. However, the difficulty is due to the low tensile strength and brittle character of infused pitch fibers. Today, the manufacturer of continuous pitch based carbon fibers generally uses the technology shown in the Figures 40 and 41. At first, pitch fibers are collected in a container or roller, through heat treated to become infused fibers (air, < 400°C), then followed at a lower temperature carbonization (N_2, < 700°C). After above treatment, the fibers possess good tensile strength (100 to 200 MPa) and flexibility (elongation about 5–8%). Those fibers are plied together and sent into carbonization furnace at between 1000 to 1600°C for a higher temperature treatment. Carbon fibers with different properties can be obtained according to the temperature engaged. In the carbonization temperature range, when at < 800°C (nitrogen atmosphere), aromatic layer inside the fibers undergo further condensation, cross-linking and cyclization, and give off H_2O, CO_2, CO, H_2, and tar. In other words, H, O, N, etc. heteroatoms are eliminated and the layer of fused ring aromatic hydrocarbon is enlarged. With the increase of temperature (1000~1600°C), non-carbon atoms are continuously eliminated, through condensation and aromatization; turbostatic graphite like structure is formed (seen in Figure 46). Tensile strength of carbon fibers changes substantially to the temperature increase as shown in Figure 47. With the increase of temperature, the tensile strength and modulus of mesophase pitch-based carbon fiber (MP) increase [307,309]; while the tensile strength of isotropic pitch-based carbon fiber (IP) decreases and modulus increases slightly. The reason of this observation is because mesophase pitch has regular aromatic hydrocarbon layers, low heteroatoms content and no impurities; while isotropic pitch contents more ash, impurities, non-carbon atoms, smaller, and poorer oriented aromatic layer. In addition, with the increase of temperature, structure of fiber converts to multi-crystaline graphite structure gradually. The relationship between structure parameter and temperature is shown in Figure 48. Those data show that whether anisotropic or isotropic pitch, with the increase of temperature, d_{002} crystalline layer distance decreases, while the microcrystalline size increases gradually, and mesophase graphite fibers are more approaching to the graphite structure.

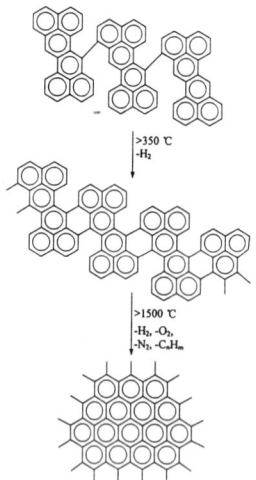

Figure 46 Schematic change of chemical structure of pitch fiber during thermal treatment process (from ref. 1).

1.4.6 Graphitization

Properties of pitch based carbon fibers changed substantially after 2000 to 3000°C heat treatment. Tensile strength and modulus of mesophase based graphite fibers increase with the increase of temperature; but at the same temperature treatment condition, tensile strength of isotropic pitch based graphite fibers decreases with the increasing of temperature, while tensile modulus increases, and PAN-based carbon fibers have the same tendency [307,309] (seen in Figure 47). The results may be due to the different molecular structure patterns; i.e., mesophase pitch has large and regular aromatic hydrocarbon layer, when temperature higher than 1800°C, the stack of aromatic plane is better than that of PAN and IPCF, and orientation degree higher, so micro-crystalline organization degree increases faster. In the process of microcrystalline growing up, anisotropic expansion of layer structure leads to a strong internal stress, so the dislocation and flaw disappear, and microcrystalline is rearranged, therefore mesophase pitch is easier to form three dimension ordered graphite-like structure. Then as seen, tensile strength and modulus increase with temperature. On the contrary, microcrystalline structure of the PAN and isotropic pitch fibers has poor orientation and increases more slowly, and their internal stress is also weaker. Therefore, tensile strength decreases with temperature, and the tensile modulus increases but the ratio is lower than that of APCF. Some of the commercially available pitch-based carbon fibers along with their mechanical properties are described in Table 6.

Table 6 Mechanical Properties of Some Commercially Available Pitch Based Carbon Fibers

Manufacturer	Product name	T.S(MPa)	Y.M(GPa)	Strain to failure %
Amoco Corp	Thornol P25	1400	140	1.0
	Thornol P555	2100	380	0.5
	Thornol P755	2000	500	0.44
	Thornol P100	2200	690	0.3
	Thornol P120	2200	820	0.2
Osaka Gas	Danacarb F140	1800	140	1.3
	Danacarb F60	3000	600	0.5
Kureka (Japan)	Kureca KCF100	900	38	2.4
	Kureca KCF200	850	42	2.1

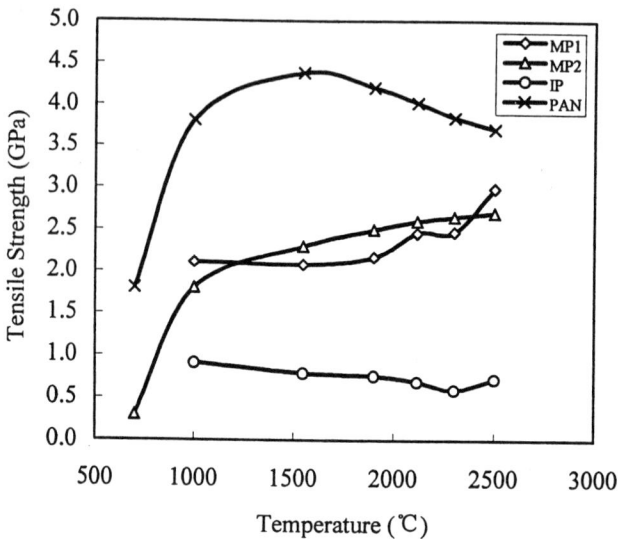

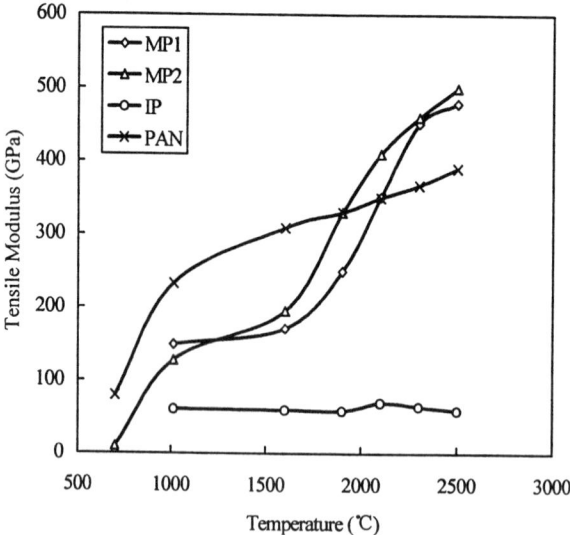

Figure 47 Effect of temperature on properties of carbon (graphite) fiber [from ref. 244 (PAN, MP2), 309 (IP,MP1)].

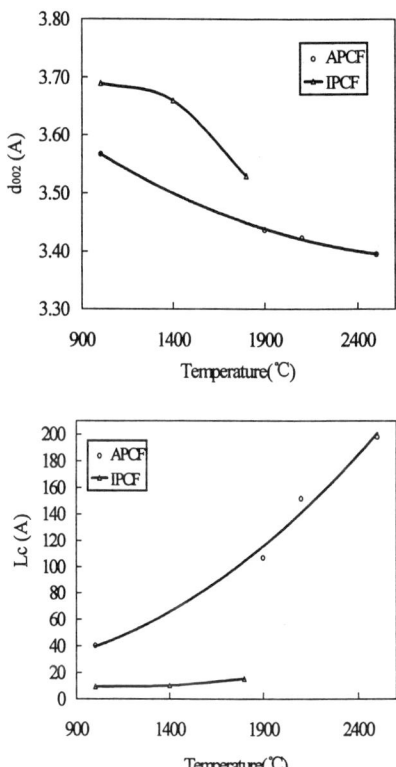
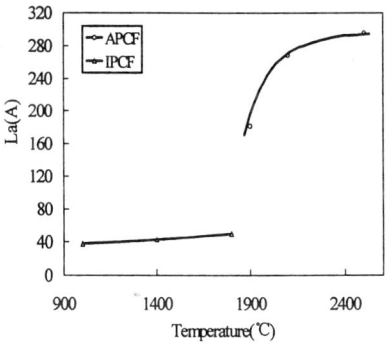

Figure 48 Effect of temperature on microcrystalline structure parameters of carbon (graphite) fiber (from ref. 250, 309).

1.5 VAPOR GROWN CARBON FIBERS

Vapor grown carbon fibers (VGCFs) are prepared by the decomposition of gaseous hydrocarbons at temperatures between 300 and 2500°C in presence of an ultra–fine metallic catalyst (e.g. Fe, Ni, and Co). VGCFs are characterized by the high preferred orientation of graphitic basal planes parallel to the fiber axis with annual ring texture in the cross-section, which give high mechanical performance, excellent electrical conductivity and high graphitizability to the fibers [325,326]. Two methods of forming VGCFs have been developed: Seeding catalysts on a substrate and fluidizing catalysts in a space. VGCFs may be grown on several types of substrates (e.g. carbon, silicon, quartz) and from many hydrocarbons (e.g., acetylene, benzene, natural gas, etc.), but in all cases, growth is favored in a hydrogen atmosphere. The apparatus used by Endo for growing vapor carbon fibers on substrate is shown in Figure 49. The conceptual schemes of VGCFs production over fluidizing catalysts methods proposed by Endo [327] are shown in Figure 50 (a) (b). Filament diameter may range from about 100 nm to several hundred micrometers. The process leads to the formation of fibers having various cross-section, namely circular, helical, twisted, or pleated fibers.

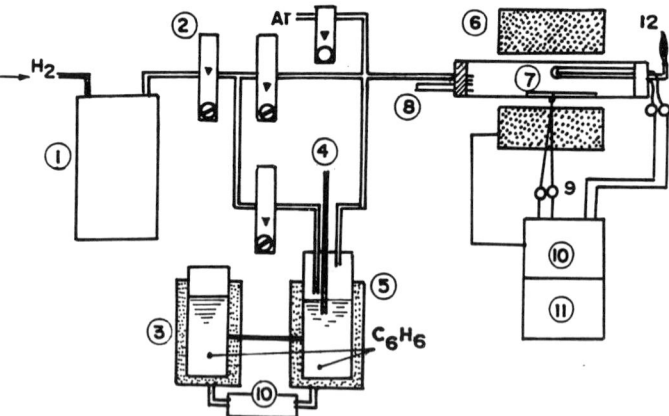

Figure 49 Schematic diagram of apparatus for preparing carbon fibers by thermal decomposition of hydrocarbons (1) Hydrogen purification unit, (2) flow meter (3) benzene reservoir, (4) thermometer (5) thermostatic bath, (6) electrical furnace (7) substrate, (8) peephole, (9) thermocouples, (10) temperature controller, (11) temperature recorder, and (12) gas exhaust [from ref. Endo, Ph.D. Thesis, University of Orleans, Orleans, France (1975)].

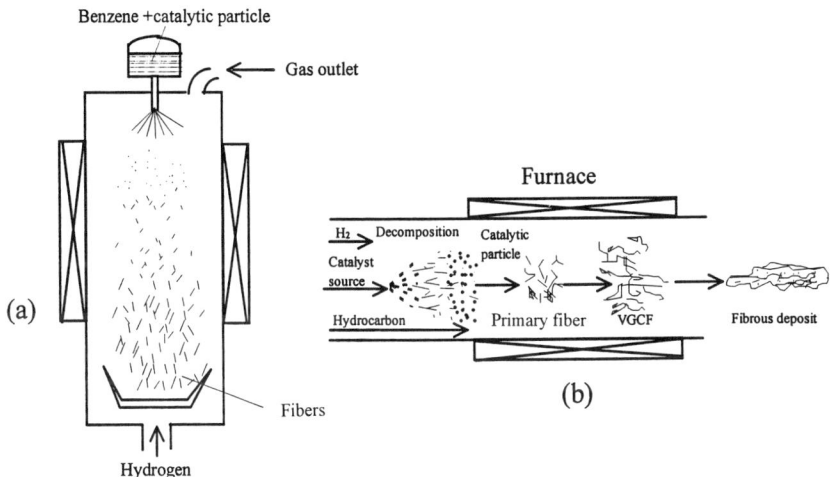

Figure 50 Conceptual scheme of VGCF production over floating catalyst particles by (a) direct, and (b) indirect methods (from ref. 327).

1.5.1 Background of VGCFs

Earliest published account of these vapor grown carbon fibers appeared in 1890 [328]. Then, Gibson et.al. [329] successfully grew the fibers from methane at 1000°C. Pittsburgh coke and chemical company [330] made an attempt to commercialize this process during the mid 1950s. However, their pilot plant operated only for a short period of time. Endo, Koyama, and co-workers produced vapor grown carbon fibers by pyrolysis of benzene [331,332]. Subsequently, Katsuki et al. [333] studied carbon fibers growth from pyrolysis of naphthalene–hydrogen mixtures, whereas Tibbetts' group reported on conditions for fiber growth from methane pyrolysis [334,335]. Kato et al. [336] studied the growth of carbon fibers on the surface of activated carbon pellets from benzene pyrolysis. It was found that VGCFs could be obtained from low cost sources such as Linz-donawitz convert gas or coal derived hydrocarbons such as 1-methylnaphthalene [337,338]. Recently, a new method, the liquid pulse injection (LPI) technique, has been announced to produce VGCFs [339,340] with the possibility of making VGCFs from a wide variety of hydrocarbons [341]. In addition, Motojima et al. obtained the micro-coiled carbon fibers using a Ni metal activated pyrolysis of acetylene under the presence of some impurities [342,343]. These fibers exhibit high crystalline perfection when subjected to high treatment temperatures. Their axial electrical conductivity is close to that of single crystal graphite [344] with an axial thermal conductivity substantially higher than that of copper [345].

The growth of fiber occurs via a catalyzed dehydrogenation reaction of a hydrocarbon in several steps. This includes the activation of the catalyst, the initial growth and elongation of a fiber in which the catalyst particle is often carried at the tip of the fiber, and a subsequent thickening process in which carbon is directly deposited on the side of the fiber [346]. It was shown by Baker and co-workers [347–349] that only a part of the catalyst particle at the tip of fiber was exposed to the hydrocarbon (Figure 49) and that a coating of carbon could cover the tip completely, arresting growth. Further growth could occur if this surface layer was removed by oxidation.

Several different kinds of vapor grown fiber can be grown. Vermicular fibers (i.e., with an irregular, wormy form) are favored at lower temperatures (e.g., < 900°C, but depending on the hydrocarbon) while straighter fibers are grown at higher temperature. Baired et al. [350,351] in their study using high resolution electron microscopy concluded that vermicular fibers tend to have an amorphous structure and straight fibers have a structure comparable to graphite. Vermicular fibers often have a low density, easily oxidized core, enclosed by a sheath of material which is resistant to oxidation [348,349]. In contrast, graphitic fibers have hollow cores and a minimum diameter [352,353].

The initial diameter of fibers growing with catalyst particles at tip is closely related to the diameter of the catalyst particles. There are differences in the structure of fibers grown from different precursors such as methane, acetylene, propane, and butadiene.

1.5.2 Growth Mechanism

Baker et al. [347] proposed the mechanism of growth of carbon fibers from pyrolysis of acetylene on isolated nickel particles as depicted in Figure 51. According to this mechanism, acetylene decomposes on the front exposed surfaces of the metal particles to release hydrogen as gas and carbon, which dissolves in the particles. The dissolved carbon diffuses through the particle and is precipitated at the trailing end from the body of the fiber. Temperature gradient across the catalyst particle due to the exothermic decomposition of hydrocarbons is one of factors that governs this mechanism. Since the solubility of carbon in a metal is temperature dependent, precipitation of excess carbon occurs at the colder zone behind the particle, thus allows the fibers to grow as

Manufacture of Carbon Fibers

depicted in Figure 51c. The fiber ceases to grow in length after the poisoning of leading tip of the catalyst (Figure 51d). Audier [354] had an alternative explanation to the mechanism of Baker et al., considering that the bulk diffusion is driven by an isothermal concentration gradient.

Baird et al. [351] and Oberlin et al. [352] proposed that the carbon fiber is formed entirely by a catalytic process involving the surface diffusion of carbon around the metal particle rather than bulk diffusion of carbon through the catalytic particle. Figure 52 illustrates the growth mechanism of the vapor grown carbon fibers. Sacco and co-workers [355] also studied the initiation and growth mechanisms of filamentous carbon over iron foils at 900K and atmospheric pressure. They suggested that the rate-determining step in filament growth was diffusion of carbon through the catalyst, but this diffusion was driven by a difference in solution of carbon at the metal-metal carbide (–iron)/(ementite) interface developed at the forward end of the crystal, and the metal–filament (–iron)/(graphite) interface at the back end (Figure 53). Several workers have established some empirical equations to calculate the growth rate of VGCFs [356–359].

1.5.3 Growth Conditions

Partial pressure of vaporized benzene, purity, and flow rate of the hydrogen gas, the residence time for thermal decomposition and the temperature of the furnace are some of the main factors that govern the growth rate as well as quality of carbon fibers.

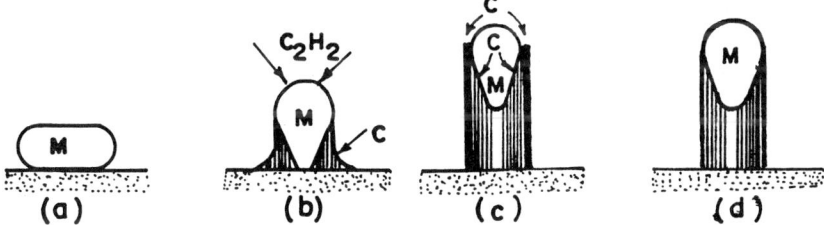

Figure 51 Schematic representation proposed for carbon filaments growth form the pyrolysis of acetylene (C_2H_2) on a metal particle (M) where C denotes carbon (from ref. 346).

Catalyst plays a vital role in the growth of vapor grown carbon fibers as already described in Figures 51 and 52. Use of a broader distribution of particle size of catalyst is known to enhance the carbon fiber yield. Further, the fiber formation is enhanced as the catalyst particle size decreases. Since fiber growth ceases when the particle surface gets covered either with carbon layers, oxygen or other impurities, high chemical purity of the catalyst is, therefore, essential for the growth of long fibers. Fiber diameter and aspect ratio of the vapor grown carbon fibers (VGCFs) are determined by the catalyst/hydrocarbon feed ratio.

1.5.4 Prospect of VGCFs

Single fiber properties of VGCFs are summarized in Table 7 [362]. Because of the VGCFs' unique structure and excellent performance, such as high mechanical, electrical, thermal properties, etc., they are expected in different fields of application as structural and functional materials. This type of carbon fibers could be sufficiently low-priced to replace ordinary carbon fibers in discontinuous yarn and thus serves as a useful filler for composites [327]. Applications have already been developed, for example, the VGCF by modifying the surface has been proved to be an excellent adsorbent with high surface area [361], VGCFs with rather thick diameter (10 μm) can be used as a host material for graphite-intercalation compounds (GICs) [362]. In addition, using nanometer-size VGCFs, thermoplastic composites with electrical conductivity have been fabricated with a microscopically smooth surface, allowing for electrostatic painting [363].

Thus VGCF is emerging in markets as a new type of carbon and graphite whisker and is under extensive study for both basic research and future applications [326].

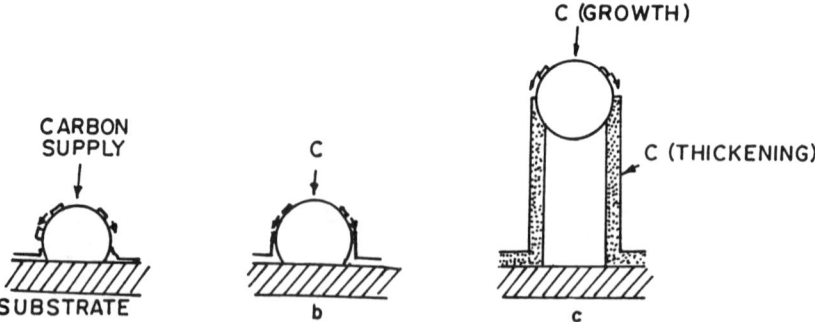

Figure 52 Schematic illustration of growth mechanism proposed for carbon fibers from benzene through a catalytic effect (from ref. 347).

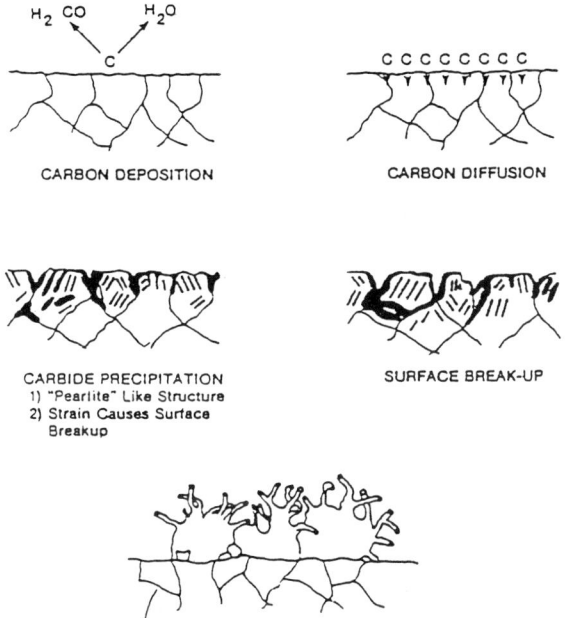

Figure 53 Proposed initiation mechanism for the development of filamentous carbon on iron foils (from ref. 355).

Table 7 Properties of Vapor Grown Carbon Fibers (VGCF)

Property	VGCF	Units
Filament diameter	< 1 to > 100	µm
Tensile strength	7.0	GPa
Tensile modulus	600	GPa
Density	2.1	gm/cm3
C.T.E.	–1.0 (Calc.)	PPm/°C
Electrical resistivity	55	µ Ω cm
Thermal conductivity	1950 at 300K	W/m-K

1.6 CURRENT SCENARIO AND FUTURE PROSPECTS OF CARBON FIBERS

As has been discussed in the chapter there are a variety of precursors for manufacturing carbon fibers. If one looks at these precursors a little bit in depth, it comes out very clearly that each precursor alone is capable of producing certain grade of carbon fibers. As shown in Figure 54 some commercial fibers, PAN-based carbon fiber have been developed both in the direction of high tensile strength (T–Type) and high tensile strength and modulus (MJ–Type), while pitch-based carbon fibers in the tendency of high modulus. High tensile strength fibers are mainly used in the airplane main structure parts; on the other hand, high tensile modulus fibers mainly used as structure materials in the aerospace applications.

Theoretically speaking it should be possible to achieve tensile strength of the order of 100 GPa, if one can control the amount of defects etc. in the PAN precursor as well as in the resulting carbon fibers. This, therefore, brings out very clearly the tremendous scope which is lying ahead of us in improving the properties of carbon fibers through processing of the precursor as well as carbon fibers therefrom. However, if one is looking for material which offers highest dimensional stability at elevated temperatures, one has to use pitch based carbon fibers. Young's modulus of the order of 900 GPa, which is very near to the theoretical value, has already been realized in pitch based carbon fibers commercially. These fibers, however, exhibit extremely poor strain to failure values because of low value of tensile strength. Tremendous effort is going on, specially in Japan, to enhance tensile strength of these types of carbon fibers by various processing routes.

The rayon based carbon fibers though offered a lot of promise in the beginning, were over shadowed by PAN based carbon fibers because the prime requirement at that time was achieving high strength carbon fibers. The other two factors which contributed to reduce the importance the phasing out of the rayon based carbon fibers was very low carbon yield and high processing cost. However, low carbon yield of rayon based carbon fibers has become a boon for producing "Activated Carbon Fibers," as such fibers offer very large surface area and uniform pore size distribution. This is the reason why such fibers are mainly used in the form of activated carbon fibers. In addition, they have low density, high purity, high deformation, and low thermal conductivity character. That is also the reason why rayon is a kind of unreplaceable material in the space industry now.

There has been increasing demand in recent years for having carbon fibers possessing ultra high thermal conductivity values. In order to achieve highest values of thermal conductivity, obviously the structure of carbon fibers should be as near as possible to that of single crystal graphite. Vapor grown carbon

Manufacture of Carbon Fibers

fibers are the only type of carbon fibers which have potential to offer the structure close to graphite. As discussed in this chapter it is impossible to realize such structures through any other precursor namely, PAN, pitch, and rayon. Value of the order of around 2000 W/m–K has already been achieved in the case of vapor grown carbon fibers. There are serious efforts going on to produce vapor grown carbon fibers commercially. For example, Allied Sciences, USA has already installed a plant to produce around 2000 kg/yr of vapor grown carbon fibers. However, there are lot of difficulties in processing these fibers. Hence there seems to be a clear need to do R & D in this vital area.

After 1989, the demand of carbon fiber in aerospace and aeronautic industry decreased. Some companies, consequently, were closed down. This lead to a world-wide re-structure of carbon fiber production [364], which is shown in Table 8.

However, into the 90s, carbon fibers are increasingly used in areas related to economy, such as sporting goods, sealing materials, automobile, absorbent and catalyst, fuel cell and electrode, building material and so on. According to the forecast by Toray company, the actual demand was 8390 tons in 1996, 16% of which in aerospace and aeronautic industry, 48% in sports and 36% in the general industry. This means that the market of carbon fibers is transferred from aerospace industry to transportation, machinery, chemical, building and sports industry. Although production capacity may exceed the demand this year, with development of science and technology, demand for carbon fiber increase, so carbon fiber is still interesting industry in future through cutting down the cost and broadening industry application. In the 21st century, following metal, ceramic, and polymer materials, carbon fiber will become the fourth generation essential industrial materials, and it is expected that from the about 10000 ton/yr demand in 1997 the production may double in the next ten years.

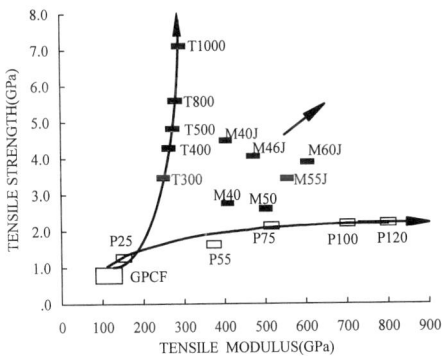

Figure 54 Developing trend of tensile strength and modulus for CF (GF).

Table 8 Production Capability of PAN Based Carbon Fibers

Zone	Manufacturer	Production name	Production capacity (T/Y)	Remarks
Asia	Toray	Torayca	2900	
	Toho Beslon	Besfite	2900	
	Mitshubishi Rayon	Pairofil	1200	
	Others		360	Taiwan, Korea
U.S.A.	Hercules	Magnamite	1715	
	Amoco	Thornel	1210	
	Akzo	Fortail	2270	
	Grafil	Grafil	710	Mitsubishi
	Zoltek	Panex	60	Rayon
Europe	Tenax Fibre	Tenax	720	Toho Rayon
	Soficar	Torayca	800	Toray
	Others		200	German, U.K.
Total			15045	

NOTE

1. Note added in proofs, A detailed review paper has recently been published (112 references), see J. Mittal, R. B. Mathur, and O. P. Bahl, *Carbon* 35, 1713–1722 (1997).

REFERENCES

1. Morida, K., and Kagaku G., (in Japanese), *126*, 6 (1981).
2. Shen, Z., Encyclopedia of Materials Science (in Chinese), p. 993, China Encyclopedia Press, 1995.
3. Edison, T. A., U. S. Patent 223,398 (1880).
4. Swan, J. W., British Patent 4933 (1880).
5. Otani, S., *Mol. Cryst. Liq. Cryst. 63*, 251 (1981).
6. Otani, S., Carbon *3*, 31 (1965).
7. Donnet, J. B., and Bahl, O. P., Encyclopedia of Physical Science and Technology *2*, 517 (1987).
8. Bacon, R., and Hoses, T. N., High Performance Polymers, Their origin and Development, edited by Sanymour R., B., and Kirshambaum, G. S., New York, Elsemer, 1986, p. 342.
9. Winter, L. L., Union Carbide Corp. National Carbon Research Laboratories, Biweekly reports 1944–45.
10. Hautz, R. C., Textile Res. J. *20*, 786 (1950).
11. Delomonte, J., Technology of Carbon and Graphite Composites, Van Nostrand Reinhold, New York, 1981, pp. 11–14.

12. Shindo, A., Studies on Graphite Fiber rept. No. 317, Gov't Ind. Res. Inst. Osaka (1961); also J. Ceram. Assoc. Japan *69*, C195 (1961).
13. Watt, W. et al., The Engineer (London) *221*, 815 (1966).
14. Watt, W., and Johnson W., in High Temperature Resistant Fibers from Organic Polymers, Preston, J., ed., Appl. Polymer Symp. *9*, 1969, pp. 229–43.
15. Muller, T., I.F.J., 46 (1988).
16. Mitsubishi Rayon Co, Ltd., Japan Kokai Tokyo Koho, 62,231,027 (1987).
17. Toray Industries, Inc., Japan Kokai Tokyo Koho, 59,168 128 (1984).
18. Toray Industries, Inc., Japan Kokai Tokyo Koho, 58,214 521 (1983).
19. Japan Exlan Co., Ltd., Japan Kokai Tokyo Koho, 58,191,704 (1984).
20. Mitsubishi Toasty Chemical Inc., Japan Kokai Tokyo Koho, 62,15, 329 (1987).
21. Platonava, N. V., Klimenko, I. B., Grachev, V. I., and Kiselev, G. A., Vysokomol Soedin Ser. A., *30* (5), 1056 (1988).
22. Mitsubishi Rayon Co., Ltd., Japan Kokai Tokyo Koho, 61,132,632 (1986).
23. Asashi Chemical Industries Ltd.. Japan Kokai Tokyo Koho, 58,163,728 (1983).
24. Nikkiso Co., Ltd., Japan Kokai Tokyo Koho, 61,75,819 (1986).
25. Bajaj, P., Chavan R. B., and Manjeet, B., J. Macromol Sc. Chem., *A23* (3), 335 (1986).
26. Grassie N., and McGuchan, R., Eur. Polym. J. *7*, 1503 (1971).
27. Gupta, A. K., Paliwal D. K., and Bajaj, P., J.M.S. Rew. Macromol Chem. Phys. *C31* (1), 1–89 (1991).
28. Guyot, A., Bert, M., Hamoudi, A., McNeil, I., and Grassie, N., Eur. Poly J. *14*, 101 (1978).
29. Bajaj P., and Padmanabam, M., Eur. Poly. J., *21* (1), 93 (1985).
30. Gupta, D. C., Agrawal J. P., and Sharma, R. C., J.Appl. Poly. Sci *38*, 265 (1989).
31. Japan Exlan Co., Ltd., British Patent, 2,161,821 (1986).
32. Jillin Chemical Industries Corp. Chinese Patent 85,1031,318 (1987).
33. Toho Beslon Co., Ltd., Japan Kokai Tokyo Koho, 61,119,720 (1986).
34. Nikkoso Co., Ltd., European Patent, 178,890 (1986).
35. Toray Ind., Inc.,Japan Kokai Tokyo Koho, 63,182,317 (1988).
36. Mitsubishi Rayon Co., Ltd., European Patent, 201,908 (1986).
37. Mitsubishi Rayon Co., Ltd., Japan Kokai Tokyo Koho, 63,59,409 (1988).
38. Mitsubishi Rayon Co., Ltd., Japan Kokai Tokyo Koho, 61,14,206 (1986).
39. Faserforsch, J. U., Textiltech, *11*, 62 (1962).
40. Grassie, N., and McGuchan, R., Eur. Polym. J., 7, 1091 (1971).
41. Patron, L., and Bastianelli, V., Appl. Polym. Symp., *25*, 105 (1974).
42. Japan Exlan Co., Ltd.,Japan Kokai Tokyo Koho, 62,299,510 (1987).
43. Stami Carbon B.V., European Patent 144,983 (1985).
44. Toray Inst. Inc., Japan Kokai Tokyo Koho, 61,119,710 (1986).
45. Asashi Chemical Industry Co., Ltd., Japan Kokai Tokyo Koho, 61,138,712

(1986).
46. Mitsubishi Rayon Co., Ltd., Japan Kokai Tokyo Koho, 63,35,820 (1988).
47. Japan Exlan Co., Ltd., German Patent 3,535,368 (1968).
48. Toyobo Co., Ltd., Japan Kokai Tokyo Koho, 60,94,613 (1985).
49. Toray Industries Inc., Japan Kokai Tokyo Koho, 62,141,111 (1987).
50. Bayor, A. G., German Patent, 3,630,244 (1988).
51. Mitsubishi Rayon Co., Ltd., Japan Kokai Tokyo Koho, 63,112,718 (1988).
52. Mitsubishi Rayon Co., Ltd., Japan Kokai Tokyo Koho, 62,78,209 (1987).
53. Asashi Chemical Industries Co., Ltd., Japan Kokai Tokyo Koho, 60,94,215 (1985).
54. Mitsubishi Rayon Co., Ltd., Japan Kokai Tokyo Koho, 62,268,812 (1987).
55. Mitsubishi Rayon Co., Ltd., Japan Kokai Tokyo Koho, 62,268,810 (1987).
56. Asashi Chemical Industries Co., Ltd., Japan Kokai Tokyo Koho, 61,138,709 (1986).
57. Asashi Chemical Industries Co., Ltd., Japan Kokai Tokyo Koho, 61,138,711 (1986).
58. Asashi Chemical Industries Co., Ltd., Japan Kokai Tokyo Koho, 61,138,713 (1986).
59. Padhya, M. R., and Karandikar, A. V., J. Appl. Polym. Sci., *33*, 1675 (1987).
60. Nimtz, F. G., Geller, A. A., and Geller, B. E., Khim. Volokna, *3*, 11 (1986).
61. Mitsubishi Rayon Co., Ltd.,Japan Kokai Tokyo Koho, 59,204,914 (1984).
62. Yua, G. R., Lin C. H., and Wang, H. H., Fang Chih Kung Ch'eng Hsueh K'an, *13*, 45 (1986).
63. East, G. C., McIntyre, J. E., and Patel, G. C., J. Text. Inst. *3*, 196 (1984).
64. Frushour, B. G., Polym. Bull. *4*, 305 (1981).
65. Frushour, B. G., Polym. Bull. *7*, 1 (1982).
66. Grove, D., Desai, P., and Abhiraman, A. S., Carbon *26* (3), 403 (1988).
67. Edie, D. D., Fox, N. K., Barnett, B. C., and Fain, C. C., Carbon *24* (4), 477 (1988).
68. Mathur, R. B., Bahl, O. P., Matta, V. K., and Nagpal, K. C., Carbon *26* (3), 295 (1988).
69. Mitsubishi Rayon Co., Ltd., Japan Kokai Tokyo Koho, 63,275,713 (1988).
70. Mitsubishi Rayon Co., Ltd., Japan Kokai Tokyo Koho, 63,99,317 (1984).
71. Morton and Watt, Carbon *12*, 543 (1974).
72. Mitsubishi Rayon Co., Ltd., Japan Kokai Tokyo Koho, 63,275,716 (1988).
73. Ferguson, J., and Mahapatro, B., Fiber Sci, Tech. *11*, 55 (1978).
74. Mitsubishi Rayon Co., Ltd., Japan Kokai Tokyo Koho, 63,59,409 (1988).
75. Celanese Corporation, European Patent, 125,905 (1984).
76. Ono Keizo, Mizuguchi Toyokazu, Tsunodo Atsushi, 02–47309 (1990)
77. Mastsuhisa Yoji, Washiyama Masayoshi, Yamane Shoji, JP 04–240221 (1992)
78. Chari, S. S., Bahl, O. P., and Mathur, R. B., Fiber Sci. Technol. *15*, 153 (1981).

79. Ko, T. H., Ting, H. Y., and Ling, C. H., J, Appl. Poly. Sci *35*, 631 (1988).
80. Donnet J. B., and Bansal, R. C., Carbon Fibers, 2nd ed. Marcel Dekker, New York, 1990.
81. Bahl, O. P., Mathur, R. B., Matta, V. K., and Sivaram, P., Carbon *27*, 494, (1989).
82. Bahl, O. P., and Manocha, L. M., Fib. Sci Tech. *9*, 77 (1976).
83. Jain, M. K., and Abhiraman, A. S., J.Mater. Sci. *22*, 278 (1987).
84. Jones, B. F., and Duncan, R. G., J. Mater. Sci. *6*, 289 (1971).
85. Bahl, O. P., Mathur, R. B., and Kundra, K. D., Fib. Sci Tech. *15*, 147 (1981).
86. Jain, M. K., and Abhiraman, A. S., J. Mater. Sci. *18*, 179 (1983).
87. Cermia, E., Man made Fibers, Science, and Technology, 3 Interscience Publishers, New York ,1968,135.
88. Japan Exlan Co., Ltd. Brit. Patent 1,499,085 (1978).
89. Toray Industries Inc., Japan Kokai Tokyo Koho, 62,184,121 (1987).
90. Mitsubishi Rayon Co., Ltd., Japan Kokai Tokyo Koho, 61,146,817 (1986).
91. Toray Industries Inc., Japan Kokai Tokyo Koho, 58,214,518 (1983).
92. Toray Industries Ltd., Japan Kokai Tokyo Koho, 58,214,518 (1983).
93. Mitsubishi Rayon Co., Ltd., Japan Kokai Tokyo Koho, 60,181,323 (1985).
94. Nikkiso Co., Ltd., Japan Kokai Tokyo Koho, 61,83,320 (1986).
95. Toray Industries Inc., Japan Kokai Tokyo Koho, 59,36,727 (1984).
96. Takemoto Oil and Fat Co., Ltd., Japan Kokai Tokyo Koho, 59,116,471 (1984).
97. Mitsubishi Acetate Co., Ltd. and Mitsubishi Rayon Co., Ltd., Japan Kokai Tokyo Koho, 58,169,516 (1983).
98. Beslon, T., German Patent De, 30,40,265 (1983).
99. Tokai Electrode Manufg. Co., Ltd, Japan Kokai Tokyo Koho, 72,29,937 (1972).
100. Popovska, N., and Malaskonov, I., Carbon, *21(3)*, 33–8 (1983).
101. Nitto Baseki Co., Ltd., Japan Kokai Tokyo Koho, 74,35,629 (1974).
102. Lyubcheva, H., Lyubchev, L., Mladenov, I., Acta Polym., *35* (10), 634–6 (1984).
103. Kalvin, I. L., DiEdwardo, A. H., Choe, I. W., Rhodes, J. M., Org. Coat. Plast. Chem., *38*, 685 (1978).
104. Mitsubishi Chemical Co., Ltd., Japan Kokai Tokyo Koho, 80,32,807 (1980).
105. Mitsubishi Rayon Co., Ltd., Japan Kokai Tokyo Koho, 76,40,432 (1976).
106. Shieldlin, A., Marom, G., Zillikha, A., Polymer, *26* (3), 447–51 (1985).
107. Lei, J., and Danghan, W., Liu Shihong, Fuhe Cailiao Yuebao, *J(1)*, 62–6 (1988).
108. Nippon carbon Co., Ltd., ,Japan Kokai Tokyo Koho, 74,00,527 (1974).
109. McCabe and Michael, U.S. Patent, 4,661,336 (1987).
110. Ko, T. H., Lin, L. H., J. Mater. Sci. Lett., *7 (6)*, 1628 (1985).
111. Robin, R. P., U.S. Patent, 3,410,874 (1968).

112. Riggs, R. P., U.S. Patent, 3,656,882 (1972).
113. Mathur, R. B., Mittal, J., Bahl, O. P., and Sandle, N. K., Carbon, *32* (1): 71 (1994).
114. Mittal, J., Ph.D. Thesis, Indian Institute of Technology, Delhi, 1994.
115. Ko, T. H., and Huang, L., J. of Material Science, *27*, 2429 (1992).
116. Mathur, R. B., Gupta, D., Bahl, O. P., Dhami, T. L., Fib. Sci. Tech., *20*, 227 (1984).
117. Simitzis, J., Atomkarnen erg/ Derntech, *38 (3)*, 205 (1981).
118. Zang, H., and Chang, G., Kao Ge, Tze Tung Hsun, *2*, 145 (1981).
119. Hartland, H. G., King, M. W., and Phillips, L. N., Brit. Patent, 1,431,883 (1976).
120. Cupp, K. H., and Stuez, D. E., U.S. Patent, 3,592, 595 (1971).
121. Bahl, O. P., Mathur, R. B., and Dhami, T. L., J. Mat. Sci. and Eng. *73*, 105 (1985).
122. Weibull, W., J. Appl. Mech. *18*, 293 (1951).
123. Peeper, R. T., U.S. Patent, 4,526,770 (1985).
124. .Bahl, O. P., Mathur, R. B., and Matta, V. K., Proc. 19th Bienn Conf. Penn State, U.S.A, 1989, p. 318.
125. Mathur, R. B., Mittal, J., and Bahl, O. P., J. Appl. Poly. Sci, *49(3)*, 469 (1993).
126. Toray Industries Inc., European Patent, 2,23,199 (1987).
127. Fitzer, E. and Muller, D. J., Carbon *13*, 63 (1975).
128. Berlin, A. A., Dubinskaya, A. M., and Moshkovskii, Y. S., Vysokomol Soedin, p. 1938 (1964).
129. Fester, W., Textil Rundschau *20*, 1 (1965).
130. Burlant, W. J., and Parsons, J. L., J. Polym. Sci., *22*, 249 (1956).
131. Grassie, N., and Hay, J. N., J. Polym. Sci, *56*, 189 (1962).
132. Friedlander, H. N., Peebles, L. H., Brandrup, J., and Kirby, J. R., Macromolecule *1*, 79 (1968).
133. Coleman, H. M., and Petearich, R. R., J. Poly. Sci., Polymer Phys. Ed., *16*, 821 (1978).
134. Coleman, M. M., and Sivy, G. T., Carbon *19*, 123 (1981).
135. Sivy, G. T., and Coleman, M. M., Carbon *19*, 127 (1981).
136. Reich, L., Macromol Rev. *3*, 49 (1968).
137. Grassie, N., and McGuchan, R., Eur. Polym J., *7*, 1503 (1971).
138. Grassie, N., and McGuchan, R., Eur. Poly. J., *6,* 1277 (1970).
139. Turner, W. N., and Hohnson, F. C., J. Appl. Polym. Sci, *13*, 2073 (1969).
140. Grassie, N. and McGuchan, R., Eur. Polym. J., *7*, 1357 (1971).
141. Johnson, J. W., and Watt, W., Polym. Prepr., 9 (2), (1968).
142. Johnson, J. W., Philips, L. N., and Watt, W., British Patent 1,110,791 (1968).
143. Rosembanm, S. J., J. Appl. Polym. Sci., *9*, 2071 (1965).
144. Layden, G. K., J. Appl. Polym. Sci., *15*, 1709 (1971).
145. Warner, S. B., Peebles, L. H., and Uhlmann, D. R., J. Mater. Sci., *14*, 565

(1979).
146. Fitzer, E., and Muller, D. J., Macromol. Chem. *144*, 117 (1971).
147. Fitzer, E., and Heym, M., Chem Ind., *16*, 663 (1976).
148. Bahl, O. P., and Manocha, L. M., Angew Makromol Chem. *48*, 145 (1975).
149. Bahl, O. P., and Mathur, R. B., Fib. Sci. Technol., *12*, 31 (1979).
150. Knovich, M. M., Evrtors, I. L., et al. Khim. Volokna, *19 (3),* 370 (1978).
151. Fitzer, E., Frohs, W., and Heins, M., Carbon, *24* (4), 387 (1986).
152. Toho Beslon Co., Ltd., European Patent, 1,79,715 (1986).
153. Ashi Chemical Industries Ltd., Japan Kokai Tokyo Koho, 62,22,826 (1987).
154. Racke, P. B., Schurmus, H., and Verhoest, J., Inorganic Fibers and Composites, Pergamon, New York, 1984.
155. Bahl, O. P., Mathur, R. B., and Kundra, K. D., Fiber Sci. Technol. *13*, 155 (1980).
156. Raskovic, V. and Marinkovic, Carbon, *16*, 351 (1978).
157. Fitzer, E., and Muller, D. J., Carbon, 13, 63 (1975).
158. Colemann, M. M., and Petcavich, R. J., J. Poly. Sci., Polym. Phys. Ed., *16*, 821 (1978).
159. Bahl, O.P., and Manocha, L. M., Carbon, *12*, 17 (1974).
160. Billmayer, F. W., Text Book of Polymer Science, 2nd. ed., Wiley–Interscience, New York, 1971.
161. Stille, J.K., Introduction to Polymer Chemistry, Wiley, New York, 1971, p. 172.
162. Fiedler, A. K.., Fitzer, E., and Muller, D. J. Am Chem. Soc. Preprints Div. of Organic Coatings and Plastic Chemistry, 161st National Meeting, Los Angeles, Calif., *31*, No.1 April, 3–5, p. 142 (1973).
163. Standage, A. E., and Matkowsky, R. D., European Polymer J., *7*, 775 (1971).
164. Watt, W., Proc. 3rd Conf. on Ind. Carbon and Graphite Soc.Chem. Ind . London, 1971, p. 431.
165. Kirby, J. R., Brandrop, J., and Peebles, L.H., Macromolecules *1*, 53 (1968).
166. Gupta, S. K., Chem. Age India, *27 (11)*, 961 (1961).
167. Goodhew, P. J., Clarke, A. J., and Bailey, J. E., Mater. Sci. Eng., *17*, 3 (1975).
168. Mittal, J., Bahl, O. P., and Mathur, R.B., Carbon, *32,(6),* 1133 (1994).
169. Mathur, R. B., Bahl, O. P., and Mittal, J., Carbon, *30(4),* 657 (1992).
170. Mathur, R. B. ,Bahl, O. P., and Mittal, J., Carbon, *29 (7),* 1145 (1991).
171. Brombley, J., Carbon Fiber–Their Composites and Applications, Pro. Intern. carbon Fiber Conf., London, p. 3 (1971).
172. Watt, W., Nature, *236*, 10 (1972).
173. Shindo, A., Carbon Fibers–Their Composites and Applications, The plastic Institute, London, p. 18 (1971).
174. Blazeuiex, S., and Chlopek, J., Loks
175. Toray Industries Inc., Japan Kokai Tokyo Koho,59,137,513 (1984).
176. Kasotochkin, V. I., and Kargin, V. A., Doklady Phys. Chem. *191*, 303

(1970).
177. Johnson, J. W., Appl. Polym. Symp. *9*, 229 (1969).
178. Morton, R., Watt, W., and Johnson, W., Nature *213*, 690 (1967).
179. Cooper, G. A., and Mayor, R. M., J. Mater. Sci. *6*, 606 (1971).
180. Union Carbide Corp., U.S.A., U.S. Patent., 3,454,362 (1969).
181. Mochids, I., Ohtsuba, R., Takeshita, K., and Marsh, H., Carbon *18*, 117 (1980).
182. Dietz, V., U. S. Patent, 3,931,392 (1976).
183. Allen, S., Cooper, G. A., and Mayer, R. M., Paper Presented at IP and PS Conference on fibers and composites, Brighton, June, 1969.
184. Mittal, J., Mathur, R. B., and Bahl, O. P., Carbon, (personal communication).
185. Moutand, G. M., and Duflos, J. L., U.S. Patent, 3,322,489 (1967).
186. Yoneshiga, I., and Teranishi, H., Nippon Carbon Patent 2775/70 (1970).
187. Bacon, P. et al., Society of the Plastics Industry, 21st technical and Management Conference section 8E (1966).
188. Bacon, R., Carbon Fibers from Rayon Precursors, Chemistry Physics of Carbon, *9*, Walker, P. L., and Thrower, P. A., eds., Marcel Dekker, New York, 1973, pp. 1–102.
189. Schmidt, D. L., and Cray, R. D., Advanced Carbon Fibric Phenolic for Thermal Protection Application, APA 116829 (1982).
190. Schmidt, D. L., Alternate Carbon Fibric, SAMPE Quarterly, *8 (4)*, 48–54 (1977)
191. Schmidt, D. L., ADA 047293.
192. Shindo, A., Nakanishi, Y., and Sema, I., Appl. Polymer Symposia, No. 9, 271 (1969).
193. Duffy, J. V., J. Appl. Polym. Sci. *15*, 715 (1971).
194. Schuyten, H. A., Weaver, J. W., and Reid, J. D., Ind. Eng. Chem. *47*, 1433 (1955).
195. Cory, M. T., U.S. Patent, 3,508,871 (1970).
196. Tamaru, K., J. Chem. Soc. Japan, *69*, 21 (1948).
197. Tamaru, K., Bull. Chem. Soc. Japan, *24*, 164 (1951).
198. Bacon, R., and Tang, M. M., Carbon, *2*, 211 (1964).
199. Kanamaju, K., and Koyamo, K., J. Appl. Polymer Sci. *3*, 143 (1969).
200. Strong, S. L., Am. Chem. Soc., Div of Organic Coatings and Plastic Chemistry prepr., *31*, 426 (1971).
201. Madrosky, S. L., Hart, V. E., and Strauss, S., J. Res. Natl. Bur. Stand. *56*, 343 (1956).
202. Shindo, A., U.S. Patent, 3,529,934 (1970).
203. Schwenker, F., and Weaver, J. W., and Reid, J. D., Ind. Eng. Chem, *47*, 1433 (1955).
204. Kilzer, F. J., and Broido, A., Pyrodynamics, *2*, 151 (1965).
205. Ross, S. E., Text Res. J., *38*, 906 (1968).
206. Peters, E. M., U.S. Patent. 3,235,353 (1966).

207. Bacon, R., Cranch, G. E., Moyer, R. O., and Watts, W. H., U.S. Patent, 3,305,315 (1967).
208. Gutzeit, C. L., U.S. Patent, 3,479,151 (1969).
209. Moore, D. R., Ross, S. E., and Tesoro, G. C., U.S. Patent, 3,527,564 (1970).
210. Diffu, J. V., Polymer Sci., *15*, 715 (1971).
211. Little, R. W., (ed.), Flameproofing Textile Fabrics, Reinhold, New York, p. 167 (1947).
212. Tang, W. K., and Neill, W. K., J. Polymer Sci., *C6*, 65 (1964).
213. Parks, W. G., Esteve, R. N., Gallis, H. M., Guescis, R., and Petrarce, A., Mechanism of pyrolytic decomposition of cellulose, 127th Meet., Am. Chem. Soc., Div. of Cell Chem., Cincinnati, Ohio, (1955).
214. Schalamon, W. A., and Bacon, British Patent 1,167,007 (1969).
215. Ruland, W. J., J. Appl. Phys., *38*, 3585 (1967).
216. Otani, S., Carbon, *3*, 213 (1965).
217. Otani, S., Yamada, K., Koitabashi, T., and Yokoyama, A., Carbon, *4*, 425 (1966).
218. Kureha Chemical Ind., Co., JP. 41–15728 (1966).
219. Kureha Chemical Ind., Co., JP. 68–4450 (1965).
220. Kureha Chemical Ind., Co., JP. 69–2511 (1969).
221. Riggs, D. M., and Diefendrof, R. J., U. S. Patent, 4208267 (1980).
222. Riggs, D. M., and Diefendrof, R. J., 3rd Int. Carbon Con., Prints, p. 330, 1980.
223. Riggs, D. M., and Diefendrof, R. J., 16th Bienn. Conf. on Carbon, p. 24 1983.
224. Riggs, D. M., Prints, Div. of Petro. Chem., *29 (2)*, P. 480 (1984).
225. Yamada, Y., Honda, H., Inouse, T., JP 58–18421 (1983).
226. Yamada, Y., Imamura, T., Inouse, T., and Honda, H., JP 58–196292 (1983).
227. Honda, H., Tanso, No. 113, 66 (1983).
228. Otani, S., JP 57–100186 (1982).
229. Tomio, A. et al. JP 59–12286 (1984).
230. Mochida, I., Nakamura, E., Malda, K., and Takeshta, K., Carbon, *14*, 123 (1976).
231. Mochida, I., Inaba, T., et al., Carbon, *21*, 543 (1984).
232. Mochida, I., Shimizu, K., et al., Carbon, *30*, 55 (1992)
233. Li, R., New Carbon Material (in Chinese), *3*, 16 (1989).
234. Exxon Research and Engineering Co., US Patent 4,208,267 (July 7, 1981).
235. DuPont, US Patent 4,915,296 (Apr. 10, 1990).
236. Hawthohe, M. et al., Industrial Carbon & Graphite 3rd Conference (1970), p.508.
237. Hawthohe, M. et al., Plastics & Polymer Conference Supplement, *5*, 81 (1970).
238. Newman, J. W., 32th SAMPE, pp. 938–944, 1987.
239. Singer, L. S., High Modulus Carbon Fiber From Pitch, Belgian Patent 797,543 (1973).

240. Singer, L. S., U. S. Patent 4,005,183 (1977).
241. Singer, L. S., Carbon, *16*, 408 (1978).
242. White, J. L., et al., Carbon, *5*, 517 (1967).
243. David, A., and Schul, Z., SAMPE J., March/April, pp. 27–31 (1987).
244. Matsumoto, T., Pure & Appl. Chem., *57,* 1533 (1985).
245. Figueiredo, J. L. et al., Carbon Fibers Filaments and Composites, Kluwer Academic Publisher, 1990.
246. Miyazaki Takamine, TANSO, *115,* 186 (1983).
247. Ueda, K., Kagaku Kogyo, *12*, 286 (1975).
248. Lisicki, Z., et al., Fuel Procesing Technology, *20*, 103 (1988).
249. Araga, Yasunar, Sagara, Hiroshi, Yamamoto, Kenza et al., JP 63–112687.
250. Shen, Z., Li, A. et al., National Symposium on GPCF, Guilin (China), 1983.
251. Matsumura, Y., Sekiyu Gakkaishi, *30 (5),* 291 (1987).
252. Fitzter, E., and Liu, G., 19th Biennial Conference on Carbon (1989).
253. Mochida et al.. Carbon, *27*, 498 (1988).
254. Otani, S., Yokoyama, A., and Nukui, A., Appl. Polymer Symp. 9, 325 (1969).
255. Otani, S., and Yokoyama, Bull. Chem. Soc. Japan *42*, 1417 (1969).
256. Donnet, J. B., Bansal, R. C., Carbon Fibers, Marcel Dekker, Inc., New York, 1984.
257. Donnet, J. B., Bansal, R. C., Carbon Fibers, Marcel Dekker, Inc., New York, 1990.
258. Sharp, J. A., Fuel, *66,* 1487 (1987).
259. Brooks, J. D., and Talyor, G. H., Carbon, *3*, 185 (1965).
260. Brooks, J. D., and Talyor, G. H., Chemistry and Physics of Carbon, Marcel Dekker, New York, *4,* 243–268 (1968).
261. Lewis, I. C., Carbon *16*, 503, 1978.
262. Riggs, D. M., and Diefendorf, R. J., "Carbon" 80, Preints, 326–333, Baden-Baden, 1980.
263. Basch, R. et al., 15th Biennial Conf. on Carbon, Extended Abstracts, 164, 1981.
264. Yamada, Y., Seni–Gakkaishi, *35(11),* 344 (1979)
265. Qian, S., Physical–Chemistry of hydrocarbon Liquid Carbonization Process (in Chinese) Sino-Academy of Science Institute of Coal Chemistry Press, 1982
266. Kovac, C. A., and Lewis, I. C., Carbon *16,* 433 (1978).
267. Imamura, T., Yamada, Y., Oi, S., and Honda, H., Carbon, *16,* 481 (1978).
268. Imamure, T., Nakamizo, M., and Honda, H., Carbon, *16*, 487 (1987)
269. White, J. L., AD–777814.
270. Mochida, I., and Korai, Y., Journal of Materials Science Society of Japan, *23 (2),* 75 (1986).
271. Peebles, L. H., Carbon Fibers Formation, Structure, and Properties, CRC Press (1995).
272. Singer, L. S., U.S. Patent 400518 (1977).

273. Mchenry, E. R., U.S. Patent 4, 026,778 (1977).
274. Chwastiak, S., British Patent 2005,298A (1977).
275. Shen, Z., Shang, Y., Wang, Y. et al., International Sympodium on the Devlopment of Carbon Fibbers and Their Application, in Korea, pp. 4–9, 1991.
276. Park, Y. D. et al., J. Matter. Sci. *21*, 424 (1989).
277. Park, Y. D. et al., Carbon, *27*, 925 (1989).
278. Rhee, B. S. et al., 19th Biennal Cof. on Carbon Extrended Abstracts, p. 212 (1989).
279. Riggs, D. M., Venner, I. G., Ext. Abst. 16th the Biennal Conf. on Carbon, p. 42 (1989).
280. Riggs, D. M., and Diefendorf, U. S. Patent 4208267 (1980).
281. Yamada, Y., Honda, H., and Inoue, T., JP 58–18421 (1983).
282. Yamada, Y., Imamura, T., Inoue, T., and Honda, H., JP 58–196292 (1983).
283. Honda, H., Tanso, 113 (1983).
284. Otani, S., JP 57–100186 (1982).
285. Lewis, I. C., Carbon, *16,* 425 (1978).
286. Lewis, I. C., Fuel, *66 (11),* 1527 (1987).
287. Otani, S., and Sannda, Y., Carbon Fiber (in Japenese), Ohm Publishing Co., Ltd.,1980.
288. Mochida, I., Sone, Y. et al., Carbon *23 (2),* 175, (1985).
289. Mochida, I. et al., Carbon *28 (2/3),* 311 (1990).
290. Mochida, I., Shimizu, K. et al., Carbon *30 (1),* 56 (1992).
291. Wang, C. Y., Zheng, J. M. et al., 22nd Bien. Conf. on Carbon, Ext. Abst. and Pro. Univ. of California, San Diego, 16–21 July 1995, pp. 254–255.
292. Yoon, S., Oka, H. et al., 22nd Bien. Conf. on Carbon, Ext. Abst. and Pro. Univ. of California, San Diego, 16–21 July 1995, pp. 46–47.
293. Hu, Z. J. and Ling, L. C. et al., 22nd Bien. Conf. on Carbon, Ext. Abst. and Pro. Univ. of Carlifornia, San Diego, 16–21 July 1995, pp. 258–259.
294. Braun, M., and Ramer, J. K. et al., Carbon, *33 (10),* 1359 (1995).
295. Braun, M., and Ramer, J. K. et al., Carbon, *32 (8),* 1073 (1994).
296. Personal Communication in Carbon'96.
297. Ryuichi, H., et al., JP 61–186520 (1986).
298. Mitsuru, U., and Hiroshi, N., JP 60–259632 (1985).
299. Kagaku Kogyo Nippo (Japan), Feb. 24th, 1986. p. 1.
300. UCC, U. S. Patent 4301135.
301. Edie, D. D., and Dunham, M. G., Carbon, *27 (5),* 647 (1989).
302. Zhang, F., and Shen, Z., 3rd Inter. Joint Sympo. between BUCT and CNNU, Oct. (1996), Beijing (China) p. 86.
303. Barr, J. B., Appl. Poly. Symp. *29*, 161 (1976).
304. Yamada, Y., Imamura, T., and Honda, H., JP 60–21911 (1985).
305. Takayaki, I., JP 61–130182 (1986).
306. Yamada, Y., and Imamura, T. et al., JP 59–53717 (1984).
307. Matsumoto, T., Pure & Appl. Chem., *57 (11)*, 1553–1562 (1985).

308. Cha, Q. et al., Carbon, *30 (5),* 739 (1992).
309. Shen, Z., and Guo, H. et al., Carbon and Carbonaceous Composite Materials Structure–Property Relationship, Abst. and Pro., Malenovice, Czech Republic, Oct. 10–13, 1995, p. 31. (preceding in the press).
310. Hideyuki, N. et al., JP 62–41320 (1987).
311. Tsutomu, N. et al., JP 01–280025(1989).
312. Ahmed, M., Textile Science and Technology, 5,174, Elsevier, New York (1982).
313. Eiji, K. et al., JP 01–156513(1989).
314. Shen, Z., and Qin, R. et al., 36th Inter. SAMPE Symp. and Exhibition, *36,* 1109–1117(1991).
315. Shen, Z., and Guo, H. et al., 21st Bienn. Conf. on Carbon, p. 352 (1993).
316. Chi, W., and Shen, Z., 2nd National Symposium on New Carbon Materials, Dalian (China) 1995 p. 83.
317. Okuda, K., Nikkakyo Geppo, 12 (1), 17 (1980).
318. Otani, S., Jidosha Gijutsu, 34 (8), 861 (1980).
319. Sakaguchi, Y., Kagaku Kogaku, *46 (3),* 145 (1982).
320. Kenichi, M., and Yoshiro, K. et al., British Patent 1426502 (1973).
321. Hawthornal, H. M., et al. Nature, *227,* 946 (1970).
322. Nippon Steel Corp., JP 58–144123 (1983).
323. Chi, W., and Shen, Z., Carbon'96 European Carbon conference, p. 395 (1996).
324. Otani, S. and Kimura, K., Carbon Fiber (in Japanese), Kindai Henshu Ltd. p. 140 (1972).
325. Dresselhaus, M. S. et al., Graphite fibers and filaments (Edited by Cardona, M.), p. 18, Springer–Verlag, Berlin(1988).
326. Oberlin, A., Endo, M., and Koyama, T., J. Cryst. Growth, *32,* 335 (1976).
327. Endo, M., CHEMTECT, 568 (Sep. ,1988).
328. Schutzenberger, L., C.R. Acad. Sci. (Paris), *111,* 774–778 (1890).
329. Gibson, J. et al., Nature, *154,* 544 (1944).
330. Anon., Chemical Engineering, October, 172–4 (1957).
331. Koyama, T., Carbon, *10,* 757 (1972).
332. Endo, M., Koyama, T., and Hishiyama, Y., Jpn. J. Apl. Phys., 15,2073 (1976).
333. Katsuki, H. et al., Carbon, *19,* 148 (1981).
334. Tibbetts, G. G., and Devour, M. G., U.S. Patent 4,565,684 (1986).
335. Tibbetts, G. G., Graphite Intercalation Compounds (Edited by Eklund, P. C. et al.), p. 186, Material Research Society, Pittsburgh, PA (1984).
336. Kato, T. and Haruta, K. et al., Carbon, *30,* 989 (1992).
337. Ishioka, M. et al., Carbon, *30,* 975 (1992).
338. Kato, T. and Mattsumoto, T. et al., Carbon, *31,* 937 (1993).
339. Masuda, T., Mukai, S. R., and Hashimoto, K., Carbon, *31,* 783 (1993).
340. Mukai, S. R., and Masuda, T. et al., Carbon, *33,* 733(1995).
341. Mukai, S. R., Masuda, T., Harada, T., and Hashimoto, K., Carbon, *34,* 645

(1996).
342. Motojima, S. and Hasegaqwa, I. et al., Appl. Phys. Lett., *62*, 2322 (1993).
343. Motojima, S. and Hasegaqwa, I. et al., Carbon, *33*, 1167 (1995).
344. Chieu, T. C. et al., Phys. Rev. B, *26*, 5867 (1982).
345. Pirauz, L. et al., Solid State Comm., *50*, 697 (1984).
346. Tesner, P. A., Robinovich, E. Y., Rafalkes, I. S., and Arefieva, E. F., Carbon *8*, 435 (1970).
347. Baker, R. T. K., Barber, M. A., Herris, P. S., and Feetes, F. S., J.J. Warte, *26*, 51 (1972).
348. Baker, R. T. K., Harris, P. S., and Thomas, R. B., J.J. Warte, *30*, 86 (1973).
349. Baker, R. T. K. and Harris, P. S., J. Phys. *ES*, 793 (1973).
350. Baired, T., Fryer, J. R., and Grant, B., Nature (London), *233*, 329 (1971).
351. Baired, T., Fryer, J. R., and Grant, B., Carbon, *12*, 591 (1974).
352. Oberlin, A., Endo, M., and Koyama, T., Carbon, *14*, 133 (1976).
353. Tibbetts, G. G., J. Cryst. Growth, 66, 632 (1984).
354. Audier, M., Carbon *23*, 317 (1985).
355. Saacco, A., Jr. et al., J. Catal., *85*, 224 (1984).
356. Tibbetts, G. G., Devour, M. G., and Rodda, E. T., Carbon, *25*, 367 (1987).
357. Tibbetts, G. G., Carbon, *30*, 399 (1992).
358. Chitrapa, P., Iund, C. R. F., and Tsamopoulous, J. A., Carbon *30*, 285 (1992).
359. Gupta, S. K., Gupta, N., and Kunzru, D., J. Anal. Appl. Pyrolysis, *26*, 131, (1993).
360. Lake, M. L., Ting, J. M., and Alig, R., Proceeding Eight Annual Conference on Material Technology, Sept.,1992, pp. 55–63.
361. Matsumoto, M., Hashimoto, T., and Murata, K., Carbon *32*, 111 (1994).
362. Dresselhaus, M. S., Dresselhaus, G., Adv. Phys. *30*, 136 (1981).
363. Dasch, C. J. et al., Extended Abstracts, 21st Bienn. Conf. on Carbon (1993), p. 82.
364. Shen, C., ACTA Material Composites SINICA, 4, 1996.

2
STRUCTURE AND TEXTURE OF CARBON FIBERS

Agnès Oberlin
Saint Martin de Londres, France

Sylvie Bonnamy
CNRS-Université, France

Khalid Lafdi
Southern Illinois University, Carbondale, Illinois

2.1 INTRODUCTION

If the term structure is not defined, it covers both atomic structure and chemical structure, etc. If the term microstructure is employed it also covers anything, including textures from nanometric and micrometric scales. The present chapter covers both crystalline structure and all kinds of textures recognized in carbon fibers (micro and nanotextures). In the first part of this chapter, the techniques used will be divided into averaging techniques (x-ray and electron diffractions, optical microscopy, Raman spectroscopy, scanning electron microscopy SEM) and nanotechniques able to "see" atoms (transmission electron microscope TEM, near field scanning microscopes). To avoid too long developments the reader is requested to consult the references (books, reviews and thesis) detailing the techniques of x-rays and electron diffraction [1–8], optical microscopy and optical properties [9,10], Raman spectroscopy [11,12], TEM [13–15] and near field scanning microscopies [16]. The second part will be devoted to modelization of PAN-based carbon fibers, then to pitch-based ones. In the same manner references are given to books, reviews and thesis [17–28].

2.2 TECHNIQUES

2.2.1 X-Ray and Electron Diffractions

The data obtained for both techniques are less and less precise as the wave length and/or the probe size decrease. For electron diffraction the angular resolution is poor and the size of the pattern is small (as an example the 002 ring of carbon is so small that no difference can be established between a heat-treated turbostratic carbon and graphite). In counter part the volume of the scattering sample decreases from x-rays to electron diffraction. The wave length for x-rays starts from 0.154 nm (copper radiation) down to 335.10^{-5} nm in a 120 kV TEM and even less as the voltage increases. Correspondingly, the volume of scattering sample decreases from a few cubic millimeters for x-rays down to 1/20 μm^3 for selected area electron diffraction (SAD) and to 10^{-6} μm^3 for microdiffraction (depending on the final size of the probe which limits the scattering area).

The data obtained are the same for both techniques, since they are deduced from the interaction between the incident beam i.e., the probe and the sample. This yields a diffraction pattern resulting from the interaction between the reciprocal incident wave (the Ewald sphere i.e., a sphere of radius $1/\lambda$) with the reciprocal space of the sample (built with $1/d_{hkl}$ distances). The larger λ the more pronounced the curvature of the reciprocal sphere, i.e., the more deformed is the section of the reciprocal space. It is more difficult to go back from x-ray pattern to the real space of the object since the Bragg law must be written $2d \sin \theta = \lambda$. It is simplified for electrons because λ, thus θ are so small that $sin\ \theta$ can be assimilated to θ and the law becomes $2d\theta = \lambda$. The reciprocal space can thus be more easily recognized. However if the relations in sine for x-rays becomes *almost* linear for electrons, the section of the reciprocal space which forms the pattern is *almost* a plane but is *not* a plane. Simultaneously the reciprocal space is always distorted both by the defects contained in the object and by the necessity to have thin samples for being transparent to electrons. All data must take that into account.

In SAD, for a thin graphite crystal [29], the mathematical nodes of the reciprocal lattice are elongated in $1/t$ due to the small thickness t of the sample. Owing to that the Ewald sphere is able to cut a few nodes and a pattern is obtained. If the single crystal is replaced by a turbostratic stack (aromatic layers rotated at random with a small angle α), the thickness becomes that of a single layer. The elongated nodes fused into continuous hk lines. Since some layers are piled up, each hk lines becomes divided into as many lines as there is layers in the stack. When such a disordered structure is produced, the diameter of the stacks also decreases tremendously down to elemental units (basic structural

Structure and Texture of Carbon Fibers 87

units i.e., BSU [15] and see [76]). However, they tend to stay in the same plane in azimutal preferred orientation (Figure 1a). The reciprocal space is made of *hk* cylinders (Figure 1b) added to *00l* elongated nodes. The patterns now obtained depend on the orientation of the sample relative to the electron beam (P_1, P_2, P_3, in Figure 1). The feature common to all fibers is a rotation of the turbostratic layer stacks at random around the fiber axis AA' (continuous rotation of Figure 1a and b around AA'). The resulting pattern (sketched in Figure 1c) is an average of all possible superimposed patterns obtained by a continuous rotation of the Figure 1a and b around AA'. Asymmetrical *hk* bands are formed along the equator. In addition to these simple sketches, one must take into account the occurrence of a slight misorientation relative to the preferred orientation plane due to a slight tilt of the stacks relative to the fiber axis AA'. Their *c* axis describes a cone and *002* forms an arc. The real pattern thus corresponds, at the best, to Figure 1c with *00l* arcs added (Figure 2a). Among the data obtained are:

1° The degree of preferred orientation i.e., the value of the half angle of the misorientation cone. It is measured by the opening of the *00l* Bragg reflection arcs [1, 30].

2° The interlayer spacing $\bar{d}_{002}$ i.e., the average value of the distances between each pair of aromatic layers.

3° The porosity which is determined by low angle scattering [31,32] (pore sizes and estimation of their shape).

4° The degree of three dimensional order. It is known [33–35] that the lowest order of almost all carbons is two-dimensional not only due to turbostratic structure but also to small random glides [36]. The best three dimensional order is that of natural graphite. From the former to the latter all intermediates are known, [15,37] as well as all intermediates can be produced by heat treatment of some peculiar turbostratic carbons (graphitization of graphitizing carbons).

Graphitization is generally obtained by heat treatment. As $\bar{d}_{002}$ decreases a progressive transformation of the reciprocal *hk* lines to elongated nodes occurs by a progressive modulation near the position of *hkl* reflections [6,38]. The longitudinal intensity profile of a turbostratic *hk* line which was initially that of $f_{(\theta)}^2$ (where $f_{(\theta)}$ is the atomic diffusion factor of carbon), begins to reinforce near *hkl* positions (Figure 2b). It is only in the x-ray pattern that, from such modulations, the degree of graphitization can be measured. It is expressed by the probability P_1 to find a pair of adjacent layers in the graphite AB positions [6,33,34,38]. P_1 increases with heat treament up to a maximum. P_{1max} is almost nul for non graphitizing materials. It is one for natural graphite. Among natural and industrial carbons all values can be found [15,37,39] (Figure 3). Two different relations between $P_{1\ max}$ and $\bar{d}_{002}$ are found for lamellar products (Figure 3a) and fibers (Figure 3b). In addition, the empirical factor

$g = 0.344 - \overline{d}_{002} / 0.344 - 0.335$, sometimes identified to $\sqrt{P_I}$ [6,7], is reported in Figure 3 as dashed line. It does not fit with experimental curves mainly for fibers. The values of P_I are thus not one to one connected to $\overline{d}_{002}$ [5, 33, 34, 39]. The term of graphitization is very often used in an improper fashion, despite the numerous remarks given in literature [4,5], by confusing the graphitization process itself with the high HTT (heat treatment temperature) necessary to obtain it. Entirely turbostratic fibers, such as most of PAN-based fibers, are often qualified as graphite fibers, only because they were heat-treated above 2000°C.

5° *Lc* and *La*. Such values should represent only the width at half maximum of a Bragg reflection. They are often assumed to be the thickness and the diameter of the carbon layer stacks, themselves qualified as crystallites. If it is so, they are meaningless [5]. However they can be used to compare series of materials if they are measured in the same conditions. Artefacts come from the gap between the coherent domain diameter *La* deduced from *hk* in the patterns and the real diameter of the corresponding aromatic layer [7,40,41]. Figure 4a illustrates what happens in a turbostratic layer stack where two reciprocal *hk* rows are slightly rotated so that parasitic interferences occurs in A due to the superimposition of the two beams (first condition of interference). In B there is no more interference. The coherent domain size measured from A is thus larger relative to B. In addition, from B to C and to the largest orders, the coherent domain size decreases again due to distorsions [7]. Since the true physical entity interesting to measure is not *La* but the diameter L_H of the <u>continuous</u> but distorted aromatic layer, it is necessary (but tedious) to extrapolate the values to the *000* reciprocal node coincident with the incident beam (Figure 4b). Similarly, stacking disorders modify *Lc* [5,7].

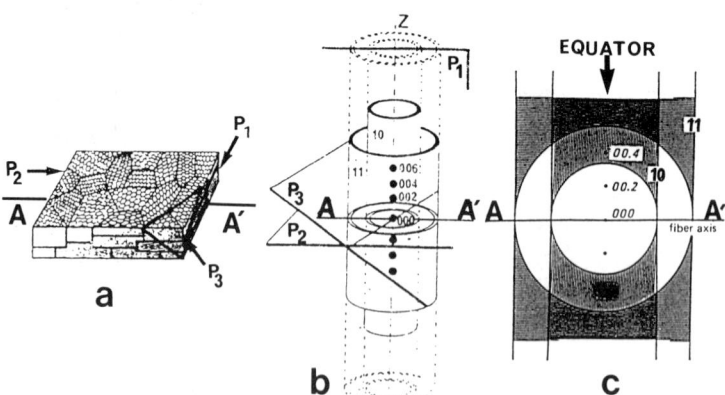

Figure 1 a. Turbostratic layer stacks in azimutal preferred orientation, b. Reciprocal space, c. Sketch of a fiber pattern.

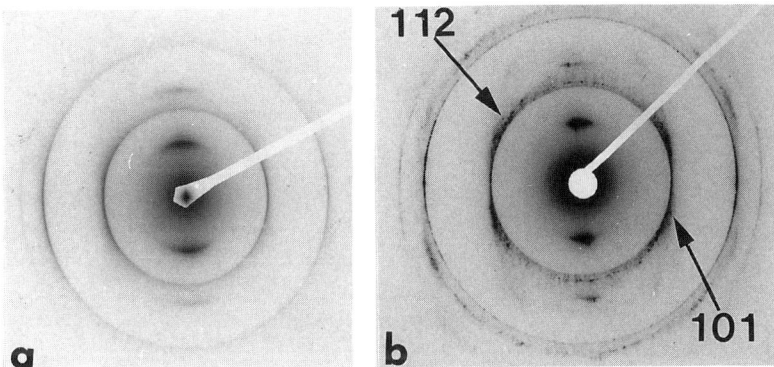

Figure 2 a. SAD pattern of a turbostratic fiber, b. SAD pattern of a partially graphitized fiber.

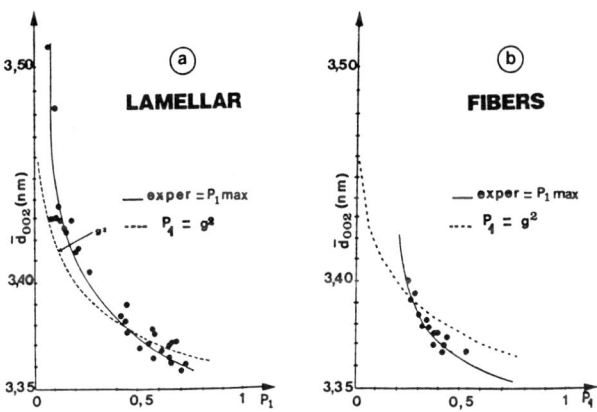

Figure 3 Relation between P_1 and $\bar{d}_{002}$, a. for lamellar products, b. for fibers. In dashed line $P_1 = g^2$ (adapted from ref. 39).

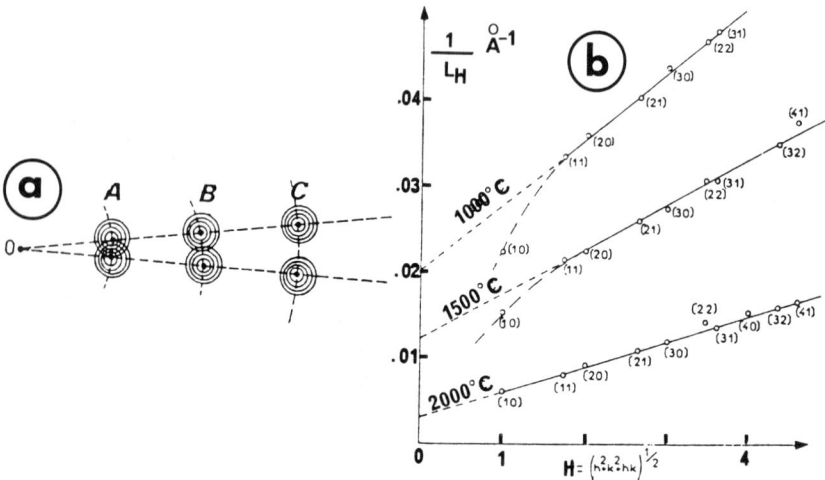

Figure 4 a. Reciprocal lattice rows of two rotated aromatic layers, b. Relation between the extrapolated size L_H of the distorted aromatic layer and the coherent domain sizes L_a for different orders of reflections (from ref. 7).

2.2.2 Optical Microscopy (OM)

Optical microscopy will be largely applied to the study of pitch-based carbon fibers, not only for characterizing the fibers themselves but, also, to characterize the pitches used as precursors. The following paragraph is thus important to understand. Since carbons are highly absorbing media, reflection microscopy must be used upon polished sections. As carbons are more or less ordered they are optically anisotropic. Both polarized light and crossed polarizers should thus be used. Optical anisotropy is expressed by the phase shift introduced by the object. A Newton chart is used to evaluate phase shift. In Figure 5a the phase shifts are marked with their associated colors. The heavy black dots mark the zero and first order of the interference colors i.e., 0 and 0.551nm.

1° Single crystal of graphite
A single crystal of graphite is an uniaxial negative crystal (Figure 5b). The ordinary refractive index n_o is n_g the largest one which is contained in the aromatic layer plane (n_g = 2.15). n_e is n_p oriented along the hexagonal axis (n_p = 1.81). Since graphite is highly absorbing, all observations have to be done

by reflection. The ellipse, section n_g, n_p of the ellipsoid is thus replaced by the indicator surface of reflectances. The higher the birefringence, the less the ovaloid differs from an ellipsoid. In the case of graphite which has strong birefringence ($\Delta n = 0.34$), the two sections can be approximated to each other so as to assimilate transmission to reflection data. Therefore the Newton chart should be used to evaluate the phase shift δ. For graphite δ is small (~230 nm) i.e., its color between crossed polarizers is almost white when the aromatic layers are almost perpendicular to the observation plane (Figure 5a). The true reflectances as well as the maximum δ value of graphite are impossible to measure [10] since it is impossible to cut or polish graphite perpendicularly to the (001) layer planes (because of the high value of cohesion due to the 0.142 nm bond). Graphite is dark i.e., isotropic when the layers are in the observation plane. In the latter case, the ellipsoid of a uniaxial crystal having a revolution symmetry, no change occurs during a rotation of the crystal. For the same reason δ_{max} is also constant i.e., 230 nm from the maximum illumination down to extinction since n_g and n_p are always seen with a constant orientation. *During the stage rotation only the intensity of the reflected light changes.* Due to its high absorption index graphite is pleochroic, i.e., when it is observed in polarized light it is dark when n_o (n_g) is parallel to the polarizer vibration plane. It is bright when n_o is perpendicular.

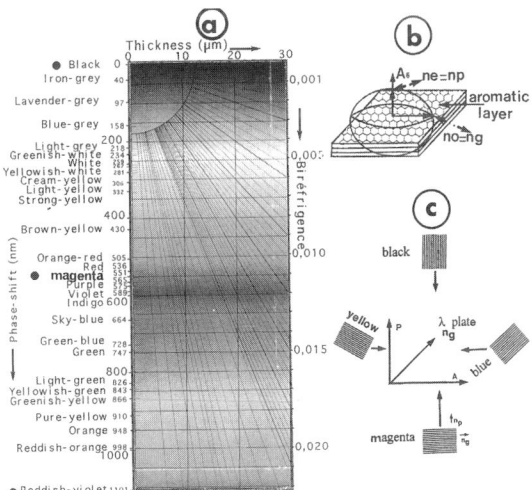

Figure 5 a. Succession of colors in a Newton chart, b. Indices ellipsoid for graphite (negative uniaxial crystal), c. Observation of a graphite crystal perpendicular to the observation plane. Crossed polarizers and λ plate added.

Since the maximum value of δ is small, it is necessary to observe graphite crystal not only between crossed polarizers, but also to add a retarder plate (λ plate) the phase shift of which is 551 nm (black dot in Figure 5a). This auxiliary plate is magenta in color. By considering the Newton chart (see Figure 5a), as n_g of graphite is parallel to n_g of the λ plate, the 230 nm phase shift of graphite adds to the 551 nm of the plate: the color turns to green blue. It remains green blue inside the whole quadrant. When n_g of graphite is perpendicular to the n_g of the λ plate the color turns to strong yellow inside the whole quadrant. During stage rotation only the intensity of the colors changes inside a given quadrant. Figure 5c summarizes the results obtained when the aromatic layers are perpendicular to the observation plane. When they are parallel, the section remains magenta during the stage rotation. *It is only when the crystal is tilted* between the above described positions that δ increases (or decreases) between zero and its maximum value. The colors observed sweep the range of the Newton chart comprized between magenta and green blue or strong yellow and magenta.

2° Less organized carbons
In less organized carbons, the elemental aromatic layer stacks (basic structural units or BSU) are much smaller than the microscope resolution. It results in averaging the data relative to each BSU considered as a nanocrystal of graphite (the turbostratic disorder inside each BSU does not modify anything except the intensity of the reflected light). The data thus obtained are not sensitive to the crystalline order but highly sensitive to the textures i.e., to the three dimensional arrangement of BSU, to the occurrence or not of a preferred orientation, and to its perfection. If there is no local orientation (local molecular orientation or LMO) or if it is largely below the resolution of the microscope (BSU at random for intance or very small local molecular orientations, themselves at random) the material is optically isotropic i.e., always magenta. If the preferred orientation is large enough to be observed, the colors depend on the degree of misorientation of the constituting BSU. Since they are usually both twisted and tilted (Figure 6a) [42], the colors are affected in their intensity by the twist (β), they change with the tilt (α). In the case of LMO (Figure 6b), the twist is ± 20° and the tilt ±10°, the colors are almost constant. If LMO are replaced by poor statistical orientations (SO) [43] as seen in Figure 6c, the green blue colors tend to indigo, the strong yellow ones tend to red whereas black and magenta are added in a certain amount.

In the peculiar case of PAN-based fibers (see Figure 17 and 25) the cross section is isotropic since BSU and LMO are twisted at random and much smaller than the resolution. The longitudinal sections are weakly anisotropic with n_p always perpendicular to the fiber axis since almost only BSU tilt is operational (obviously it is meaningless to describe such an arrangement as a

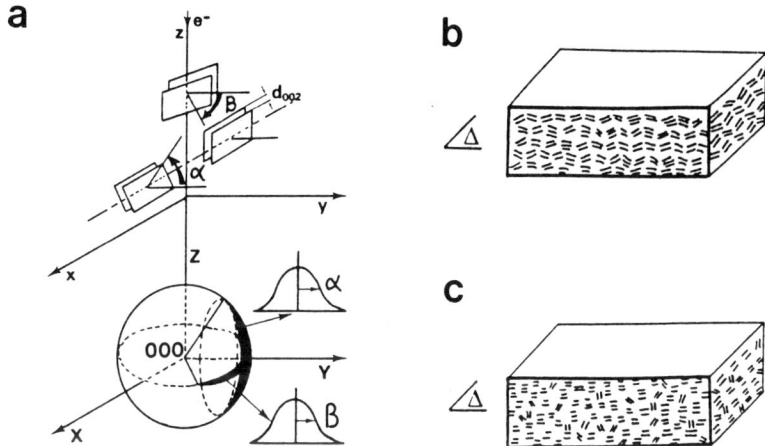

Figure 6 a. Possible misorientations of BSU inside a LMO, b. Sketch of local molecular orientation (LMO), c. Sketch of statistical orientation (SO) including BSU misoriented up to 90°.

positive uniaxial crystal since the latter has a revolution symmetry around a unique axis n_g). In the case of anisotropic pitch-based fibers where the texture scale is above the microscope resolution, concentric, radial or random textures can be studied (see Section 2.4).

2.2.3 Raman Spectroscopy

The sample interacts with a monochromatic beam of photons (laser), the frequency of which is υ_0. Then the sample scatters a polychromatic light adding to the major Rayleigh scattering υ_0, much weaker scattered waves with frequencies $\upsilon_0 \pm \Delta\nu$ (Raman effect). The $\pm\Delta\nu$ (stokes and antistokes) are characteristic of the molecules or ions present in the sample. Raman spectroscopy can thus be used as a tool for chemical analysis by determining the wave number $1/\lambda$ of the peaks present in the spectrum. To the normal Raman spectrometers can be added microprobe spectrometers in which an optical microscope is associated so as to focus the laser beam to a few μm^2 through the objective. As an example the MOLE (Molecular Optical Laser Examiner) [11] equipped with a 514.5 nm laser was used to study a few carbonaceous materials [11,12,44–47].

1° Graphite

For graphite the first order spectrum shows only a sharp peak at 1580cm^{-1} (the E_{2g} carbon-carbon vibration in the aromatic layer plane).

2° Less organized carbons

For less organized carbons the 1580 cm^{-1} band is sometimes displaced to 1610cm^{-1} and an additional peak (more or less broad) is added at 1350–1380 cm^{-1}. It is attributed to "defects." As an example [12] thermal graphitization of thin films of carbon can be followed by MOLE (Figure 7). From the initial amorphous stage (bottom of the figure) to the most graphitic sample (top) the two peaks develop at first, then only the graphite one is present at 2700°C. From a Raman spectra one can take out the occurrence of various bands, their wave number, their width at half maximum, their relative surfaces, so as to correlate these data to the purity and perfection of the specimen.

2.2.4 Scanning Electron Microscopy (SEM)

It is still an averaging technique for all modes using the total penetration depth of the probe with the associated secondary effects (x-ray photons or Auger electrons). It approaches nanotechniques in the secondary electron mode i.e., morphology, since the resolution of modern SEM is almost nanometric. SEM is a *pseudo-imaging* system since no lens is used. The electron probe scans a

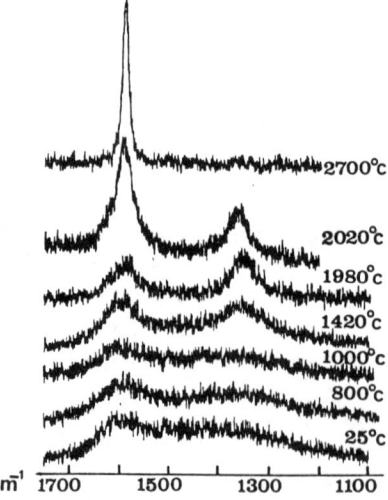

Figure 7 Graphitization of thin films followed by MOLE Raman spectroscopy (from ref. 12).

Structure and Texture of Carbon Fibers

given x^2 surface of the sample whereas a suitable collector gathers the secondary emissions and send them point by point to a monitor screen standardized in size (10 cm). If secondary electrons issued from the surface (depth lower than 5 nm) are collected through a positively charged collector, the magnification of the pseudo-image is 10 cm divided by x. This morphological mode yields details of the surface with a pseudo relief due to the heterogeneous attraction of secondary electrons by the collector. SEM is a very valuable and widely used tool since the preparation of the sample is almost negligible: just put it in the microscope after fracture.

2.2.5 Transmission Electron Microscope (TEM)

With TEM, imaging of atoms is approached, joined to the priceless advantage to exploit the same crystallographic data as x-ray or electron diffractions through a double Fourier transform. Correspondingly, a direct relationship can be established with these techniques as well as with another imaging technique: optical microscopy. At least electron diffraction, TEM and OM images can be obtained exactly on the same area of the sample [43] by observing a thin section and its residual block used as a polished section.

1° Sample preparation
The first point to discuss briefly is the preparation of the sample. The simplest one is *grinding*, often blamed for its supposed degradation of the specimen. It is in fact one of the less disrupting for carbons even for a prolonged action [48]. The main disadvantage is the loss of the three dimensional microtexture. *Ion thinning* is more controversial [49] since it has three major defects. The sample observed has a wedge profile, strongly disrupting the lattice fringes. Then thinning destroys preferentially the fragile phases: disordered carbons, amorphous ones and resins are increasingly destroyed yielding a lot of artefacts in the interpretation of the images. In addition, thinning also destroys the texture of the sample by mechanical weakening of the thinnest areas or by chemical and thermal etching (sublimation followed by recondensation). As an example, a matrix of resin carbonizes under the ion beam and is redeposited upon fiber reinforcements. A pure mechanical thinning process (tripod polisher) was also developed which could be less harmful [50]. Thin sectioning (when it is possible) is the best technique, at first because the sample has an almost constant thickness instead of being a wedge. The main disadvantage is the vibration of the microtome diamond knife as it touches the carbon sample. The section is broken quasi-periodically into fragments which are displaced. However this happens only on highly carbonized or graphitized samples. It may be often overcome by suitable tricks such as changing the embedding media, cut the sample more obliquely, re-embed it after cutting or sharpen only a part

Figure 8 Bright-field image. A part of a thin section (cross-section) of a DuPont E105 pitch-based carbon fiber.

of the block before cutting again [51]. Usually there is no artifacts introduced in the number and nature of the components. Also if the displacements of the fragments of the section are small enough, they do not prevent recognition of the texture (radial in Figure 8).

<u>2° Choice of the mode</u>
The second point to review is the choice of the modes employed. Basically TEM is a true imaging technique which restores the object real space through a double Fourier transform operated by the objective lens. The first Fourier transform is the diffraction pattern occurring at the back focal plane of the lens. The second one is the image produced at the image plane of the lens. TEM thus exploits the same data than x-rays but yields automatically the real space with a resolution of a few tenth of nanometer. The efficiency of TEM thus depends on the good choice of the modes to use, so as to describe the three dimensional arrangement of the components of the object. As an example, let us consider a thin carbon film on the way to graphitize [52]. In such films all aromatic layers are oriented parallel to the film plane but in azimutal disorder. Their reciprocal space is that of Figure 1b if the film is still turbostratic. However cracks are very common which cause the film to tilt or even roll around an axis (sketch in inset of Figure 9). The reciprocal space is thus that of Figure 1c i.e., that of a

Structure and Texture of Carbon Fibers

fiber. The folded film emits *00l* beams in the areas where the aromatic layers are almost parallel to the incident beam. It emits *10* and *11* beams when the layers are perpendicular or oblique to the incident beam. If one *hkl* beam at a time is employed to form the image, the corresponding regions of the object would appear bright upon a dark-field (*hkl* dark-field images or DF). By successively using all the beams the three dimensional arrangement in the film would be determined. The parts of a fold almost perpendicular to the observation plane would appear bright in *002* dark-field. If graphitization is marked enough, individual aromatic layer stacks would be produced and superimposed with random rotations along the film thickness. Rotational moirés fringes would occur in *10* or *11* dark-fields (Figure 9a). In the areas strongly folded, the more and more oblique stacks would appear more and more elongated with more and more strong disruption in the periods of their moirés. The more oblique ones appear as ribbons (arrow). From such analyses, in addition to nanotexture, the diameter La of the stacks and their thickness Lc would be estimated. If one beam is allowed to interfere with the incident beam or with a higher order of the same reflection, a high resolution image of the dense lattice planes (or atoms rows) would be obtained (*lattice fringes* or LF). As an example Figure 9b (from box in Figure 9a) shows *002* lattice fringes in contrast in a fold of the film. Since TEM images are only the projection of various features on the observation plane, the *002* fringes disappear suddenly as the aromatic layers cease to be under the Bragg angle. The film, despite its continuity, appears as a ribbon in Figure 9b. Without Figure 9a an artefact should have produced wrong data. Up to now, it is the case for all ribbons models established for carbons and fibers [31,53–55].

The main disadvantage of TEM is the spherical aberration Cs of the objective lens, which introduces a variable phase shift in the scattered beams (transfer function of the lens). Correspondingly, all modes of TEM are affected. In selected area electron diffraction (SAD), the aperture used to limit the scattering area must necessarily be large enough to avoid to obtain the pattern of an unknown region of the object (the smallest area reliably used must be at least 1μm in diameter). In dark-field, the effect of Cs is avoided by using high resolution tilted dark-field [15,42]. In lattice fringes mode, it is necessary to suitably defocus the lens. Unfortunately for that, only the eye forms the starting point. In addition, if the specimen is not homogeneous in thickness (wedge or surface steps) the fringes are either curved or with a variable spacing despite the lattice planes are flat and perfect [14].

Two associated techniques must be mentioned as very convenient: image analysis (see ref. 202) whether coupled or not to the TEM and laser optical diffraction bench (OD data). Owing to them the Fourier transforms of the lattice fringe image are performed, allowing to determine the interfringe spacing and its eventual fluctuation, as well as measurements of preferred orientation if any.

98
Oberlin et al.

Since in composites, the mechanical properties are highly sensitive to the first nanometers at the interface fiber-matrix (see ref. 109 and 115), studies of areas of a few square nanometers are without price.

2.2.6 Near-Field Scanning Microscopes (see Chap. 4)

To visualize the surface of a solid with a resolution better than 0.2 nm, it is necessary to detect and explore the field emitted by the surface along a thickness of a few nanometers. The probe apex must thus interact with the surface with a tight coupling. The latter can be electronic. It was the case when Binnig and Rohrer [56] used tunnel effect of electrons to built the first scanning tunneling microscope (STM). The coupling can be mechanical as in atomic force microscope (AFM). It can be also electromagnetic which is realized in the photon scanning tunneling microscope (PSTM). More recently a large range of

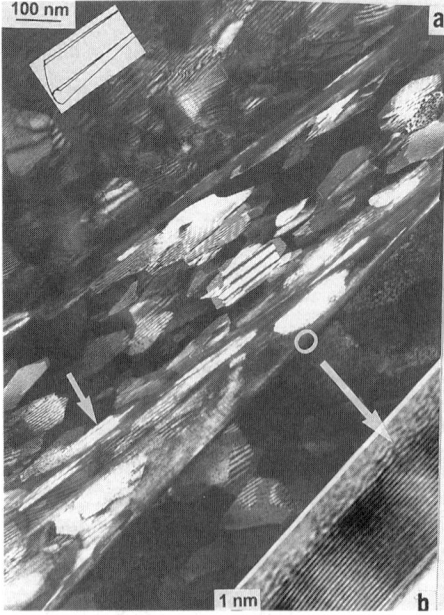

Figure 9 Thin carbon film in the way to graphitize (from ref. 52), a. 11DF of the folded part (the fold is parallel to AA'). The bright stacks are those whose $\vec{a}$ axis is parallel to AA', b. 002 lattice fringes (LF) in the fold. In inset sketch of the fold.

variants appear. Only STM and AFM are commercial apparatus applied respectively to electrical conductors and insulators. Only STM was widely applied to fibers. In STM the probe is a nanoelectrode measuring the current I_e at a distance a of the surface. A reference servocurrent displaces the electrode in Z to avoid touching the rugosity a. The time constant is about one millisecond. The data obtained are not really atom imaging but local electronic state density. STM allows to distinguish between three and bi-dimensional order in carbons [16,57–59] in a very interesting manner. Graphite is three dimensionally ordered with ABAB stacking (Figure 10a). This is responsible for a lack of equivalence between the six carbon atoms of a cycle. The α atoms (crosses) superimposed to those of the adjacent aromatic layer are less visible than the β ones (circles) which have no neighbours immediately below. Only a ternary symmetry is observed in the patterns with a point to point distance of 0.246 nm (Figure 10b). In the cases where underlayers are not coherent with the upper ones (two dimensional order) the six carbon atoms are visible (triangles in Figure 10c). The symmetry is hexagonal. Figure 10d shows a transition zone between the trigonal symmetry of graphite in the upper part of the figure and the hexagonal symmetry of a two dimensional aromatic layer in the lower part of the figure. Figure 10e is a sketch of Figure 10d inside which α atoms are still crosses β ones circles and the six identical carbon atoms are triangles. All kinds of defects are supposed to be recognized such as vacancies and disclinations. However the cleanliness of the surface to be observed is drastic since any insulating ad-atom or molecule and peculiarly oxygen entirely blurrs the image. Cleavages are the most safe way when they are possible. Unfortunately they are not for fibers, the surface of which being particularly covered with contaminants. In addition, for fibers other methods (etching or peeling) easily generate artefacts. Lack of something better, not treated and not sized fibers must be used.

2.3 PAN-BASED CARBON FIBERS

As other fibers such as pitch-based ones, PAN-based carbon fibers are spun, stabilized by air then carbonized (high tensile strength fibers i.e., HTS fibers). Eventually they are heat-treated in the temperature range of graphitization (high modulus or ultra high modulus fibers i.e., HM fibers). However almost all of them are non-graphitizing carbons. For detailed bibliography the reader may consult ref. 60 and 61. As any other organic matters [15,62,63] polyacrylonitrile (PAN) cyclizes. BSU occurs which subsequently associate within oriented volumes (LMO). After association into preferred orientation areas, BSU coalesced into distorted layers, the distorsions of which may anneal more or less completely as HTT increases.

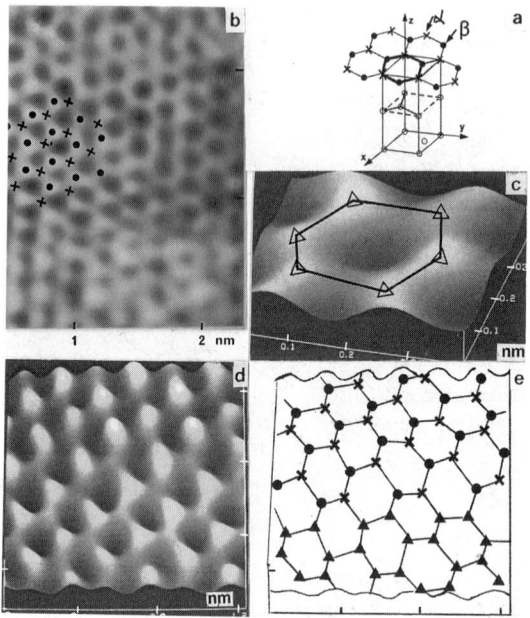

Figure 10 a. Graphite lattice showing the unit-cell, the three atoms α (crosses) and the three atoms β (circles), b. STM image of trigonal symmetry of graphite (from ref. 59), c. STM image of hexagonal symmetry of a two-dimensional aromatic layer (from ref. 58), d. Transition between graphite and two-dimensional order (from ref. 58), e. Schematic representation of 10d.

2.3.1 Polymerization of Acrylonitrile

The monomer is a mixture of acrylonitrile with a few percents of additives. After polymerization, the PAN is dissolved and extruded under tension into a coagulating bath (wet spinning). The resulting polymer is oriented along the fiber axis and characterized by x-rays. The most intense reflection is 0.52 nm and a weaker reflection at 0.305 nm is sometimes, but not always, observed [64].

PAN decomposes exothermally from 180°C with a maximum at 250–300°C [65]. By heating in sealed tubes it softens and behaves as a graphitizing carbon [66]. To avoid these phenomenons which destroy the fibers, stabilization is performed.

2.3.2 Stabilization

The fiber is submitted to air between 200 and 300°C still under stretching. This step is of utmost importance in industrial fabrication. The exact parameters are thus unknown, except that the treatment must exceed 5 hours near 220°C and must yield a suitable oxygen fixation (about 6 wt.%). The polymer obtained (ladder polymer) is often improperly called precursor.

Oxygen and copolymer additives combine during stabilization, partly to control some of the volatiles release, partly to favor cyclization in the ladder polymer. The nature and amount of volatiles released (HCN, NH_3, N_2, CO_2, H_2O, CH_4, and hydrocarbons) have a tremendous importance since they govern the ultimate chemical composition of the final carbon fiber (C, H, O, N). The latter itself regulates the fiber properties and textures. In fact, stabilization corresponds to three different reactions: cyclization, dehydrogenation, and oxidation. It is very difficult to know the exact succession of the reactions but they are probably very close to each other and cause the formation of the elemental units of carbon (BSU). At first it is known that only five nitrogenated heterocycles are stable [67]. Then it is known that only coronene and larger molecules get a stable face to face configuration [68] so that a carbon BSU has the minimum size of dicoronene (0.71nm) with a molecular weight of 600 amu. Its maximum size measured by TEM is around 1nm i.e., between ovalene (800 amu) and diekacoronene (1300 amu–1.2 nm in size). It is plausible [60] to expect a transition from cyclized parts of the oxidized ladder polymer through dehydrogenation (Figure 11a). Then lateral growth forms polyaromatic BSU components. A large difference occurs between such a planar molecule and the ones constituting BSU of other carbons. At the edges are located nitrogen and also pentagons responsible for disclinations (Figure 11b). Such defects are two dimensional analogues of dislocations, either in a liquid crystal [69] or in a solid [70]. They are usually very stable in solids. They thus maintain curvatures in aromatic layers after subsequent coalescence of BSU [71], preventing or deleting graphitization [72]. When submitted to stretching, the planar molecules would have a tendency to pile up with one or two of their neighbors. However, due to stresses, the units formed (nitrogenated BSU) are turbostratic. BSU can be imaged as bright dots in *002* dark-field images (see Figure 14a).

It is questionable whether or not, there is an advantage to develop many aromatic BSU at such an early stage. This question is partially answered by considering x-ray patterns of increasingly oxidized PAN [61,73]. Beyond the 0.52 nm sharp and intense reflection of PAN occurs a wide and faint one situated either at 0.31 or at 0.35 nm depending on the authors [73,74] (Figure 12a). The increase in intensity of the additional reflection accompanying the decrease of the 0.52 nm one (Figure 12b) corresponds to an increase of oxidation joined to a failure in the mechanical properties. The reflection at

0.31-0.35 nm was attributed to BSU which were thus ill-timed. In fact, due to the inaccuracy of the measurements it seems to be difficult to choose between a BSU reflection (too small value) or the 0.31nm reflection of non oxidized PAN sometimes mentioned [64].

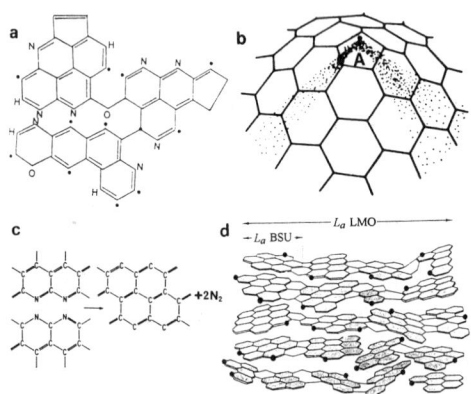

Figure 11 a. Possible model of dehydrogenation of an oxidized PAN-based fiber (from ref. 60), b. Disclination formation from a pentagon (from ref. 22), c. N_2 release according to Watt mechanism, d. Sketch of a small LMO containing nitrogenated BSU. (●) = N atoms.

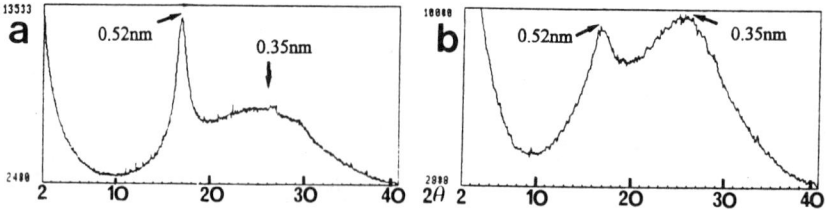

Figure 12 X-ray diffractograms of increasingly oxidized PAN fiber, a. The 0.52nm sharp reflection is associated to very faint additional halo (0.35nm), b. The 0.35nm halo overcomes the 0.52nm reflection (from ref. 61).

2.3.3 Carbonization

The industrial process is performed in continuity with stabilization under constant length but without stretching and under nitrogen. The fibers obtained are high tensile strength (HTS) fibers. It is a short duration operation carried on at a medium HTT. As in stabilization, the sequence, the nature and the amount of volatiles released, depend on additives used in the precursor. They depend also on oxygen and nitrogen content still present at the end of stabilization and of the HTT of carbonization. Roughly NH_3 leaves first, then HCN and N_2, probably by Watt mechanism (Figure 11c) [75].

1° BSU occurrence
Since the starting point of carbonization plausibly contains nitrogenated BSU of carbon, a comparison must be done with carbonization of most of the other organic matters (tars, pitches, kerogens and coals [15,37,62,63,76], (see also ref. 140 and 141) for which the main data will be recall. As HTT increases, H_2O, CO_2, and hydrocarbons lighter and lighter release. Correspondingly, BSU get a certain mobility (softening). It is due to the breakage of the aliphatic groups grafted upon BSU edges which form a light suspensive medium before releasing. Softening is increasingly prevented by cross-linking if any. Oxygen and some kind of sulfur are responsible for. All intermediate behaviors are known forming a continuous series between maximum plasticity and an almost negligible one. Near the limit, only microhardness could recognize and measure softening. During the stage where they are mobile, BSU tend to align themselves with their aromatic layers parallel to a plane. The oriented volumes are LMO (Figure 11d) visible in *002* dark-field images by a clustering of the bright BSU into clouds (see Figure 14b). Association of BSU into LMO is favored by the bubbles due to volatile release [77,78]. As cross-linking increases, the less marked the plasticity, the slowest the bubbles movement and the smallest the LMO. After the total release of aliphatic CH groups, there is no more suspensive medium, the material is solid and LMO development stops. However, the persistence of aromatic CH groups around BSU delete their coalescence into continuous aromatic layers, insuring to them a certain flexibility. At this point the primary carbonization is over and only elimination of CH_4 and H_2 occurs accompanying progressive organization of LMO into aromatic layer stacks (see Figure 15). Since during the softening stage hydrogen and cross linkers act as antagonistic parameters, it is necessary now to include nitrogen in the process and to determine its relations with O and H.

2° LMO occurrence
For doing that, increasingly oxidized PAN were spun and carbonized into a semi-industrial pilot unit allowing to study the samples at various HTT. The

more oxidized precursor corresponds to fiber C. The occurrence of LMO was determined by TEM and connected to elemental analyses. The mechanical properties of the final HTS fibers (D and C) were also measured (Table 1) [61, 73]. For each initial oxidized PAN-based fiber, $(H/C)_{at.}$ was plotted versus $(O/C)_{at.}$ and $(N/C)_{at.}$. Then the displacement of the representative point was plotted in the Van Krevelen pattern for each HTT and for the two fibers (Figure 13). During carbonization, $(H/C)_{at.}$ decreases first, down to 0.4–0.5, then $(H/C)_{at.}$ and $(O/C)_{at.}$ decrease altogether. Correspondingly, $(N/C)_{at.}$ decreases also, but more steadily. Table 1 indicates the initial elemental composition, the temperature to which LMO occurs, its size and the elemental composition at LMO. The last two columns are the tensile strength σ and the Young's modulus E. If the LMO occurrence is reported on Figure 13 (arrow), it corresponds systematically to the inflection point of the curve $(H/C)_{at.}$–$(O/C)_{at.}$ i.e., 550°C for D (Figure 13a) and 400°C for C (Figure 13b) i.e., maximum of dehydrogenation. The rule of other carbons is obeyed since with increasing cross-linking, i.e., $(O/H)_{at.}$ increasing, the LMO temperature decreases. Figure 14a and b shows the *002* DF images for C fiber respectively before LMO (300°C) and after [73]. At the LMO occurrence for D and C fibers $(O/H)_{at.}$ increases from 0.34 up to 0.48 whereas $(N/C)_{at.}$ decreases from 0.28 to 0.23. By considering Figure 13 and Table 1 the nitrogen role becomes more obvious. LMO corresponds to the brittle solid state. At that point the carbon skeleton is formed. If reticulation by oxygen is more pronounced LMO occurs at a lower HTT, $(O/H)_{LMO}$ is larger and there is less "lubrication" by aromatic CH groups. If it is admitted that nitrogen plays the same role as CH aromatics, it would be necessary to keep the highest amount of nitrogen available after LMO so as to insure flexibility of the LMO. Table 1 already shows that a limited oxidation and a maximum $(N/C)_{at.}$ value is favorable to better mechanical properties, mainly to a better tensile strength σ. On the contrary, under-oxidation favors softening of the materials, destroying the fiber texture, whereas over-oxidation yields more and more fragile materials [79]. However the mechanism of nitrogen effect is not yet known. It will be clear up owing to modelization after a careful study of commercial fibers.

Table 1 Elemental Analyses and Mechanical Data for Pilot PAN-Based Carbon Fibers

	Initial			LMO occurrence			Mechanical properties	
	$(O/H)_{at.}$	$(N/C)_{at.}$	HTT °C	Size (nm)	$(O/H)_{at.}$	$(N/C)_{at.}$	σ Gpa	E Gpa
D	0.20	0.31	550	4-9	0.34	0.28	3.7	245
C	0.24	0.29	400	4-9	0.48	0.23	3.1	238

Source: from ref. 73.

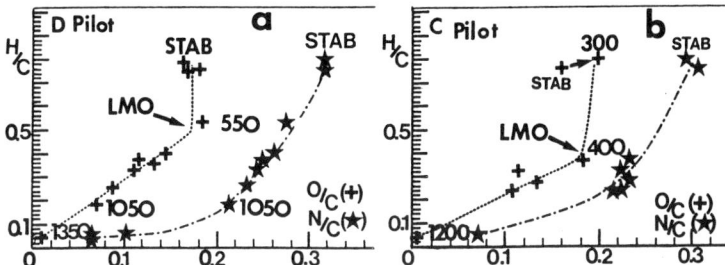

Figure 13 Van Krevelen patterns, a. Oxidized PAN transforming into a carbonized fiber *D*, b. Increasingly oxidized PAN transforming into fiber *C*. Arrows indicate LMO occurrence (from ref. 73).

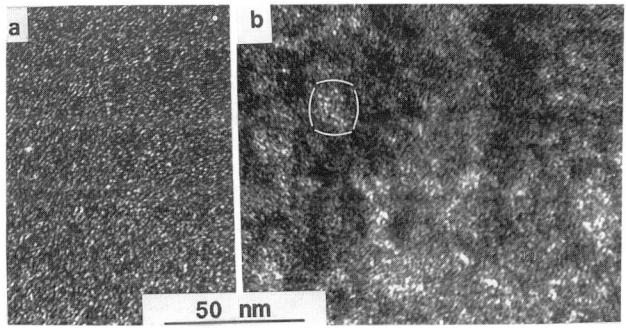

Figure 14 002DF images of cross-sections of fiber *C* (the more oxidized), a. before LMO (300°C), b. after LMO (400°C). One LMO is circled (from ref. 73).

3° Chemical data

Table 2 (columns 1, 2, 3) shows the elemental analyses of some commercial fibers [22,60]. In the same table (column 4) the number N_b of nitrogen atoms occupying an available site on a BSU edge is calculated on the basis of nitrogen fixation instead of aromatic CH groups. The minimum size of a BSU (dicoronene) is 2 × 7 i.e., 14 cycles i.e., 24 available sites. The maximum size (1 nm piled up by 3) is 45 cycles i.e., 51 available sites. For T300, in the initial fiber there is 1.5-3 nitrogen atom per BSU. In Table 2 HTT_{N2} (column 5) is an equivalent HTT usable to compare with other heat-treated carbons. Since no heating rate is really defined in fiber fabrication, the kinetics of the carbonization reaction is modified relative to the usual conditions (4°C min^{-1} heating rate). A direct comparison with other carbons heat-treated at 4°C min^{-1} could be established only by retreating the fibers at this heating rate, up to the temperature where the elemental composition changes [22,60] i.e., up to the moment where the fibers begin to lose nitrogen. Then the nitrogen residual content $(N/C)_{at.}$ is measured at 1100°C (Table 2 column 6) which is the denitrogenation temperature of the more stable fibers (Courtaulds HTS). σ and E (Table 2 columns 7 and 8) are also reported (commercial values). Table 2 shows that a large amount of N_2 present in the commercial fibers is connected to σ since all Toray fibers have a large number N_b of nitrogen per BSU (max. 2–3) and high σ (3.19–7.06 GPa), whereas Courtaulds fibers have a low N_b (0.8-1.4) with poor σ (2.99–3.43 GPa). In addition [22,60,80], among fibers belonging to the same series (Toray fibers) as $(N/C)_{at.}$ decreases σ increases, i.e., a large amount of N_2 release (at a low temperature) improves σ. When $(N/C)_{at.}$ is too low and the equivalent HTT_{N2} carbonization temperature too high (Courtaulds fibers) σ is low. On the contrary T300 heat-treated at 2800°C [81] gets an increase of 45% in σ as compared to the 20% loss of the Courtaulds fibers. Also it was shown that all PAN-based fibers get their maximum value of σ at 1500–1600°C [82] which is the temperature where all fibers are devoid of nitrogen. This means that the highest σ fibers have more nitrogen to lose than the others and lose it at a lower temperature. At that point a strong correlation is established between N_2 and σ whereas it is less clear for E.

Table 2 Elemental Analyses and Mechanical Data for PAN-Based HTS Carbon Fibers

	1	2	3	4	5	6	7	8
	W% C	W% N_2	$(N/C)_{at.}$ 10^{-2}	N_b	$HTT\,N_2$ °C	$(N/C)_{at.}$ $.10^{-2}$ 1100°C	σ GPa	E GPa
T 300	92.6	6.6	6.4	1.5–3	750	3.5	3.19	230
T 400	93.0	6.6	6.1	1.5–3	850	4.0	4.61	233
T 600	93.6	5.3	4.8	1.2–2.4	850	3.0	5.32	265
T 800	95.2	4.9	4.4	1.1–2.2	nd	nd	5.59	294
T 1000	96.0	3.5	3.1	1.1–2.2	nd	nd	7.06	294
Courtaulds XAS	96.2	4.0	2.9	0.7–1.4	1000	1.5	3.43	237
Courtaulds HTS	97.8	2.7	1.6	0.4–0.8	1100	1.6	2.99	237

nd = not determined.
Source: from ref. 22 and 92.

Structure and Texture of Carbon Fibers

Why is the correlation so strong between σ and $(N/C)_{at.}$ for the pilot fibers (where $(N/C)_{at.}$ is measured after LMO [61,73]) and the Toray fibers where it is measured after carbonization? A plausible hypothesis arises from chemical data which attributes to N_2 a role similar to that of CH aromatic groups in other carbons [76,83]. The role of aromatic CH is to prevent coalescence between BSU inside a given LMO as long as they remain fixed upon their edges. Their release begins after LMO occurrence and stabilization (Figure 15a). BSU become at first free radicals in increasing concentration as HTT increases. After a maximum around 600°C, free radicals annihilate, i.e., BSU tightly associate [84], a little after the disappearance of aromatic CH groups [85] (Figure 15b and c). For doing that, all sites available on the edges of BSU, i.e., 24 to 51, have been used progressively and at random. In the case of fibers which contain nitrogenated BSU, nitrogen can be used to tighten BSU inside a LMO only by Watt mechanism [75] (see Figure 11c). This necessitates that two BSU put their nitrogen atom face to face so as to eliminate N_2. By considering Figure 11d, and knowing that per BSU the maximum number of nitrogen sites is three, the chances of the latter are low to be face to face so as to contribute to rigidification. The nitrogen already present well after solidification in HTS fibers has thus an other role which is not that simple.

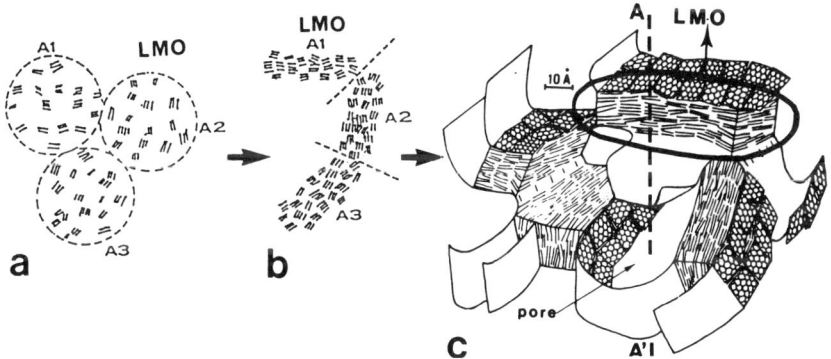

Figure 15 Arrangements of BSU inside LMO as carbonization progresses, a. LMO occurrence, b. Ordering of BSU inside each LMO (from ref. 15), c. Formation of pores by LMO association (adapted from ref. 83).

4° Structural data

Since HTS fibers are all turbostratic, with a poor organization, most of the x-ray data given in literature concern high modulus fibers. They were used to establish models not applicable to high tensile fibers. The very few data obtained on HTS fibers [54,86,87] yield $\bar{d}_{002}= 0.36$nm, almost isometrical domains 3–4 nm in diameter (La) and 1.7 nm in thickness (Lc) misoriented of 43°. Low angle X-ray scattering gives pore sizes of 1-2 nm in diameter and 20–30 nm in length [32,88] or 0.5–1.4 nm in diameter [31,89].

Table 3 gives nanostructural values measured upon longitudinal thin sections in SAD (La // SAD column 1), 002DF (La // DF column 2 and Lc) and lattice fringes. In the latter case are given the interfringe spacing $D002$ (column 3) and Lc (column 4). At last the BSU misorientation inside a given LMO (β_{OD} column 5) was measured upon optical diffraction patterns of LF images (arc opening). The total misorientation β_{SAD} (column 6) was measured in SAD patterns (002 arc opening). The data in the two first columns are in conflict with x-rays since the SAD values for La // approach BSU sizes wheras the La // DF values correspond to the real length of an LMO in longitudinal thin sections (Figure 16 to compare to Figure 14b). On the contrary, Lc and β given in Table 3 are in agreement with those deduced from x-rays. The relative lack of x-ray data makes difficult to establish a model so that only one is given in literature from TEM data [22,60,80]. All other models apply to HM fibers.

Table 3 Crystallographical Data for HTS Fibers, Longitudinal Thin Sections

	La // SAD nm	La // DF nm	D_{002} OD nm	Lc nm	β_{OD}	β_{SAD}
T 300	1.2	10	0.358	1.5–2.3	39	40
T 400	nd	nd	0.360	nd	nd	35
T 600	nd	17.5	0.364	2.3	nd	35
Courtaulds XAS	1.6	12.5	0.357	1.5–1.9	37	37
Courtaulds HTS	2.6	17.5	0.355	2.2–2.5	30	35
	1	2	3	4	5	6

nd = not determined.
Source: from ref. 60 and 61.

Structure and Texture of Carbon Fibers 109

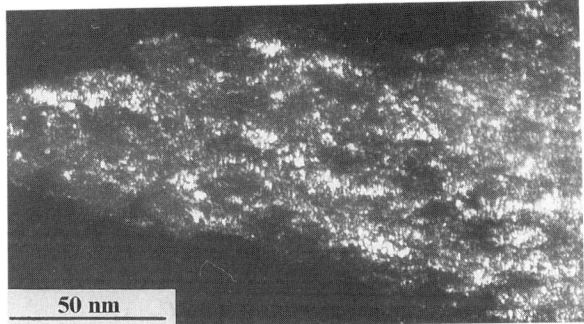

Figure 16 002DF image of a longitudinal section of LMO in Courtaulds XAS.

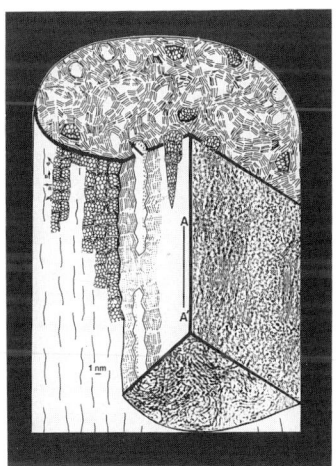

Figure 17 Model of HTS carbon fibers, sketch partially filled-up with 002LF images (adapted from ref. 15). The single arrow illustrates the bonding between adjacent BSU sheets. The $\vec{a}$ axis of BSU (small arrows) are at random.

5° Lamellar model for HTS fibers

At that point, it is necessary to recall that a real satisfactory model must explain the role played by nitrogen, justify the values of mechanical properties and account for all other data. Moreover it must be easily transformed to fit also with high modulus fibers which are known to be derived from HTS fibers only by heat-treatment. The model here described was adapted from the more ancient one elaborated for HM fibers through TEM and SAD data [90,91]. However, since carbon content of HTS fibers ranges between 93 and 98% of carbon, the model was built of individual BSU with $\vec{a}$ axis at random (small arrows in Figure 17) instead of continuous layers. Figure 17 is the sketch partially filled-up with lattice fringe images. LMO are arranged in strongly distorted and entangled sheets including pores elongated along the fiber axis AA'. This model corresponds to that of a porous carbons [83] (Figure 15c) to which stretching was applied along an axis AA'. The crumpled sheets of BSU thus elongate along AA' to yield the lamellar model of Figure 17.

From this model was first defined a compactness factor [22,60,80] taking in account the opening of the nanotexture as observed by eye on the cross-sections images. The higher this factor, the better the tensile strength. Unfortunately, a misconception was introduced by integrating to the pore size (transverse and longitudinal) a parasitic factor due to the temperature of treatment i.e., the perfection of the fringes and that of their stacking. As a result a high index of compactness was attributed to all fibers heat-treated at or above 1000°C such as Courtaulds fibers (Table 2, column 5) which have clearly defined stacks. In cross section (Figure 18), the porosity is almost entirely masked by the contrast of the fringes. It is only in longitudinal sections (Figure 19a) that the open porosity of Courtaulds HTS was recently recognized. If Figure 19a is compared to 19b (T400), the increase in compactness is obvious. It is the reason for which, here, only fibers heat-treated in the same range of HTT will be compared [92]. Figure 20 shows cross sections of increasingly compact HTS Toray fibers for which σ increases from 3.19 to 5.59 GPa. At first Figure 20a (T300) and Figure 20b (T800) show the continuity of the folded BSU sheets not compatible with ribbons. They also show the decreasing size of the pores accompanying the increasing compactness. However, individual BSU are still detectable everywhere. The longitudinal sections (Figure 21a and b) also show the decrease in pore size. In addition, in T800 fiber (Figure 21b), the different sheets of BSU are so close that only one common stack seems to occur at their contact point (box in Figure 21b). After this more precise definition eliminating the perfection of the layers, it can be emphasize that a high index of compactness favors a high tensile strength. The low modulus common to all HTS fibers (230–294 GPa) could be justified by the flexibility of the BSU

Structure and Texture of Carbon Fibers 111

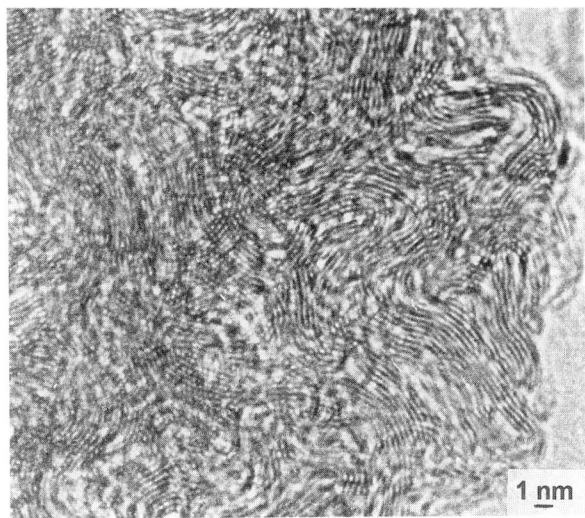

Figure 18 002LF image of Courtaulds HTS cross-section (from ref. 80).

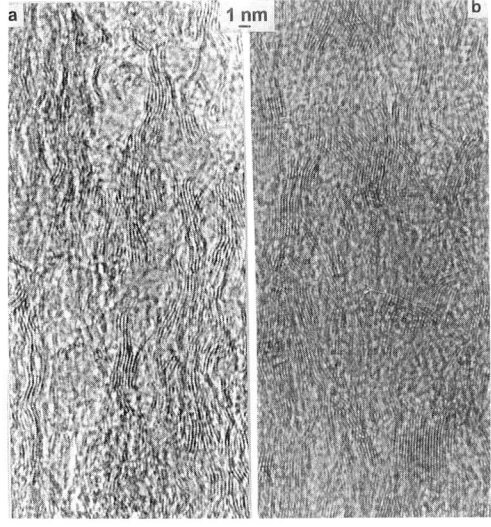

Figure 19 002LF images of longitudinal sections, a. Courtaulds HTS, b. Toray 400.

sheets insured by the misorientations of nitrogenated BSU inside each LMO (Table 3 β_{OD}) to which could be added the misorientation of LMO themselves along the fiber axis (Table 3 β_{SAD}).

The model here described not only justifies the mechanical properties but it also explains the role of nitrogen. As carbonization progresses, stretching favors the face to face of two nitrogenated BSU belonging to different but adjacent sheets (arrow in Figure 17). The elimination of N_2 between these sheets by Watt mechanism [75] (see Figure 11c) creates a tight connection i.e., a strong lateral cohesion in the fiber thus increasing σ. The probability of this event increases with the number N_b of nitrogenated sites around a BSU edge (see Table 2 column 4). The high σ is due to precursors able to keep, after stabilization, a still high potential of N_2. The latter is thus usable all along the carbonization to produce binding between two adjacent LMO sheets. The highest chances, but not yet realized, are in T300. Since it has the largest available supply of nitrogen, it improves by 45% the tensile strength [81] during further heat treatment eliminating N_2. Considering the Toray series (Table 2) in which σ increases, an increasing part of the nitrogen initially available has been consumed respectively by T400, T600, T800, and T1000. The Courtaulds fibers though devoid of nitrogen have low σ probably because they had not enough nitrogen available at the beginning of carbonization. These considerations justify the great importance attributed in 2.3.2 to the nature and amount of volatiles released during stabilization which governs the final nitrogen content of the precursor. sp^3-sp^3 bonds were also assumed to bind adjacent sheets of layers, insuring the tensile strength [24,93]. In opposition to the inter LMO nitrogen, the nitrogenated sites still present inside a LMO favor its flexibility in the same manner as CH aromatic groups.

6° Microtexture and texture heterogeneities

They are very changeable not only from one type of fiber to an other but even for the same kind, depending on uncertain fabrication parameters. As an example, T300 was described as entirely random throughout its cross-sections [60]. Some of them are now random outside, with a concentric 0.3μm thick internal shell at about 3μm of the surface. It is again random in the core. During 150 hours oxidation by air at 350°C, the fiber loses 60% of its weight and then remains stable. However since it is reduced to the core and its concentric shell, it is only 4μm in diameter. It is thus so thin that it is spaghetti-like i.e., flexible and folded. Similar features were found through plasma oxidation [94]. Parasitic skin and core texture was also observed as an external better organized and fully stabilized sheath is formed, protecting the core against further oxidation [94]. Since such features are not constant they must be recognized but they present only a temporary interest.

Structure and Texture of Carbon Fibers

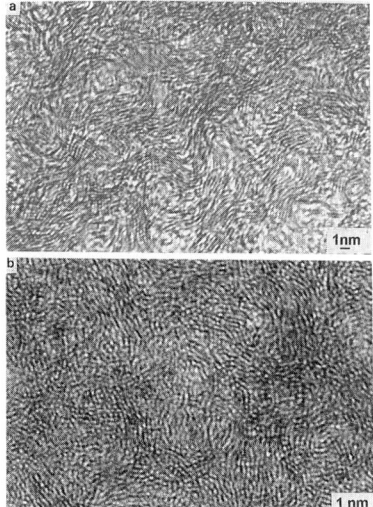

Figure 20 002LF images. Cross-sections of increasingly compact fibers, a. T300, b. T800 (from ref. 73).

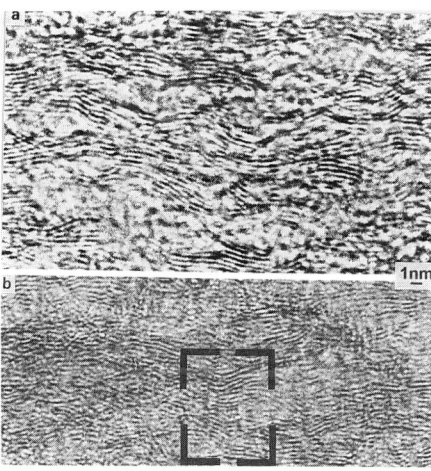

Figure 21 002LF images. Longitudinal sections of increasingly compact fibers, a. T300, b. T800. The box illustrates the lateral cohesion between two adjacent sheets of BSU (from ref. 61).

In the same manner, catalytic effects introduce heterogeneities in the fibers [95] which cause flaws. Voids are produced surrounded by a well organized carbon shell. This is due to the formation- decomposition of an unstable carbide (Fe, Ni, Co). Since these metals are problematical impurities, such voids are also incidental.

2.3.4 High Modulus Fibers

HM fibers are industrially obtained from HTS ones by a short heat-treatment at or above 2000°C under nitrogen, while keeping a constant length. They are almost pure carbon. They derive from HTS fibers by heat treating some of them at 2800°C [96] since identical data were obtained.

1° Nanotextures
Here also, to compare with other carbons, heat-treatment at $4°C.min^{-1}$ must be performed on various initially HTS fibers [22,60,96]. The first transformation to occur is common to all carbons including fibers [15]. The BSU contained in one LMO coalesce into a distorted layer stack. The latter keeps the initial diameter of the LMO but becomes thinner because of the BSU ordering, improving their parallel orientation (see Figure 15b and c). Above 2000°C, in carbons other than fibers [15,52], the distorsions kept inside the aromatic layers of a single LMO anneal and the *002* lattice fringes become perfect in the grains. If the grains are lamellar, the carbon is entirely graphitizing, because the lamella derives from a single almost infinite LMO (mosaic isochromatic domains). If the texture of the carbon is either spherical (carbon blacks [97]) or cylindrical (filaments [98]), above 2000°C as the layers become stiff and perfect, a sudden polygonization occurs. It transforms the black into polyhedrons and the filaments into prismatic columns. In porous carbons also [15,37] the pores become polyhedral. The lamellae, the flat faces of the polyhedron and the columnar filaments, as well as the flat pore walls, are considered as the grains able to graphitize. A general rule is thus established connecting exponentially the decreasing grain size to the decreasing graphitization degree P_{1max}. The grains do not have disclinations since they are gathered at the grain boundaries.

On the contrary, in the case of HM fibers, pentagons fixed on the nitrogenated BSU edges are responsible for stable disclinations (as already mentioned in 2.3.2 and Figure 11b). Since they remain frozen-in after heat-treatment, they maintain permanent curvatures, which are systematically found in all commercial HM fibers (Figure 22) [81,99]. Since curvatures are produced instead of polygonization, very partial or non-existent graphitization can be expected for HM fibers (see Figure 2b) as in some pitch-based carbon fibers [72]. The cross-sections of Figure 22 show continuous *002* fringes with positive

Structure and Texture of Carbon Fibers

curvatures not compatible with ribbons. The same conclusions were already given for HTS fibers (compare with Figure 20). Entangled pores are evident everywhere and the radius of curvature r_t of their walls is measurable. The longitudinal sections show elongated pores limited by walls already curved, with radius of curvature r_l (Figure 23). The pores usually end by approaching their walls as it is sketched in inset of Figure 23. Table 4 allows to compare various commercial HM fibers (Courtaulds HMS and Toray M40, M40J, M50J) [22,60,61,81,99]. In this table σ is given in column 1, E in column 2 and r_t in column 3. The longitudinal radius of curvature r_l (column 4) is sometimes so large that it cannot be measured directly on lattice fringe images. It is either approximated by L the pore length, or indirectly measured upon *hk* dark-field images. The layer diameter $L_a//$ was assimilated to the length of *002DF* bright bands or *002LF* (column 5).

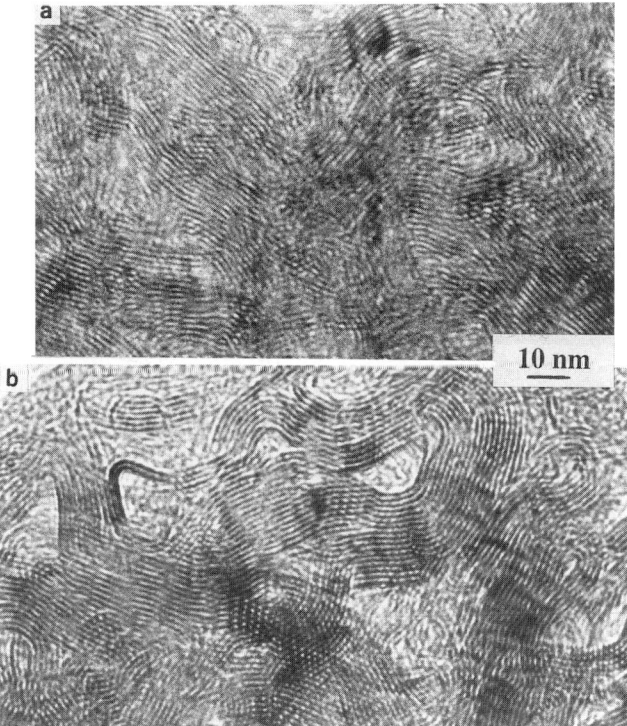

Figure 22 002 LF images. Cross-sections of HM fibers with increasing r_t, a. Toray M40J, b. Toray M50J (from ref. 99).

In the ancient generation of HM fibers as E increased σ decreased [100]. By considering Table 4 it is verified and more marked in the two last lines of the Table which recall the results formerly obtained [22,60,81]. M40J and M50J, where σ is always better than expected, are an exception to the rule [99]. For all fibers mentioned in Table 4, E increases with r_t and r_l suggesting a direct relationship between the diameter of the stacks and E. Such a linear relation was already proposed [60]. Other authors proposed also to identify the coherent domain measured by x-rays to the stiffness of the fiber [54,87,95,101]. On the contrary, others [3,32,54,86] connect E to the preferred orientation of the stacks along the fiber axis as measured by q (x-ray parameter of orientation). However so many deviations [55] were found that the validity of the exponential law proposed is questionable. Concerning σ the only relation proposed is also deduced from old data [90,91] and derived from HTS model [22,60]. σ is assumed to increase as lateral cohesion increases. The latter being due to tight bonding between defective adjacent BSU sheets, it has to decrease for HM fibers. It is in linear relationship with the density of defective areas [22, 81] (see Chap. 5). No relation is actually available for the more recent fibers.

Table 4 Comparison Between Some HM Fibers

	1	2	3	4	5
	σ GPa	E GPa	r_t nm	r_l or L nm	L_a nm
Courtaulds HMS	2.62	345	1.5–2.5	7.6	10–15
M 40 B Toray	2.74	392	3	8	/
M 40 J	4.69	386	2–3	10	10–15
M 50 J	3.92	475	2–10	13	15–20
Celion GY 70	1.86	517	5–8 core 10–30 skin	69	46
Serofim AG 58	1.6	600	6–10 core 20–50 skin	250	71

Source: from ref. 22, 60, 61, 99.

Structure and Texture of Carbon Fibers

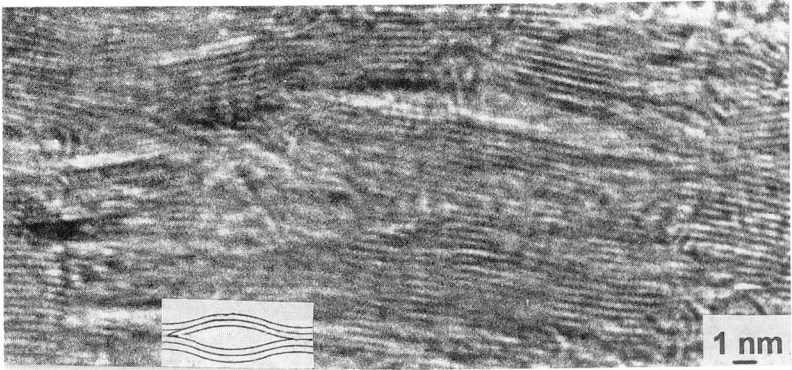

Figure 23 002 LF image. Longitudinal section of Courtaulds HMS. Inset shape of a pore (from ref. 22).

2° Models of high modulus fibers
The first model established from x-ray and electron diffraction data was the ribbon model [3,32,54,86] (Figure 24a). It is based on the fact that measurements yield an elongated shape for the coherent domains, L_a // measured along the fiber axis is larger than $L_a \perp$ [54, 86, 102-104]. Nevertheless the maximum elongation given is 2 (L_a // = 11.9 nm and $L_a \perp$ = 6.4 nm) for PAN-based fibers heat-treated at 2700°C [54]. Even isometric domains without elongation (L_a = 6nm) were found [86]. Firstly these data do not really suggest the occurrence of ribbons. Then they do not fit with TEM

data showing continuous folded layer stacks. In 2.2.1 5° (see Figure 4a and b) it was shown that the real physical entity (distorted aromatic layer stack) is not accessible by measuring the coherent domain size diameter L_a from *10* or *11* bands. The distorsions make promptly disappear the coherence of the layer stack which is thus restricted in size in the distorsion direction. If the basic layer stack is folded parallel to the fiber axis its length would be true, but its width is increasingly decreased as r_t decreases. If the reader refers to that, he may immediatlely guess that the so-called ribbons are the coherent part of a layer stack distorted by folding around the fiber axis. In addition, lattice fringe images of longitudinal sections which were used to try to confirm the ribbon model are peculiarly unsuitable for doing that, since the fringes are issued only from the part of the stacks fulfilling the *002* Bragg angle. Even by taking in account the interference error, when the stack is inclined ± 10° it disappears, so that only a negligible part is visible. A narrow ribbon or an infinitely large but folded layer stack should give identical images. It is necessary not only to link cross sections images to longitudinal ones but also to use *hk* dark-fields to see the stacks projected along their diameter. Mechanical properties cannot be explained by the ribbon model which has no lateral cohesion. To improve it, the fibrils were basket weaved (Figure 24b) [55]. It does not bring any additional explanation to mechanical properties.

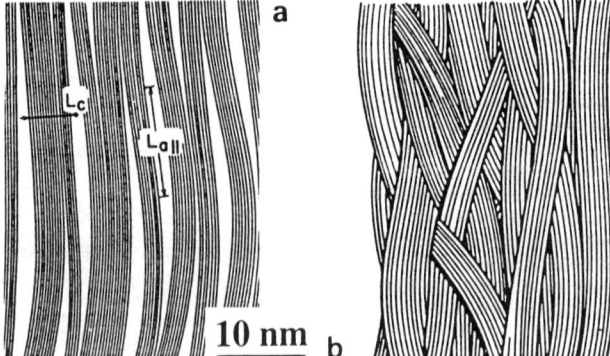

Figure 24 a. Ribbon model, b. Basket weaved model.

A convenient model satisfying to all data (x-ray and electron diffractions, TEM, mechanical and STM data) is the lamellar model established by Oberlin et al. in 1977 [90,91]. It is illustrated by Figure 25 (sketch partially filled up with lattice fringes images). Instead of LMO made of distinct BSU, distorted sheets of continuous layers stacks are entangled and bound from place to place by their defective areas. This model was established by a careful comparison between *002 DF, 10 DF,* and *11 DF* images of commercial Serofim fibers. It was then retaken and verified in the study of some other commercial fibers [22,60,81] by using the same techniques. The lamellar model is confirmed by exfoliation experiments. They were performed by making an intercalation compound with H_2SO_4 + $K_2Cr_2O_7$ exfoliated by H_2O_2 [91] or by forming graphitic oxide with H_2SO_4, $NaNO_3$, and $KMnO_4$ deflagrated above 200°C [105, 106]. Lamellar isometrical fragments were always produced and not ribbons. The TEM data fit well with x-rays [54,102,103,107] by the values of fibrous orientation which is much better than the HTS one. The data obtained by TEM (Table 4 columns 3 and 4) upon r_t and r_l [81,99] are also consistent with the pore sizes given by x-rays [31,32,88,89] i.e., a width of about 1–3 nm for a length of 15–30 nm for x-rays. It is remarkable that the L_a // given by x-rays (9–12 nm) well agree with the L_a // given by TEM (column 5) and with the L_a // of the LMO in HTS fibers (10–17.5 nm) (Table 3). Obviously this means that one may derive the HM model of Figure 25 from the HTS model of Figure 17 only by heat-treatment since BSU are expected to coalesce into each individual LMO to give a single layer stack. One of the two models proves the validity of the other. A model, very similar to the lamellar model of Oberlin et al., was proposed one year latter by Johnson in 1978 [19,87,101] (Figure 26). The two models well agree.

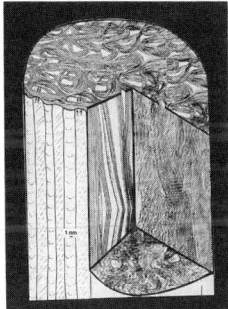

Figure 25 First lamellar model of HM carbon fibers. Sketch partially filled-up with 002LF images (adapted from ref. 90). The skin effect is marked by the double arrows.

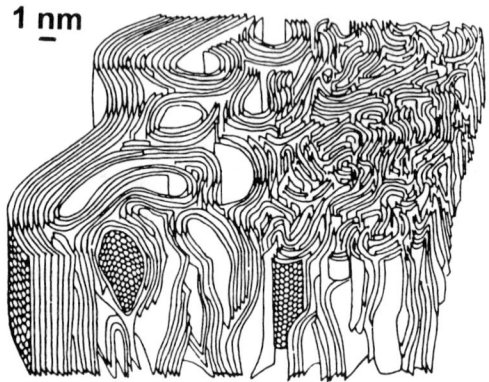

Figure 26 Lamellar model of HM carbon fibers (from ref. 87).

3° Microtextures (heterogeneities)

A concentric skin is very often formed around a random core in HM fibers (double arrows in Figure 25). It is sometimes explained by a stabilization gradient. The skin is supposed to be fully stabilized and protect the core which is thus understabilized [87,95,101,108]. Nevertheless such skins are found also in fully stabilized fibers [22]. More plausibly, they can be attributed to the stress relaxation occurring near the fiber surface. They correspond to the largest transverse radius of curvature. The skin is almost always partially graphitized [22,60] (see also 2.4.4.2° ref. 208).

2.3.5. Fiber Surfaces

In composites, the mechanical properties among others, are determined by the adhesion between the fibers and the matrix. If it is too strong the composite is fragil, if it is weaker it becomes resilient due to the pull-out of the fibers. It was pointed out by many authors [109–113] that the fiber surface itself (treated, coated or sized) is responsible for adhesion through chemical or physical coupling. To study such delicate mechanisms, convenient tools are necessary. Ten years ago only Raman and TEM were available. Raman was not

Structure and Texture of Carbon Fibers

extensively employed [44] since it gave limited data. It shows [114] that the fiber surface is highly defective. However recent results will be discussed in 2.4.4 2° (see ref. 208). TEM was applied successfully [115] to study the adhesion between PAN fibers and epoxy matrix in composites. In the case of PAN fibers the basic idea was the role played by surface aminated functional groups [116] able to react with epoxy. At first it was shown [115] that surface rugosity (1 to 10 nm) favors good adhesion. Then the occurrence of accessible layer edges also favors good adhesion. As an example, Courtaulds HMS fibers treated with hexamethylene tetramine show both nanometric rugosities and accessible edges joined to a strong adhesion with epoxy matrix. On the contrary, treated with urea, Courtaulds HMS has a poor adhesion to epoxy because of the smooth surface produced by the concentric skin. A micro-crack develops at the interface. An interesting result is the good adhesion between Toray 300 and epoxy without any additional surface treatment. By using TEM and electron energy loss spectroscopy, this was attributed to the occurrence of an excess of nitrogen near the surface [117]. However these results are in conflict with those reported by Boehm et al. [118], who found the reverse by ESCA on the same fiber. As an other example, adhesion between a pyrocarbon deposit and reinforcing fibers is also submitted to the fiber rugosity and to the eventual occurrence of accessible edges [109,113]. Figure 27a illustrates the adhesion between a HTS fiber and a pyrocarbon deposit. It is strong only in the place where accessible edges occur in the fiber (arrow). In this region the aromatic layers of the deposit follow those of the fiber. In other places decohesions occur. Figure 27b illustrates the same features for a HM fiber. The arrow also indicates a place of strong adhesion [113]. In conclusion, chemical coupling between matrix and fibers in a composite must always be associated with nano-rugosity of the fiber surface and connected to the active sites present on the aromatic layer edges if some are available. Indirect methods are still used to determine surface roughness such as contact angle measurements [119]. To evidence sizing, decoration by heavy metals, followed by TEM, is also employed [120].

Finally, to have a reasonable chance to explain and then predict the behaviour of the composites reinforced by PAN-based fibers it is necessary to know the surface state of the fiber at the atomic level. The progresses of STM or AFM make this possible. Nevertheless surface contamination and very poor organization of the fibers are two overwhelming problems which were not overcome at first [121]. However more recent papers got much better results [122–124]. HTS fibers such as T300 and Courtaulds X-AU were examined by STM within micrometric and nanometric scales. Micrometric striations and roughness of the surface were attributed to the spinning and stretching processes. However the most important data for T300 [123] is the occurrence of numerous nanometric domains (Figure 28a) irregular in shape (one is circled) entirely devoid from the graphite triangular pattern with frequent missing atoms and with their $\vec{a}$ axis at random. They are thus turbostratic and represent the stacks, the basal planes of which are sufficiently parallel to the surface. They are separated by areas without contrast probably due to stacks protruding by their edges or at least oblique. This is an image of individual BSU. Their size is about 1nm in diameter and 0.7–1.5 nm in the areas without contrast. In Courtaulds X-AU (Figure 28b) the domains are associated into more elongated units, inside which the $\vec{a}$ axis are also at random (LMO). This well agrees with the higher organization of Courtaulds fibers (see Figure 18) as compared to T300 (see Figure 20a). Figure 28b is an image of the surface of the model of Figure 17. The surface of HM fibers (Toray M40 [122]) are also striated but less rough. They show (Figure 29) periodically (2–3nm) wavy domains, markedly elongated parallel to the fiber axis. Their organization is improved in the basal planes but they are still turbostratic. They probably represent the folds having the same level. This is an image of the surface of the model of Figure 25 (see also ref. 217 and 218).

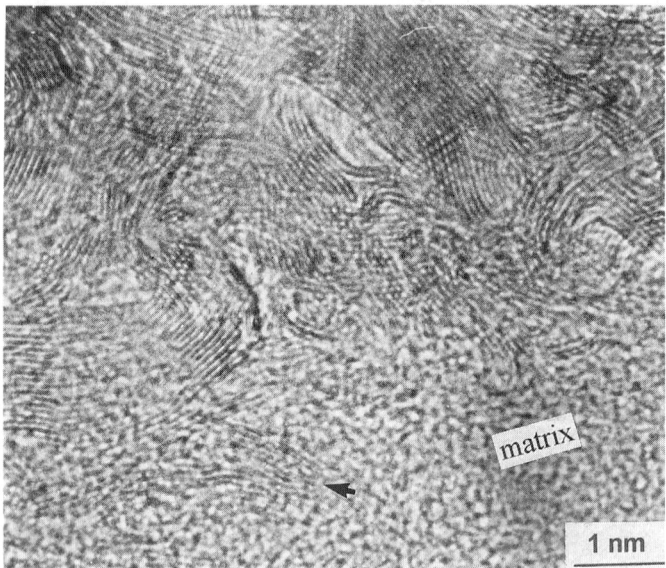

Figure 27 002 LF images. Thin sections of composites pyrocarbon PAN-based fibers showing the strong adhesion (arrows) and decohesions, a. HTS fiber, b. HM fiber (from ref. 113).

2.3.6 Conclusion

By combining mostly the non-averaging techniques such as all modes of TEM and STM, the PAN-based carbon fibers are now fully described in the bulk and at their surface. HTS fibers can be represented by individual BSU about 1nm or less in diameter (Figure 14a and 28a), associated in azimutal disorder into oriented volumes or LMO. LMO themselves tend to elongate parallel to the fiber axis in wrinkled, folded and entangled sheets (see Figure 16, 17 and 28b). Pores elongated parallel to the fiber axis are thus produced. Since BSU are turbostratic and are not coherent into a given LMO because their $\vec{a}$ axis are at random (see Figure 17 and 28b), the fiber themselves are turbostratic. The adjacent sheets when they touch each other with nitrogenated protruding edges in contact (misoriented BSU) are strongly bonded by N_2 elimination (Watt mechanism). However the nitrogen remaining insures each LMO flexibility. Therefrom lateral cohesion is responsible for high tensile strength, whereas flexibility of LMO is responsible for low modulus. Spinning causes striations on the surface (STM). Stretching causes marked preferred orientation along the fiber axis contributing to the modulus. From T300 to T800 and Courtaulds HTS the increasing carbonization temperature both makes N_2 to release and increases the perfection of the aromatic layers. The N_2 release (when enough is available) is responsible for the increasing compactness and thus increasing σ from T300 to T800. The tendency to anneal distorsions of the layers and the lack of potential N_2 makes Courtaulds HTS lower in σ but higher in E. Nevertheless HTS fibers are not thermally graphitizable since stable disclinations maintain permanent curvatures of the layers.

The picture is not very different for HM fibers except that BSU have coalesced into distorted continuous layer stacks (see Figures 22b, 25, and 29) having roughly kept the initial LMO size. The entangled sheets are thus made of individual stacks isometrical but folded and entangled, parallel to the fiber axis. They are oriented at random in the core but with a tendency to a concentric skin arrangement due to the increase of their transverse radius of curvature near the surface. r_t measured by TEM [60] (see Table 4) as 1.5–3 nm is responsible for a wavy appearance of the fiber surface having the same period. This well agree with the wavy domain periods measured by STM [122] as 2–3 nm. The stable disclinations are still present, insuring the permanent curvature of the layers and preventing graphitization, except partial in the skin. Since the sheets are better organized, they have less defective areas so that there is less chances of lateral bonding, i. e., σ often decreases. For the same reason, the increase in diameter of the stacks (from one BSU to the equivalent of one LMO) increases tremendously E. The improved fibrous orientation also contributes to increase E.

The ideal tensile performances of a fiber should reach simultaneously the calculated graphite Young's modulus (1030 GPa) [125] and the strength of graphite whiskers (20 GPa). However the paradox lies in the fact that crystalline graphite is prohibited. If, at first, each single aromatic layer should have an almost infinite modulus without strain (minimum C–C bond of 0.142 nm), then their piling-up should have an infinite strain but no tensile strength (Van der Waals spacing of 0.335 nm). The only way to approach graphite theoretical values is to block the easy glide present between the (001) planes. The solution adopted in PAN-based carbon fibers favors the locking to the detriment of crystallinity so that the precursor is spinnable and not graphitizable. The tensile strength is high because of the lateral cohesion, the modulus is medium because of the tremendous amount of stable disclinations.

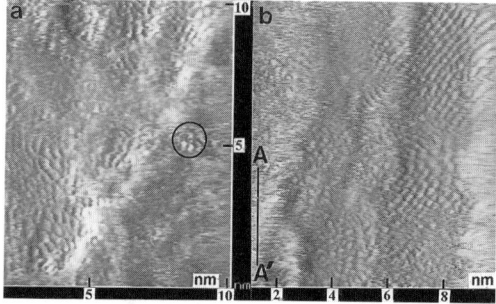

Figure 28 STM top view images of HTS PAN-based fiber surface, a. T300 9nm × 9nm, b. Coutaulds X-AU 10nm × 10nm (from ref. 123).

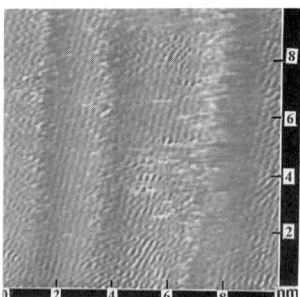

Figure 29 STM top view image of HM PAN-based M40 fiber surface, 9nm × 9nm (from ref. 122).

2.4 PITCH-BASED CARBON FIBERS

In the present chapter only high performances fibers obtained from either mesophase or anisotropic pitches will be described in the purpose to determine the relations between structure, textures and properties. Consequently, only the data connecting processes and characterization (at least by OM) will be discussed.

The solution chosen to obtain pitch-based fibers is exactly the reverse of the one chosen for PAN-based ones, since crystallinity is favored. The precursor is thus spinnable and graphitizable. The tensile strength is medium and the modulus high. The way to prepare such fibers is similar to PAN-based ones since after spinning the fibers are stabilized by air, then carbonized and graphitized. Nevertheless no tension is applied. All properties of the final fiber depend on three parameters as in PAN-based fibers:

- the pitch precursor primarily acting upon subsequent nanotexture,
- the spinning process including the spinneret configuration which mainly governs the microtexture,
- the stabilization which influences textures and graphitizability.

2.4.1 Processes to Prepare Precursors

The leading thread of this paragraph lies in the utmost importance of the various principles of preparation providing more or less homogeneous and/or defective pitches suitable for spinning. They must be characterized by their physicochemistry and their textures before discussing their nature and their part in fiber properties. All fiber precursors are issued from isotropic coal-tar or petroleum pitches. Themselves derive respectively from coal-tars or heavy petroleum products, made heavier by distillation, heat-soaking or combination of both. All pitches are complex organic matters which cannot be reliably analyzed since after separation into various components they cannot be restored by mixing them. Nevertheless the pitches are somewhat arbitrarily defined by:

- their elemental composition often given as atomic ratio H/C, O/C,
- their softening point (SP or KS i.e., Kraemer Sarnov point),
- their molecular weight range. The latter is frequently roughly expressed by solvent fractionation into three kind of increasingly heavier fractions: γ resins, the lightest part, which are toluene soluble (TS), β resins which are toluene insoluble but quinoline soluble (QS, TI), α resins, the heaviest part, which are anthracene-oil or quinoline insoluble (QI),
- their aromaticity factor F_a often measured by infra-red or NMR [62,63].

Coal-tar and petroleum pitches (Table 5) [126,127] do not differ much in elemental composition and softening points, but by their molecular weight range and aromaticity. The former contain up to a few percent of QI inherited from tars and a noticeable amount of β resins. Coal-tar pitches are more aromatic than the petroleum ones which are always devoid of QI and rarely contain a small amount of β resins. The QI of coal-tar pitches are carbon-blacks (free carbon) and heteroatom rich plastic phases [128]. Both are anisotropic in TEM. They can be eliminated only by filtration or centrifugation. The β resins of coal-tar pitches are also partly inherited from coals. They get very easily long range orientation under weak stresses in coals [129] and pitches [130] or through flow orientations produced by filtration [131]. They thus become OM or TEM anisotropic (Figure 30). Figure 30a and b are orthogonal 002 DF inside which the filtration direction is represented by the white arrow. Figure 30c is the sketch of the flows deduced from a and b. The γ resins are always isotropic at any scale [131].

In the same manner as tars, coals or heavy petroleum products, all isotropic pitches are almost unanimously considered as colloidal suspensions (sols) of micelles [63,127,132–138]. As an example (Figure 31), the core of a micelle is occupied by something like a BSU surrounded by a shell of lighter molecules: mono, di or triaromatic and chain-like (aliphatic). The suspensive medium is made of the lightest molecules. The sol turns to gel by elimination of suspensive medium either by volatile release (heat-soaking or carbonization) or by fractionation. After the total withdrawal of suspensive medium the gel flocculates irreversibly into a solid. Before that point all sol-gel transformations are reversible (thermotropy) by further dispersion of the micelles. Depending on the thermodynamical conditions, liquid crystal phases may appear in the phase equilibrium diagram.

Table 5 Physicochemical Data for Pitches

	A 240	Coal tar pitch
Softening point	120	110
$(H/C)_{at.}$	0.8	0.6
$(O/C)_{at.}$	/	0.01
Fa	0.7	0.8
α resins %	nil	1-2
β resins %	8	16-20

Source: from ref. 125, 126, and commercial data.

Structure and Texture of Carbon Fibers

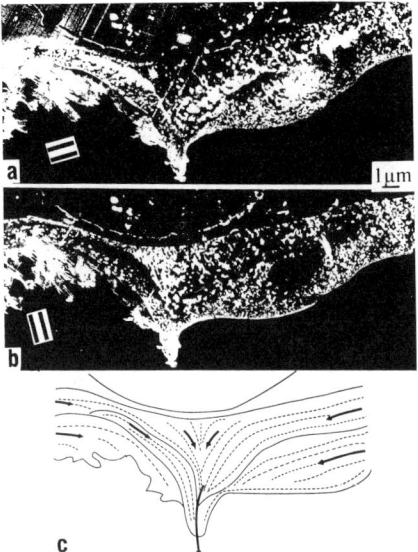

Figure 30 Flow orientation produced in a tar by filtration, a and b: orthogonal 002DF, c: sketch of the flows (from ref. 131).

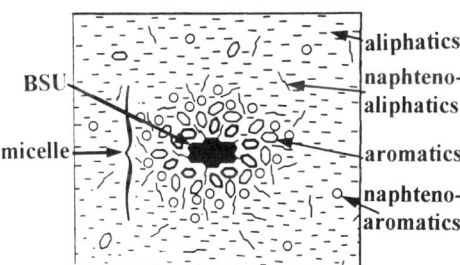

Figure 31 Sketch of a colloidal suspension. Example: heavy petroleum product (from ref. 136).

1° Mesophase pitches (1970–1977)

During quiescent carbonization of isotropic pitches, anisotropic α resin spheres (Brooks and Taylor mesophase) [139] precipitate all of a sudden and increase in size by coalescence only [140,141]. Then islands of mosaic, disclination rich [142,143], develop at the expense of the pitch. Correspondingly, the apparent viscosity increases, measured when it is too high by Vickers microhardness (HV) [63]. Figure 32 (Ashland 240 i.e., A240) shows that HV jumps to the infinity at the brittle solid state [141], also marked by maximum dehydrogenation and stabilization at the maximum size of the oriented domains (LMO, mosaics or bands) [140]. This point marks the end of primary carbonization defined by Brooks and Taylor [139] as the semi-coke stage. The pitch is made of 100% α resins or bulk mesophase [139]. This product is not spinnable despite it is graphitizable.

In the first patents and papers of Singer [144,145] isotropic pitches were used as A240 and acenaphtylene pitches both characterized as conventional pitches. In these pitches, 100% anisotropy due to Brooks and Taylor mesophase mixed with the β resins of the pitch is the operative phase [146]. Since it was understood that spinning necessitates to get plasticity, A240 composition was chosen at the secondary minimum of viscosity [147]. However, it was also necessary to have extended anisotropic grains to approach graphite. The authors tend to the end of primary carbonization without reaching it. They use at the most 50–90 weight% α resin mesophase with 100% of anisotropic domains larger than 200µm, H/C is about 0.5. The convenient viscosity is between 1–20pa.s. To obtain that, the precursor is heated at about 400°C up to several days. It is spun between 450 and 480°C. In all these experiments the mesophase content was equal to the QI content since mesophase was the Brooks and Taylor mesophase [139,148,149]. Nevertheless, the pitches obtained are not convenient for making high performance fibers because of their high spinning temperature and of their heterogeneity (since the problem caused by the latter will arise at the stabilization stage it will thus be explained in Section 2.4.3).

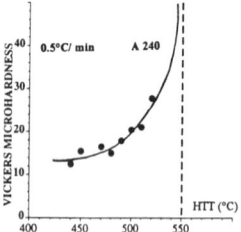

Figure 32 Vickers microhardness of A240 plotted versus HTT (from ref. 141).

Structure and Texture of Carbon Fibers

2° Gas sparge preparation
To overcome the numerous defects preventing progresses in fiber properties, a strong mechanical stirring was introduced during heat-soaking of a conventional pitch (A240). The gas sparge preparation (1977–1980) [150,151] introduces a gradient of pressure associated to the gradient of temperature by bubbling nitrogen under pressure. This enhances the release of low molecular weight volatiles, accelerates the anisotropy development and provides a softening point of 309°C. This process is claimed to have the exclusive result to narrow the molecular weight spectrum of the final pitch: 90% below 1500amu. All further Union Carbide patents were devoted to reduce the spectra of the pitches as 100% mesophase pitches. At this point it is necessary to elucidate what are the real parameters measured and on what kind of sample.

a. Molecular weight spectra   Mochida et al. [152] in 1978 evaluated the spectrum of single spheres of Brooks and Taylor mesophase (issued from an isotropic petroleum pitch), to be between 400 and 4000amu. Greinke and Singer [153] in 1988 evaluated the spectrum of a gas sparge pitch not characterized issued from A240, to be between 400 and 2200amu with two maxima at 600 and 900amu. The two authors treated their pitches with the same technics and at the same HTT (400°C). Both were obliged to solubilize their mesophases by a drastic chemical treatment which transform them into some kind of γ resins (they were benzene or trichlorobenzene soluble). As a result of the 400°C heat-treatment all volatiles of molecular weight less than 400 amu were released. The light fractions of the initial isotropic pitches are thus unknown. However molecular weights as low as 130amu are present in isotropic pitch and those as high as 6000amu are found in coalesced mesophase [154]. The regression down to γ resins due to the chemical treatment prevents any recognition of the true solubility of the mesophase obtained.

A tentative treatment was further performed to roughly evaluate the molecular weight range of a single sphere of Brooks and Taylor mesophase without any solubilization treatment and without extracting it from its pitch [155]. Thin sections of mesophase single spheres already embedded in their pitch were examined in TEM using 002 lattice fringes mode. Figure 33a shows that the sphere is heterogeneous at the nanometric scale and somewhat anarchic. This picture is different from the more ordered one (Figure 33b) corresponding to a sphere extracted from the pitch by anthracene oil (see ref. 174). In Figure 33b, the fringes form distorted columns comparable to piles of plates very close to each other. To understand why the spheres are so different whether extracted or not, image analysis was carried out on Figure 33a [173]. The first Fourier transform is given in inset. Apparently three components appear. At first a strong diffuse scattering reinforced into two sectors. Then a diffuse 002 ring superimposed to two 002 arcs. At first only the small angle scattering was used to form the image excluding 002 and the incident beam (Figure 33c). Almost slit shaped domains appear preferently oriented parallel to a plane. Their thickness is about that of a single lattice fringe, their length is about 1nm. They are homogeneously distributed. Their average distance parallel to the slit approaches the size of a BSU.

Then only the interference image contribution is used to form the image (Figure 33d). An interconnected array of distorted columns appears. The micrograph is obviously similar to the image of the extracted sphere of Figure 33b. However the areas devoid of columns are larger and fill-up with single BSU distributed at random. If the BSU associated into a column are numbered (a single BSU is assimilated to dicoronene (molecular weight 600amu)) then 4000 to 6000amu are obtained. The regions fill-up with random BSU correspond to 600amu. The bright domains of Figure 33c are either pores or disordered single molecules lower in molecular weight. Finally the molecular weight spectrum deduced from the image of Figure 33a is in the order of magnitude given in ref. 152. To estimate what could be the "mesophase" studied in ref. 153 it is necessary to consider now the solubilities of anisotropic products.

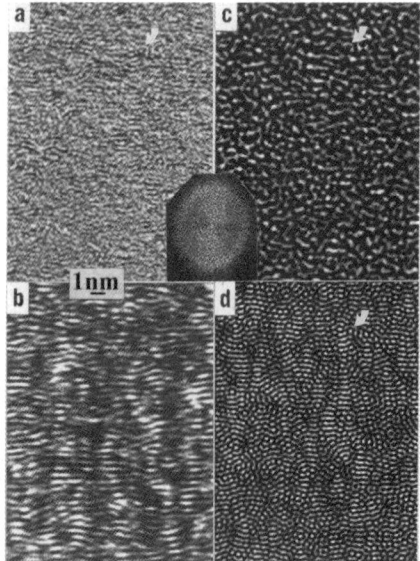

Figure 33 Thin section of a single mesophase sphere: a. In situ in its pitch, TEM 002 LF image (from ref. 155) inset its Fourier transform, b. Extracted by anthracene oil, TEM 002 LF image (from ref. 174), c and d Image analysis of image a (from ref. 173), c. Small angle diffusion image, d. 002 interference image. The white arrow in Figure a, c, and d marks a point common to the three micrographs.

b. Solubility of mesophases   As long as conventional isotropic pitches were heat-treated in quiescent conditions (see 1°) the mesophase content was equal to the α resin content. However, as early as 1976 [156] and more recently in 1978 [157] it was found that anisotropic areas of a pitch happens to be far more important than the QI content. As an example [158] in some peculiar pitch 95% of anisotropic areas corresponds only to 30% of QI, whereas A240 carbonized in a usual way contains 60% of anisotropic areas for about 40% of QI and 100% anisotropic areas for 80% of QI. For a gas-sparge petroleum pitch [157] the amount of anisotropic areas is also much larger than the QI content i.e., 100% of anisotropic areas for 30% of QI and 70-80% of TI. The so-called mesophases are thus sometimes almost entirely QI (Brooks and Taylor mesophase), sometimes noticeably soluble in quinoline (70%) and even partly soluble in toluene (20–30%). It is clearly pointed out from these data, that Brooks and Taylor mesophase is not alone or perhaps is not at all present in the anisotropic areas of gas-sparge pitches. Even in Brooks and Taylor mesophase spheres TEM data [155] joined to solubility measurements [152] show that parts of the spheres (probably regions devoid of fringes in Figure 33) are not QI. This could explain the 20% discrepancy between anisotropic areas and QI content in A240 [158].

It is possible now to come back to the measurements of molecular weights [153] with some better understanding. After gas sparging A240 has lost all volatiles lower than 400amu and progressively develops anisotropy whereas the residual pitch increases in molecular weight. There is a constant molecular weight of 900amu for the anisotropic phase to which tends the pitch at the end of the process. However it is instructive to consider the whole curves for anisotropic fraction (64%), isotropic fraction and whole pitch [153]. All of them contain a very thin peak at 600amu and a very broad one at 900amu which extends up to 2200amu. It seems that the observation extracted from ref. 153 sums up quite well the situation: the coexisting phases contain similar size molecules but in different proportions. The peak at 600 amu could coincide with single BSU of the minimum size i.e., made of dicoronene whereas the broad peak at 900amu could be attributed to labile associations of BSU. In addition, the first peak could be connected to the more easy to disperse fraction (TS) the second to the more difficult to disperse one (QS-TI). From these data it is clear that the anisotropic product is not Brooks and Taylor mesophase. The situation will be somethat clear up after textural characterization (2.4.2).

3° Fractionation process
To further narrow the molecular weight spectrum of the pitch, fractionation process was carried out (1980 and later) by Riggs and Diefendorf in Exxon patents [159]. An insoluble fraction of a heat-soaked pitch is separated either by solvent, solvent mixtures or super-critical fluid, then heat-soaked again.

Something similar to β resins is obtained having only one peak of molecular weight in the range 800-900amu. 100% mesophase was also claimed to be obtained despite the product is not α resins.

4° Direct preparation

Many tentatives were successfully performed to obtain directly 100% of anisotropic areas without gas-sparging or fractionating. Almost all trials used catalytic processes [160,161] or additives such as hydrogen donors. In the same category entered the dormant mesophase prepared by hydrogenation of A240 [162]. As early as 1980 [163] and more recently [164] anthracene was supposed to give 100% mesophase pitch when carbonized under pressure with or without catalysts. However, it was also pointed out that no sphere at all occurs when pure anthracene is carbonized under pressure [165]. A homogeneous product fills-up the gold tube used in these experiments. It is full of bubbles and systematically oriented parallel to the bubbles or the gold tube walls. This could be explained by the retention of volatiles inside the sealed tubes, which yields some kind of homogeneous β resin rich material, without production of Brooks and Taylor α resins spheres. As other β resins, the ones produced under pressure easily get flow orientations along bubbles and tube walls. On the contrary a large amount of spheres, but radial in texture, are produced by carbonization under pressure of more or less aliphatic products [165,166]. When the processes already mentioned are carried out upon coal-tar pitches instead of petroleum ones, it is necessary to compensate first their excess of aromaticity by hydrogenation before narrowing their molecular weight spectra [43,167]. It is also necessary to filter them before treatment so as to eliminate their primary QI (free carbon and anisotropic phases). The latter are responsible for flaws in the final fiber. Whether they are prepared from coal-tar or petroleum pitches, by gas-sparge, fractionation or any other technique the 100% optically anisotropic pitches have similar physicochemical characters:

- their H/C values are 0.5–1 [126],
- their softening point is high (250-300°C),
- their molecular weight is narrow (down to 800–900amu),
- they are approaching β resins and are thus, in average, QS-TI,
- they are thermotropic i.e., after softening they become reversibly isotropic-anisotropic.

Without foreseeing what their texture are (it will be seen in Section 2.4.2. that it is specific), it is obviously illogical to assimilate the pitches above described to 100% mesophase since the term mesophase was up to now restricted to Brooks and Taylor mesophase. In addition, in the liquid crystal

Structure and Texture of Carbon Fibers

branch of learning, the term mesophase is not employed alone because not accurate. It is usually clarified by adding the name of the mesogenic molecule. As mesophasic and anisotropic pitches are so different, different names have to be employed for fear of misconceptions. All pitches used for high performances fibers except those described in 1° will be named anisotropic pitches and not mesophase pitches even if they were demonstrated to be liquid crystals [168]. The specificity of their textures added to the specificity of their physicochemistry render this distinction absolutely inevitable. Whatever the interpretations *in-fine* given to the nature of those pitches, their original individuality must be recognized.

2.4.2 Pitch Characterization

The above considerations explain why processes or products not reliably characterized, cannot and will not be discussed in this chapter since physicochemical data become almost meaninglesss if nobody knows upon what material they have been measured.

1° Feature common to gas-sparge or fractionation prepared pitches
It was studied mainly by Lafdi et al [43,72,169–171] then discussed by Fitzgerald et al [172] and at last verified by Bonnamy and Oberlin [173]. OM and TEM were employed by all authors but only Lafdi et al. measured microhardness and local solubilities (toluene, quinoline or mixtures of both). Solubilities were tested at a micrometric scale (OM upon polished section or thin sections residues) and at a nanometric scale (TEM upon thin sections). Such experiments yield so much data that they should be employed more often. In true mesophase pitches the mesophase part has a marked preferred orientation (LMO) sketched in Figure 6b. The misorientation of BSU is limited to ± 20° in twist and ± 10° in tilt [174] (see 2° in Section 2.2.2). On the contrary, anisotropic pitches were claimed to have a poor long range orientation which is only statistical (SO) sketched in Figure 6c. BSU are in majority parallel to a plane as in mesophase but a noticeable amount of them is misoriented statistically i.e., at random up to 45° or even 90°. Such opposite characters, if they are confirmed, are the expression of a decrease of molecular weight and an increase in solubility, i.e., the change from α to β resins since it indicates a more labile BSU association easier to disperse.

The differences between LMO and SO should be easily detected and even measured by OM and TEM. In OM they are immediately evidenced by eye and with the help of Newton chart (see Section 2.2.2 and Figure 5a which must be used to follow the reasoning). The Brooks and Taylor mesophase spheres or mosaics and graphite appear almost white between crossed polarizers (near greenish white). Their phase shift δ is thus near 234nm. By adding the λ

retarder plate [175] the colors are green blue and strong yellow. On the contrary, between crossed polarisers, anisotropic pitches appear between lavender grey and blue grey [173]. Their phase shift δ is thus in the range 120-150 nm (at the most) [43,172,173]. The opposition between the SO of anisotropic pitches and the LMO of Brooks and Taylor mesophase is obviously confirmed by TEM 002DF images [43,72,169,173] taken with a progressive rotation of the objective aperture [42]. The Brooks and Taylor mesophases are lit off well before a 30° rotation whereas anisotropic pitches remain noticeably bright for a 45° rotation (Figure 34) and sometimes even for a 90° one. The statistical orientations observed in anisotropic pitches are indeed very similar to those observed in β resins (see Figure 30). The first specific character of anisotropic pitches thus lies in their poor anisotropy due to their large misorientations.

2° Feature specific to gas sparge preparation
The second feature is specific to gas-sparge preparation or laboratory preparation made by bubbling [171]. It was also debated [172] then verified by precise measurements [173]. Gas-sparge pitches [43,169,171] are made of a major component devoid of disclinations at TEM scale and characterized by its poor SO (Figure 34). Then it contains droplets ranging between 10nm up to 1μm more or less polydispersed in size. They are small enough to be scarcely observed in OM. The major component is QS-TI whereas the droplets are lighter since they are partially soluble in toluene [43]. All of them are weakly anisotropic in comparison with the major component (Figure 34). By image analysis and optical densitometry performed upon a thin section the anisotropy of the droplets is estimated to be 25 graduations on the grey scale versus 63 for the main component with a constant zero reference taken upon a supporting film [173].

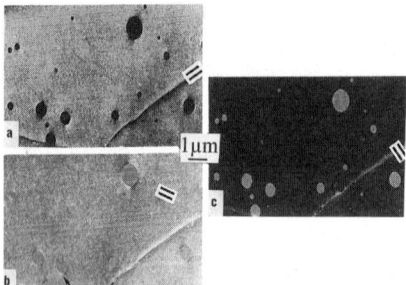

Figure 34 Thin section of a gas-sparge anisotropic pitch. 002DF images (a) 0°, (b) 45°, and (c) 90° rotation of the objective aperture, the aromatic layers are sketched by the double bar (from ref. 169).

3° Features specific to fractionation preparation
The pitches prepared by fractionation [72] are still poorly oriented but they are entirely devoid of anisotropic droplets. This is consistent with their molecular weight spectrum reduced to one peak at 800–900amu. However, they are often very rich in disclinations (Figure 35) [142,143].

4° Heterogeneities
A large amount of anisotropic pitches are heterogeneous [171]. They correspond to mixtures of all possible permutations of true mesophase pitch domains described in 1° (mosaics plus isotropic pitch drops containing Brooks and Taylor mesophase spheres), anisotropic pitches containing anisotropic droplets described in 2° and anisotropic pitches without droplets described in 3°. Figure 36 is an example of heterogeneous pitch made of large drops of true mesophase pitch with spheres (white arrow) and a continuum of anisotropic pitch containing anisotropic droplets (black arrows) [43]. As a result, tremendously confusing data restrict possible synthesis to a few well characterized series. As an example of misconceptions the percentage of optically anisotropic areas, whatever they are, is attributed to mesophase percentage. As an other example, large drops of residual isotropic pitch are intermingled with the small anisotropic droplets already described.

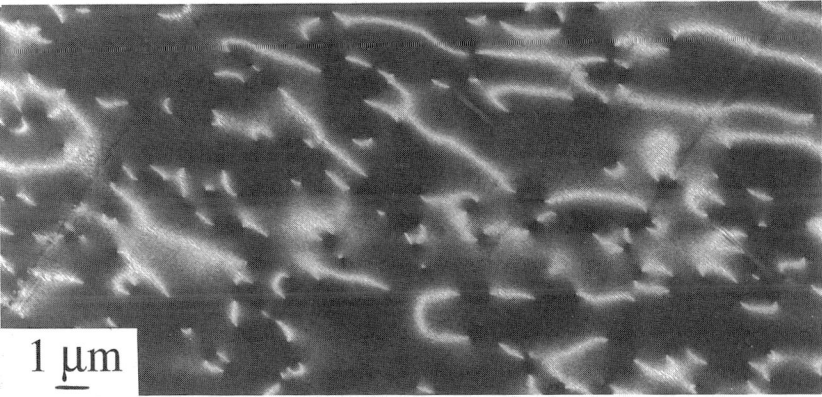

Figure 35 002DF image. Thin section of a pitch prepared by fractionation. showing the large amount of disclinations (from ref. 72).

In summary, there is a physicochemistry peculiar to anisotropic pitches (at the most β resins) i.e., relatively high H/C values, high softening points, narrow and low molecular weight ranges, thermotropy. To that, are tightly connected the textures i.e., labile associations of poorly oriented BSU forming long range flow orientations. The largest molecular weight range products (gas-sparge) corresponds to a continuum made of a major strongly anisotropic component containing weakly anisotropic droplets. The narrowest molecular weight range product (fractionation) is reduced to a single anisotropic component more or less disclination rich.

These strongly interdependent specific characters will be shown to be partly responsible for the pitch-based carbon fibers properties.

2.4.3 Effect of the Spinning Process

Melt spinning involves three steps: melting the precursor in a quite narrow temperature range [176], extrusion through the spinneret capillary and drawing down the fibers as they cool by winding. An important parameter is the pitch flow combining shear and extensional flows. The rheological properties can be studied experimentally then modelized [177]. At the spinneret exit, the spinneret die tends to align the flows [178]. At last, outside from the spinneret occurs a die-swell region followed by a thinning of the fiber which contributes

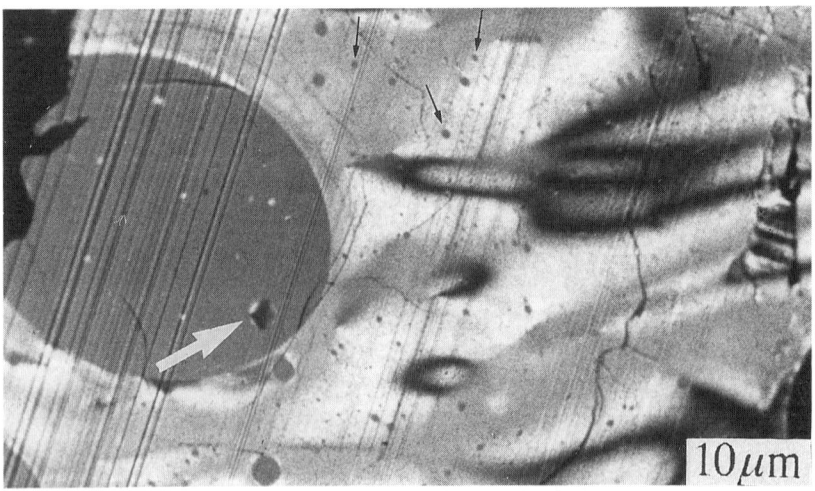

Figure 36 Optical microscopy image. Heterogeneous pitch mixing mesophase pitch and anisotropic pitch (from ref. 43).

Structure and Texture of Carbon Fibers

to modify the orientations. The shape of the spinneret or more generally its overall configuration is also highly important [179–183]. More recent papers more or less reproduce or emphazize the results of earlier papers [184–186]. However ref. 187 is an interesting tentative to correlate the HTT melting range, the β resin content of the pitch and the spinneret configuration. In addition, trial to image by OM the type of flows, was performed by forcing a highly viscous anisotropic pitch through nozzles having different shapes [188]. As the flow is laminar the cross and longitudinal sections of the extruded filament are well ordered along a long range. As the flow is turbulent the textures are random.

The effect of the spinning process upon fibers was studied indifferently upon as-spun, carbonized or graphitized fibers by using OM, SEM and more rarely TEM. A part of the above data is discarded here because either they are not new, or the origin and characterization of the samples are not given, the way to prepare the specimen not convenient or even experiments not clearly described.

1° Fiber shape
The shape of the fibers is the first character attributable to spinning [189,190]. A large variety of shapes bi or multilobal, C-shape, ribbon shape, etc.was produced and related to fiber properties.

2° Cross-sectional microtexture
The cross-sectional microtextures can be divided into a limited number of categories, if the multiple subdivisions further introduced are neglected:

- radial textures include radial and radial with open wedge
- oriented core (pan-am texture)
- concentric
- random

Figure 37 illustrates three of the major microtextures: radial with wedge, radial and concentric, seen by SEM. Figure 37a corresponds to the radial with wedge microtexture. Figure 37b corresponds to the radial microtexture and Figure 37c to the concentric one [191].

A great majority of authors attribute the cross-sectional microtextures to spinning [25,28,179–186]. The most comprehensive experimental study is that of Endo's group [179–183]. A variable spinneret design and variable melting temperature range are related to the occurrence of various microtextures and to the preferred orientation degree of the final graphitized fiber. When the flow through the spinneret capillary is laminar, radial arrangements are obtained (spinneret devoid of stirrer). The concentric or random microtextures are attributed to the increasingly turbulent flow due to the mobile stirrer increasingly approaching the spinneret die. This is consistent with ref. 188. A decreasing viscosity of the pitch only increases the preferred orientation of the fiber. A new idea also arises from ref. 180. SEM images show wavy (or folded) sheets of aromatic layers, almost periodic at the micrometric scale. The larger the wave periodicity i.e., the flater the sheets observed in SEM, the better the graphitizability as well as electrical and thermal conductivities, but the lower the tensile strength (since nothing prevents crack propagation).

A complementary view will be given further [171,191–195] but at a much smaller scale, since distinction will be introduced between micro and nanodistorsions, both inducing different properties but more or less predominantly. The two concepts are not in conflict since flatness at the micrometric scale is not detrimental to properties (except tensile strength). Nevertheless the ordering into graphite as well as electrical and thermal properties are also known to be predominantly dependent upon atomic scale. These concepts will be discussed in Section 2.4.5 (see also Figure 42 and 43).

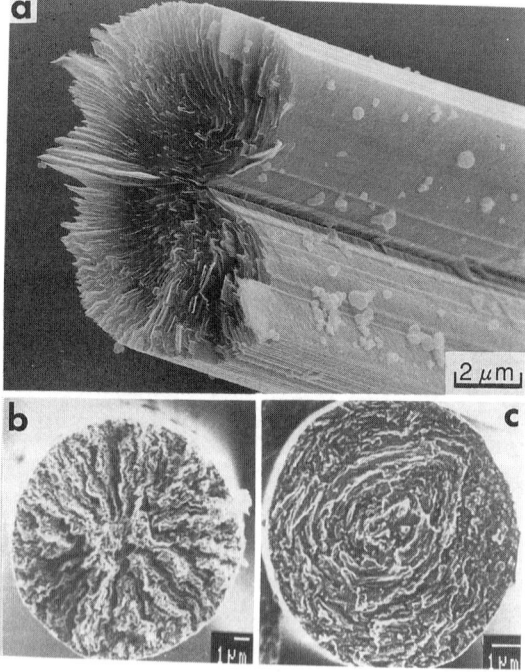

Figure 37 Microtextures of pitch-based carbon fibers, S.E.M. images: a. radial with wedge fiber (courtesy of M. Endo), b. radial fiber, c. concentric fiber (from ref. 191).

2.4.4 Stabilization

A great deal of work was done upon stabilization without a clear understanding of the process even in recent publications [196,197]. As a matter of fact it should be necessary to tie up elemental or infrared analysis with nanotexture to better understand the process. Only a limited tentative was done [198]. As spun fibers remain able to soften, whereas after stabilization they do not soften anymore because of oxygen fixation. A strong possibility of cross-linking has thus to be account for (see Section 2.3.3 1°). It was shown upon many carbonaceous materials [15,37,76,83,140,141] that cross-linking decreases the LMO size and also the graphitizability, so that either $\bar{d}_{002}$ minimum and P_1 maximum graphitization degree vary according to the atomic ratio $(O/H)_{at.}$, i.e., the atomic ratio of cross-linkers (oxygen) over their antagonist (hydrogen). Figure 38 [198] shows the variation of $\bar{d}_{002}$ versus $(O/H)_{at.}$ for five fibers increasingly stabilized then heat treated at 2800°C (black circles). A240 (as a reference), the precursor pitches of the fibers and the as-spun fibers were graphitized in the same conditions as a comparison (star). Their $(O/H)_{at.}$ ratio is negligible and their $\bar{d}_{002}$ stays at 0.336nm. After heat treatment, a Du Pont E105 fiber (d_{002} initial 0.342nm) gets 0.337 nm for $(O/H)_{at.} = 0.13$ (star). One of the precursor pitch was oxidized then heat treated at 2800°C (open triangle), its $\bar{d}_{002}$ is 0.337nm for $(O/H)_{at.} = 0.125$. It is clear from Figure 38 that all the fibers studied are increasingly oxidized without any benefit. However with a milder treatment of stabilization (plasma) one of the fibers instead of $\bar{d}_{002} = 0.339$nm stays at 0.336nm. Overstabilization has thus to be suspected but to a small extent for E105 and the fiber having a radial with wedge microtexture (the $\bar{d}_{002}$ of which being 0.337nm).

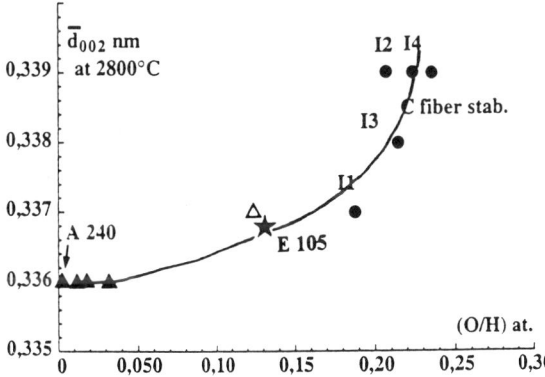

Figure 38 Variation of $\bar{d}_{002}$ versus $(O/H)_{at.}$ for pitches, oxidized pitches and stabilized fibers. Black circles: stabilized fibers, black triangles: pitches and as-spun fibers, star: Du Pont E105 fiber, open triangle: oxidized pitch, (from ref. 198).

2.4.5 Fiber Textures (Micro and Nano)

As in any other carbons the high temperature behaviour of the fibers is fixed up by the nanotexture of the stabilized fiber. Since no change have been recognized between as-spun and stabilized fibers all steps will be treated altogether in the paragraph. Among the abundant literature, the same never-ending remarks should be done : almost nothing is known about the past of the fibers. However three well defined series of fibers have been extensively studied: commercial ones, among which are the DuPont E, and Amoco's P series, then laboratory fibers [43,171,191–195]. To the credit of the latter, it must be recall that the precursor pitch preparation is known, the product is characterized at micrometric (OM) and nanometric (TEM) scales. The four fibers studied from as-spun to carbonized and graphitized, belong to classical microtextures (radial with wedge, radial, concentric, and random) and are also characterized at all scales. The two other commercial series are only studied as received. It can only be assumed that DuPont could be originating from fractionated pitches because of the Exxon patents and Amoco from gas-sparge because of the Union Carbide patents, but this is not scientific data. Anyway, the recent papers upon DuPont fibers [199–203] and Amoco ones being less comprehensive [204,209], will be discussed before those of laboratory fibers.

1° DuPont fibers

The first study appears in 1988 [199] upon a fiber having an oriented core microtexture. After graphitization the fiber is made of thin graphite platelets evidenced by a careful *hkl* DF study. In particular the mosaic grains viewed in 110DF are similar to those of Figure 9. A little further [200] a laboratory made fiber with a radial microtexture was cross-thin sectioned after graphitization. 002 lattice fringes images evidenced a high amount of disclinations. Smooth waves were observed but no graphite platelets in mosaic arrangement obviously incompatible with the smooth curvatures due to disclinations. The persistence of disclinations was consistent with the low graphitizability of the fiber. Later on, papers devoted to the DuPont E series (E01 to E130) appear, [201,202]. For E130 fiber having the highest modulus the microtexture was proposed as oriented core [201] or radial [202]. Anyway all authors agree to find a tremendous amount of disclinations in E130. Correspondingly the graphitizability is poor. At that step, it is necessary to point out an important remark concerning the scale at which the observations are made. Ref. 202 insists that DuPont, intentionally makes the zigzag radial texture so as to accomodate easily the residual stresses causing failure in all other textures. The zigzag radial thus described have the same periodicity than the smooth waves described in ref. 179–183, i.e., about micrometric (see also Figure 8). Recall that the disclination smooth curvatures are about ten times smaller and cannot

Structure and Texture of Carbon Fibers 141

be detailed by SEM but only by TEM. It will be understood further how important for properties are the micrometric and nanometric characters altogether but clearly differenciated.

The last paper [203] is an attempt to use image analysis to quantify the TEM data for fibers and compare them to x-rays. The 002 DF and lattice fringes images were binarized. Then the interfringe spacing and its distribution were measured as well as L_c deduced from 002DF. A tortuosity index described the fringe distorsions at the atomic scale. In addition, a crystallite imperfection factor and a stacking imperfection factor were determined to compare with x-rays data. For E130, $\overline{d}_{002}$ is 0.338nm with a long tail of the histograms near larger spacings. At this occasion, early data obtained upon other carbons by 002 DF got the same results [210]. If the distortion factor is negligible for E130 and the stacking imperfection factor low, the crystallite imperfection factor is relatively high (41%). This is consistent with the poor graphitization and high amount of disclinations. The great interest of this paper is the improvement in precision joined to the replacement of the tedious measurements made by eye in the early papers. The comments and conclusions support them [22,60,61].

In conclusion, the data gathered upon E130 sums up to a microtexture favorable to tensile strength (micrometric zigzags) and a nanotexture unfavorable to graphitization and electrical or thermal conductivities as well as disclinations.

2° Amoco (and Union Carbide) fibers

One of the earliest paper [204] almost immediately follows the Union Carbide patent [144]. Two kind of fibers were heat treated by steps up to 3000°C: one radial with wedge, the other random. Electrical resistivity, magnetoresistance, ESR, X-rays and OM data were compared to mechanical properties. The random fibers seem to have a tendency to be less oriented than the radial but have a similar crystallographic behavior. Both are graphitizing. Electrical resistivity and tensile strength improve as $\overline{d}_{002}$ decreases. Correspondingly the Young's modulus and the density also increase. The important conclusion of this early work is the certitude to have strong correlations between electrical, crystallographical, electronic, and mechanical properties as it was already demonstrated for PAN fibers.

Later on [205] Amoco P fibers (P55S to P100) were studied by SEM and TEM. P55S and P75 were oriented core and P100 radial with wedge (Figure 39a). P100 was entirely made of flat and stiff lamellae easily separated from each other. It was the first time that this character very peculiar to Amoco fibers was mentioned. Neither waves, nor micrometric zigzags were observed. P100 was the best crystalline material as compared to DuPont and other Amoco fibers studied (Figure 39b). P100 was the only fiber in the series to be homogeneous. All others are made of alternate bands made of microporous carbon, folded sheets of turbostratic carbon (disclination rich), and graphitic material (Figure 40). The occurrence of microporous non-graphitizing carbon was attributed to the persistence of the isotropic pitch in the precursor pitch (see also [191–195] and the following paragraph for the explanation). The same data were reproduced later [206] using the same techniques to characterize the same fibers. Additional data were obtained upon the new P120 and P130X which strongly reinforce the early data upon the lamellar constitution of the best Amoco fibers (radial microtexture). The graphitization degree of P100 also recognized as very high is much more marked in the newest P130X fiber. The lamellar graphite of P130X is so much developed that SAD are single crystal patterns. The most graphitizable fibers are those of Amoco due to the flatness of their aromatic layer sheets [205,206]. Correspondingly, the P130X modulus is higher than the E130, whereas the tensile strength is a little lower. The density is the same (2.19) [202]. The occurrence of microporous carbon in alternate bands with graphite is rediscovered in ref. 206 from P20 to P75. It is also attributed to a two phase precursor pitch (see Section 2.4.1 1°).

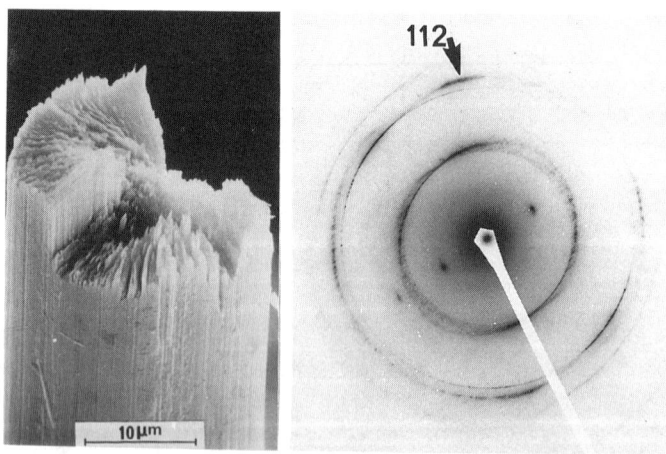

Figure 39 a. Amoco P100 fiber microtexture, SEM image (from ref. 205), b. SAD pattern of Amoco P100 (from ref. 205).

Structure and Texture of Carbon Fibers

Raman spectroscopy may be applied not only to the fiber surface, but upon polished sections at various level [207]. It is thus possible to characterize the whole fiber. A Raman microprobe was used to compare PAN-based HTS fibers (T300, T800) to HM ones (GM70), to Amoco P25 to P100 [208], after characterization by OM, SEM, TEM and x-rays. After Sakata et al. [211] the authors measured the shift to lower wave number of the graphite band versus strain applied to the fibers. E_{2g} band frequency decreases linearly with strain starting from a lower initial value for the higher modulus fibers. The shift decreases from HTS PAN-based fibers to HM PAN-based fibers and moreover to pitch-based fibers. The common limit is that of P100 and graphite (HOPG) i.e., 1580cm^{-1} [212]. The authors correlate Raman data to the microtexture of the fibers, since they attribute the decrease of the shift to the increase of order from HTS PAN-based fibers (entirely microporous and non-graphitizing), to HM fibers (having a partially graphitized skin joined to a porous core [22]), then to pitch-based fibers approaching single crystals of graphite such as P100 [205]. In addition, TEM was combined with SAD to estimate the preferred orientation so as to correlate the data with the Young's modulus. It was good for both, PAN and pitch-based fibers.

The data gathered upon Amoco P100 up to P130X sums up to microtexture and nanotexture as being favorable to electrical and thermal conductivities (perfect planar sheets of lamellae with graphite grains almost infinite), but unfavorable to tensile strength (easy glides and cleavages).

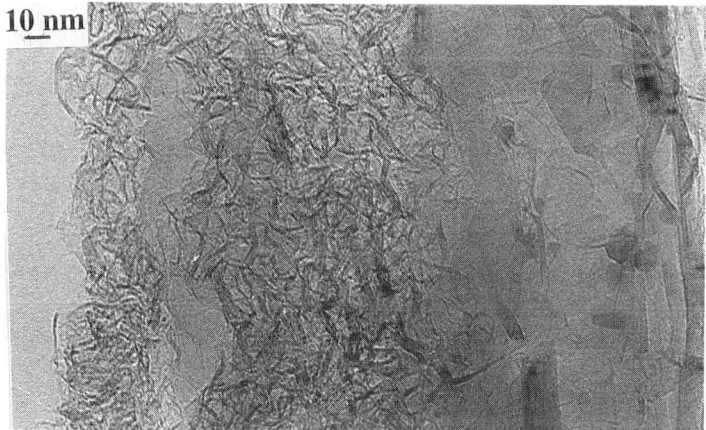

Figure 40 Bright field image. Alternate bands of graphite and microporous carbon in Amoco P75 (from ref. 205).

3° Laboratory fibers

a. Gas-sparge pitch-based fibers. The precursor pitch (described in detail in ref. 43) is heterogeneous. One part is made of a classical gas-sparge pitch, i.e., a combination of anisotropic component and anisotropic droplets (see Figure 34), and the other part is made of remainders of isotropic pitch present as large drops, and containing Brooks and Taylor mesophase single spheres (see Figure 36). The as-spun fibers are very convenient to study since thin-sectioning is so easy that no knife induced slip-lines are produced, so that the micro- and nanotexture are not disrupted. Figure 41 [194] illustrates the cross-section of such a fiber. The microtexture is radial with a slit which will become a wedge. The anisotropic droplets systematically present in the pitch have disappeared. Then, inside the radial major anisotropic component occurs a dense array of channels running radially. They are dark in the bright sectors and bright in the dark sectors. The authors [170,192–195] attribute the channels to the coalescence between adjacent pitch droplets, since when the droplets were observed, using a hot stage [170], their optical anisotropy increases and they fuse in line forming channels. In addition, Figure 41 shows that the heterogeneity of the pitch is maintained in the as-spun fiber. An isotropic drop of residual pitch (arrow) contains a Brooks and Taylor mesophase sphere. In the longitudinal section these drops are recognized as isotropic fibrils, since they were elongated by spinning. They are better seen in graphitized carbon fibers (see Figure 44). All other references of the same authors point out the persistence of both the heterogeneity initially described and the channels in all gas-sparge pitch-based fibers, whatever their microtexture is, i.e., radial with wedge [194], radial [195], concentric, or random [191,192]. Moreover, the channels are also visible in longitudinal sections as the isotropic fibrils are. In the radial with wedge microtexture (laminar flow) the channels are not only radially elongated but almost perfectly elongated along the fiber axis having thus a slit shape. They are distorted and short in all other textures (turbulent flow) and are not oriented in cross-sections. It is difficult to avoid to relate channels to the initial pitch anisotropic droplets, having been shaped by the type of flow during spinning. A marked memory of that texture will be retained even after carbonization and graphitization [171,191–193] indicating how strong is the persistence of the characters inherited from the precursor pitch. The channels derived from the initial anisotropic droplets give rise to nano-zigzags (30-100nm at the most) superimposed to the wavy microtexture (about 1μm) described by ref. 179–183. This double scale effect is particularly clear in Figure 42 (cross-section), not only in bright field but also in lattice fringes images (Figure 43). The flat part of each nano-zigzags represents the individual grains separated from each other by grain boundaries (GB) where all defects are gathered. Each grain is entirely devoid of defects (disclinations or layer tortuosities). They are equivalent to the graphite platelets described in ref. 199.

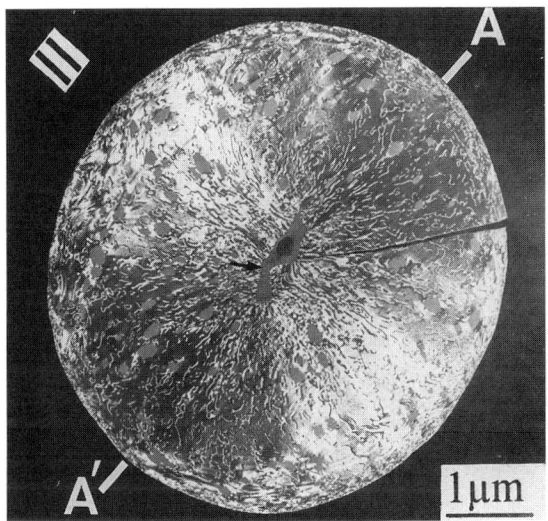

Figure 41 002DF image. Cross-section (thin section) of an as-spun fiber issued from a gas-sparge pitch. (from ref. 194).

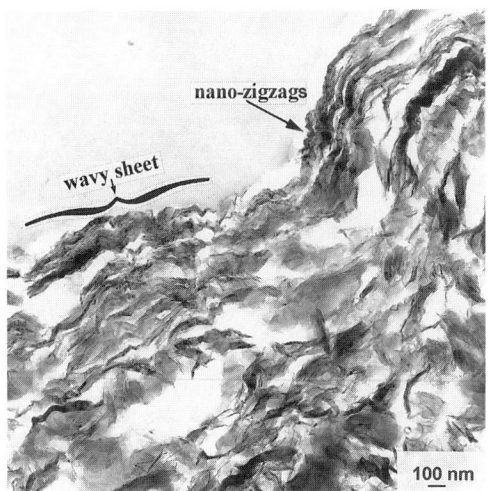

Figure 42 Bright field image. Cross-section (thin section) of a graphitized fiber issued from a gas-sparge pitch. (from ref. 171).

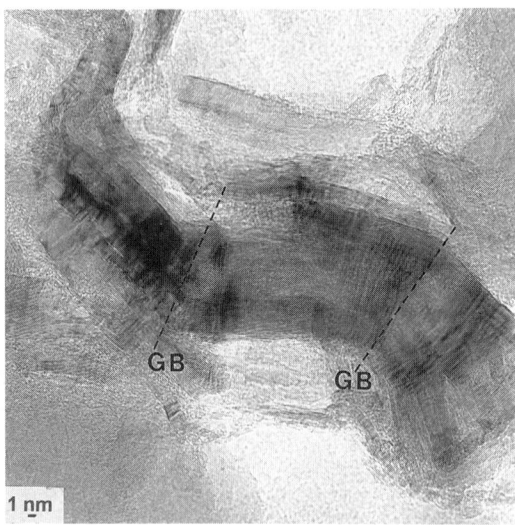

Figure 43 Cross-section (thin section) of a graphitized fiber issued from a gas-sparge pitch. Lattice fringes image illustrates the nano-zigzags and the grain boundaries (GB) (from ref. 193). In a the nano-zigzags have a period 30–100nm versus about 1μm for the wavy sheets.

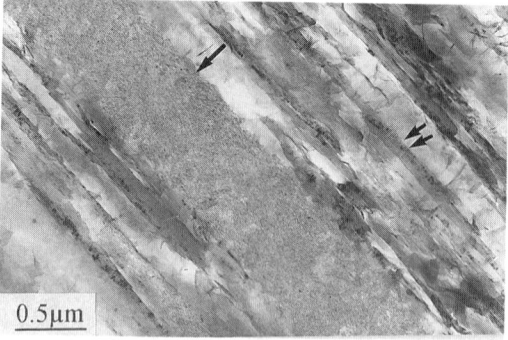

Figure 44 Bright field image. Longitudinal section (thin section) of a graphitized fiber issued from a gas-sparge pitch. The isotropic fibril is arrowed. The graphitized component is double arrowed (from ref. 193).

The aromatic layers there are perfect. It is clear that there is no comparison between the periods of the waves and that of the nano-zigzags. It is their flat part size which rules the graphitizability [193] as in any other carbons [15,37,76,83,140,141]. Here the best graphitizability also occurs for the radial with wedge fiber with a high positive magnetoresistance [191]. The combination of microwaves and nano-zigzags were found again under the name of domains and micro-domains in ref. 221.

Isotropic pitches usually yield lamellar graphitizing carbons. Unexpectedly, the isotropic fibrils contained in the as-spun fibers are transformed into microporous non graphitizing fibrils (Figure 44, single arrow). It is because the stabilization process is applied to two different components (anisotropic and isotropic pitches) that this transformation occurs. It is known that isotropic pitches are more able to fix oxygen than anisotropic ones [213,214]. Correspondingly the O/H atomic ratio of the former increases. The large LMO shrink down to small ones, producing micropores during carbonization. Nothing changes in the anisotropic component so that it is still lamellar and graphitizing (double arrow in Figure 44). This interpretation also applies to the Amoco P25 to P75 series [205] (see Figure 40). Such heterogeneities are ill-timed for two reasons:

- the microporous fibrils are fragile.
- they have poor adhesion to the major component which is harmful to mechanical properties.

b. Fractionated pitch-based fibers The as-spun fibers issued from fractionated pitches devoid of droplets do not contain any channel [72]. However they are often very rich in disclinations and thus poorly graphitizable.

The data of Lafdi et al. are comforted by the comparison between the molecular weight spectra and the solubilities of the two kind of pitches. The double peak of the gas-sparge pitch spectrum and the double solubility of the two components could be in accordance with the combination anisotropic droplets-anisotropic pitch, responsible for the couple channels-major component. On the contrary, the single peak of the QS-TI fractionated pitch could explain the absence of channels (see Section 2.4.1 2° a and b) in the as-spun fibers.

2.4.6 Fiber Surfaces

An experiment to determine the entire fiber microtexture as a pleat texture from SEM surface observations [215] appears very controversial since the experimental fibers studied are almost all radial (5 over 6). The images observed can be thus more simply explained by the protrusion of the layer sheets radially oriented or by spinning striation. In any case, nothing can be inferred by internal arrangement from surface observation alone.

A much more reliable technique for surface studies is STM, provided the images are clear enough, which is strangely difficult for fibers [216]. However very clear micrographs are given in ref. 122, 217, and 218 of Amoco fibers (P55, P120) as compared to PAN-based fibers and to unknown pitch-based carbon fibers. In these papers various oxidation treatments and coatings were also studied. Except striations due to spinning, the surfaces of the as-received fibers are smooth and highly graphitic. P120, among others, shows developed areas with trigonal symmetry typical of graphite. At the most, the hexagonal structure is locally distorted by twist. Nodular nanotexture is visible everywhere (Figure 45). However it is surprizing to see, that even in the disrupted regions the aromatic layer planes are almost systematically parallel to the surface and not protruding by their edges as it is suggested by SEM images (see Figure 37b and c and Figure 39a). It should be noted that the smaller the observation scale, the smaller the surface observed. Not long ago TEM scientists were tempted to only consider their data. Despite being helpful, it is necessary to compare to other data obtained at a larger scale. In the above references [122,217,218], the cross sectional microtexture is not mentioned so that it is not known if the pitch-based fibers studied are radial, concentric, or random. Knowing that P120 is radial, it is difficult to make a connection between SEM and STM. However, since the data obtained by these different authors agree, it suggests the occurrence of a very thin surface concentric skin not visible by SEM or even TEM.

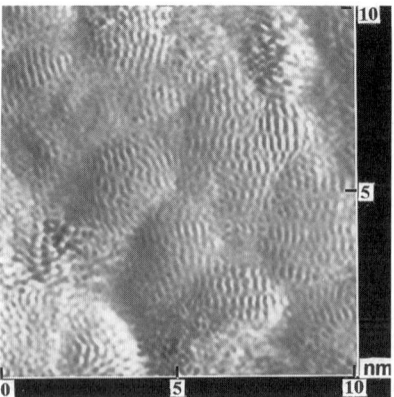

Figure 45 STM top view image of pitch-based graphitized fiber 10 nm × 10 nm (from ref. 122).

Structure and Texture of Carbon Fibers

2.4.7 Conclusion: Relations Between Textures and Properties

At the occasion of a new fiber preparation, a comprehensive discussion of the properties of most of the above described commercial fibers was performed [219]. This reference will be used to clear up the respective influences of micro- and nanotextures of properties, using the DuPont and Amoco fiber performances measured by the authors. Figure 46 a and b respectively represent the electrical resistivity versus the Young's modulus and the tensile strength versus the modulus. As well as Lavin et al. [220], the authors consider the electrical resistivity as a reliable predictor of thermal properties, i.e., too low resistivity corresponds a high thermal conductivity. It is clear from the two figures that the lowest electrical resistivity is that of Amoco P130 and the highest modulus as well. On the contrary, the tensile strength of DuPont, for the same modulus, is better than Amoco: a little more than 3GPa versus either 2.2GPa or 2.7GPa. The graphitizability of the top fibers in each series was already discussed and recognized much better for Amoco than DuPont.

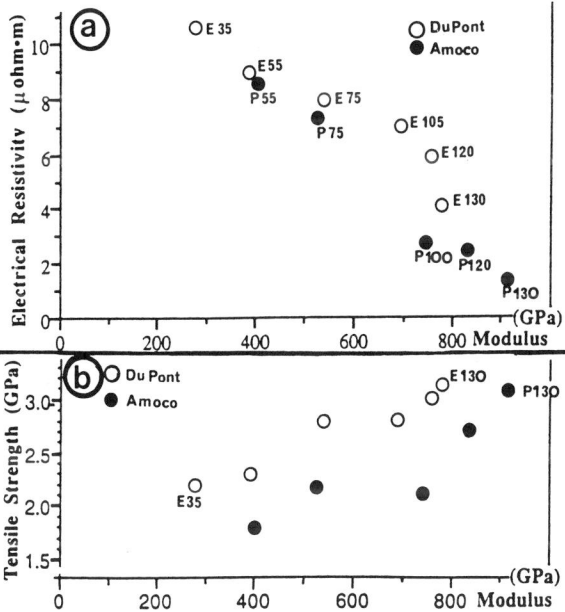

Figure 46 Comparison between DuPont and Amoco pitch-based carbon fibers (adapted from ref. 219), a. Electrical resistivity versus Young's modulus, b. Tensile strength versus Young's modulus.

In conclusion such results could be expected since:

–The micrometric zigzags of DuPont fibers (as well as the smooth waves described in ref. 179–183) favor entanglement of the adjacent layer sheets, i.e., favor the tensile strength. However the combination of micrometric distorsions and nanometric disclinations insure a permanent curvature of the layers that HTT (even high) cannot anneal. The flat parts able to yield graphitizable grains are thus small enough to almost prevent or lower graphitization. In the same manner, the electrical resistivity is high. The thermal conductivity is thus low, since the charge carriers cannot jump over disclinations or strong distorsions. The modulus which is predominantly connected to the stiffness i.e., to the graphite grains diameter cannot be maximum.

–The total flatness of Amoco fibers is not convenient for insuring a high tensile strength, but is so strongly in favor of optimum electrical and thermal conductivities that this is one of the most basic needs researched. In fact, the heat dissipation is so important in many composites, that, despite their high cost, pitch-based carbon fibers seem to come first and foremost in that purpose.

–The nano-zigzags contained in the laboratory fibers restrict the size of the grains almost as drastically as the disclinations do for DuPont fibers. The graphitizability is about the same for the two series i.e., 0.337nm for $\overline{d}_{002}$ at 2800°C for E130 and for the radial with wedge laboratory fiber. However, if there is almost no hope to completely anneal disclinations, at a reasonable cost, it is always possible to increase the grain size by a convenient choice of the precursor. It is probably the way chosen by Amoco.

–The last but not least remark concerns the precursor homogeneity since an heterogeneous precursor always yield heterogeneous fibers, which is rarely good for the properties.

ACKNOWLEDGMENT

The authors thank Thomas Cacciaguerra for his help to illustrate this chapter.

REFERENCES

1. Guinier, A., Théorie et technique de la radiocristallographie, p. 694, Dunod, Paris (1956).
2. James, R. W., The optical principles of the diffraction of X-rays, p. 661, G. Bell and sons, London (1962).
3. Ruland, W., in Chemistry and Physics of Carbons, Walker Jr., P. L., ed. vol. 4, pp. 1–84, Marcel Dekker, New York (1968).
4. Maire, J., and Mering, J., in Chemistry and Physics of Carbons, Walker Jr., P. L., ed. vol. 6, pp. 125–190, Marcel Dekker, New York (1970).

5. Fischbach, D. B., in Chemistry and Physics of Carbons, Walker Jr., P. L., ed. vol. 7 pp. 1–105, Marcel Dekker, New York (1971).
6. Maire, J., Recherche sur le phénomène de graphitation, Doctorat d'Etat Thesis, AO 1350 Schiffer, Paris (1967).
7. Bouraoui, A., Bull. Soc. Franç. Minér. Crist. 88, 633 (1965).
8. Thomson, G. P., and Cochrane, W., Theory and Practice of Electron Diffraction, p. 334, MacMillan, London (1939).
9. Bowie, S. H. V., in Physical Methods in Mineralogy–Microscopy in reflected light, Zussman, Y., ed. pp. 103–159, Acad. Press, New York (1967).
10. Ergun, S., in Chemistry and Physics of Carbons, Walker Jr., P. L., ed. vol. 3 pp. 45–120, Marcel Dekker, New York (1968).
11. Beny-Bassez, C., and Rouzaud, J. N., in Scanning Electron Microscopy, pp. 119–132, SEM inc. AMF O'Hare, Chicago (1985).
12. Rouzaud, J. N., Oberlin, A. and Beny-Bassez, C., Thin Solid Films, 105, 75 (1983).
13. Magnan, C., Traité de Microscopie Electronique, vol. 1, p. 649, Hermann, Paris (1961).
14. Hirsch, P. B., Howie, A., Nicholson, R. B., Pashley, D. W., and Whelan, M. J., Electron Microscopy of Thin Crystals, p. 549, Butterworths, London (1965).
15. Oberlin, A., in Chemistry and Physics of Carbons, Thrower, P. A., ed., vol. 22, pp. 1–141, Marcel Dekker, New York (1989).
16. Colloque: la microscopie à effet tunnel, Mulhouse, Juillet 1991, L'actualité chimique n°2 mars-avril (1992).
17. Reynolds, W. N., in Chemistry and Physics of Carbons, Walker, P. L., Jr. and Thrower, P. A., eds., vol. 11, pp. 1–67, Marcel Dekker, New York (1981).
18. Delmonte, J., Technology of Carbon and Graphite Fiber Composites, p. 452, Van Nostrand Reinhold Co., New York (1981).
19. Johnson, D. J., in Chemistry and Physics of Carbons, Thrower, P. A., ed., vol. 20, pp. 1–58, Marcel Dekker, New York (1987).
20. Dresselhaus, M. S., Dresselhaus, G., Sugihara, K., Spain, I. L., and Goldberg, H. A., Graphite Fibers and Filaments, Springer Series in Materials Science 5, p. 382, Springer-Verlag, Berlin (1988).
21. Strong Fibres, vol.1, Watt, W., and Perov, B. V., eds. Handbook of Composites Series, Kelly, A., and Rabotnov, Y. N., eds., vol. 1, p. 752, North-Holland, Amsterdam (1988).
22. Oberlin, A., and Guigon, M., in Fibre Reinforcements for Composite Materials, Bunsell, A. R., ed., Composite Materials Series, Pipes, R. B., ed., pp. 149–210, Elsevier Science Publishers, Amsterdam (1988).

23. Donnet, J. B., and Bansal, R. C., Carbon Fibers, vol.3, 2nd ed., Lewin, M., series ed., p. 291, Marcel Dekker, New York (1991).
24. Fitzer, E., Carbon 27, 621 (1989).
25. Rand, B., in Strong Fibers, Watt, W., and Perov, B. V., eds. Handbook of Composite Series, Kelly, A., and Rabotnov, Y. N., eds., pp. 495–575, North Holland, Amsterdam (1988).
26. Petroleum derived carbons, ACS Symposium series no. 303, Bacha, J. D., Newman, J. W., and White, J. L., eds. p. 406, ACS New York (1986).
27. Diefendorf, R. J., in Carbon Fibres and their Composites, Fitzer, E., ed. p.46, Springer-Verlag, Berlin (1985).
28. Edie, D. D., in Carbon Fibers, Filaments and Composites, Figueiredo, J. L., Bernardo, C. A., Baker, R. T. K., and Huttinger, K., eds., p. 43, Kluwwer Academic Publishers, Dordrecht, the Netherlands (1990).
29. Bacon, G. E., Acta Cryst. 3, 137 (1950).
30. Tombrel, F., and Rappeneau, J., in Les carbones, ed. Groupe Français d'Etude des Carbones, vol. II, pp. 805–811, Masson Paris, (1965).
31. Johnson, D. J., and Tyson, G. N., Brit. J. Appl. Phys. D2, 787 (1969).
32. Perret, R., and Ruland, W., J. Appl. Cryst. 3, 525 (1970).
33. Warren, B. E., Phys. Rev. 59, 693 (1941).
34. Franklin, R. E., Acta Cryst. 4, 253 (1951).
35. Rannou, I., Bayot, V., and Lelaurain, M., Carbon, 32, 833 (1994).
36. Endo, M., Takeuchi, K., Takahashi, K., Oshida, K., Dresselhaus, M. S., and Dresselhaus, G., Ext. Abstr. 22nd Biennial Conf. on Carbon, San Diego, p. 340 (1995).
37. Oberlin, A., Bonnamy, S., Bourrat, X., Monthioux, M., and Rouzaud, J. N., in Petroleum derived carbons ACS Symposium series no. 303, Bacha, J. D., Newman, J. W., and White, J. L., eds., pp. 85–98, ACS New York (1986).
38. Brindley, G. W. and Mering, J., Acta Cryst. 4, 441 (1951).
39. Iwashita, N. and Inagaki, M., Carbon 31, 1107 (1993).
40. Mering, J. and Longuet-Escard, J., J. Chim. Phys. 51, 416 (1954).
41. Warren, B. E. and Bodenstein, P., Acta Cryst. 18, 282 (1965).
42. Oberlin, A., Carbon 17, 7, (1979).
43. Lafdi, K., Bonnamy, S., and Oberlin, A., Carbon 29, 831 (1991).
44. Tuinstra, F. and Koenig, J. L., J. Chem. Phys. 53, 1126 (1976).
45. Vidano, R. and Fischbach, D. B., J. Am. Ceram. Soc. 61, 13 (1978).
46. Marchand, A., Lespade, P., and Couzi, M., Ext. Abstr. 15th Biennial Conf. on Carbon, Philadelphia, p. 282 (1981).
47. Cuesta, A., Dhamelincourt, P., Laureyns, J., Martinez-Alonso, A., and Tascon, J. M. D., Carbon 32, 1523 (1994).
48. Oberlin, A. and Rousseaux, F., C. R. Acad. Sc. C274, 108 (1972).

49. Elkind, J. L., J. Vac. Sci. Technol. B 10, 1460 (1992).
50. Benedict, J. P., Klepeis, S. J., Vandygriflt, W. G., and Anderson, R., EMSA Bull. 19, 74 (1989).
51. Maniette, Y., J. Mater. Sci. Letters 9, 48 (1990).
52. Goma, J., Oberlin, M., and Oberlin, A., Thin solid films 65, 221 (1980).
53. Jenkins, G. M. and Kawamura, K., Nature 231, 175 (1971).
54. Fourdeux, A., Perret, R., and Ruland, W. J., in Carbon fibres. Their composites and applications, Proceed. Intern. Carbon Fibres Conf., p. 57, The Plastic Institute London (1971).
55. Diefendorf, R. J. and Tokarsky, E. W., Polymer Engineer. Sci. 15, 150 (1975).
56. Binnig, G. and Rohrer, H., Helv. Phys. Acta 55, 726 (1982).
57. Nysten, B., Roux, J. C., Flandrois, S., Daulan, C., and Saadaoui, H., Phys. Rev. B48, 12527 (1993).
58. Biensan, P., Roux, J. C., Saadaoui, H. and Flandrois, S., Microsc. Microanal. Microstruct. 1, 103 (1990).
59. Saadaoui, A., Roux, J. C., and Flandrois, S., Carbon 31, 481 (1993).
60. Guigon, M., Relation entre la microtexture et les propriétés mécaniques et électriques des fibres de carbone ex-PAN, Doctorat d'Etat Thesis, Université de Technologie de Compiègne, Novembre 1985.
61. Deurbergue, A., Fibres de carbone ex-PAN: stabilisation, carbonisation. Relations avec les propriétés mécaniques, Ph. D. Thesis, Université de Pau et des Pays de l'Adour, Juin 1989.
62. Loison, R., Foch, P., and Boyer, A., Coke quality and production, p. 567, Butterworths, London (1989).
63. Van Krevelen, D. W., Coal: Typology, Chemistry, Physics, Constitution, 3rd ed., p. 979, Elsevier, Amsterdam (1993).
64. Chari, S. S., Bahl, O. P., and Mathur, R. B., Fibre Sc. Technol. 15, 153 (1981).
65. Grassie, N. and Mc Guchan, R., Eur. Polymer J. 6, 1277 (1970).
66. Frushour, B., Polymer Bull. 7, 1 (1982).
67. Muller, D. J., Fitzer, E., and Fiedler, A. K., in Carbon Fibres. Their composites and applications, Proceed. Intern. Carbon Fibres Conf., p. 10, The Plastic Institute London (1971).
68. Vorpagel, E. R. and Lavin, J. G., Carbon 30, 1033 (1992).
69. Friedel, G., Ann. Phys. 18, 273 (1922).
70. Sadoc, J. F. and Mosseri, R., Phil. Mag. B45, 467 (1982).
71. Zimmer, J. E. and White, J. L., Advances in Liquid Crystals 5, 157 (1982).
72. Lafdi, K. and Oberlin, A., Carbon 32, 61 (1994).
73. Deurbergue, A. and Oberlin, A., Carbon 29, 621 (1991).
74. Bahl, O. P. and Manocha, L. M., Carbon 12, 417 (1974).

75. Watt, W., Johnson, D. J., and Parker, E., in Carbon fibres. Their place in modern technology, Proceed. Intern. Carbon Fibres Conf., p. 3, The Plastic Institute London (1974).
76. Oberlin, A., Boulmier, J. L., and Villey, M., in Kerogen, Durand, B., ed. pp. 191–241, Technip, Paris (1980).
77. Mrozowski, S., Proceed. 1st and 2nd Carbon Conf., Buffalo, p. 31 (1956).
78. White, J. L., Dubois, J., and Souillart, C., J. Chim. Phys., numéro spécial, 33 (1969).
79. Mathur, R. B., Bahl, O. P., and Mittal, J., Carbon 30, 657 (1992).
80. Guigon, M., Oberlin, A., and Desarmot, G., Fibre Sc. and Technol. 20, 55 (1984).
81. Guigon, M., Oberlin, A., and Desarmot, G., Composites Sc. and Technol. 20, 177 (1984).
82. Reynolds, W. N. and Moreton, R., Phil. Trans. Roy. Soc. London A 294, 451 (1980).
83. Oberlin, A., Villey, M., and Combaz,A., Carbon 18, 347 (1980).
84. Ingram, D. J. E., Ext. Abstr. 3nd Biennial Conf. on Carbon, Buffalo, p. 97 (1959).
85. Rouxhet, P. G., Robin, P. L., and Nicaise, G., in Kerogens, Durand, B., ed., pp. 163-190, Technip, Paris (1980).
86. Johnson, D. J., in New fibres and their composites, pp. 36–41, The Royal Soc. of London (1980).
87. Bennett, S. C., and Johnson, D. J., Proceed. 5th Industrial Carbon and Graphite Conf., vol. 1, p. 377, Soc. Chem. Ind. London (1978).
88. Perret, R. and Ruland, W., J. Appl. Cryst. 2, 209 (1969).
89. Johnson, D. J., in Carbon Fibres. Their composites and applications, Proceed. Intern. Carbon Fibres Conf., p. 52, The Plastic Institute London (1971).
90. Oberlin, A., Molleyre, F., and Bastick, M., Ext. Abstr. 13th Biennial Conf. on Carbon, Irvine, p. 371 (1977).
91. Oberlin, A. and Oberlin, M., Rev. Chim. Miner. 18, 442 (1981).
92. Deurbergue, A., and Oberlin, A., Proceed. Intern. Carbon Conf., Tsukuba, vol. 1, p. 280 (1990).
93. Northold, M. G., Veldhuizen, L. H., and Jansen, H., Carbon 29, 1267 (1991).
94. Barnet, F. R. and Norr, M. K., Carbon 11, 281 (1973).
95. Johnson, D. J., Phil. Trans. Roy. Soc. London A294, 443 (1980).
96. Guigon, M., Oberlin, A., and Desarmot, G., Fibre Sc. and Technol. 27, 1 (1986).
97. Heidenreich, R. D., Hess, N. M., and Ban, L. L., J. Appl. Cryst. 1, 1 (1968).

98. Oberlin, A., Endo, M., and Koyama, T., J. Cryst. Growth 32, 335 (1976).
99. Deurbergue, A., and Oberlin, A., Carbon 30, 981 (1992).
100. Dorey, G., Phys. Technol. 11P, 56 (1980).
101. Bennett, S. C. and Johnson, D. J., Carbon 17, 25 (1979).
102. Johnson, D. J. and Tyson, C. N., Brit. J. Appl. Phys. D3, 526 (1970).
103. Johnson, D. J., Nature 226, 750 (1970).
104. Bacon, R., in Chemistry and Physics of Carbons, Walker Jr., P. L., and Thrower, P. A., eds., vol. 9, pp. 1–102, Marcel Dekker, New York (1973).
105. Donnet, J. B., Voet, A., Ehrburger, P., Dauksch, H., and Marsh, P. A., Carbon 11, 431 (1973).
106. Donnet, J. B. and Ehrburger, P., Carbon 15, 143 (1977).
107. Fourdeux, A., Perret, R., and Ruland, W., J. Appl. Cryst. 1, 252 (1968).
108. Watt, W. and Johnson, J. W., Proceed. 3rd Industrial Carbon and Graphite Conf., vol. 1, p. 417, Soc. Chem. Ind. London (1971).
109. Després, J. F., Les interfaces de carbone pyrolytique dans les composites carbone/ carbure de silicium, Ph. D. Thesis, Université de Pau et des Pays de l'Adour, Septembre 1993.
110. Donnet, J. B., Papierer, E., and Dauksch, M., in Carbon fibres. Their place in modern technology, Proceed. Intern. Carbon Fibres Conf., p. 9, The Plastic Institute London (1974).
111. Fitzer, E., Geige, K. H., Huttner, W., and Weiss, R., Carbon 18, 389 (1980).
112. Drzal, L. T., Rich, M. J., and Lloyd, P. F., ACS Polym. Preprints 22 (1981).
113. Narcy, B. and Bruneton, E., submitted to Carbon (1996).
114. Matsui, J., Critical Reviews in Surface Chemistry 1, 71 (1990).
115. Oberlin, A. and Guigon, M., Revue Scient. et Techn. de la Défense 89 (Oct. 1988).
116. Sanchez, M., Desarmot, G., Mérienne, M. C., and B. Barbier, C.R. 5ème Journées Nationales des Composites, p. 472, AMAC Paris (1986).
117. Serin, V., Fourneaux, R., Kihn, Y., Sevely, J., and Guigon, M., Carbon 28, 573 (1990).
118. Boehm, H. P., Diehl, E., and Heck, W., Symp. on Carbon, Tokyo, Japan VIII-10-1 (1964).
119. Dilsiz, N., Erinc, N. K., Bayramli, E., and Akovali, G., Carbon 33, 853 (1995).
120. Coustumer, P. Le, Lafdi, K., and Oberlin, A., Carbon 30, 1127 (1992).
121. Marshall, P. and Price, J., Composites, 22, 388 (1991).
122. Donnet, J. B. and Qin, R. Y., Carbon 30, 787 (1992).
123. Donnet, J. B. and Qin, R. Y., Carbon 31, 7 (1993).

124. Qin, R. Y. and Donnet, J. B., Carbon 32, 323 (1994).
125. Blackslee, O. L., Proctor, D. G., Seldin, E. J., and Weng, T., J. Appl. Phys. 1, 3373 (1970).
126. Barr, J. B. and Lewis, I. C., Carbon 16, 439 (1978).
127. Lafdi, K., Elaboration et caractérisation des fractions lourdes de goudron de houille, Ph. D. Thesis, Université de Pau et des Pays de l'Adour, Décembre 1989.
128. Lafdi, K., Bonnamy, S., and Oberlin, A., Carbon 28, 57 (1990).
129. Rouzaud, J. N., and Oberlin, A., C. R. Acad. Sc., Paris, 757 (1983).
130. Auguié, D., Mésophase carbonée et interaction entre coke de pétrole et brai de houille, Thèse Ingénieur-Docteur, Université d'Orléans, Décembre, 1979.
131. Lafdi, K., Bonnamy, S., and Oberlin, A., Carbon 28, 617 (1990).
132. Pfeiffer, J. P. and Saal, R. N. J., J. Phys. Chem. 44, 139 (1940).
133. Sack, H., C. R. Acad. Sci., 224, 833 (1947).
134. Yen, T. F., Erdman, J. G., and Pollack, S. S., Anal. Chem. 33, 1587 (1961).
135. Tissot, B. and Welte, D. H., Petroleum formation and occurrence, 2nd ed., p. 538, Springer-Verlag, Berlin (1984).
136. Speight, J. G., in The chemistry and technology of petroleum, vol. 1, pp. 247–252, Marcel Dekker, New York (1980).
137. Pelet, R., Behar, F., and Monin, J. C., in Advances in Organic Geochem., Leythaeuser, D., and Rullkötter, J., eds., pp. 481–498, Pergamon (1985).
138. Espinat, D., and Ravey, J. C., Proceed. SPE Intern. Symp. on Oilfield Chemistry, New Orleans, p. 365, Soc. Petr. Eng. Inc. (1993).
139. Brooks, J. D., and Taylor, G., in Chemistry and Physics of Carbons, Walker Jr., P. L., ed., vol. 4, pp. 243–286, Marcel Dekker, New York (1968).
140. Bonnamy, S., part I, submitted to Carbon (1996).
141. Bonnamy, S., part II, submitted to Carbon (1996).
142. White, J. and Zimmer, J., Carbon 16, 469 (1978).
143. White, J. and Buechler, M., in Petroleum derived carbons ACS Symposium series no. 303, Bacha, J. D., Newman, J. W., and White, J. L., eds., pp. 62–84, ACS New York (1986).
144. Singer, L. S., US Patent 4.005.183 (Union Carbide) (1977).
145. Singer, L. S., Carbon 16, 409 (1978).
146. Oberlin, A., Bonnamy, S., and Rouxhet, P., in Chemistry and Physics of Carbon, to be printed (1998).
147. Balduhn, R. and Fitzer, E., Carbon 18, 155 (1980).
148. Ihnatowicz, M., Chiche, P., Deduit, J., Pregermain, S., and Tournant, R., Carbon 4, 41 (1966).

149. Honda, H., Kimura, H., Sanada, Y., Sugawara, L., and Futura, T., Carbon 8, 181 (1970).
150. Lewis, I. C., McHenry, E. R., and Singer, L. S., US Patent 3.976.729 (1976), 4.026.788 (1977), 4.017.327 (1977).
151. Chwastiak, S., US Patent 4.209.500 (1980).
152. Mochida, I., Maeda, K. and Takeshita, K., Carbon 16, 459 (1978).
153. Greinke, R. A. and Singer, L. S., Carbon 26, 665 (1988).
154. Mochida, I., Maeda, K., and Takeshita, K., Carbon 15, 17 (1977).
155. Lafdi, K. and Oberlin, A., Ext. Abstr. 20th Biennial Conf. on Carbon, Santa Barbara, p. 150 (1991).
156. Otani, S., Endo, T., Ota, E., and Oya, A., Tanso 87, (1976).
157. Chwastiak, S. and Lewis, I. C., Carbon 16, 156 (1978).
158. Otani, S. and Oya, A., in Petroleum derived carbons ACS Symposium series no. 303, Bacha, J. D., Newman, J. W., and White, J. L., eds., pp 323–335, ACS New York (1986).
159. Diefendorf, R. J. and Riggs, D. M., US Patent 4.208.267 (1980).
160. Mochida, I., Zeng, S. M., Korai, Y., and Toshima, H., Carbon 28, 311 (1990).
161. Moldenaers, P., and Mewis, J., J. Non Newtonian Fluid Mechanics 34, 359 (1990).
162. Otani, S., Japan patent 156022 (1983).
163. Lewis, I. C., Carbon 18, 191 (1980).
164. Mochida, I., Shimizu, K., Korai, Y., Sakai, Y., Fujiyama, S., Toshima, H., and Hono, T., Carbon 30, 55 (1992).
165. Ayache, J., Oberlin, A., and Inagaki, M., Carbon 28, 337 (1990).
166. Washiyama, M., Sakai, M., and Inagaki, M., Carbon 26, 303 (1988).
167. Yamada, Y., and Honda, H., Japan Patent 58.18421 (1983).
168. Lewis, J. C. and Lewis, R. T., Carbon 26, 757 (1988).
169. Lafdi, K., Bonnamy, S., and Oberlin, A., Carbon 29, 849 (1991).
170. Lafdi, K., Bonnamy, S., and Oberlin, A., Carbon 29, 857 (1991).
171. Lafdi, K. and Oberlin, A., Carbon 32, 11 (1994).
172. Fitzgerald, J. D., Taylor, G. H., and Pennock, G. M., Carbon 32, 1389 (1994).
173. Bonnamy, S., Clinard, C., Lafdi, K., and Oberlin, A., Ext. Abstr. 23rd Biennial Conf. on Carbon, PennState, USA (1997).
174. Auguié, D., Oberlin, M., Oberlin, A., and Hyvernat, P., Carbon 19, 227 (1981).
175. Honda, H., Kimura, H., and Sanada, Y., Carbon 9, 695 (1971).
176. Edie, D. D. and Dunham, M. G., Carbon 27, 647 (1989).
177. Park, N. A., Cho, Y. I., and Irvine, T. F., J. Non Newtonian Fluid Mechanics 34, 351 (1990).
178. Nazem, F. F., Carbon 20, 345 (1982).

179. Hamada, T., Nishida, T., Sajiki, Y., Matsumoto, M., and Endo, M., J. Mater. Res. 2, 603 (1987).
180. Endo, M., J. Mater. Sci. 23, 598 (1988).
181. Hamada, T., Nishida, T., Furuyama, M., and Tomioka, T., Carbon 26, 837 (1988).
182. Hamada, T., Sajiki, Y., Furuyama, M., Tomioka, T., and Endo, M., J. Mater. Res. 4, 1027 (1989).
183. Hamada, T., Furuyama, M., Sajiki, Y., Tomioka, T., and Endo, M., J. Mater. Res. 5, 1271 (1990).
184. Yoon, S. H., Korai, Y., and Mochida, I., Carbon 31, 849 (1993).
185. Edie, D. D., Fain, C. C., Robinson, K. E., Harper, A. M., and Rogers, K., Carbon 31, 941 (1993).
186. McHugh, J. J. and Edie, D. D., Ext. Abstr. 22nd Biennial Conf. on Carbon, San Diego, p. 2 (1995).
187. Zha, Q. F., Shi, J. L., Ji, Y., Lin, L., and Qian, S. A., Carbon 30, 739 (1992).
188. Lafdi, K., Oberlin, A., and Rand, B., Ext. Abstr. 20th Biennial Conf. on Carbon, Santa Barbara, p. 172 (1991).
189. Anderson, S. H., and Chung, D. D. L., Carbon 22, 613 (1984).
190. Thesis of Clemson University: Fox, N. K. (1984), Harrison, M. G., (1985), Schikner, R. C., (1989), Harper, A. M., (1990), Fleurot, O., (1992).
191. Inagaki, M., Iwashita, N., Hishiyama, Y., Kaburagi, Y., Yoshida, A., Oberlin, A., Lafdi, K., Bonnamy, S., and Yamada, Y., Tanso 147, 57 (1991).
192. Lafdi, K., Bonnamy, S., and Oberlin, A., Carbon 31, 29 (1993).
193. Lafdi, K., Bonnamy, S., and Oberlin, A., Carbon 30, 533 (1992).
194. Lafdi, K., Bonnamy, S., and Oberlin, A., Carbon 30, 551 (1992).
195. Lafdi, K., Bonnamy, S., and Oberlin, A., Carbon 30, 569 (1992).
196. Yanagida, K., Sasaki, T., Tate, K., Sakanishi, A., Korai, Y., and Mochida, I., Carbon 31, 577 (1993).
197. Drbohlav, J., and Stevenson, W. T. K., Carbon 33, 693 (1995).
198. Lafdi, K. and Oberlin, A., Ext. Abstr. 21st Biennial Conf. on Carbon, Buffalo, p. 280 (1993).
199. Roche, E. J., Lavin, J. G., and Parrish, R. G., Carbon 26, 911 (1988).
200. Bourrat, X., Roche, E. J., and Lavin, J. G., Carbon 28, 236 (1990).
201. Pennock, G. M., Taylor, G. H., and Fitzgerald, J. D., Carbon 31, 591 (1993).
202. Kogure, K., Sines, G., and Lavin, J. G., Carbon 32, 1469 (1994).
203. Doble, M. G., Guo, H., and Johnson, D. J., Carbon 33, 1115 (1995).
204. Bright, A. A., and Singer, L. S., Carbon 17, 59 (1979).

205. Guigon, M., and Oberlin, A., Composites Science and Technology 25, 231 (1986).
206. Fitzgerald, J. D., Pennock, G. M., and Taylor, G. H., Carbon 29, 139 (1991).
207. Morita, K., Murata, Y., Ishitani, A., Murayama, K., Ono, T., and Nakajima, A., J. Pure Appl. Chem. 58, 455 (1986).
208. Yuang, Y., and Young, R. J., Carbon 33, 97 (1995).
209. Endo, M., personal communication.
210. Oberlin, A., Terriere, G., and Boulmier, J. L., Tanso 83, 153 (1975).
211. Sakata, H., Dresselhaus, G., Dresselhaus, M. S., and Endo, M., J. Appl. Phys. 63, 2769 (1988).
212. Chien, T. C., Dresselhaus, M. S., and Endo, M., Phys. Rev. B26, 5867 (1982).
213. Joseph and, D., and Oberlin, A., Carbon 21, 559 (1983).
214. Joseph and, D., and Oberlin, A., Carbon 21, 565 (1983).
215. Yoon, S. H., Korai, Y., and Mochida, I., Carbon 32, 1182 (1994).
216. Effler, L. J., Fellers, J. F., and Annis, B. K., Carbon 30, 631 (1992).
217. Hoffman, W. P., Elings, V. B., and Gurley, J. A., Carbon 26, 754 (1988).
218. Hoffman, W. P., Carbon 30, 315 (1992).
219. Edie, D. D., Robinson, K. E., Fleurot, O., Jones, S. P., and Fain, C. C., Carbon 32, 1045 (1994).
220. Lavin, J. G., Boyington, D. R., Lahijani, J., Nysten, B., and Issi, J. P., Carbon 31, 1001 (1993).
221. Mochida, I., Yoon, S. H., Takano, N., Fortin, F., Korai, Y., and Yokogawa, K., Carbon 34, 941 (1996).

3
SURFACE TREATMENT OF CARBON FIBERS

Jimmy C. M. Peng, Jean-Baptiste Donnet, and Tong Kuan Wang
Ecole Nationale Supérieure de Chimie and Université de Haute-Alsace, Mulhouse, France

Serge Rebouillat
DuPont de Nemours International S. A., Geneva, Switzerland

3.1 INTRODUCTION

The development of strong and stiff carbon fiber and its use as the reinforcing element in lightweight structural parts is one of the major technological achievements of the past decade. As the use of various CF increases, so does the understanding of those CF properties that are needed for a good composite performance. This performance often depends on the degree of adhesion between the fiber and the binder. Adhesion is thought to be controlled by chemical bonding due to functional groups and by mechanical interlocking due to surface morphology. These hypotheses were proposed and investigated by several research groups [1–5] in the past decade.

Carbon fibers are supplied to the customers with and without surface treatment, and are normally coated with a thin layer of sizing compounds. Carbon fibers, when used without any surface treatment, produce CF-epoxy composites with low interlaminar shear strength (ILSS). This has been attributed largely to weak bonding between the fiber and the epoxy matrix. Kaelble et al. [6], Larsen et al. [7], and Daukeys [8] have shown that ILSS is directly related to the fiber-matrix bonding. It was observed that as the temperature of graphitization of the carbon fiber increased, ILSS decreased, although the Young's modulus showed an increase. These observations led investigators to develop a number of surface treatments that could improve the fiber matrix interfacial bonding. Note that the interfacial shear behavior in a are produced by polymerization, oxidation, carbonization, and graphitization processes from acrylonitrile molecules [9], have many graphitic edge planes exposed at the surface. On the other hand, pitch-based fibers, which are produced by heat treatment, extrusion, spinning followed by stabilization, carbonization, and graphitization processes from petroleum asphalt or coal tar materials [10], have a larger fraction of the basal plane of graphite exposed at the surface. PAN-based CF-epoxy composites typically exhibit much better adhesion, a result that has been attributed to the higher chemical reactivity of graphitic edge planes compared to the basal plane.

Surface treatments may be classified into oxidative treatments and non-oxidative treatments. The oxidative treatments can further be subdivided into gas-phase oxidations at lower or at elevated temperatures, liquid-phase oxidations carried out chemically or electrochemically, and catalytic oxidations. The non-oxidative treatments that improve the fiber-resin bonding consist of depositing more active forms of carbon, such as the highly effective whiskerization, the deposition of pyrolytic carbon, or the grafting of polymers on the carbon fiber surface. Some average improvements in composite shear strength which can be achieved by the various surface treatments are shown in Table 1.

Table 1 Surface Treatment of Carbon Fiber and Improvement of Composite

Treatment	Improvement in ILSS(%)
Gaseous oxidation (air, ozone, RF plasma)	10–15
Liquid-phase oxidation (HNO_3, NaClO, electrolytic)	100–200
Whiskerization (Si_4N_4, TiO_4, SiC)	200–300
Pyrolitic carbon coating (CH_4, FeC, SiC)	60–100
Polymer grafting	80–100

3.2 GAS-PHASE OXIDATIVE TREATMENTS

The gas-phase oxidative treatments are carried out with air, oxygen, or oxygen-containing gases such as ozone and CO_2. The oxidation may be carried out either as such or in the presence of a catalyst. Air is a very commonly used gaseous oxidizing agent; oxidation is also sometimes carried out in the presence of certain other gaseous species, such as CO_2, CO, or water vapor. The gaseous treatments may be carried out at low or elevated temperatures. The treatments at elevated temperatures are, however, generally very drastic and cause severe degradation and excessive pitting of the carbon fiber surface, which reduces the fiber strength.

3.2.1 Oxidative Treatment in Air or Oxygen

Oxidative treatment in air, although very convenient, has not been very efficient in improving the composite shear properties. In order to obtain good results, the treatment has to be carried out at elevated temperatures where degradation and a large weight loss are likely to occur. For example, Herrick et al. [11] observed almost no improvement in the composite shear properties of rayon based carbon fibers after the treatment in air for 16 hours at 500°C. However, when treatment temperature increased during the same period the result was an increase of the composite shear strength by about 45%. Because 600°C is very close to the ignition temperature of the carbon fiber and, thus, the treatment results in excessive weight loss. On the other hand, Sach and Basche [12] heat-treated carbon fibers in air by using small amounts of several oxidation inhibitors such as halogens, SO_2, and halogenated hydrocarbon (CCl_4), and found that the oxidation behavior was considerably slower, milder, and did not cause excessive pitting of the fiber surface. As a result, the adhesion of carbon fiber and the resin matrix was notably enhanced.

The effect of oxidation in air on the mechanical properties of carbon fibers has been variously interpreted and has been shown to depend on the experimental conditions under which the oxidation is carried out. The oxidation causes marked pitting of the carbon fiber surface at elevated temperatures, which reduces the fiber strength [13]. Kucera et al. [25] has recently reached a similar conclusion that when fibers were oxidized in air at various temperatures (600-900°C) for a very short time (several seconds) and a marked decrease of the tensile modulus and strength were observed. Mckee and Mimeault [17] indicated that the etch pits were produced only during the initial stages of the oxidation process. Concentric layers of carbon were continuously eliminated during the processes of oxidation, the surface becoming more and more smooth and uniform, with a considerable decrease in the diameter of the carbon fiber.

Transmission electron microscopy (TEM) of type I and type II carbon fibers, after oxidation between 400 and 600°C, showed that the pits formed in lines which concentrate into channels [18]. The points attack were parallel to the fiber axis at places where the carbon fibers are aligned approximately parallel to the fiber axis in such a way that the edge planes rather than basal planes and the imperfections such as point defects in basal planes are exposed. The measurements of surface area by BET (N_2) showed a rapid increase with burn-off followed by a slow decrease. The rapid increase is due to the formation of pits, and the slow decrease, to the formation of channels by the coalescing of the pits. Interlaminar shear strength of the fiber composites increased on oxidation, which has been attributed partly to an increase in surface area and partly to the formation of surface groups.

Ahearn and Rand [26] have recently published the results on the oxidation of PAN-based carbon fibers in air at 420 and 730°C to low extents of burn-off (less than 3.5%). The authors discussed the fracture behavior of a brittle carbon-carbon composite by controlled oxidation. They observed by SEM that the fiber loss occurred in the areas close to internal pores and cracks in the case of 420°C treatment. The mechanical property changes of the low temperature oxidized composites are shown as a function of weight loss in Figure 1 [26]. The material strength declined to be almost equal to that of the unoxidized composite after a weight loss of 2.2%. On the other hand, at a higher oxidation temperature (730°C), the strength was found to decrease with increasing weight loss, although not as significantly as for the lower temperature treated materials. A 7% decrease in strength at 730°C compared to 26% for the lower temperature (420°C) when the sample oxidized at the same extent to 2.4% weight loss. A weight loss of 3.5% was observed and corresponding to a strength reduction of 26% in the higher temperature sample, the modulus also suffered a decrease with increasing amounts of oxidation by a maximum of 26% after 3.5% weight loss.

The use of oxygen for an efficient oxidative treatment of carbon fibers needs to be carried out at comparatively higher temperature (>400°C), where the probability of the spontaneous ignition of the carbon fiber is quite high. By the thermogravimetry technique, Molleyre and Bastick [14] studied the oxidation behavior of rayon and polyacrylonitrile (PAN) carbon fibers in oxygen at 550°C, as well as examined the external and internal structures of the treated and untreated carbon fibers, using adsorption and scanning electron microscopy (SEM) techniques. Surface area (surface rugosity) and pore diameter of carbon fibers were increased after the oxidative treatment. They found the increase in surface rugosity resulted in an increase of the fiber-resin composite shear strength. Ko and Li [23] reached a similar conclusion from a recent study. The properties of three kinds of carbon fibers, which were pre-carbonized at 500°C, 550°C, and 600°C, were measured after being oxidized in air for 1-6 minutes at 550°C. The pre-carbonization process strongly affected the surface properties (elemental compositions) and the mechanical properties of the final carbon fibers, as measured after air oxidation. After one to six minutes air oxidation carbon fibers showed a different oxidation behavior in the surface morphology for each pre-carbonization temperature. Optimum conditions not only enhanced the fiber tensile strength and modulus by over 50%, but also increased the oxygen content.

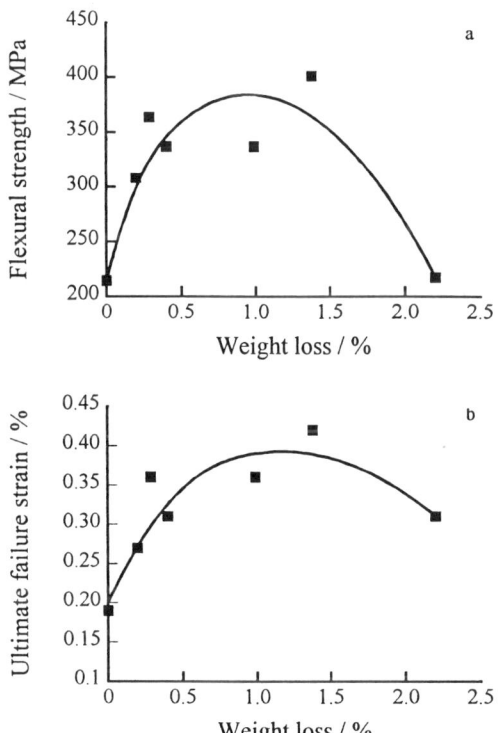

Figure 1 Graphs illustrating the changes in (a) flexural strength and (b) ultimate failure strain as a function of weight loss owing to oxidation [26]. (Reprinted with permission from publisher.)

High-strength, untreated and unsized Toray T300 carbon fibers were oxidized by Vukov [22] in an air atmosphere at 420°C for different period of time. The measured results of carbon fiber weight loss are presented in Figure 2. Weight loss increases with treatment time and reaches the value of 0.9% after 7 hour treatment. The concentration of surface acidic groups was measured by 1M NaOH volumetric titration as shown in Figure 3 curve a. An increase in the concentration of acidic surface groups with treatment time is clearly seen. However, when the treated fibers were heated up to 950°C in argon atmosphere for one hour, those functional groups decreased dramatically. The change of surface area with treatment time is presented in Figure 4. After seven hours of oxidation in air, surface area 0.5 m^2g^{-1} of untreated fibers increases to the value of about 29 m^2g^{-1}. On the other hand, the surface area measurements of post-treated fibers yielded almost identical values. The author claimed that such a result was expected because the post-treatment process should only cause the decomposition of surface oxides while surface area remains unchanged. The influence of oxidation treatment on the adhesion of carbon fiber/epoxy composites is shown in Figure 5a. After one hour oxidation treatment, the interlaminar shear strength by a short-bean test shows an increase from 53 MPa of untreated fibers to 72 MPa. Further oxidation after seven hours exhibits a slow increase in ILSS and reaches the value of 75 MPa. In contrast, composites made of post-treated fibers results in a little change of ILSS when oxidized in air. Vukov concluded that acid surface groups play an important role at the fiber/epoxy interface, acting in three ways: (1) forming chemical bonds with the matrix molecules, (2) enhancing the fiber wettability, and (3) the surface area increase is an important effect for the adhesion.

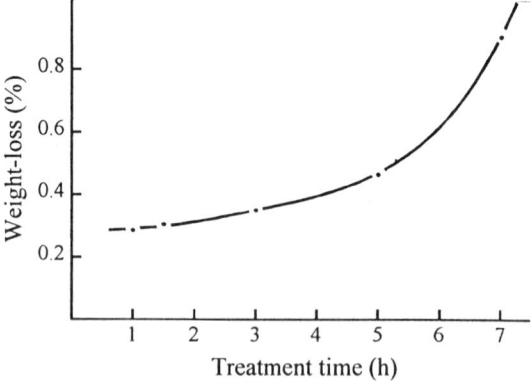

Figure 2 Carbon fiber weight loss as a function of the treatment time [22] (Reprinted with permission from *Journal of the Serbian Chemical Society*).

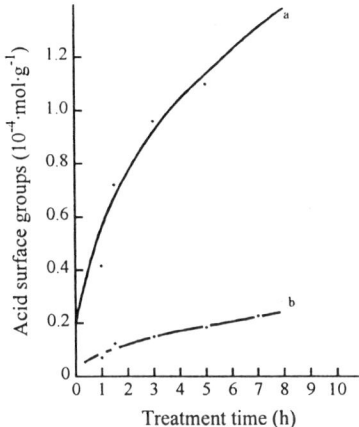

Figure 3 Concentration of acid surface groups vs. treatment time: a) treated fibers, b) post-treated fibers [22]. (Reprinted with permission from *Journal of the Serbian Chemical Society*.)

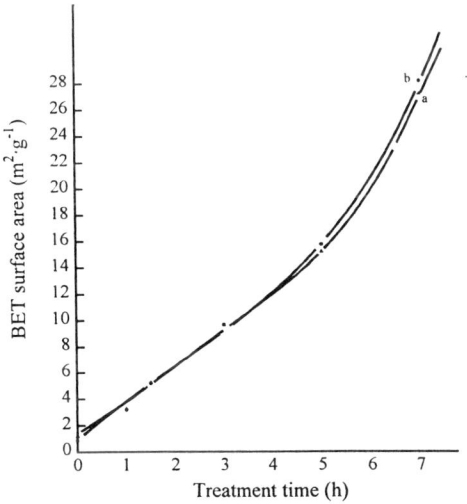

Figure 4 BET surface area of carbon fibers vs. treatment time: a) treated fibers, b) post-treated fibers [22]. (Reprinted with permission from *Journal of the Serbian Chemical Society*.)

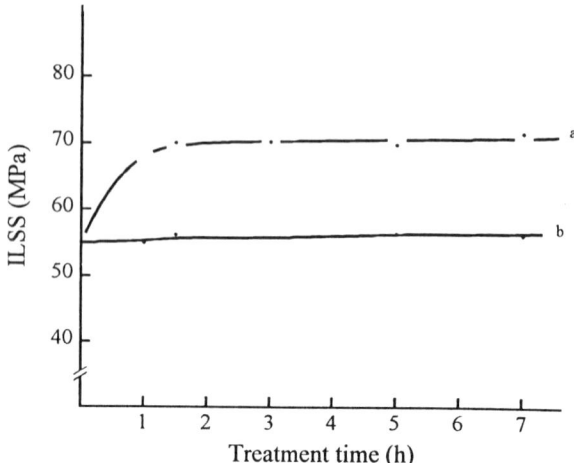

Figure 5 ILSS of carbon fiber/epoxy resin composites as a function of the carbon fiber treatment time: a) composites made of treated fibers, b) composites made of post-treated fibers [22]. (Reprinted with permission from *Journal of the Serbian Chemical Society*.)

3.2.2 Oxidation in Oxygen-Containing Gases

Several oxygen-containing gases, such as CO_2 and ozone, have also been used for oxidative treatment of carbon fibers. Scola and Basche [12], Druin et al. [19] oxidized carbon fibers in an inert atmosphere containing small amounts of oxygen and observed that the oxidation increased the specific surface area and the ILSS of the fiber-resin composite produced by using the modified fiber. Molleyre et al. [21] treated carbon fibers with CO_2 in the temperature range 850-925°C and observed changes in surface rugosity similar to that observed on treatment with oxygen diluted with nitrogen. The surface topography of the carbon fiber, however, was modified dramatically when the fibers were contaminated with metallic impurities. The presence of iron impurities produced holes which were randomly distributed on the carbon fiber surface.

Alcaniz-Monge et al. [24] examined the effect of the activating gas on the tensile strength and pore structure of pitch-based carbon fibers. The activation of carbon fibers (1 g of sample in each experiment) was achieved at 1160 K with both CO_2 (0.1 MPa, 80 ml/min) (CFC) and a steam/N_2 mixture (1/1 by volume, 0.1 MPa, 80 ml/min) (CFS) in a horizontal furnace. Figure 6 shows the experimental results of fiber diameter with the burn-off percentage tested for each series by Scanning Electron Microscopy. It is evident that the diameters of fibers decrease notably for the steam activated specimens (CFS), however, in the CO_2-activated samples, the diameter remains almost constant after a small initial decrease. The difference of the fiber tensile strength with the burn-off percentage during the activation was also presented in Figure 6. There is a significant decrease in the tensile strength of the CFC samples as the burn-off percentage increases. In contrast, the CFS series exhibits an initial decrease in tensile strength for low burn-off percentages (15%), whereas it remains constant at higher burn-off percentages. The steam/N_2 mixture treatment causes an external fiber burn-off, leading to an expansion of microporosity and to a fiber diameter reduction based on the evaluation of porous texture, fiber diameter, and tensile strength with the burn-off percentages. On the other hand, the authors explained that CO_2 activation creates microporosity without varying fiber diameter, which suggests that instead of on the external surface the reaction takes place within the fibers. The fact that tensile strength decreases with burn-off in the case of CO_2 suggests that CO_2 activates not only by generating a small size of porosity, but also porosity growing deeply towards the inside of the fibers. Therefore, they concluded that defects generated in this way within the material will have a greater influence on a decrease of the tensile strength.

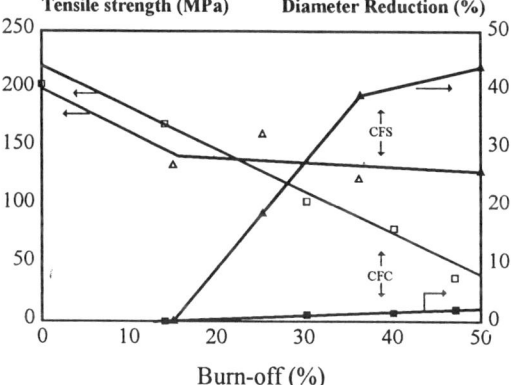

Figure 6 Variation of the activated carbon fiber diameter and tensile strength with burn-off [24]. (Reprinted with permission from Elsevier Science, Ltd.)

3.2.3 Catalytic Oxidation

Oxidation of graphite has been found to be greatly enhanced by the presence of small amounts of metallic impurities such as oxides of Cu, Pb, V, and other transition metals [15,16]. The presence of these metal oxides causes rapid pitting of the entire graphite surface at considerably lower temperatures. Thus, McKee and Mimeault [17] treated carbon fibers with solutions of copper acetate and lead acetate in order to incorporate metallic impurities into the fiber surface. The catalytic-treated carbon fibers were oxidized in air and composites were prepared from the oxidized carbon fiber. The composites showed considerable improvement in the shear properties, which suggests that a mild catalytic oxidation may be more suitable than the more drastic gaseous oxidation.

Hoffman [20] developed a process for treating carbon fiber surfaces which increases the surface area with a total weight loss <2%. The carbon fiber surface was deposited with a single coating of materials which is capable of catalyzing carbon gasification and results in either pitting or channeling upon heating in air. The coating materials selected by the author are from the oxides or metals of the group consisting of Pt, Ni, Ir, Re, V, Pb, W, Pd, Co, Fe, Mo, Cu, Cd, Cr, Mn, Ru, Ag, Au, and mixtures. The fiber surface is heated when air is applied to a certain temperature at which the coating promotes localized oxidation to cause pitting or channeling to increase the surface area, and then removing all of the coating remaining after heating and cooling the carbon fiber. The treatment improves fiber to matrix bonding.

3.3 LIQUID-PHASE OXIDATIVE TREATMENTS

Solution-phase oxidation of carbon fibers has been found to be quite effective in improving the shear properties of carbon fiber-resin composites. Several types of oxidizing agents, such as nitric acid, acidic potassium permanganate, acidic potassium dichromate, hydrogen peroxide, and potassium persulfate have been used with varying degrees of success. These liquid phase oxidative treatments are milder compared to the gaseous phase oxidative treatments and generally do not cause excessive pitting and degradation of the carbon fiber.

3.3.1 Nitric Acid Solution Oxidation

Herrick et al. [11] studied the effect of concentration of nitric acid and the treatment time under reflex on the oxidation of WyB and Thornel type I carbon fibers, and found that both the surface functionality and the surface area increased on oxidation. More recently, Pittman and his co-workers [34] oxidized Thornel T300 PAN-based carbon fibers in 70% nitric acid solution at 115°C for periods of time from 20 to 90 minutes. They determined the acidic functionality of oxidized fibers by neutralization with NaOH and the acidic capacity by dye (methylene blue) adsorption. The acidic capacities of these treated fibers increased almost linearly with the oxidation time as shown in Table 2. More than 85% of phenol and 99.9% of carboxylic acid could be neutralized by using a 5×10^{-4} M sodium hydroxide solution. The basic solutions chosen efficiently neutralize one gram of carbon fibers with acidic capacities up to 100 μeq/g. In addition, a methylene blue (cations) solution of *pH* 10.5 would adsorb those surface sites with negative charges, where more than 99.9% of the carboxylic acid and 85% of phenolic hydroxyl would be ionized. The tensile strength testing data of PAN-based low modulus (THORNEL T300) fibers, before and after nitric acid oxidation treatment, are listed in Table 3. As shown, an increase in oxidation time resulted in the tensile strength decrease. The authors suggested that the less crystallized regions were more reactive toward nitric acid and the surface of untreated fibers were pitted and fragmented by the acidic solution. Such defects may be the reason causing the decrease in the tensile strength.

Table 2 The Acidic Capacities of Nitric Acid-Oxidized Carbon Fibers[a] Determined by Both NaOH Uptake and Methylene Blue Adsorption

	Oxidation time (min.)				
	0	20	40	60	90
Acidic capacity (μeq/g)					
via NaOH uptake	2.0	10.5	30.7	55.8	89.2
via methylene blue adsorption	1.4	3.60	8.06	16.6	26.4

[a]Ex-PAN T-300 high-strength fibers were oxidized in 70% nitric acid at 115°C.
Source: from ref. 34. (Reprinted with permission from Elsevier Science, Ltd.)

Table 3 The Effect of HNO_3 Oxidation on the Tensile Strength of THORNEL T-300 Carbon Fibers[a]

	Oxidation time (min.)				
	0	20	40	60	90
Acidic capacity (µEq/g)	2.0	10.5	30.7	55.8	89.2
Tensile strength (MPa)	3077	3039	2781	2773	2616
Strength loss (%)	0	1.2	9.6	9.9	15.0

[a]Tensile strength values are the average of more than 10 measurements for each oxidation time. The relative standard deviation was about 10%.
Source: from ref. 34. (Reprinted with permission from Elsevier Science, Ltd.)

The influence of HNO_3 treated-carbon fiber on the mechanical properties of composites was evaluated by Ma and his co-workers [36]. A lower tensile strength with broad distribution was observed when the carbon fiber was treated by HNO_3. They observed that the tensile strength is strongly related with the flow patterns on the fiber surface. The surface morphology changes considerably after acid treatment, hence, the distribution of fiber strength becomes broader. C/C composites made with treated fibers exhibit lower flexural strength than those made with untreated fibers due to the formation of cracks during carbonization process. Due to the formation of perfect graphite lattice in the matrix after a higher temperature graphitization, the C/C composites made with treated fibers show higher textural strength than those made with untreated fibers. Bansal and Chabra [27] oxidized PAN-based carbon fibers with aqueous solutions of mild oxidizing agents such as hydrogen peroxide and potassium persulfate by contacting the carbon fiber with their aqueous solutions in closed bottles for periods up to 24 hours. Both treatments resulted in the chemisorption of appreciable amounts of oxygen. Similar results were obtained by Druin and Dix [28] who oxidized carbon fibers in an aqueous hypochlorite solution at moderate temperatures, and observed that the surface characteristics of the treated carbon fibers were considerably modified. During the aqueous oxidation, carbon fibers picked up considerable amounts of oxygen, which resulted in an improvement of the adhesion of the fiber/resin matrix composites [28].

Surface treatment of type I (high-modulus HM) as well as type II (high-strength HT) carbon fibers was carried out by Bahl et al. [29] with dilute (34%) and concentrated (68%) HNO_3 solutions for different durations. Oxygen-containing functional groups on the type II fiber surface are produced after a concentrated HNO_3 treatment. However, this treatment is so severe that it

Surface Treatment of Carbon Fibers

damages the fibers and results in composites with poor mechanical properties. On the other hand, oxidation in diluted nitric acid for short period of time increases the mechanical properties of fibers and improves the bonding between fibers and matrix molecules. They suggested the poor mechanical properties of carbon fibers when treated with concentrated HNO_3 for >1 h may be caused by the incorporation of nitrate ions into the graphitic planes of the fibers and thus decreases the mechanical properties of composites. Surface treatment of the fibers improves the mechanical properties of the composites made with furan resin to a greater extent than those of the composites made with phenolic resin. In Desimoni et al. [30] work, high modulus carbon fibers were chemically treated with potassium permanganate in concentrated sulphuric acid. The pre- and post-oxidation surface status of the fibers was characterized by x-ray photoelectron spectroscopy, x-ray-excited Auger spectroscopy and scanning electron microscopy.

Donnet et al. [31] treated type I and type II acrylic carbon fibers in boiling 68% nitric acid and observed a significant increase in the oxygenated group content in both cases. The weight loss after 24 hours of treatment was 32% in type II, and 0.1% in type I carbon fibers. The specific surface of the type I carbon fiber, however, increased markedly. Type I carbon fibers were also treated with Hummer's reagent (which is a mixture of $NaNO_3$, H_2SO_4, and $KMnO_4$). This treatment caused appreciable chemisorption of oxygen with formation of functional groups such as -COOH and -OH, identified by infrared (IR) spectroscopy. The oxygen content of the fiber surface increased with increase in time and temperature of treatment. The surface area, however, remained almost unchanged. The high oxygen pickup is due primarily to the formation of graphitic oxide having been shown by IR and ESCA measurements.

Fitzer et al. [133] confirmed Donnet's observation that oxidation with nitric acid results in the formation of considerable amounts of the surface oxygen structures. Both high-modulus (HM) and high-strength (HT) carbon fibers shown in Figure 7 indicate the oxygen content is proportional to the increase of oxidation time. The BET surface area increased initially but decreased with time of oxidation most likely due to the degradation of carbon fiber caused by excessive oxidation. The translation of fiber properties into the composite also increased with time of oxidation, but more significantly for HM carbon fibers (Figure 8). In the ILSS properties, While the surface-treated HM carbon fibers increased with increase in surface area and surface functionality, those of HT carbon fibers illustrated an increase and then a decrease (Figure 9). This decrease in ILSS for HT carbon fibers was attributed to the degradation of the HT carbon fiber caused by excessive oxidation.

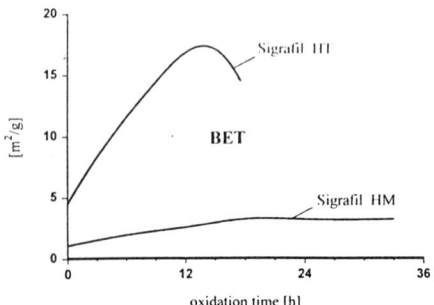

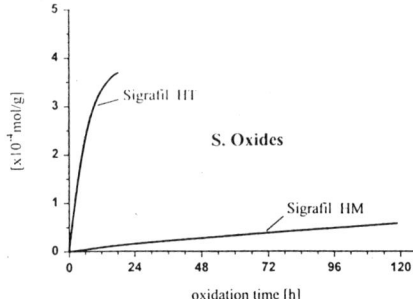

Figure 7 Effect of surface oxidation in HNO_3 on surface area and surface oxide buildup on HT and HM carbon fibers [133].

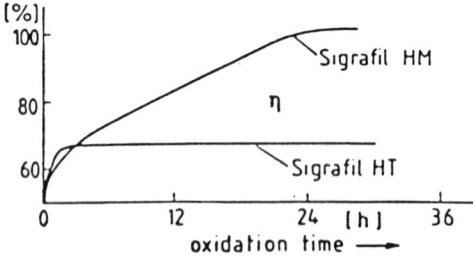

Figure 8 Effect of oxidation in HNO_3 on translation of carbon fiber properties into composite properties [133].

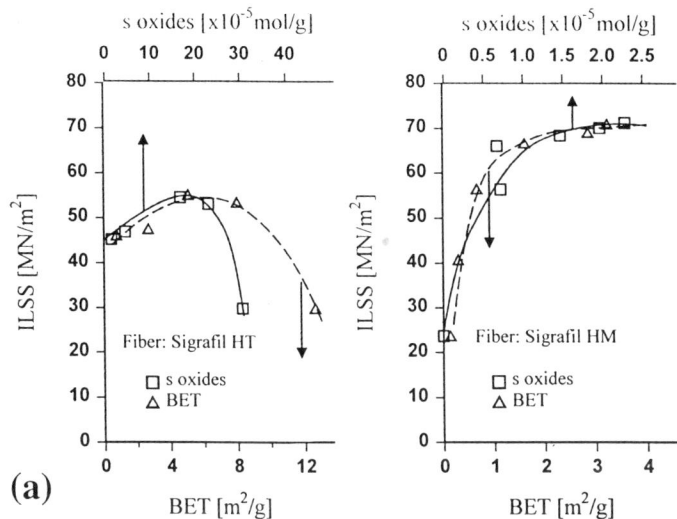

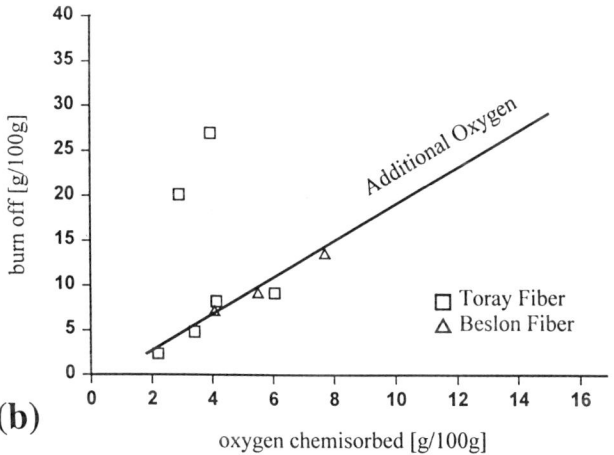

Figure 9 (a) Relationship between ILSS and surface area and surface oxides on HT and HM oxidized carbon fibers. (b) Relationship between burn-off and the oxygen chemisorbed on oxidation of carbon fibers [133].

3.3.2 Other Oxidizing Solutions

Several other oxidizing solutions with varying oxidizing power have been used for the oxidation of carbon fibers, but the details of many of these treatments are part of the patent literature. Well and Colclough [32], and Druin and Dix [28] oxidized carbon fibers by contacting with aqueous solutions of hypochlorides at moderate temperatures, and observed that the surface characteristics of the treated carbon fibers were considerably modified. The carbon fibers picked up appreciable amounts of oxygen, which resulted in enhancing the carbon-resin interfacial bond [32] and improving the adhesion of the fiber with the resin matrix [27,28], producing composites with better shear strengths. Compared with nitric acid Donnet et al. [33] treated type I PAN–based carbon fibers with Hummer's reagent and observed the formation of acid groups on the fiber surface even after a very short treatment time, but no significant change in the surface area as shown in Table 4. On the other hand, Nakanishi and Fujita [35] discovered that continuous anodic oxidation of carbon fiber was more efficient in an alkali bath than in acidic (H_2SO_4) bath. A small charge of $\sim 10^{-4}$ C/mm^2 of aqueous NaOH gives the best results for anodic oxidation of the fibers. The matrix was selected from 2 epoxy resins or polybismalemide resin, and the fabricated fiber composites showed an increased strength with higher active (functional groups involved) surface area. The authors suggested that the interfacial bonding strength between carbon fibers and the resin matrix was influenced by the surface microroughness, as well as the contributions of –COOH, –OH, and other functional groups.

Table 4 Acidic Groups (COOH + OH) Determined on Type 1 Fibers Treated with Hummer's Reagent

Time (min)	Temperature (°C)	Acidic groups (µEq/g)	Specific surface area (m²/g)
Nontreated	–	3	0.38
3	35	200	–
3	40	270	–
10	40	400	0.23
40	40	835	0.37

Source: data from ref. 33.

3.4 ELECTROCHEMICAL OXIDATION METHOD

The possibility of oxidation of carbon fibers in electrolytic baths of acid and alkaline aqueous solutions has been indicated [37] and industrially used. The electrochemical oxidative treatment offers more control over surface chemistry and has become the preferred industrial process. This treatment is preferred because it can allow continuous processing of the carbon fiber.

The pioneer work in the field was carried out by Donnet et al. [38,39], who oxidized type I (2500°C) and type II (1100°C) acrylic carbon fibers anodically in solution of dilute nitric acid (10 wt%) containing 2.5% potassium dichromate and sodium hydroxide (1N), using a graphite plate as the cathode. The current density about 1 mA/cm^2 and a work potential of 2V was employed. Oxidation in nitric acid solution increased it considerably, although oxidation in alkaline medium did not appreciably change the oxygen content of the carbon fibers. The treatment of type I carbon fibers generated graphitic oxides, while no graphitic oxide was detected in the case of the type II fibers. However, the type II fibers underwent appreciable weight loss due to the evolution of CO_2. In addition, small amounts of polycarboxylic acids, which are degradation products, were also formed. The topography of the treated fibers by SEM showed a smoothing of the surface by anodic etching for acidic and alkaline treatments. No appreciable change in surface area was noticed as a result of the treatment. Composite shear strength did not show any appreciable improvement when nonpurified carbon fibers were used. However, the shear properties showed significant increase when the composites were prepared by using the purified (extracted) fiber (Table 5). This has been attributed to the presence of degradation products in the nonpurified fibers, which tend to decrease cohesion of the fiber-resin interface. Fitzer [40] has shown that removal of these degradation products by washing the surface with alkaline aqueous solution markedly improves the performance of the treated fiber. The efficiency of the washing process has been demonstrated by Donnet and his co-workers [41,42].

The electrochemical treatment of the carbon fiber was extensively studied by Sherwood et al. [43]. For example, they studied Type I (high-modulus HM) and Type II (high-strength HT) carbon fibers which were electrochemically treated in a variety of electrolytes. By XPS, they detected that the amount of carbon/oxygen functionality is greater for type II for both the untreated and electrochemically treated fibers. There are more edge sites present on type II fibers than type I whose edge sites contain carbons with unsaturated bonds which can be reacted with a variety of oxidizing agents. The edge site functionalities can be further oxidized in the electrochemical oxidation processes, most possibly to carboxyl functionalities and/or CO_2. Sherwood and his co-workers [44–49] carried out electrochemical oxidations of PAN and pitch based carbon fibers using NaOH, HNO_3, H_2SO_4, H_3PO_4, and a number of ammonium salts as electrolytes. For example, at different anodic potentials high-modulus pitch-based carbon fibers were anodically oxidized in 1M HNO_3 solution, leading to a change both the surface and bulk chemistry of the fiber. When fibers were oxidized at low anodic potentials (lower than 2V), mainly hydroxyl-type and some carbonyl-type functional groups were created on the surface. However, the oxidation degree was largely accelerated and the fibers are considerably oxidized when the potential was above 2.5V [47]. The oxidation of fiber surfaces was quite remarkable using solutions containing ammonium salts of strong acids, producing similar surface functionalities as those from using the hard acid alone [44]. They also found out that galvanostatic treatments gave more reproducible changes to the surface and bulk chemistry of the carbon fibers than did potentiostatic treatments. When the fibers were treated for a short period of time (less than one minute), the difference was more evident [48]. Both galvanostatic and potentiostatic treatments led to treated fibers which produced composites with high interlaminar shear strengths (80 to 90 MPa). The increase in ILSS with surface treatment is not dependent upon the O-1s/C-1s ratios or the amount of carbonyl/carboxyl functionalities present on the surface. Therefore, they suggested that mechanical keying of the resin to the fiber surface might play an important role in the adhesion increase of composites [46].

The effect of an electrochemical surface treatment on the surface microstructure of high tensile strength carbon fibers was investigated by Mahy et al. [50]. They observed that the surface area did not increase significantly followed by the treatment intensity increase. Both basic and acidic surface functionalities were formed, and their relative ratios changed with the treatment intensity. Nakao et al. [51] electrochemically oxidized carbon fibers in different basic/acidic solutions and studied the effect of treatment on the interfacial adhesive strength, i.e. interlaminar shear strength (ILSS) and transverse flexural strength, of epoxy composites. The nature of the electrolyte used had a great influence on the surface properties of carbon fibers. The authors compared the effect of basic and acidic solutions on the mechanical properties and concluded (1) a basic electrolyte resulted in an improvement of composite transverse flexural strength and ILSS due to an increase in the number of surface functional groups; (2) an acid electrolyte resulted in the ILSS increasing while the transverse flexural strength decreased for composites because the carbon fibers had a high surface content.

Bader and Baillie [52] comprehensively investigated the interface adhesion of electrolytically oxidized carbon fibers in an epoxy-resin matrix. An ammonium-bicarbonate electrolyte was electrolytically applied to a high-strength PAN-based carbon fiber. The current density varied from 1 to 100 Cm^{-2} (total charge density with a constant residence time of the fibers in the electrolyte). The process was found to be quite effective in increasing the

Surface Treatment of Carbon Fibers

interfacial strength of composites as evidenced by the measurements of a single-fiber pull-out or an embedded-fiber fragmentation test. Figure 10 illustrates the relation between the shear strengths and the acidity levels and as seen the higher acidity levels have a clear tendency towards higher strengths. The acidity groups on the fibers are mainly the –COOH group which has a good correlation with the measured interface, was found to increase with the electrochemical treatment. Jones [53], however, indicated that there was no correlation between the fiber/resin bond strength and the amount of chemical functionality on the fiber surface, i.e., chemical bonding did not play a major role in fiber-resin adhesion. He believed that chemical bonding between carbon fiber and resin was not a key factor but was possible for the adhesion mechanism. It was suggested that the surface concentration of chemical groups is too low to make a significant contribution.

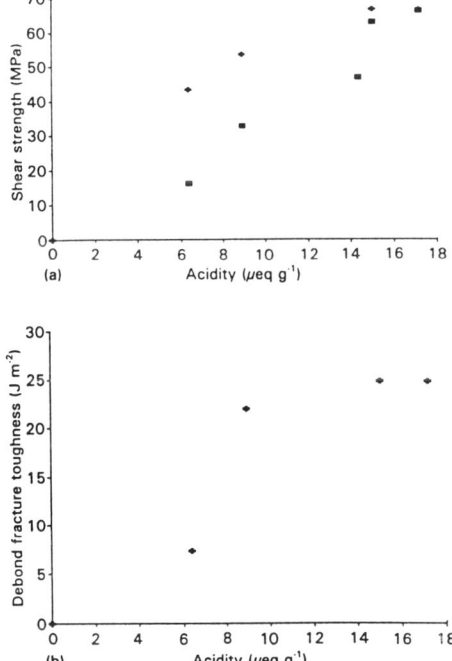

Figure 10 (a) The interfacial shear strength, τ_1, from both (✦) pull-out and (■) fragmentation tests, and (b) the debond fracture toughness, G_{ic}, from pull-out tests as a function of the surface acidity of carbon fibers [52] (Reprinted with permission from Chapman & Hall).

Table 5 Influence of Degradation Products on the Shear Strength of Composite

	Strength (MPa)	
Fibers[a]	Nonpurified fiber	Purified fiber
AC nontreated	—	55
AC HNO$_3$ 1 min	53	55
AC HNO$_3$ 5 min	49	70
AC HNO$_3$ 10 min	48	68

[a]AC means acrylic carbon fiber heat treated to 1100°C.
Source: from refs. 38 and 39.

3.5 PLASMA TREATMENT

In the carbon fiber composites, good adhesion between the carbon fiber surface and the matrix materials happens when the surface energy of fibers is higher or equal to the surface energy of the matrix. Untreated carbon fiber surface usually has a low surface energy and consequently is difficult to form a strong adhesive bond with the matrix. Thus, surface treatments are applied to carbon fibers to increase their surface activity and surface energy. Among those many possible means, low pressure plasma processing offers an attractive and efficient route to modify fiber surface chemistry without affecting bulk properties. In this method, an operating pressure range from about 0.1~10 Torr, which can be achieved easily by conventional mechanical vacuum pumps. Free electrons in the plasma, accelerated by the applied field, have kinetic energies of several electron volts, equivalent to thousands of degrees Kelvin. The bulk of the gas remains near ambient temperature (below 100°C) because the degree of ionization is quite low (about 1 part in 10^6). A variety of active species such as ions, radicals, and various electronically excited molecules, as well as UV and vacuum UV radiation from radiative relaxations are present in the plasma gas. This possess an unique environment for performing high energy chemistry on the fiber surfaces at near-ambient processing temperatures. This feature is particularly important for fibers which, because of their small cross-sectional area, are very susceptible to degradation by conventional processing in aggressive environment. Under normal processing conditions in a plasma, extensive modifications may occur on the outmost few atomic layers of a substrate, leaving bulk properties intact.

3.5.1 Oxygen-Containing Compound Plasmas

Donnet and co-workers [54–57] applied plasma treatments onto different types of carbon fibers to enhance the polar component of the surface energy, which could improve the interfacial adhesive interactions between the fiber and the matrix. Similar results were obtained recently by Blackman et al. [65] who examined the effect of oxygen plasma treatment on the thermoplastic PEEK and PPS fiber composites. The thermoplastic composite joints fail at the adhesive/composite interface at a very low applied load for the samples without plasma modification. However, the authors observed that the oxygen-plasma and corona treatments led to an increase in the surface concentration of polar (oxygen-containing) groups, leading to both higher epoxy adhesive wettability on the substrate and stronger intrinsic adhesion across the adhesive/composite interface. The improvement in intrinsic adhesion is reflected in the center of failure of the joints moving away from the adhesive/composite interface, with far higher values than the adhesive fracture energy (Gc).

Two types of carbon fibers, high-modulus fiber-M40 and high-strength fiber-T300, were surface treated by using a microwave plasma under reduced pressure in an atmosphere of air and argon. The carbon fibers, fixed on a glass holder, were placed in a glass tube that was introduced into the middle of a microwave cavity as shown in Figure 11. A reduced pressure of air or argon in the range of 0.5–1 torr was kept constant during the treatment by the combination of continuous evacuation and gas inlet through a microleak. A quadrupole mass spectrometer was used to control the gas composition. The input power to the microwave generator was maintained at 50 W while the reflected power varied between 2 and 7 W. The duration of treatment was varied from a few second to several minutes in air and argon. The surface energy of the carbon fibers before and after plasma treatment was determined from the wetting contact angle within two liquids.

The microwave plasma treatment results in an increase in the polar component γ_s^p and a decrease in the dispersive component γ_s^d of the surface energy shown in Figure 12. The polar component increases significantly from 5–8 mJ/m^2 for the untreated M40 fiber to 25 mJ/m^2 after a 10-second plasma treatment and indicates the generation of polar oxygen-containing groups on the carbon fiber surface. The presence of atomic oxygen in the plasma leads to these surface groups be incorporated at this low temperature (room temperature). Surprisingly, more oxygen groups were presented on the carbon fibers when plasma treated in argon was used. This could be caused by the fact that the bombardment with argon induces the formation of radicals on the fiber surface which are very reactive with air during exposure after treatment to form the surface oxygen groups. The treatment results in the creation of some pores which are oriented transversely to the fiber axis, suggesting that these carbon fibers may have a peculiar carbon structure. A comparison of the tensile breaking load of a single filament of T300 untreated and modified by 5 minutes of plasma treatment in Table 6 indicates that plasma treatment affects the mechanical properties less than do the other surface treatments.

Table 6 Mechanical Properties of T300 Fibers

	Original untreated, unsized	Just treated	Fully treated	Treated
Breaking load (mN)	98.1	89.3	70.6	90.2
%	100	91	72	92

Source: from ref. 54.

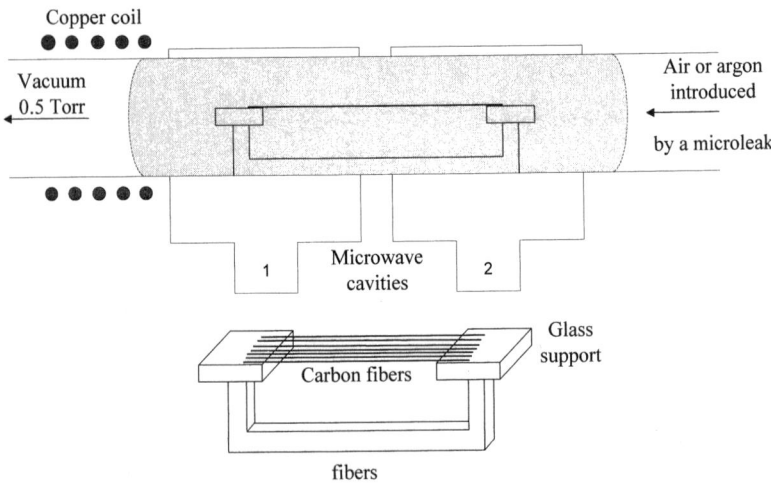

Figure 11 Schematic of plasma treatment equipment [54]. (Reprinted with permission from Pergamon Press.)

Surface Treatment of Carbon Fibers

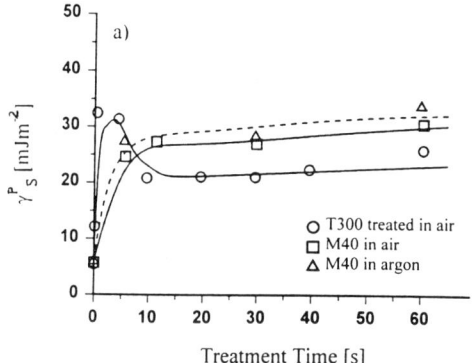

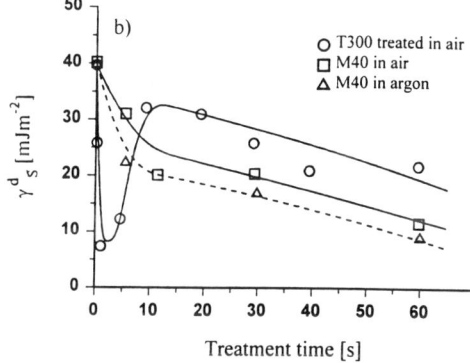

Figure 12 Variation of γ_s^p and γ_s^d with microwave plasma treatment for M40 and T300 carbon fibers. (o) T300 treated in air, (□) M40 in air, (Δ) M40 in argon [55].

Examination of the carbon fiber fracture surfaces by SEM before and after microwave plasma treatment shows that the spinning grooves present in the untreated M40 fiber are almost completely removed by a 12-second microwave plasma treatment in air (Figure 13a), which indicates the removal of disorganized material from the outer layer. This outer-layer material removal occurs preferentially and often leads to the formation of longitudinal grooves (Figure 13b). Microwave plasma treatment in argon caused deposits of low-density material around the fiber (Figure 13c) and its thickness along the fiber in the same direction as the gas flow suggests that it was due to the redeposition of the carbon sputtered by argon ions from other parts of the fiber. High-magnification SEM micrographs of M40 and T300 carbon fibers after microwave plasma treatment under argon indicate roughening of the surface, but no deep crevices are observed. The erosion of the carbon results in the formation of flat or concave longitudinal structures. Plasma treatment in argon does not appreciably change the topography of the fiber surface, and thus the mechanical properties are not expected to change appreciably. In the case of plasma treatment in air, however, the diameter of the fiber is decreased, and some crevices are created by the oxidative processes, and consequently there is a fall in the strength of the carbon fibers. But when plasma treatment in air is carried out for an interval of time (a few seconds), the mechanical properties remain more or less unchanged. It has been observed [57] that the fall in mechanical properties of a carbon fiber on microwave plasma surface treatment in air or argon appears to be a function of the nature of the carbon fiber structure because pitch-based KCF-100 carbon fibers showed only a slight fall in strength on microwave plasma treatment, even in air.

Jones [59] gave a low power nitrogen plasma treatment on Type II (high-strength) carbon fibers, resulting in a significant improvement of fiber/resin adhesion. He proposed that this improvement was due to two reasons 1) chemical interaction via amine/epoxy bonding at the edge sides of fiber surface; 2) together with the interaction of activated basal planes present on the fiber surface with the epoxy. This improvement is only achieved if the fibers are immersed in resin before being exposed to air. Otherwise, due to the adsorption of moisture from the environment, if exposing the plasma-treated fibers to air dramatically reduces the fiber/matrix adhesion. Heating these air-exposed fibers in a vacuum at 130°C for one hour finds some recovery of the interfacial strength. It was also demonstrated that the interfacial shear strength falls substantially when the nitrogen-containing functional groups were removed completely from the fiber surface. Farrow and Jones [63] also examined the effect of low-power air plasma treatment on the mechanical properties of carbon fiber/epoxy composites. The wide scan XPS spectrum for XAU carbon fibers after air plasma treatment is illustrated in Figure 14a. It is evident that both oxygen and nitrogen containing groups are incorporated onto the fiber surface. Then, they heated the plasma treated fibers to 1000°C under a high vacuum by removing the chemical functionalities to examine the temperature effect. The XPS spectrum in Figure 14b confirms that most of the functional groups had been desorbed. The single-fiber fragmentation testing (IFSS) results of the plasma treated fibers for different periods of time are shown in Table 7. The improvement of fiber-matrix adhesion was obtained considerably even only after 10 seconds of plasma treatment. However, plasma exposed to a longer time did not result in higher values of IFSS but remained within the range of 44-55 MPa. In order to improve the adhesion capability, the inherently slippery carbon fiber was treated using ultra high vacuum plasma treatment. This surface treatment effectively improves the interfacial bond strength, evidenced by the Short Beam Shear Strength (SBSS) of fiber-reinforced polymer matrix composites. The testing results [69] of composite unibars at 60% fiber volume of untreated and plasma treated fibers are shown in Table 8. Ranging from a 16% interfacial strength increase for the E35 fiber to a 110% increase for E125 fiber after the plasma treatment was observed. Similar results (Table 9) were obtained from other types of commercial fibers [70], i.e., Hercules–AU4 PAN-based fibers were treated in various O_2 plasma environments. The plasma treated fibers show remarkable adhesion improvement, even better than the commercial surface-treated AS4 fiber. These plasma treatments seem to be reproducible without any loss in fiber tensile strength and good chemical stability.

Table 7 Results for Interfacial Shear Strength (IFSS) from the Single-Fiber Fragmentation Test for Air Plasma-Treated Fibers

Treated time (s)	IFSS (MPa)
0	21
10	51
30	49
60	46
300	55
600	44

Source: from ref. 63. (Reprinted with permission from John Wiley & Sons.)

Table 8 The Composite Interfacial Shear Strength of Carbon Fibers with Various Modulus by Surface Plasma Treatment

Fiber type	Interfacial shear strength (MPa)
Surface untreated	
DuPont E-35 fiber	79
DuPont E-75 fiber	59
DuPont E-125 fiber	33
Surface plasma treated	
DuPont E-35 fiber	92
DuPont E-75 fiber	78
DuPont E-125 fiber	69

Source: from ref. 69.

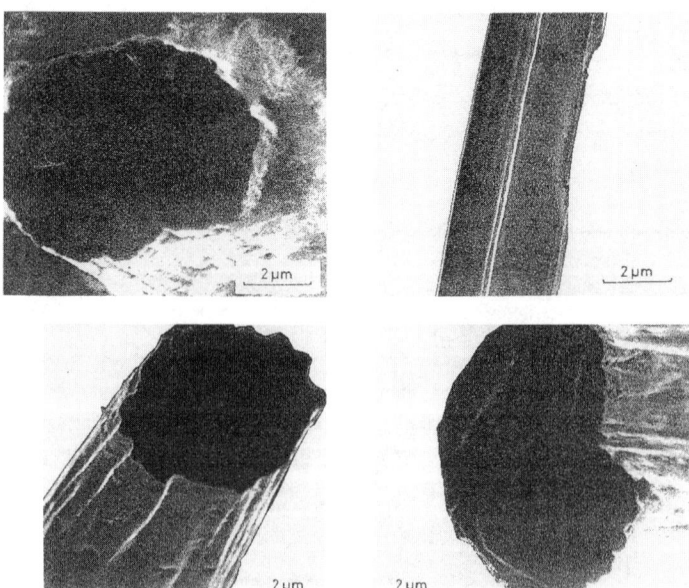

Figure 13 SEM micrographs of untreated and microwave plasma-treated M40 and T300 carbon fibers [55]. (Reprinted with permission from Pergamon Press.)

Surface Treatment of Carbon Fibers 187

Table 9 The Effect of Surface Treatment on the Interfacial Shear Strength of Hercules PAN-Based Carbon Fibers

Hercules PAN-based Carbon Fiber	Interfacial Shear Strength Rating (%)
AU4 untreated fiber (Control sample- index 100%)	100
AU4 plasma treated fiber (Ar/O_2, O_2)	243
AS4 commercially treated fiber	166

Source: from ref. 70.

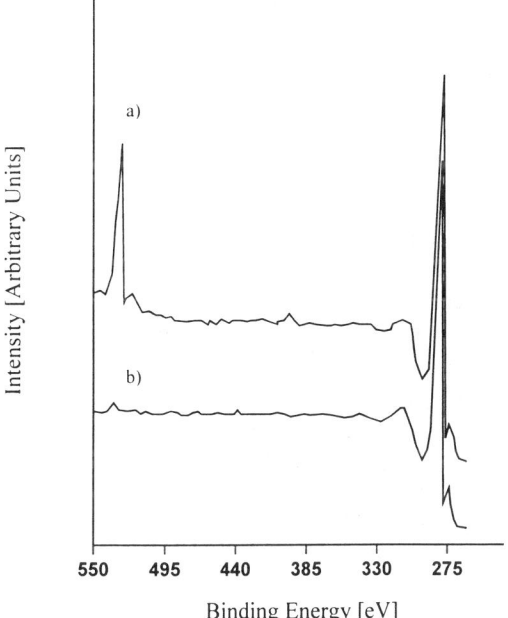

Figure 14 Widescan XPS spectra of: (a) XAU fibers after air plasma treatment, (b) XAU fibers treated with air plasma and heated at 1000°C in a vacuum for 1 hour [63]. (Reprinted with permission from John Wiley & Sons.)

3.5.2 Organic Compounds Plasmas

Besides the oxygen-containing compounds, organic gas plasma, such as methane, ethylene, trifluoromethane and tetrafluoromethane plasmas [68] were also used to treat fiber surface. XPS showed that a layer of fluorocarbon was deposited on the surface of the fibers from the trifluoromethane plasma, as well as a thin layer of hydrocarbon from the methane and ethylene plasmas. The tetrafluoromethane plasma etched the fibers and introduced a significant amount of fluorine on the surface. The adhesion between untreated and plasma-treated carbon fibers (IM7) and polyethersulfone (PES) and an epoxy resin (Epon 828 with Jeffamine Du-700 curing agent) was measured by the microbond pull-out test. The results indicated that an etching plasma, such as the tetrafluoromethane plasma, improved the adhesion between carbon fibers and PES. The authors believed that the adhesion was primarily due to the differential thermal shrinkage between the fiber and the matrix. The fiber chemical composition plays a role in the fiber-matrix adhesion in the case of a reactive matrix such as an epoxy resin. However, the cleaning effect of the surface plasma treatment is more superior to this chemical effect.

The plasma polymerization of allylamine onto PAN-based carbon fibers were investigated by Smiley and Delgass [58] using XPS and SEM. Poly(allylamine) compounds were grafted onto fibers accompanied with a monomer-saturated stream of Ar gas into the plasma chamber and the reaction lasted for 2–30 minutes. The deposited N on the surface was revealed by XPS in the form of amine and not amide species. In addition, from the SEM pictures it was clearly seen that the polyamine coatings were filled in the fiber grooves and left clumps of material on the surface.

A radio-frequency (RF) plasma processing technique, known as plasma-induced polymerization (PIP), was employed by Nay et al. [60] in a multistep plasma process to deposit thin, uniform layer of viscoelastic material on a high-temperature polyimide film in order to enhance adhesion between the film and the resin in carbon fiber/epoxy composites. In a three-step study, plasma pressure, the processing parameters of residence time, and plasma power were varied to investigate adhesion improvement. To examine the maximum condition of RF process for a thermoplastic (polyimide) interleaf composite, compressive mode II strain energy release rate (GIIc) was characterized by end-notched flexure testing. Adhesion was improved by as much as 140% over the adhesive performance of RF plasma untreated composites. However, this adhesive performance was tested at a relatively low temperature (177°C) processing system, and therefore this plasma deposition may not be stable at the higher processing temperatures needed for fabricating polyimide matrix systems.

Ebert and Weisweiler [61,62] also applied the RF sputtering technique to control deposition of polymeric films to create a reactive fiber surface. The authors described plasma polymerization of benzene, aniline, or pyridine that form polymeric films on different carbon fibers. Plasma polymerization of a given type of aromatic monomer was performed with a radio frequency sputtering apparatus using argon as carrier gas and additional gases like sulfur dioxide, water vapor, ammonia, or air as reactive component for special experiments. Several surface sensitive techniques such as XPS, IR-spectroscopy, mass spectroscopy, elemental analysis, ESR, as well as contact angle measurement were employed to characterize and verify the plasma polymeric films. Composites of different polymeric films-treated carbon were manufactured with to examine the fiber/matrix adhesion. The interlaminar shear strength is increased by about 35% for the coated fibers in the composites. Similar results were obtained by Waldman et al. [66], who studied the plasma polymerization of ethylene and ammonia gas mixtures to form polymer coatings on the surface of AS4 carbon fibers. Plasma-deposition rate of 100% ethylene was the smallest and increased by three- to four-fold when ammonia was added to the monomer mixture. The polymer films were coated with an uniform thickness and exhibited a complex crosslinked structure. The 100% ethylene plasma polymer was mainly composed of hydrocarbon but with some additional oxygen- and nitrogen- containing groups. On the other hand, plasma polymers deposited from an ethylene/ammonia mixture contained more polar groups, such as hydroxyl, carbonyl groups, primary, and secondary amines. The plasma polymer deposition process has a little effect on the carbon fiber tensile strength but affect the adhesion of fiber/matrix composite. The single fiber composite test results [67] indicated that the interfacial shear strength (IFSS) between ethylene/ammonia gas mixture plasma treated carbon fibers and PEEK resin increased by about 84%. However, in the case of 100% ethylene plasma deposition, which was strongly hydrocarbon in nature, had almost no effect on the IFSS. Dilsiz et al. [68] reached a different conclusion that the plasma treatment of the PAN-based carbon fibers with allylcyanide and xylene/air/argon gas mixtures resulted in an increase in the fiber strength and elongation capacity.

3.6 NONOXIDATIVE SURFACE TREATMENTS

Besides those oxidative methods previously described, several non-oxidative carbon fiber surface treatments have been developed and used with good results for improving the mechanical properties of the composites. These treatments include whiskerization, grafting of different kinds of polymers and vapor-phase deposition of pyrolytic carbon.

3.6.1 Whiskerization

Whiskerization involves a nucleation process and the growth of very thin and high-strength single crystals of other chemical compounds, such as silicon carbide (SiC), titanium dioxide (TiO_2), and silicon nitride (Si_3N_4) on the fiber surface perpendicular to the fiber axis [71]. The whiskers grow from individual fibers and their growth on the fiber surface generally start at points of defects such as structural irregularities, compositional heterogeneities, and imperfections. Scanning electron microscopic examination [71] has shown that in the case of PAN carbon fibers the growth nucleus of the whisker is typically located in the valley or trough of the fiber. As shown in Figure 15, the whiskers may be single crystals, or aggregates in the form of hedgehogs or balls (Figure 16) which was formed by twisting of single crystals or by secondary nucleated crystals [71]. Sun et al. [72] examined the carbon fiber whiskers mechanism by chemical vapor growth from acetylene using Fe salt solution as a catalyst. While the concentration of catalytic solution is appropriate, the carbon whiskers could deposit uniformly and link each other on the carbon fiber surface at an optimal growing temperature at about 1000°C. However, a higher temperature or concentration, might lead to a coarse carbon cluster deposited on the fiber surface. They suggested that the growing mechanism was the following: the carbon atoms in vapor phase dissolve first in the Fe salt drops over fibers, then the solid carbon grains participate on the film surface.

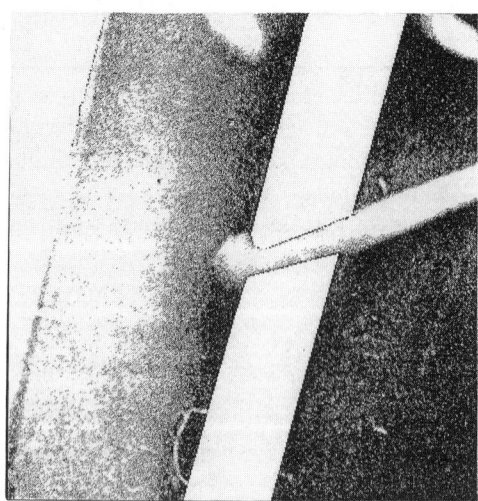

Figure 15 Single whiskers on carbon fiber ribbon (× 10,000) [71].

Surface Treatment of Carbon Fibers

Figure 16 Hedgehogs (a) and balls (b) of whiskers on carbon fiber ribbon (a: × 33; b: × 700) [71].

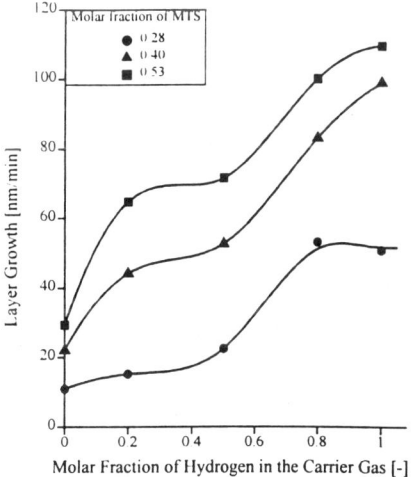

Figure 17 Influence of the molar fraction of hydrogen in the carrier gas on the growth rate of SiC layers (T = 950°C) [90]. (Reprinted with permission from Elsevier Sciences, S.A.)

Schoch et al. [90] deposited SiC from methyltrichlorosilane (MTS) onto mesophase pitch-based carbon fibers. The influence of the concentration of the reactants and composition of the carrier gas on the kinetics of the deposition of SiC was investigated. The effect of the molar fraction of hydrogen in a carrier gas on the layer growth from different molar fractions of MTS is shown in Figure 17, in which the experiments were conducted at a deposition temperature of 950°C. The layer growth is quite low with a typical value 10~30 nm/min in argon gas without hydrogen, but an increase of the molar fraction of hydrogen in the carrier gas leads to a considerable increase of the deposition rate to as high as 110 nm/min. The hydrogen in the carrier gas acts as a catalyst on the reduction of silicon chlorides [91], this effect becomes more significant when the molar fraction of MTS increases (XMTS > 0.28). The morphology of the pitch-based carbon fibers coated with SiC was investigated by scanning electron microscope. As illustrated in Figure 18a, the uncoated fiber surface is quite smooth with the fibrilar structure. However, after the fiber surface was coated at 950°C with SiC layers using carrier gases with a hydrogen content ranging from 0 to 0.8, the morphology changed dramatically as shown in the micrographs of Figures 18b-d. It is clearly seen in Figure 18b that the fine SiC dust from thermal decomposition of MTS in argon atmosphere is incorporated into the growing SiC layer by gas-phase nucleation, resulting in the formation of small tubers on the fiber surface. The adherence of the coating to the fiber is quite poor but improved with increasing hydrogen content in the carrier gas. When the content of hydrogen gas is up to 50%, the gas-phase nucleation of dust into the growing layer is then observed. However, compared with the SiC growing in Figure 18c, the gas-phase nucleation becomes less important as shown in Figure 18d when further increasing the ratio of hydrogen in the carrier gas. The relative mass change of SiC coated fibers from the oxidation test, which was measured in a thermal balance and expressed as the mass change during heating in air with a constant heating rate of 10 K min^{-1}, is given in Figure 19. A decrease of the mass due to a burn-off of carbon (starting at about 600°C) was observed in all cases, and the microstructure or thickness of the SiC layer is an important factor on the protection against oxidation. A thin layer of SiC (such as 43 nm) does not efficiently protect the fiber against oxidation. When the SiC layer is about 100 nm thick also has a poor effect because of a porous microstructure. In their experiments, a smooth surface deposited with about 150 nm thickness of SiC offered the maximized protection ability against oxidation.

The whiskerization of a carbon fiber has been found to increase the interlaminar shear strength of the resulting carbon fiber composite significantly. The improvement in mechanical properties on whiskerization is attributed to a strong bonding of the whiskers to the graphite substrate and to an increase in the interfacial area rather than to any increase in bonding between the fiber and

Surface Treatment of Carbon Fibers

the resin matrix. Japanese Yamashita et al. [3] proved that the orientation of the whiskers is magnetically controlled to improve the reinforcement efficiency. The cleavage interlaminar fracture toughness (Mode I testing) is improved as a result of the addition of oriented whiskers to the direction of the plate thickness. However, the inplane shear test (Mode II) of interlaminar fracture toughness, as well as interlaminar shear strength is not affected.

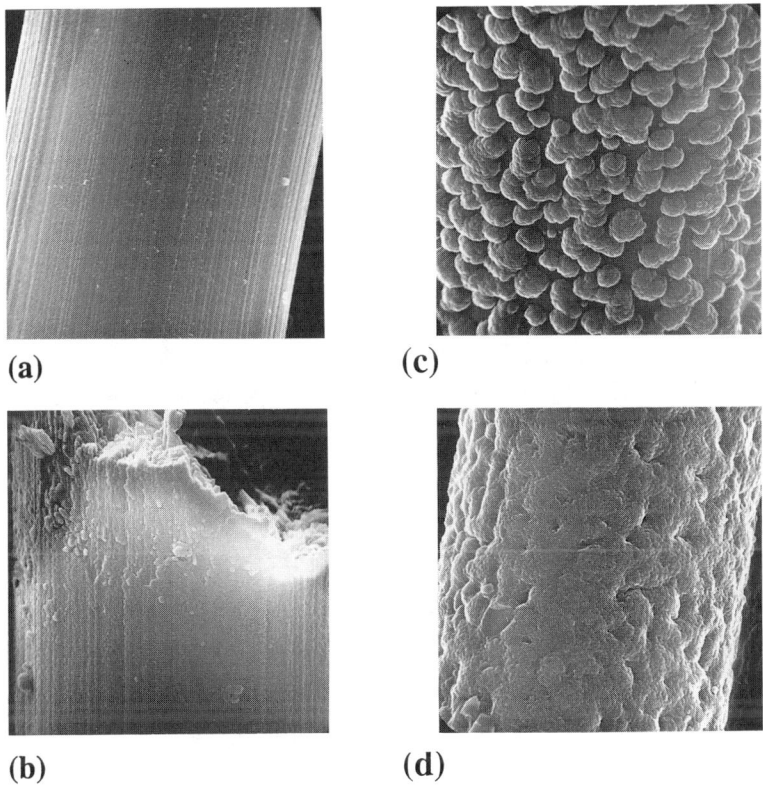

Figure 18 Scanning electron micrographs of mesophase pitch-based carbon fibers as received and coated with SiC (950°C, X_{MTS} = 0.4) using a carrier gas with different molar fractions of hydrogen: (a) uncoated fiber; (b) X_{H2} = 0 (Ar); (c) X_{H2} = 0.5; (d) X_{H2} = 0.8. [90] (Original photos from authors reprinted with permission from Elsevier Sciences, S.A).

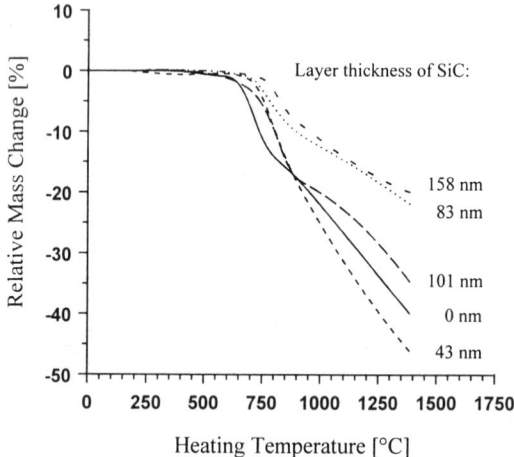

Figure 19 Relative mass change of mesophase pitch-based carbon fibers as received and coated with SiC [90]. (Reprinted with permission from Elsevier Sciences, S.A.)

3.6.2 Electropolymerization

Organic polymeric materials such as polyvinyl benzene [82], polyamide [83], polyimide [84], and organosilanes [85] could be coated on the fiber surface to enhance the fiber-resin interfacial bonding. The requirement for a polymer to be used as a coating is that it should be capable of being fixed on the carbon fiber surface by covalent or ionic bonds, and be compatible with the resin to produce a strong composite. To achieve better bonding between the resin and the polymer coating, it is better if the polymer coating has functional groups capable of reacting with the resin.

Subramanian et al. [73] investigated the effect on composite properties of coating/grafting polymers onto the carbon fiber surface during electroinitiated polymerization of a variety of monomers. Carbon fibers as anodes passing continuously through an electrolytic cell, were coated with a number of polymers containing carboxyl functional groups. After the electrodeposition, carbon fibers were washed and the weight of the deposit was largely reduced and reached to a constant value after 5 minutes washing, leading to the residual polymer coating with a structure more tightly held and more uniform. The composites made of polymers treated carbon fibers showed a significant improvement both in interlaminar shear and impact strength. In a similar manner, Scola et al. [74] electropolymerized several types of poly(carboxyphenylmethacryl-amide), such as 2-carboxyphenylmethacrylamide (2-CPM), 4-carboxyphenylmethacrylamide (4-CPM), 4-carboxyphenyl-methacrylamide /methylmethacrylate (4-CPM/MMA), and 4-carboxyphenyl-methacrylamide/N-phenyl maleimide (4-CPM/NPMI) matrixes onto AS4 carbon fiber composites to study the crosslinking effect. In the electropolymerization experiment, carbon fiber was used as cathode electrode and the metal as anode electrode, controlled by passing a constant current of 30 mA per gram of fiber. It was found that 2-CPM polymers underwent an intramolcular imidization and anhydride formation on the fiber surface, which result in a small amount of crosslinked network. On the other hand, a significant increase in gel fraction was found on the thermally cured 4-CPM polymers. The notched Izod impact strength data of different matrix composition at a 67% fiber volume fraction are displayed in Figure 20. The 4-CPM/MMA composites exhibit a slight better impact strength compared to the 2-CPM/MMA composites at the same composition. Upon heating at 280°C, 4-CPM/MMA composites maintained a higher Izod impact strength than a typical epoxy composite (200 kJ/m^2 vs 100 kJ/m^2). Figure 21 indicates the interlaminar shear strength (ILSS) of 4-CPM/MMA and 2-CPM/MMA fiber composite plates as a function of CPM polymer composition. The same tendency as the impact strength, the ILSS of 4-CPM/MMA is slightly higher than for the 2-CPM/MMA, approximately 60 MPa as compared to 50 MPa. Possible explanation for the increase of ILSS could be due to a larger amount of intermolecular hydrogen bonding of 4-CPM/MMA molecules.

The synthesis of a series of graphite fiber-polyacrylamide composites was performed electrochemically by Labes et al. [75] in diluted sulfuric acid (0.125 M)-acrylamide (2M) solution, 1:1 sulfuric acid (0.25 M)/2-propanol-acrylamide (2M) solution, and 1:1 sulfuric acid (0.25 M)/acetone-acrylamide (2M) solution, respectively. The graphite fiber-polyacrylamide composites, those synthesized from dilute sulfuric acid-acrylamide or acetone involved solution were less easily characterized than the one synthesized in a 1:1 2-propanol: sulfuric acid-acrylamide solution. But, composites that were synthesized in a diluted sulfuric acid solution were more crosslinked. FTIR confirmed the formation of interchain and intrachain imide functional groups after the electrochemical-treated fibers were cured with resin at 200°C. Both surface morphology from SEM and polymer weight gain analysis data proved that the thickness of the coatings and the volume fraction of the resin in the composites varied linearly with the time of electropolymerization. Papaspyrides et al. [86] explained the role of a polyamide interphase on carbon fibers reinforcing an epoxy matrix. This behavior was attributed to the improved affinity between matrix and asbestos fibers as a result of the well-known compatibility between polyamide and the epoxy phase.

A method of improving the interfacial bonding between carbon fibers and an epoxy resin matrix was developed by Chiu [87] in a continuous electrochemical deposition (ECD) of polypyrrole on carbon fibers. The ECD-treated carbon fibers are characterized by porous structure analysis, SEM, ESCA, and wettability measurements. Furthermore, mechanical evaluation is used to assess the macrointerfacial bonding capability of carbon fiber-epoxy resin composites. He proved that there is a good interfacial bonding between ECD-treated carbon fibers and the epoxy matrix.

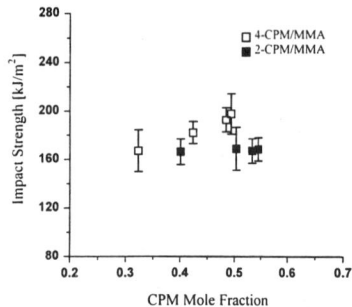

Figure 20 Izod impact strength of composite plates of different matrix composition at a 67% fiber volume fraction [74].

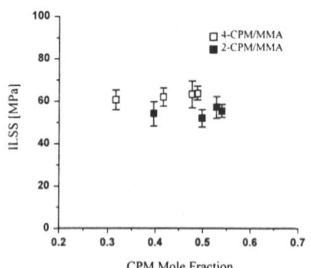

Figure 21 Interlaminar shear strength of composite plates of different matrix composition at a 67% fiber volume fraction [74].

3.6.3 Pyrolytic Surface Coatings

Because of their low cost [88] due to the simplicity of their manufacturing process [89], vapor-grown carbon fibers (VGCF) could be potential substitutes for PAN-based carbon fibers in some applications that require only a reasonable strength level. The chemical vapor deposition (CVD) process for carbon deposition uses volatile hydrocarbon compounds such as methane, propane, or benzene as precursor gases. Thermal decomposition of any of these gases is achieved on the hot surfaces of a carbon filament substrate, resulting in the deposition of pyrolytic carbon and the emission of volatile by-products. Several methods have been used to manufacture vapor grown carbon fibers. Fibers are produced and grown in a hot methane and hydrogen atmosphere chamber containing catalysts such as nickel, cobalt, and/or iron-nickel powder at 1000°C–1300°C. Vapor growth of carbon fibers initiates and emanates from the catalytic powder. A more recent patented process allows for a continuous method for producing vapor-grown carbon fibers. In this process the catalyst particles are either incorporated in the feedstock of, or produced in, a reactor by the decomposition of organometallics [81].

There are two methods of producing VGCFs that have been developed: seeding catalysts on a substrate and floating catalysts in a space. VGCFs produced from the seeding catalyst method are heterogeneously ranging in diameter (several tens microns) and in length (several centimeters), highly dependent on the geometrical factors of the reactor. This method produces a low yield of carbon fibers, and therefore is not favored from the industrial point of view. On the other hand, the floating catalyst method has advantages of a rather uniform size of a high yield of VGCFs because the catalytic particles derived from pyrolysis of organometallic compounds in a reactor chamber have a three-dimensional dispersion [80]. Therefore, the fiber structures produced from the floating catalyst method are more consistent and homogeneously ranging than those from the seeding catalyst method, although the fiber from the previous method has a shorter length and a smaller diameter.

The effect of improving the interfacial adhesion with the resin matrix or increasing the oxidation resistance of the fiber by coating carbon fiber surface via vapor-phase deposition of pyrolytic carbon, has also been studied by several workers [16,76]. Schmidt and Hawkins [77] deposited pyrolytic carbon coatings on carbon fibers by passing the fiber through a mixture of methane and hydrogen at 1500°C. The hydrogen was added to control the thickness of the coating and the rate of deposition. Although the oxidation resistance of the fiber improved considerably, an improvement in the composite shear properties was limited.

Vapor-grown carbon fibers (VGCFs) were prepared by Okada et al. [78] using floating catalysts in Linz-Donawitz converter gas (LDG) under several conditions for supplying a benzene solution containing ultra-fine transition metal particles as catalysts. The LDG used had the composition of 67% CO, 16% CO_2, 15% N_2, and 1.2% H_2, and the benzene solution contained 0.4 wt% ferrocene, 0.1 wt% cobalt acetylacetonate, and 0.2 wt% thiophene. The yield of VGCFs depended on the experimental conditions, i.e., the temperature history of the catalyst particles. The size of the catalyst particles remaining inside the VGCFs was determined by transmission electron microscope. They found that by using the catalyst growth model based on Brownian collision-coalescence motion, the average catalyst size was estimated as 20 nm, and this size was appropriate for preparing VGCFs in LDG.

In the industrial manufacturing process for VGCFs, using a high purity hydrogen as a carrier gas will raise the cost of the resulting fibers. Therefore, Ishioka et al. [79] examined the possibility of the use of less expensive Linz-Donawitz converter gas (LDG), instead of pure hydrogen, as a carrier for the formation of vapor-grown carbon fibers (VGCFs). The effect of various compositions of CO, H_2, and CO_2 on the formation of VGCFs was investigated. A small amount of hydrogen (5–7%) was added in carbon monoxide to accelerate the formation of VGCFs and gave a high yield. Same as the addition of carbon dioxide to the CO-H_2 mixtures did increase the VGCF yield. A similar composition to that of LDG a mixture of 77% CO, 17% CO_2, and 4% H_2 was used to obtain the highest yield, 45 wt.%. Figure 22 shows the scanning electron micrograph of VGCFs prepared in this maximized composition. Both VGCFs produced in pure hydrogen and a mixture of 95% CO and 5% H_2 are illustrated in Figure 22b and 22c, respectively for comparison. As seen the VGCFs prepared from the pure hydrogen are shorter and have smaller values of fiber diameter and density than those made from the CO-H_2-CO_2 and also CO-H_2 mixtures which have several hundred microns in length, 2–3 mm in diameter, and 2.05 g/cm^3 in density. As also noticed, there is no visible difference in surface morphology between fibers prepared in these CO-H_2-CO_2 and CO-H_2 mixtures.

Influence of catalyst on the formation of vapor-grown carbon fibers in carbon monoxide-carbon dioxide-hydrogen mixtures was further evaluated by Ishioka et al. [80]. The authors added several types of catalysts such as metal acetylacetonates, ferrocene and additives, to increase the growth rate of vapor-grown carbon fibers in a mixture of 77% CO, 19% CO_2, and 4% H_2. The decomposition behavior of ferrocene in the CO-CO_2-H_2 mixture was similar to that in pure hydrogen. The growth rate along with the yield were accelerated by the addition of acetylacetonates of Fe, Co, and Mn to ferrocene due to the reduction of particles size. Among those combinations they found, the best results were obtained by using 80 ferrocene and 20 wt.% cobalt acetylacetonate, leading to the growth rate and the yield of fibers as high as 40 µm/s and 70 wt.%, respectively.

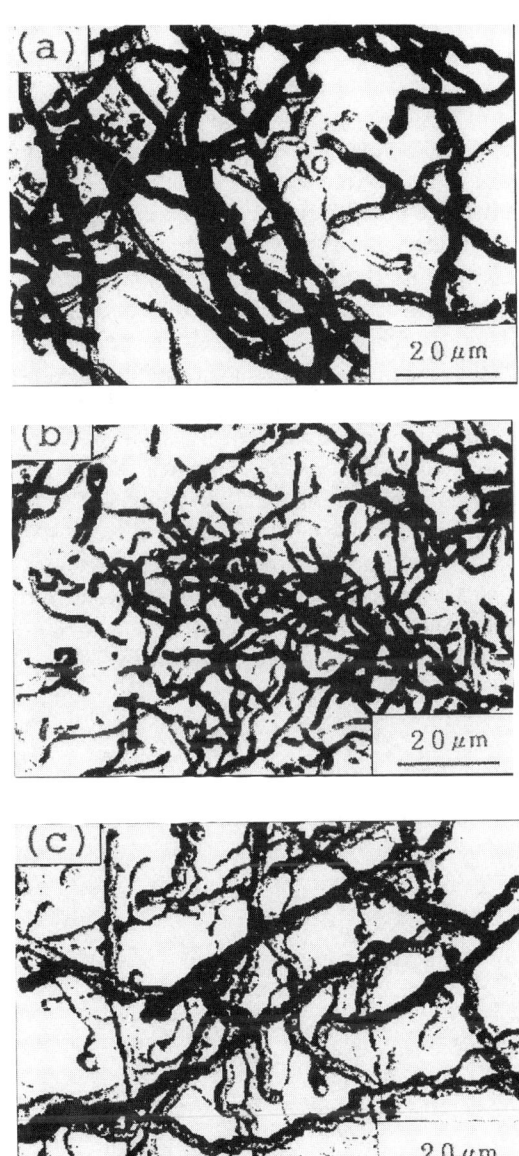

Figure 22 Scanning electron micrographs of VGCFs; (a) prepared in the CO-CO_2-H_2 mixture, (b) in pure hydrogen, and (c) in the CO-H_2 mixture [79]. (Reprinted with permission from Elsevier Sciences, Ltd.)

Besides the LDG method, another possibility of producing vapor grown carbon fibers from the light paraffins, methane, ethane, propane, iso-butane, n-pentane, n-hexane, and cyclohexane, can also be realized by a liquid pulse injection (LPI) technique [92,93], according to Mukai et al. [94]. This new method as illustrated in Figure 23 [92] provides a way to produce long vapor grown carbon fibers at high growth rates compared with traditional methods. Vapors of the carbon source and the catalyst material are introduced into the reactor chamber by the carrier gas in the usual floating catalyst technique. The temperature at the inlet of the reactor, however, is not high enough to immediately decompose the catalyst material; in other words, the catalyst material is gradually heated and slowly decomposed, leading to an insufficient growing process. On the other hand, in the LPI technique, when the catalyst material is heated simultaneously above the decomposition temperature they hit the hot wall of reactor as soon as they are injected into the reactor. Therefore, the catalyst material decomposes rapidly and release primary metal particles (nuclei) which are important intermediates for the fiber growth. When the nuclei grows through the coalescence among the particles to a proper size (a few to a few hundred nanometers), and then the fiber growth is initiated. As illustrated in the VGCF growing mechanism in Figure 23, carbon source material, hydrocarbon, dissolves into the catalyst particle on one side of it, and precipitates on the other side of it in the form of a fiber.

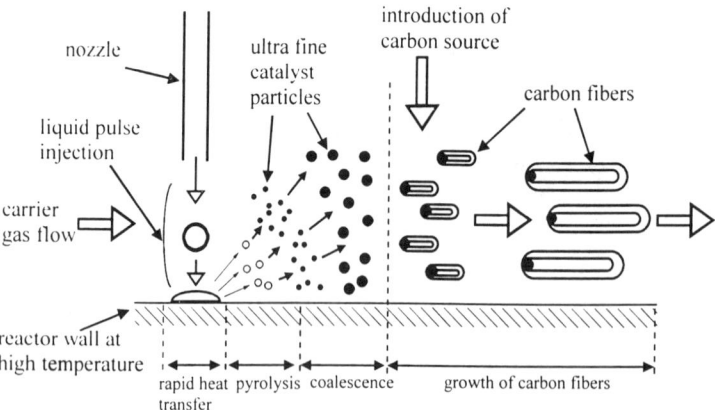

Figure 23 Demonstration of the effect of liquid pulse injection [92]. (Reprinted with permission from Elsevier Science, Ltd.)

Surface Treatment of Carbon Fibers

Figure 24 [94] present some of the SEM images of the VGCFs grown by the LPI technique from different carbon sources, a) i-butane, b) n-pentane, c) n-hexane, d) cyclohexane. It is clearly seen that straight fibers with the diameters in the range 1–4 µm up to 10 mm long could be easily obtained by this technique. The growth curves of VGCFs obtained from different carbon sources are illustrated in Figure 25 [94], where t is reaction time, and L fiber length. The trends of each curve were similar to those for the aromatic compounds source. The VGCFs grew rapidly at the beginning of the reaction (< 1 min), then the growth rate decreases after 1 min and finally stops.

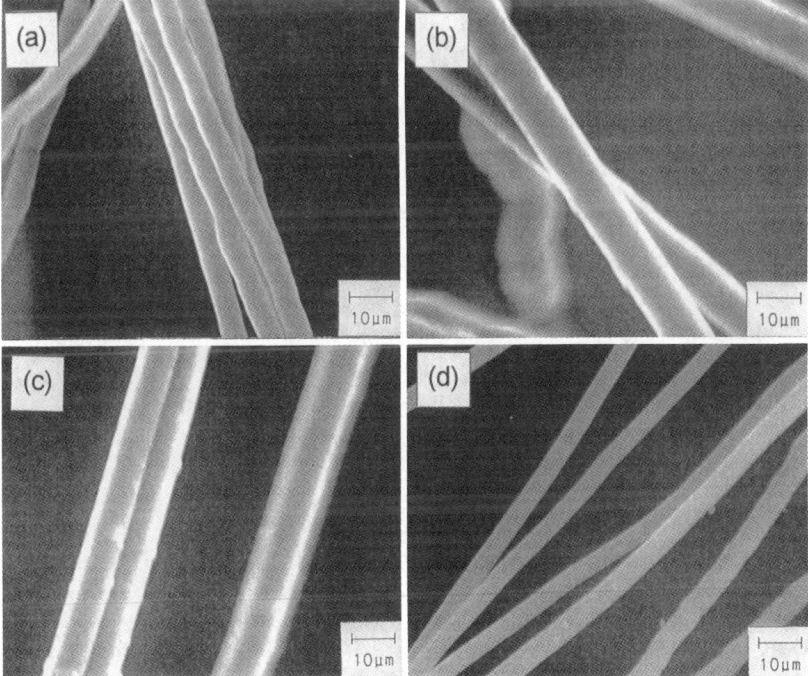

Figure 24 SEM images of typical VGCFs. Carbon source: (a) i-butane, (b) n-pentane, (c) n-hexane, (d) cyclohexane [94]. (Original photos from authors and reprinted with permission from Elsevier Science, Ltd.)

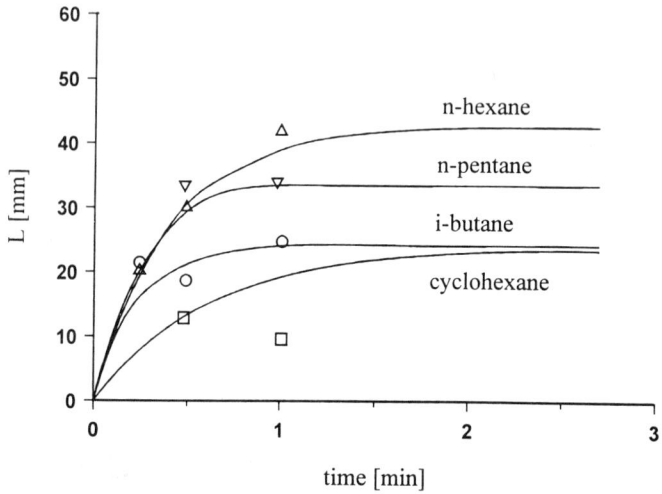

Figure 25 Growth curves of VGCFs [94]. (Reprinted with permission from Elsevier Science Ltd.)

3.7 COATING OF CARBON FIBERS

Carbon fiber/matrix (C-C) composites are being utilized for high-temperature structural applications. The most important of these applications require the C-C components to operate in oxidizing environments. Current applications for carbon fiber composites at higher temperatures include rocket propulsion components, re-entry thermal protection, and aircraft brakes [95]. At elevated temperatures in oxidizing environments carbon fiber composite materials usually need protection. Thus, research work was initially developed in the early 1970s on oxidation protection for carbon fiber composites for the thermal protection of shuttle orbiter materials. In many cases, external coatings are the most effective method of protecting materials from chemical attack. An external coating that effectively inhibits contact with oxidizing species is essential to C-C oxidation protection at all but the lowest temperatures, due to the oxides of carbon are gases and provide no physical barrier to the progression of oxidation. Chemical vapor deposited or infiltrated oxidation resistant coatings deposit protective inorganic coatings onto carbon-carbon

Surface Treatment of Carbon Fibers 203

composites through the decomposition and reaction of gases in a chamber. This process involves the diffusion of gases on the surface and into a carbon-carbon composite structure. This decomposition or reaction of the gases produces a solid oxidation resistant material deposited on the pore walls and fibers of the composite structure.

A potentially effective approach for achieving higher temperature protection is to start with a nonoxide coating that needs to be oxidized first before oxygen can diffuse to the underlying carbon. Because SiO_2 has a low oxygen permeability compared with the other oxides, silicon-based ceramics are commonly employed in the application of oxidation resistance. For example, in 100 hours at 1650°C only about 10 μm of oxide would be expected to form on high purity SiC or Si_3N_4 coatings [96]. Besides silicon-based compounds, it should also be mentioned that the platinum group metals, iridium and rhodium, have been considered as coatings [97] due to their ultra low permeabilities at very high temperatures. However, the utility of these materials is considerably limited due to the fact of extremely high cost, high thermal expansion, and volatile oxide formation. On the other hand, many phosphate coatings are also added to the oxidation protection systems, such as aluminum, magnesium, alkali and alkaline metals [109], in order to enhance the oxidation resistance. They are usually either mixed with ceramic powders or applied on a transition layer such as SiC. The disadvantage is that the phosphate coating materials can only be used at low temperatures (up to 850°C) in a dry atmosphere for short period of time applications.

3.7.1 Boron-Based Coatings

Boron-based coatings are other attractive compounds used in the application of oxidation protection [112-115]. Economy and Cofer [112] examined the effect of a boron nitride (BN) coating on the oxidation resistance of carbon fiber composites. A layer of BN coating onto carbon fiber composites could offer comparable or better mechanical properties than those observed in C-C composites [113]. The authors used vacuum infiltration to impregnate the borazine oligomers into a porous carbon fiber composite (1.41 g/cc with 23.3% open porosity). The SEM images shown in Figure 26 clearly indicate there is a great difference of surface porosity before and after the BN coatings; the sample surface before coating full of pores was densified by a layer of BN. To examine the effect of BN coating on the oxidation resistance, samples were heated in air up to 1000°C. Figure 27 display the weight loss curves versus temperature for the uncoated carbon fiber composite and BN densified C-C. The uncoated fiber with porous structures was very sensitive to oxidative attack, and the initial weight loss at about 500°C could be caused by a rapid oxidation of the carbon matrix. On the other hand, after one cycle of BN densification improved oxidation protection was observed from curve b which showed a less degree loss of weight. They tried to minimize the residual porosity by a second BN densification step. As a result, in Figure 27c, the rate of weight lost was observed to decrease and above that of the single densified C-C. Piquero et al. [114] reported that B_4C and B_4C/SiC multilayers provided better protection on carbon fibers both in a dry oxygen and in room air atmosphere than SiC. Figure 28 [114] shows that all the four type of coatings slow down the carbon fiber gasification more effectively than the uncoated T300 fiber sample, and the thermogravimetric weight loss curves confirm that boron-coatings, with a mixed layer or a double layer of B_4C/SiC, have a better effect on the oxidation protection than SiC coating. The authors explained that boronsilicate and boron oxide have a glassy structure, and they could act as a diffusion barrier, providing protection by inhibition of oxygen diffusion, thereby slowing down the carbon gasification more efficiently.

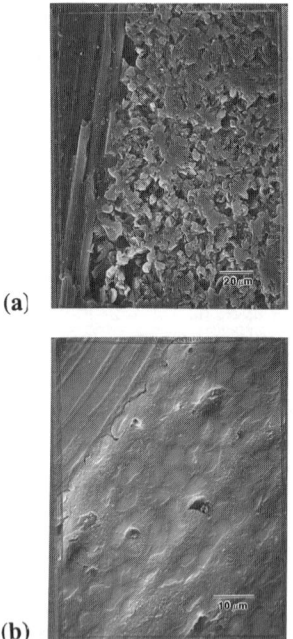

(a)

(b)

Figure 26 Polished cross-sections of (a) porous C-C, and (b) porous C-C after one densification with BN [112]. (Original photos from authors and reprinted with permission from authors and Elsevier Sciences, Ltd.)

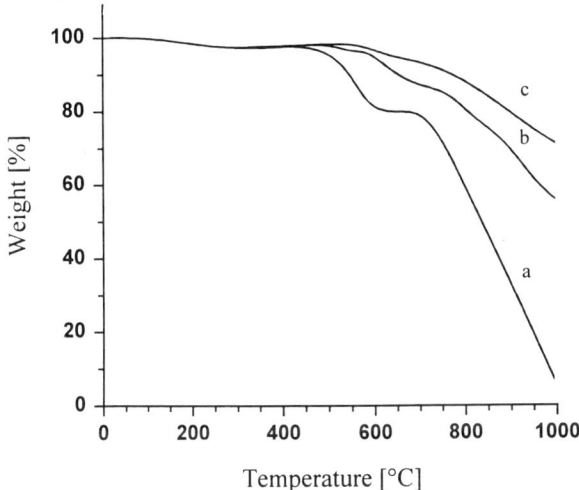

Figure 27 Weight loss versus temperature heated 10°C/min in air: (a) porous C-C, (b) porous C-C after one densification with BN, and (c) porous C-C after two densifications with BN [112]. (Reprinted with permission from Elsevier Sciences, Ltd.)

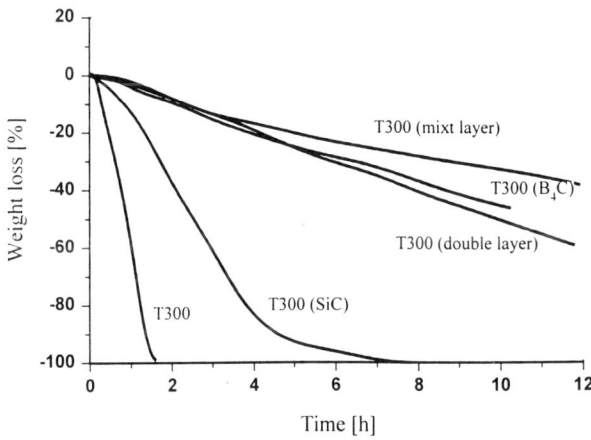

Figure 28 Weight loss changes as a function of time. T = 600°C, ambient air [114]. (Reprinted with permission from Elsevier Sciences, Ltd.)

3.7.2 Silicon and Alumina-Based Coatings

SiC and Si_3N_4 are common external coatings for C-C oxidation protection by providing inherently low oxygen permeation. The temperature limitations of SiC and Si_3N_4 are not based on the rate of oxide formation, but rather on the ability of the oxide to release gaseous oxidation products. The protective SiO_2 that forms on SiC and Si_3N_4 at lower temperatures is disrupted by gaseous reaction products in the 1500 to 1800°C range [98,99]. This leads to a rapid destruction of thin coatings by the accelerated formation of nonprotective SiO_2 and by material loss caused by the production of gaseous SiO. The thermal expansion characteristics of prominent coating candidates and those of a high-performance fabric laminate C-C composite are compared in Figure 29 [100]. The laminate C-C usually has two defects: a) expansion across the fibers is much greater than along the fibers, and b) a very low average thermal expansion coefficient (TEC) along the fiber. This anisotropy character can be avoided by the use of multidirectional weaves, but the cost increases and still has the problem of a low TEC. Several ultra low expansion oxides, such as SiO_2, match fairly well with the TEC of C-C along the fiber axis, while SiC and Si_3N_4 exhibit thermal expansion characteristics that fall between the C-C extremes.

SiC coatings prepared from polycarbosilane (PCS) solution, and SiO_2 and Al_2O_3 coatings prepared by a sol-gel method were investigated [110] to improve the oxidation resistance of carbon fibers. The authors prepared the coating by several approaches: CH_3SiCl_3 and H_2 mixtures were pyrolysed at 1373-1573°K to deposit a SiC layer on carbon fibers, and the coating thickness could be controlled according to the requirements in practice. A fine SiC/SiO_2 coating was also prepared by immersion and pyrolysis of PCS solutions in argon. The third method, from silica and alumina solutions, fine and dense SiO_2 and Al_2O_3 coatings could be obtained, respectively, under exact control of process parameters. The effect on the oxidation resistance of carbon fibers from the above coatings is quite effective but in the sequential order of CVD SiC, PCS-SiC, SiO_2, and Al_2O_3. An explanation was suggested to decrease defects like pores and microcracks in individual coatings to obtain the best oxidation protection for carbon fibers.

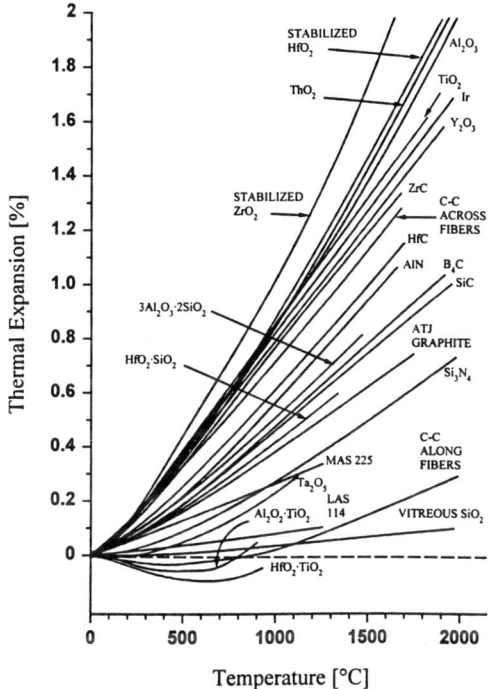

Figure 29 Thermal expansion characteristics of a high-performance fabric laminate C-C and external coating candidates [100]. (Reprinted with permission from Annual Reviews, Inc.)

3.7.3 Other Orgametallic Coatings

A multi-layer of SiC-(Si/ZrSi$_2$)-ZrSi$_2$ was coated on carbon-carbon (C-C) composite to improve resistance to oxidation [101]. The coating layer after controlled oxidation, which converted its surface to oxides, provided a passive function on exposure to an oxidizing atmosphere at high temperature. The oxidation properties of the composite were tested by TGA up to 1400°C. Static, cyclic oxidation tests, and a thermal ablation test with a plasma jet in the atmosphere were also performed. Microstructural analysis, including SEM and x-ray mapping, was used to reveal the features of the coating layer with and without oxidation. TGA measurement revealed that C/C coated with SiC-

(Si/ZrSi$_2$)-ZrSi$_2$ underwent moderate oxidation up to 1000°C, but experienced less loss of mass above 1000°C than uncoated samples. Zircon (ZrSiO$_4$), cristobalite (SiO$_2$), and zirconia (ZrO$_2$) were found in the oxidized layer of the coating. Cycles of oxidation between room temperature and 1600°C were carried out to explore the effects of the coating on the oxidation protection of C/C composites. Four C/C samples were tested at 1600°C, and the variation of the mass change was shown in Figure 30. The data showed that the as-received C/C and the C/C with plain SiC coating were not good enough to withstand oxygen attack for an oxidation test longer than 15 h. However, the multi-layer or SiC-Si coatings significantly improved the oxidation resistance of C/C, the resistance to thermal ablation and cyclic oxidation of C/C composites, and was superior to single-layer SiC-coated C/C and copper-infiltrated tungsten composites.

Manocha and his co-workers [111] made a similar conclusion that Zircon coatings were found to have better oxidation resistance than silica coatings. Harding et al. [102] used plasma-enhanced chemical vapor deposition to deposit silicon nitride on graphite fiber-reinforced polyimide composites to protect against oxidation at elevated temperatures. The adhesion and integrity of the coating were evaluated by isothermal aging (371°C for 500 h) and thermal cycling. The amorphous silicon nitride coating could withstand stresses ranging from approximately 0.18 GPa (tensile) to −1.6 GPa (compressive) and provided a 30 to 80 percent reduction in oxidation-induced weight loss.

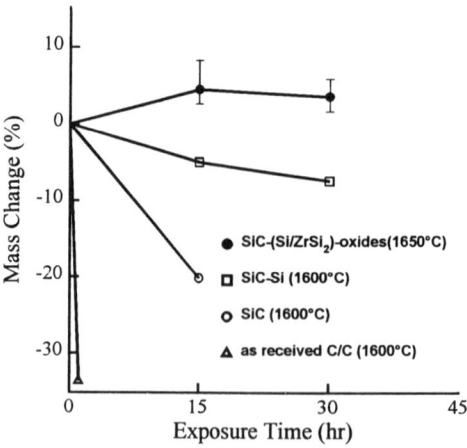

Figure 30 Weight change (%) of various C/C samples that have been tested at 1600 or 1650°C in air furnace up to 30 h [101]. (Reprinted with permission from Elsevier Sciences, Ltd.)

Kennedy et al. [103] studied carbon fiber composite interfaces with and without poly(vinyl alcohol) (PVAL) coating (both fibers being surface-treated and sized before coating) by high-resolution XPS. The main difference in the fracture surface of the fiber composites with and without coating is that the latter has a significant amount of Si (6 %) associated with the epoxy matrix, but Si is almost absent in the PVAL-coated fiber composites. This suggests that debonding mechanism in the uncoated fiber composite, which has a strong interfacial bonding, is controlled by the combination of cohesive failure of the matrix material and adhesive failure at the interface. In contrast, the PVAL coating promotes adhesive failure due to the weak bonding at the fiber-matrix interface. This observation is consistent with SEM observations in that the uncoated fiber composite consists of significant deformation of matrix material, whereas the coated fiber composite shows a less amount of matrix deformation with relatively clean fiber surface.

Interfacial bond properties were investigated by Yamada et al. [104] between carbon fibers treated with epoxy emulsion containing fine SiO_2 (10–80 nm) particles. Interfacial bond strength, evaluated by the fiber bundle pull-out test, was improved significantly when finer particles were dispersed in the epoxy resin and also increased proportional to SiO_2 concentration. TEM observation confirmed the good dispersion of SiO_2 particles. It was suggested that the SiO_2 particles in the epoxy resin enhance its affinity to cement matrix when finely dispersed and not aggregated.

As discussed earlier, these coating materials, such as SiC or Si_3N_4, were selected because of the inherently low oxygen permeability of their oxidation product, SiO_2, and average CTE values that are intermediate between the extremes of the C-C. The concept that appears to have the best potential for providing protection while accommodating the inevitable coating cracks from CTE mismatch and primary in-service stresses is to seal the cracks with a compliant glass. Borate glasses appear to have viscosity and wetting characteristics that are appropriate, and perhaps uniquely suited for this task [105,106]. Buchanan [107] used the technique of low pressure chemical vapor deposition (LPCVD) to deposit silicon carbide coatings to a carbon-carbon (C/C) composite using methyltrichloro-silane as the precursor. When cooling from the deposition temperature, the SiC coatings develop cracks due to thermal expansion mismatches between the coating and substrate. The cracks require sealing at temperatures between 600 and 1100°C to prevent oxidation of the underlying carbon. Two compositions of a borosilicate glass have been tested as crack sealants, being applied either as an outer glaze or between two layers of the CVD SiC.

3.8 SIZING OF CARBON FIBERS

In the manufacturing process, carbon fibers are usually coated with a very thin layer of epoxy resin or so-called "matrix-compatible sizings" after the proprietary surface treatments. These sizings are applied to carbon fibers for the protection of possible damage during transportation or handling. Carbon fibers are brittle materials and are therefore susceptible to strength degradation due to the presence of surface flows. It is believed that the fiber sizing is beneficial because it prevents fiber to fiber contact and hence the introduction of surface flaws. Other advantages are to improve the wetting ability of the fiber by the matrix, as well as to prevent the surface reactivity by blocking the surface functional groups [116]. From the composite manufacturing viewpoint, complete and thorough wetting of the tows with several thousand filaments is a necessary condition for good composite properties. The application of a matrix compatible finish could enhance wetting of the fibers by the matrix and therefore improve the interfacial shear strength.

Drzal et al. [116] examined the effect of the sizings on the fiber-matrix adhesion. The authors used the embedded single-fiber critical length test to measure the effect on adhesion of surface-treated carbon fibers which were coated with 100 *nm* epoxy resin (without curing agent), and found the interfacial shear strength increase by about 25%. Drzal [117] proposed that the sizing layer interacts with the bulk matrix and causes a change in local properties in the fiber-matrix interphase. The properties of this sizing layer itself are imparted to the interphase and can affect adhesion. They suggested [116] mechanisms for the function of sizing on the adhesion of carbon fiber composites. First of all, application of a fiber sizing could contribute to fiber-matrix adhesion through the creation of a protective environment for the reactive surface groups which were added to the fiber surface with treatment. Secondly, the failure mode was changed from an interfacial crack propagation to a matrix crack growth perpendicular to the fiber axis. Yumitori et al. [121] reached a similar conclusion from their study on the role of sizing resin in carbon fiber-reinforced poly(ether sulfone) composites. Sized carbon fibers show a higher interfacial shear strength than the unsized ones. They suggested from the analysis of time-of-flight secondary ion mass spectrometry that this adhesion enhancement arises from a strong interaction between sizing resin, the fiber and the matrix.

Sherwood et al. [118] studied the effect of sizings on PAN and pitch-based carbon fibers by XPS, scanning electron and x-ray wavelength dispersive microscopies. A sizing compound, bisphenol A epoxy resin, was added to the carbon fibers were previously subjected to various levels of surface treatment. By using solvents or heat-treatment to remove the size from the fiber surface, they examined the size compounds on the effect of adhesion. The surface chemistry of the sized fibers was dominated by the sizing and also more sizing compounds were found on the fibers which received greater prior surface treatment. They claimed that some remarkable interaction probably occurs between the size and the surface treated fiber, and this interaction leads to a greater stability of the size on the fiber surface. Sizing resin structure and interphase formation in carbon fiber composites were investigated by Anderson et al. [119]. The presence of a sizing resin on the as-received fibers reduced the interfacial shear strength of the composite. They suggested that compatibility of the deposited size with the matrix determines the adhesive bond between fiber and matrix and the formation of an interfacial region. On the other hand, deposition of a sizing resin from solution led to a different conclusion that chemical interaction with the fiber surface had occurred. During composite fabrication these sizing resins will therefore have to act as "coupling agents" with the matrix. Solvent extension of emulsion-deposited sizing resins, particularly at elevated temperatures, appeared to promote their interaction with the fiber surface. Wu and his co-workers [122] also discovered that the sizing treatment of carbon fiber has a great effect on the properties of 3D C/C composites. Furfuryl alcohol-maleic-anhydride-ethylene-glycol resin was found to be one of the best sizing agents because of its ability to form a firm layer to protect fiber from attack by a matrix precursor. The sizing improves the mechanical properties of 3D C/C composites, in particular the toughness and flexural strength. An almost two-fold increase of flexural strength has been obtained by this process.

Sato and Kurauchi studied the effect of fiber sizing on composite interfacial deformation by thermo-acoustic emission measurement [120]. The system was run by heating the furnace to 200°C at a rate of $2°C\ min^{-1}$ and then to cool to room temperature at the same rate. The acoustic emission from the specimen under the thermal cycle was detected by a transducer. Figure 31 shows the change of the emission activity with increasing fiber sizing. The acoustic emission was hardly observed during heating, whereas a lot of emission was observed during cooling, particularly below the glass transition temperature of the matrix resin (140°C). The thermo-acoustic emission activity was found to decrease dramatically with the increase of sizing. Accordingly, the mechanical properties of the composites with different sizings are listed in Table 10 where the samples were evaluated by a bend test at room temperature and 120°C, respectively. The tested composite is a randomly dispersed short carbon fiber-resin reinforced sheet-moulding compound made by hot-compression moulding under 10 MPa at 140°C for 10 min. The matrix was a modified vinyl ester resin and the volume fraction of the fiber for the composites was 40%. As shown, the bending modulus and bending strength decreased with the increase of sizing, particularly at 120°C.

Possible reasons were proposed by the authors to explain the change of the emission activity and mechanical properties. A larger sizing generates a thicker interphase which acts as a buffer layer for the stress transformation. The generated thermal stress concentration is relaxed through the interphase when the composite is subjected to a thermal cycle. A thicker interphase can prevent the thermal cracking because it provides a larger relaxation. That is the reason why the emission activity decreased with an increase in the sizing. In contrast, when the composite is subjected to an external load the efficiency of stress transfer from the matrix to the fiber is decreased by the presence of a thicker interphase, therefore the bending strength decreases.

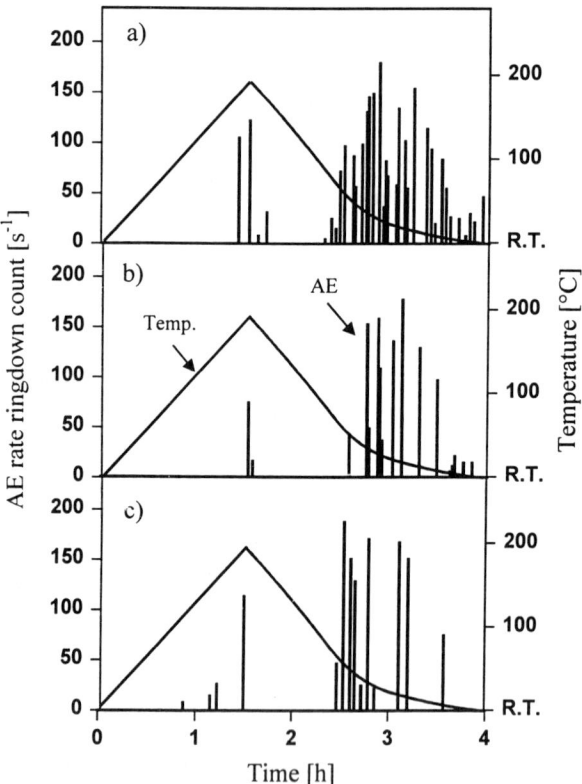

Figure 31 Thermo-acoustic emission behavior of the specimens, showing the decrease of the emission activity with the increase in the fiber sizing. AE, Acoustic emission; R.T., room temperature. Sizing (a) 0.2%, (b) 1% and (c) 3% [120] (Reprinted with permission from Chapman & Hall).

Table 10 Change of the Mechanical Properties of the Specimens with Increasing Fiber Sizing

	Room temperature		120°C	
	Bending modulus (GPa)	Bending strength (MPa)	Bending modulus (GPa)	Bending strength (MPa)
Sizing 0.2%	35	410	23	170
Sizing 1%	35	310	16	110
Sizing 3%	28	270	8.0	46

Source: from ref. 120. (Reprinted with permission from Chapman & Hall.)

3.9 Influence of Surface Treatments on Carbon Fiber Composite Properties

Carbon fiber composite properties could be affected by the surface treatment in the fiber direction. Two effects play an important role in the longitudinal tensile strength. First, the carbon fiber surface treatment has some influence on the individual fiber strength. Secondly, an improved interface strength leads to better load transfer in the composite. The effect of surface treatments and fiber sizings on the stress transfer characteristics and composite properties of AS-4 carbon fiber/epoxy composites were evaluated by Okhuysen et al. [123]. Fiber surface functional groups were experimentally varied from acidic to basic with RF glow discharge plasmas of CO_2 and NH_3 and the products were characterized by XPS techniques. Sizings composed of diglycidyl ether of bisphenol-A (DGEBA) were applied to some of the treated fibers. The interfacial shear strength of carbon fiber composites made with DGEBA/*m*-phenylene diamine resin was measured by the single fiber tension tests. Short beam shear and transverse flexural tests were used to examine the composite properties of modified materials. Results showed that the surface chemistry of the fiber was efficiently affected by the plasma treatments but the changes had surprisingly little effect on the critical stress transfer length. Transverse flexural tests were more sensitive to the changes in surface characteristics, however, the interlaminar shear strength of the composites were unaffected by the treatments.

Blackketter et al. [124] examined the surface treatment on the transverse tensile and shear strengths of carbon fiber-reinforced thermoset and thermoplastic matrix composites. The mechanical properties of high performance composites depend extensively on the adhesion of fiber and matrix. It is particularly true for maximizing the strength of unidirectional composites in off-axis directions. The composites were made by PAN-based carbon fibers (XA and A4) with EPON 828 epoxy (thermoset) and liquid crystal polymer (LCP, thermoplastic) matrix compounds. Four types of XA fibers were studied, untreated and treated fibers without sizing (XAU & XAS), and with sizing levels of 0.7% (XA-C1) and 1.2% (XA-C2), all in 12K filament tows. The A4 fibers were supplied in 6K tows, of which untreated (AU4), treated but unsized (AS4), and treated with an epoxy sizing (AS4-C) were studied. The influence of surface treatment was investigated by measuring the transverse flexural tensile (TF) and the short-beam shear (SBS) strengths of unidirectional composites. The shear and transverse tensile strengths illustrated in Table 11 were improved by the surface treatment for both the thermoset and thermoplastic reinforced fiber composites.

Table 11 Fiber Volume, Short Beam Shear (SBS) Strength, Transverse Flexural (TF) Strength, and the Ratio of SBS to TF Strength

	Fiber volume (%)	SBS strength (MPa)	TF strength (MPa)	Ratio SBS/TF
AU4/EPON 282	59	39.6	29.6	1.34
AS4/EPON 828	61	71.2	39.7	1.80
AS4-C /EPON 828	62	73.1	78.2	0.94
XAU/EPON 828	57	65.0	40.0	1.63
XAU-C1/EPON 828	55	59.3	41.0	1.45
XAU-C2/EPON 828	61	67.0	38.0	1.76
XAS/EPON 828	55	73.0	63.0	1.16
XAS-C1/EPON 828	55	80.0	77.0	1.04
XASS-C2/EPON 828	55	75.0	64.0	1.17
AS4/LCP	60	49.0	43.7	1.12
XAU/LCP	60	39.6	43.2	0.92
XAS/LCP	61	51.9	56.8	0.91

Source: from ref. 124. (Reprinted with the permission from Society of Plastics Engineers.)

Peng and Buttry [133,134] have recently proposed a possible mechanism of producing good adhesion in the carbon fiber/epoxy composite system which involve the attack of amine groups as nucleophiles from the hardener directly at non-oxygen-containing reactive sites at the carbon fiber surface. As illustrated in Figure 32, they proposed that a Michael-like reaction [135,136] might occur on the carbon fiber surface in which amine groups as nucleophiles and the vinylic C = C bonds at the carbon fiber surface act as electrophiles. In support of this proposal, they selected a variety of N-alkylamines to thermally react with carbon fibers. This means of manipulating adhesion is schematically presented in Figure 33. In the case of reaction A, an interfacial immobilization of simple N-alkylamines, if the amine immobilized species consume most of the reactive sites at the carbon fiber surface, then the interfacial layer will not have enough functionalities to react with epoxy resins to produce good adhesion. On the other hand, reaction B gives an example of interfacial immobilization of a reagent which could offer a direct chemical bonding between the fiber surface and the matrix, and the adhesion is predicted to be improved relative to the case for reaction A. A set of Tonen HMU (high modulus untreated) pitch-based carbon fibers were reacted with didecylamine, dodecylamine, 1,6-diaminohexane, and 1,9-diaminononane, respectively, at 180°C for 15 hours under the same reaction conditions. Then, the amine treated fibers were mixed with EPON828-mPDA epoxy resin for the embedded single fiber tests to measure the amine immobilization effect on the interfacial shear strength between fiber and matrix. The apparent fiber-matrix interfacial shear strength, τ_y, was measured at 83, 59, 21, 29, and 28 MPa for Tonen HMU + 1,9-diaminononane, Tonen HMU + 1,6-diaminohexane, Tonen + didecylamine, Tonen + dodecylamine, and Tonen HMU samples, respectively. As expected, the interfacial shear strength of the 1,9-diaminononane or 1,6-diaminohexane immobilized fiber shows ≥100% improvement over the control sample, presumably due to the chemical bonding between the free end of the immobilized amines and the epoxy groups in the resin. For the didecylamine sample, the authors speculated that the long alkyl chains of didecylamine blocked some of the resin molecules from reacting with carbon fiber, leading to a poor interfacial shear strength, about 22% less than that of the HMU control experiments.

When sizing was applied to the fibers after surface treatment, there was an additional improvement in the SBS and TF strengths for all the fiber/epoxy composites. Also, the TF strength was improved to a greater degree than the SBS strength by the surface treatment and sizing. Furthermore, TF strength appeared to be more sensitive to sizing than the SBS strength. The low sensitivity of the SBS strength could be explained by the fact that the shear failure mode is more dependent on friction or mechanical interlocking between the fiber and resin molecules. To achieve the maximum SBS and TF strengths for XA fibers in the epoxy systems with respect to sizing level effect on the properties, the 0.7% level seems to be a better approach than the 1.2% level. Possible reasons for the observed difference could be a reduced effect of mechanical interlocking or the lack of hardener in the interface region [126] for the higher level of sizing. Drzal et al. [126] suggested that the interface becomes more brittle as the sizing levels increase, and therefore the hardener molecules are more difficult to penetrate the sizing layer.

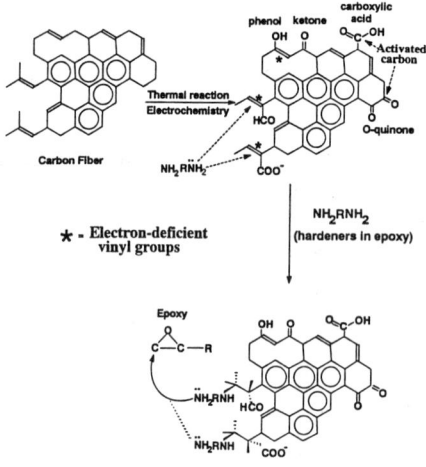

Figure 32 A proposed mechanism of nucleophilic reaction of amine compounds with vinylic groups at the carbon fiber surface [133].

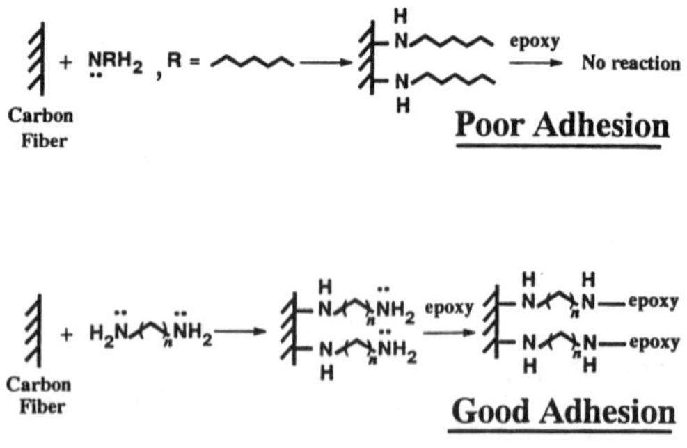

Figure 33 Schematic presentation of N-alkylamines as adhesion controlling reagents to manipulate the interfacial adhesion in carbon fiber/epoxy composites [133].

Surface Treatment of Carbon Fibers

The effect of surface treatment was investigated on pitch-based carbon fiber composites reinforced with engineering thermoplastic resin-polyethernitrile, possessing high heat and chemical resistance [125]. Both nitric acid (60%) and hydrogen peroxide (15%) solutions were used as oxidizing agents to treat fibers at 60°C for different lengths of time (12 and 24 h). The effect of the fiber oxidative treatment on the interlaminar shear strength (ILSS) is presented in Figure 34 [125]. The ILSS values of the oxidative treated fibers were about two times greater than that of the untreated CF. It is evident that the surface treatment in nitric acid and hydrogen peroxide solutions on the carbon fiber/polyethernitrile powder-impregnated composite (IPCF) is effective to improving the interlaminar shear strength. As seen in Figure 35, the oxidative treatment on IPCF resulted in an increase in the transverse flexural strength as well as in the ILSS. On the other hand, the longitudinal flexural strength of the oxidative treated IPCF composite varied slightly compared with the untreated samples. The composite prepared after a 60% nitric acid solution 24 h treatment illustrated slightly lower strength than composites from the other two treatments. It was thought that the strong bonding between fiber and matrix molecules, of which the nitric acid 24 h treated samples had the highest values of both the ILSS and the transverse flexural strength, caused the decrease of the longitudinal flexural strength.

The effect of varied wet oxidation treatments (68% HNO_3, 110°C, 10 min–150 h) of high modulus fibers on the properties of the fiber and on the properties of carbon fiber-epoxy composites are presented in Figure 36 [127]. It is evident that the surface oxides and the BET surface area are increased after oxidation. However, the fiber tensile strength is decreased because of the oxidative degradation. Both the surface oxides and surface area affect the mechanical properties of composites. The observed slow BET increase could hardly be correlated with the steep improvement of ILSS and flexural strength after a short-time (< 24 h) oxidation of fibers. Instead, the increasing amount of surface oxides could better explain the improvement of ILSS. The improvement of the adhesion with continued oxidation treatment is illustrated by the translation of the fiber strength into the composite. This translation relates to the values of fiber strength after the oxidation treatment [127]:

$$\eta = \tau_b / \chi_F \tau_{Z,F}$$

Where $\tau_{Z,F}$ is the fiber tensile strength, τ_b is the composite flexural strength and χ_F is the fiber volume content. The translation of the fiber strength is further limited after a short oxidation time. This indicates that increased oxidation by the extended treatment produces an overoxidizing effect which strongly influences the fracture behavior.

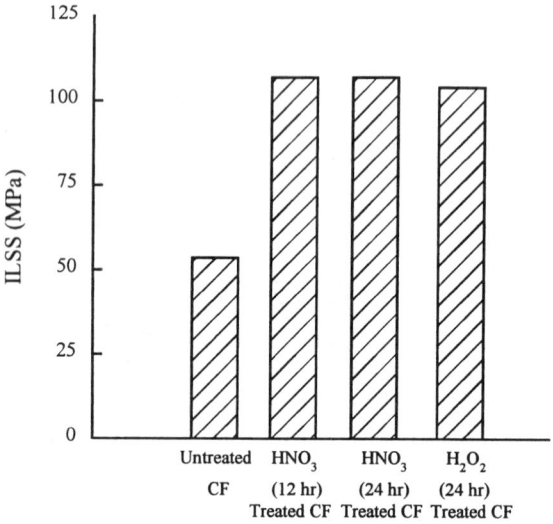

Figure 34 Interlaminar shear strength and surface treatment conditions [125]. (Reprinted with permission from Society of Plastics Engineers.)

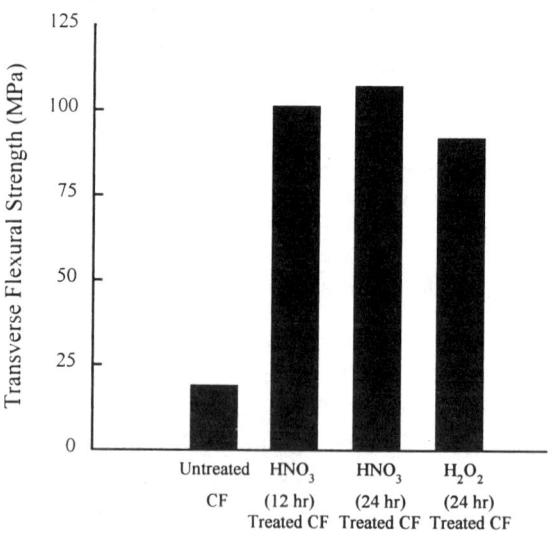

Figure 35 Transverse flexural strength and surface treatment conditions [125]. (Reprinted with permission from Society of Plastics Engineers.)

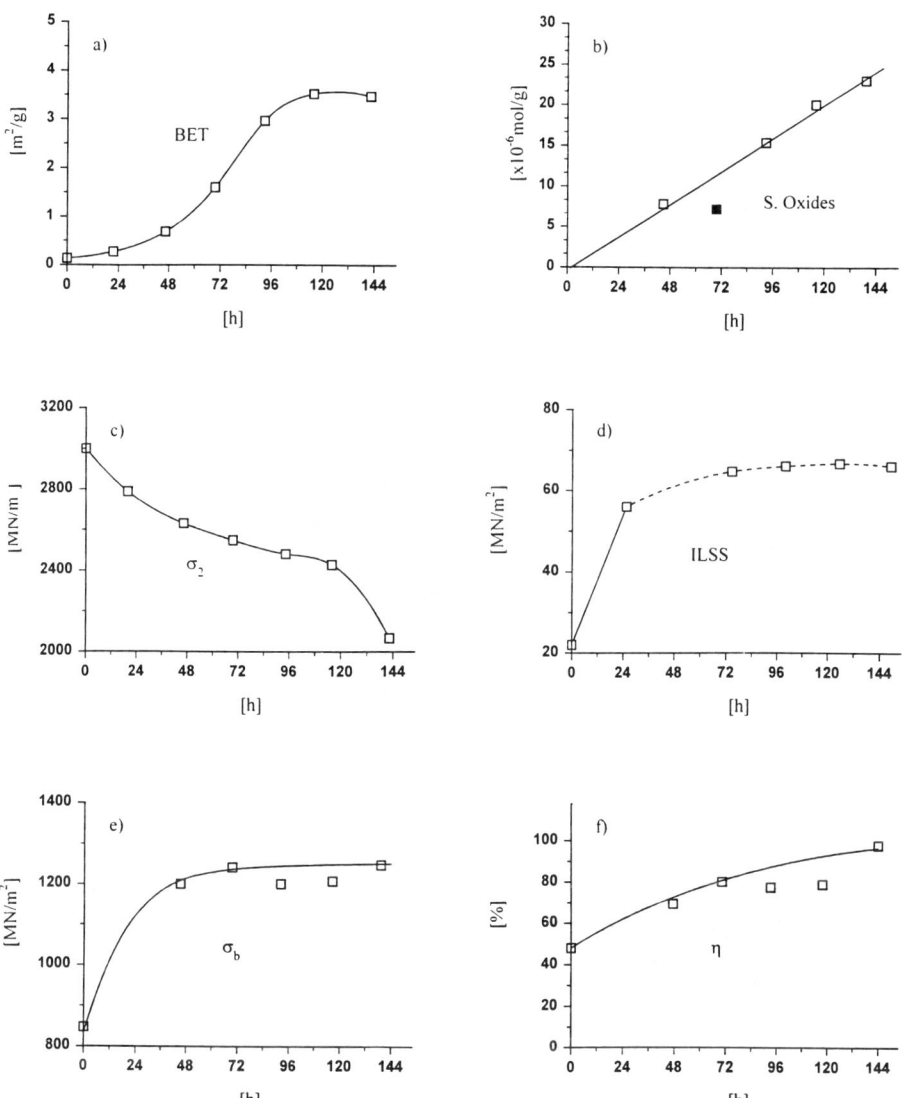

Figure 36 Influence of the wet oxidation procedure of Sigrafil HM carbon fibers on fiber and composite properties. Fiber properties: (a) BET surface area, (b) amount of surface oxides and (c) tensile strength (σ_2) of C-fibers. Composite properties: (d) Interlaminar shear strength (ILSS), (e) flexural strength and (f) translation of fiber properties into the composite (η) [127]. (Reprinted with permission from Elsevier Science, Ltd.)

Drzal and Madhukar [128] comprehensively discussed the relationship between the fiber-matrix adhesion and its composite mechanical properties. Three types of PAN-based fibers, untreated (AU4), surface treated (AS4), and surface treated with sizing were studied. Average values of the on-axis and off-axis properties for the three composite systems are listed in Table 12 and Table 13. A discussion on the results is provided below. The composite 0° tensile strength is proportional to the interfacial shear strength in the intermediate region. If the interfacial strength is too weak, the composite fails too soon due to a cumulative weakening of the material. However, the failure mode shifts from interfacial to matrix and the composite becomes more brittle when the interfacial bond strength is excessive. Composite compressive strength is enhanced by increasing the fiber-matrix adhesion. However, when the ISS was increased from low to intermediate values the flexural properties in the longitudinal are little affected. In regard to the off-axis properties, the transverse flexural test results illustrate that fiber-matrix adhesion has a stronger effect on the transverse flexural strength than on the transverse tensile strength. The authors suggested that a strong sensitivity of the ISS on the transverse flexural strength should be expected. Because the bending stresses are directly borne by the interface and the specimen failure occurs either due to tensile (or compressive) failure of the interface or matrix or a combination of the two.

Table 12 Summary of an On-Axis Property Result for Carbon/Epoxy Composite

Test	AU4/epoxy	AS4/epoxy	AS4C/epoxy
$[O]_{12}$ tensile modulus, E_{11} (GPa)	130 ± 9	138 ± 5	150 ± 9
$[O]_{12}$ tensile strength, σ_1^f (MPa)	1403 ± 107	1890 ± 143	2044 ± 256
$[O]_{12}$ compressive modulus, E_{1c} (GPa)	131 ± 8	126 ± 9	153 ± 8
$[O]_{12}$ compressive strength, σ_{1C}^f (MPa)	679 ± 116	911 ± 180	1174 ± 207
$[O]_{12}$ three-point flexural modulus, E_{1B} (GPa)	154 ± 6	136 ± 11	147 ± 5
$[O]_{12}$ three-point flexural strength, σ_{1B}^f (MPa)	1662 ± 92	1557 ± 102	1827 ± 52

Source: from ref. 128. (Reprinted with permission from Chapman & Hall).

Table 13 Summary of Off-Axis Property Results for Carbon Fiber/Epoxy Composites

Measurement	AU4/ epoxy	AS4/ epoxy	AS4C/ epoxy
[90]$_{12}$ tensile modulus, E_{22} (GPa)	8.9 ± 0.6	9.8 ± 0.6	10.3 ± 0.6
[90]$_{12}$ tensile strength, σ_2^f (MPa)	18.0 ± 3.9	34.2 ± 6.2	41.2 ± 4.7
[90]$_{12}$ flexural modulus, E_{2B} (GPa)	10.2 ± 1.5	9.9 ± 0.5	10.7 ± 0.6
[90]$_{12}$ flexural strength, σ_{2B}^f (MPa)	21.4 ± 5.8	50.2 ± 3.4	75.6 ± 14.0
[± 45]$_{3s}$ in-plane shear modulus, G_{12} (GPa)	9.1 ± 1.5	6.2 ± 0.5	6.0 ± 0.2
[± 45]$_{3s}$ in plane shear strength, τ_{12}^f (MPa)	37.2 ± 1.8	72.2 ± 12.4	97.5 ± 7.4
Iosipescu in-plane shear modulus, G_{12} (GPa)	7.2 ± 0.5	6.4 ± 1.0	7.9 ± 0.4
Iosipescu in-plane shear strength, τ_{12}^f (MPa)	55.0 ± 3.0	95.6 ± 5.1	93.8 ± 3.3
Short-beam interlaminar shear strength, τ_{13}^f (MPa)	47.5 ± 5.4	84.0 ± 7.0	93.2 ± 3.8

Source: from ref. 128. (Reprinted with permission from Chapman & Hall.)

Several other investigators [38,129,130] carried out detailed studies of the influence of various surface oxidation on the mechanical properties of carbon fiber-reinforced polymer (CFRP) composites. They estimated the carbon fiber composition, the amount of oxygen-containing functional groups, and surface areas of carbon fibers by measuring the content of two types of surface groups (Table 14): (1) strongly acidic groups such as carboxylic and phenolic, which could be neutralized by sodium hydroxide; and (2) weakly acidic groups (hydroxyl), neutralized by sodium ethoxide. The tensile failure mode of the composite is determined by the presence of strong acidic groups, which is indicative of a strong interfacial bond. Introduction of weak acidic groups enhanced the ILSS without a considerable alternation of the mode of failure. In other words, the nature of the interfacial bond remains almost unaltered. To examine the effect of surface groups more closely, these workers degassed oxidized carbon fibers by heat treatment in argon at 1000°C. This heat treatment removed the acidic groups almost completely. ILSS studies of the composites prepared by using these carbon fibers showed a slight decrease, although the values remained much higher than in the case of the untreated fibers. This clearly indicates that the presence of acid groups on carbon fibers strongly influenced the shear strength of the resulting composites.

To differentiate between the influence of increased surface area and increased surface functionality of the oxidized carbon fibers on the improvement in ILSS Fitzer and co-workers [131] selected a carbon fiber (Torayca 300) which does not undergo any marked change in surface area on surface oxidation or during subsequent heat treatment. The commercial finish of the carbon fiber was removed by extraction in boiling acetone for several days followed washing with boiling water. The carbon-oxygen surface compounds formed during oxidation were removed by gradual heating between 500 and 1400°C. The surface oxides decomposed into CO_2 and CO. In addition methane was also evolved. The CO_2 was evolved mainly between 300 and 850°C, while CO was evolved largely between 350 and 1000°C. The amount of gases evolved were different for untreated (T300 U) and surface-oxidized (T300 S) samples (Table 15). The surface treated carbon fiber composites showed higher ILSS values than those of untreated fiber composites. When surface-treated carbon fiber was heated at gradually increasing temperatures, the ILSS of the fiber-composite decreased. The decrease in ILSS occurred mainly on heat treatment in the temperature range 300–450°C (Figure 37), which can be correlated directly with the removal of the surface oxides which evolve CO_2 on heating. Only a small decrease in ILSS occurred in the temperature range 500–1000°C, where most of the carbon monoxide is evolved, indicating that CO-evolving surface oxygen groups have only a minor effect on ILSS. A drastic decrease in the ILSS of the composite occurred when the carbon fibers, treated as well as untreated, were heated at temperatures above 1000°C. This decrease in ILSS at high HTT values was found to be related directly to the elimination of nitrogen from the carbon fiber samples as shown in Figure 37.

Table 14 Composite Properties in Relation to Surface Acidic Groups on Carbon Fibers

Fiber[a]	Surface group neutralized by group		ILSS	Failure mode
	NaOH ($\mu Eq/g$)	$NaOC_2H_5$ ($\mu Eq/g$)		
AC nontreated	7	10	55	shear
AC HNO_3	14	20	85	intermediate
AC HNO_3 h.t.	0	15	77	intermediate
AC NaOH	16	16	92	tensile
AC NaOH h.t.	0	13	77	intermediate
AG nontreated	3	2	14	shear
AG NaOH	16	17	50	tensile
AG NaOH h.t.	2	21	51	intermediate

[a]h.t., heat-treated; AC, heat-treated at temperatures below 1500°C; AG, heat-treated at 2000°C.
Source: from ref. 130.

Table 15 Amount of Desorbed Gases CO_2, CO, and CH_4 During Heat Treatment of HT Carbon Fiber T300 up to Final Temperature, Fiber Nitrogen Content, and Resulting ILSS

Fiber type	Desorption treatment (°C)	Desorbed gases (μmol/g)			Fiber analysis nitrogen (%)	Adhesion in LY556 ILSS (MPa)
		CO_2	CO	CH_4		
T300 U	Nil	—	—	—	—	78.6
	1000	5.16	5.93	2.76	5.6	78.7
T300 S	—	—	—	—	—	91.9
	930	11.01	5.12	1.97	—	82.0
	1000	17.69	12.53	4.59	5.4	78.9
	1200	17.50	12.82	4.92	2.6	63.5
	1400	17.82	13.91	3.97	1.1	32.6

Source: from ref. 131.

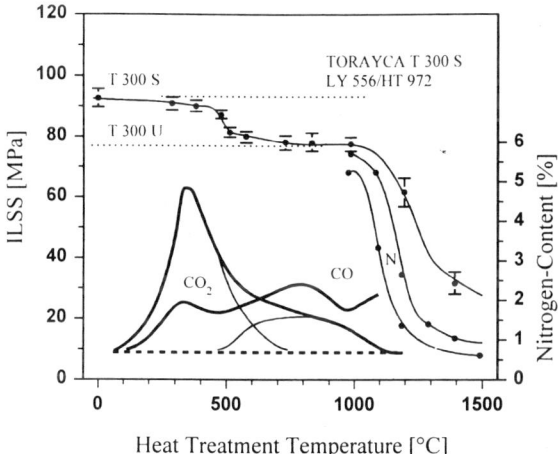

Figure 37 Relationship between ILSS and gases desorbed from carbon fiber surface [132].

REFERENCES

1. Nakayama Y., Soeda, F., and Ishitani, A., *Carbon* 28, 21 (1990).
2. Guoxiang L., Daozhi, L., and Huafu, W., *Carbon* 30, 961 (1992).
3. Davis, C. Q., He, Q., and Gustafson, R. R., *Carbon* 30, 177 (1992).
4. Barbier, B., Pinson, J., and Sanchez, M., *J. Electrochem. Soc.* 137, 1757 (1990).
5. Desimoni, E., Casella, G. I., Salvi, A. M., Cataldi, T. R. I., Morone, A., *Carbon* 30, 527 (1992).
6. Kaelble, D. H., Dynes, P. J., Crane, L. W., Maus, L., *J. Adhesion* 7, 25 (1975).
7. Larsen J. V., Smith, T. G., Erickson, P. W., "Carbon Fiber Surface Treatment," 1971, NOLTR, pp. 71-165.
8. Daukeys, R. J., *J. Adhesion* 5, 211 (1973).
9. Donnet, J. B., Bansal, R. C., "*Carbon Fibers*," 1984, Marcel Dekker Inc., New York.
10. Kureha Chemical Industry Co., Brit. Patent 1,293,900 (October 15, 1972).
11. Herrick, J. W., Grüber, P. E., and Mansur, F. T., *Surface Treatments for Fibrous Carbon Reinforcements*, AFML-TR-66-178 Part I, Air Force Materials Laboratory (July 1966).
12. Sach, R. S., and Basche, M., *U.S. Patent 3, 720, 536* (March 1973).
13. Novak, R. C. In Composite Materials: *Testing and Design, ASTM Special Tech. Publ.* No. 460 (1969), p. 540.
14. Molleyre, F., and Bastick, M., *Proc. 4th London Intern. Conf. on Carbon and Graphite*, Soc. Chem. Ind., London, 1974, p190.
15. Mckee, D. W., *Carbon* 8, 131(1970).
16. Mckee, D. W., *Carbon* 8, 623 (1970).
17. Mckee, D. W., and Mimeault, V. J., *J. Chemistry and Physics of Carbon*, Vol.88, Marcel Dekker, New York, 1973.
18. Clark D., Wadsworth, N. J., and Watt, W. *Carbon Fibers-Their Place in Modern Technology*, The Plastics Institute, London (1974), p. 44.
19. Druin, M. L., Ferment, G. R., and Rad, V. N. P. *U.S. Patent 3,754,957* (August 1973).
20. Hoffman, W. P. *U. S. Patent US 5,271,917*, 21 December 1993.
21. Molleyre, F., and Bastick, M. *High Temp. -High Pressures* 9, 237 (1977).
22. Vukov, A. J., J. Serb. Chem. Soc. 55, 333 (1990).
23. Ko, T. H., and Li, C. H., Polym. Compos. 16, 224 (1995).
24. Alcaniz-Monge, J., Cazorla-Amoros, D., Linares-Solano, A., Yoshida, S., and Oya, A., Carbon 32, 1277 (1994)

25. Kucera, M., Glogar, P., Balik, K., and Tomanova, A., Acta Mont., Ser. B 3, 91 (1994).
26. Ahearn, C. and Rand, B. *Carbon* 34, 239 (1996).
27. Bansal, R. C. and Chabra, P., *Indian J. Chem.* 20A. 449 (1981).
28. Druin, M. L. and Dix, R., *U.S. Patent 3, 894, 884* (July 1975).
29. Jain, R. K., Manocha, L. M., and Bahl, O. P. *Indian J. Technol.* 29, 163 (1991).
30. Desimoni, E., Salvi, A. M., Casella, I. G., and Damiano, D. *Surf. Interface Anal.* 15, 627 (1990).
31. Donnet, J. B., Papirer, E., and Dauksch, H. *Carbon Fibres-Their Place in Modern Technology*, The Plastics Institute, London, (1974), p. 58.
32. Well, H. and Coleclough, W. F., *U. S. Patent 3,657,082* (April 1972).
33. Herqué, J. J. Ph.D. thesis, Mulhouse and Strasboug (1975).
34. Wu, Z., Pittman, C. U., and Gardner, S. D., *Carbon* 33, 597 (1995).
35. Nakanishi, Y., and Fujita, K., *Curr. Jpn. Mater. Res.* 12, 213 (1994).
36. Ma, C. C. M., Tai, N. H., Chang, W. C., Fang, C. K., and Chao, H. T., *Annu. Tech. Conf.- Soc. Plast. Eng.* 1994, 52nd (Vol. 2), 2411.
37. Rieux, J. P., Lehurreau, P., *J. Fr. Patent 72.14.950* (1972).
38. Ehrburger, P., Herque, J. J., and Donnet, J. B. in *Petroleum Derived Carbons* (edited by M. L. Deviney and T. M. O'Grady), ACS Symposium Series No. 21, Amer. Chem. Soc., Washington, D. C. (1975), p. 324.
39. Ehrburger, P., Herque, J. J., and Donnet, J. B. *Proc. 4th Intern. Conf. Carbon Graphite*, Soc. Chem. Ind., London (1976), P. 201.
40. Fitzer, E., and Weiss, R. *Intern. Symp., Soc. Plastic. Engi.*, University of Leige, 1983.
41. Donnet, J. B., Ehrburger, P., and Ilvoas, A. M. *Bull. Soc. Chim. Fr.* 6, 2239 (1972).
42. Donnet, J. B. *Carbon* 20, 268 (1982).
43. Kozlowski, C. and Sherwood, P. M. A. *Carbon* 25, 751 (1987).
44. Kozlowski, C., and Sherwood, P. M. A. *J. Chem. Soc. Fraday Trans.* 81 (2), 45 (1985).
45. Kozlowski, C. and Sherwood, P. M. A. *Carbon* 24, 357 (1986).
46. Harvey, J., Kozlowski, C, and Sherwood, P. M. A. *J. Mater. Sci.* 22, 1585 (1987).
47. Xie, Y. and Sherwood, P. M. A. *Appl. Spectrosc.* 44, 1621 (1990).
48. Xie, Y and Sherwood, P. M. A. *Appl. Spectrosc.* 45, 1158 (1991).
49. Xie, Y., Wang T., Franklin O., and Sherwood, P. M. A. *Appl. Spectrosc.* 46, 645 (1992).
50. Mahy, J., Jenneskens, L. W., Grabandt, O., Venema, A., and Houwelingen, G. D. B. van *Surf. Interface Anal.* 21, 1 (1994).
51. Nakao, F., Takenaka, Y., Asai, H. *Composites* 23, 365 (1992).

54. Donnet, J. B., Brendle, M., Dhami, T. L., and Bahl, O. P. *Carbon* 24, 757 (1986).
55. Donnet, J. B., Dhami, T. L., Dong, S., and Brendle, M. *J. Phys. D: Appl. Phys.* 20, 269 (1987).
56. Brendle, M., Dong, S., and Donnet, J. B. *6th Intern. Conf. Ion Plasmic Assist. Tech.*, IPAT, Brighton, U.K., May 1987.
57. Brendle, M., Dong, S., and Donnet, J. B. *18th Bienn. Conf. Carbon*, 1987, extended abstracts and program, p. 7.
58. Smiley, R. J. and Delgass, W. N. *Mater. Res. Soc. Symp. Proc.* 305, 129 (1993).
59. Farrow, G. J. and Jones, C. *J. Adhes.* 45, 29 (1994).
60. Nay, J. C., Pitt, W. G., and Armstrong-Carroll, E. *J. Appl. Polym. Sci.* 56, 461 (1995).
61. Weisweiler, W. *NATO ASI Ser., Ser.* E 230, 269 (1993).
62. Ebert, E. and Weisweiler, W. *NATO ASI Ser., Ser.* E 230, 287 (1993).
63. Farrow, G. J., Atkinson, K. E., Fluck, N., and Jones, C., *Surf. Interface Anal.* 23, 313 (1995).
64. Commercon, P. and Wightman, J. P., J. Adhes. 47, 257 (1994).
65. Blackman, B. R. K., Kinloch, A. J., and Watts, J. F., *Composites* 25, 332 (1994).
66. Waldman, D. A., Zou, Y. L., and Netravali, A. N., *J. Adhes. Sci. Technol.* 9, 1475 (1995).
67. Zou, Y. L. and Netravali, A. N., *J. Adhes. Sci. Technol.* 9, 1505 (1995).
68. Dilsiz, N., Erinc, N. K., Bayramli, E., and Akovali, G., *Carbon* 33, 853 (1995).
69. Edie, D. D., Cano, R. J., and Ross, R. A., *20th Biennial Conference on Carbon*, Santa Barbara, USA, June 1991.
70. Personal communication.
71. Rabotnov, J. N., Perov, B. V., Lustan, V. G., and Stepanitsov, I. E. *Carbon Fibers-Their Place in Modern Technology*, The Plastics Institute, London, 1974, p.65.
72. Sun, S., Wei, Y., Liu, M., and Li, M. *Jinshu Xuebao* 29, B158 (1993).
73. Subrsmanian, R. V. and Jakubowski, J. *J. Polymer Eng. and Sci.* 18, 590 (1978).
74. Liang, J. L., Bell, J. P., and Scola, D. A. *Polym. Eng. Sci.* 3, 341 (1994).
75. Iroh, J. O., Suhng, Y., and Labes, M. M. *J. Appl. Polym. Sci.* 52, 1203 (1994).
76. Fitzer, E., Huttner, W., and Wolter, D. *13th Bienn. Conf. on Carbon*, Irvine, Calif., 1977, extended abstracts, p. 180.
77. Schmidt, D. L. and Hawkins, H. T., *Filamentous Carbon and Graphite*, AFML-TR-65-160, Airforce Materials Laboratory, August 1965.
78. Ishioka, M., Okada, T., and Matsubara, K. *Carbon* 31, 699 (1993).

79. Ishioka, M., Okada, T., Matsubara, K., and Endo, M., *Carbon* 30, 865 (1992).
80. Ishioka, M., Okada, T., and Matsubara, K. *Carbon* 30, 859 (1992).
81. Buckley, J. D. *Proc. Int. Conf. Compos. Mater.*, 9th, 3, 675 (1993).
82. Trostyanskaya, E. B. and Kobets, L. P. *Plast. Massen Wiss. Tech.* 1, 53 (1970).
83. Herrick, J. W. *12th Natl. Symp., Soc. Aerosp. Mater. Process Eng.*, Anaheim, Calif., October 1967, paper AC-8.
84. Marks, B. S., Mauri, R. E., and Bradshaw, W. G. *12th Bienn. Conf. Carbon*, Pittsburgh, Pa., 1975, extended abstracts, p. 337.
85. Prosen, S. P., Duffy, J. V., Erickson, P. W., and Kinua, M. A. *21st Ann. Tech. Conf.*, Soc. Plastics Ind., Section 8D, 1966.
86. Skourlis, T., Duvis, T., and Papaspyrides, C. D. *Compos. Sci. Technol.* 48, 119 (1993).
87. Chiu, H. T. and Lin, J. S. *J. Mater. Sci.* 27, 319 (1992).
88. Beck, S., In *How to apply advanced composites technology*. ASM International Congress, Dearborn, MI (1988).
89. Endo, M., *Chemtech* 18, 568 (1988).
90. Emig, G., Popovska, N., and Schoch, G., *Thin Solid Films* 241, 361 (1993).
91. Christin, F., Naslain, R., and Bernard, C., *Proc. 7th Int. Conf. on Chemical Vapor Deposition*, Electrochem. Soc., Princeton, NJ, 1979, P. 499.
92. Masuda, T., Mukai, S. R., and Hashimoto, K., *Carbon* 31, 783 (1993).
93. Masuda, T., Mukai, S. R., Fujikawa, H., Fujikata, Y., and Hashimoto, K., *Mater. Manufac. Proces.* 9, 237 (1994).
94. Mukai, S. R., Masuda, T., Fujikata, Y., Harada, T., and Hashimoto, K., *Carbon* 33, 733 (1995).
95. Fitzer, E. and Gkogkidis, A. In *Petroleum Derived Carbons* (edited by J. D. Bacha, J. W. Newman, and J. L. White), pp. 346-379. American Chemical Society (1986).
96. Costello, J. A. and Tressler R. E. *J. Am. Ceram. Soc.* 69, 674 (1986).
97. Criscione, J. M., Mercuri, R. A., Schram, E. P., Smith, A. W., and Volk, H. F., *Air Force Tech. Rep. ML-TDR-64-173.* Dayton, OH: air Force Mater. Lab. (1965).
98. Strife, J. R., and Sheehan, J. E., *Am. Ceram. Soc. Bull.* 67, 369 (1988).
99. Luthra, K. L., *J. Am. Ceram. Soc.* 74, 1095 (1991).
100. Sheehan, J. E., Buesking, K. W., and Sullivan, B. *J. Annu. Rev. Mater. Sci.* 24, 19 (1994).
101. Wei, W. C. and Wu, T. M. *Carbon* 32, 605 (1994).
102. Harding, D. R., Sutter, J. K., and Papadopoulos, D. S. *Int. SAMPE Tech. Conf.* 25, 610 (1993).

101. Wei, W. C. and Wu, T. M. *Carbon* 32, 605 (1994).
102. Harding, D. R., Sutter, J. K., and Papadopoulos, D. S. *Int. SAMPE Tech. Conf.* 25, 610 (1993).
103. Kim, J. K., Mai, Y. W., and Kennedy, B. J. *J. Mater. Sci.* 27, 6811 (1992).
104. Yamada, T., Yamada, K., Hayashi, R., and Herai, T. *Int. SAMPE Symp. Exhib.* 36, 362 (1991).
105. Mckee, D. W. *Carbon* 25, 551 (1987).
106. Sheehan, J. E. *Carbon* 27, 709 (1989).
107. Buchanan, F. J. and Little, J. A. *Corros. Sci.* 35, 1243 (1993).
108. Criscione, J. M., *ML-TDR-64-173*, part II (1974).
109. Chakraborty S., *J. Mater. Sci. Lett.* 8, 1358 (1989).
110. Wang, Y. Q., Zhou, B. L., and Wang, Z. M. *Carbon* 33, 427 (1995).
111. Manocha, L. M. and Manocha, S. M., *Carbon* 33, 435 (1995).
112. Cofer, C. G. and Economy, J. *Carbon* 33, 389 (1995).
113. Kim, D. P. and Economy, J. *J. Chem. Mater.* 5, 1216 (1993).
114. Piquero, T., Vincent, H., Vincent, C., and Bouix, J., *Carbon* 33, 455 (1995).
115. Kobayashi, K., Maeda, K., Sano, H., and Uchiyama, Y., *Carbon* 33, 397 (1995).
116. Drzal, L. T., Rich, M. J., Koenig, M. F., and Lloyd, P. M. *J. Adhesion* 16, 133 (1983).
117. Drzal, L. T. and Madhukar, M. *J. Mater. Sci.* 28, 569 (1993).
118. Weitzsacker, C. L., Bellamy, M., and Sherwood, P. M. A. *J. Vac. Sci. Technol., A* 12, 2392 (1994).
119. Cheng, T. H., Zhang, J., Yumitori, S., Jones, F. R., and Anderson, C. W. *Composites* 25, 661 (1994).
120. Sato, N., Kurauchi, T. *J. Mater. Sci. Lett.* 11, 362 (1992).
121. Yumitori, S., Wang, D., Jones, F. R. *Composites* 25, 698 (1994).
122. Wu, H. D., Chen, C. I., Chou, S. T., Wang, K. L., and Hsu, S. E. *Mater. Manuf. Processes* 8, 549 (1993).
123. Okhuysen, B., Cochran, R. C., Allred, R. E., Sposili, R., and Donnellan, T. M., *J. Adhes.* 45, 3 (1994).
124. Blackketter, D. M. and Upadhyaya, D., *Polym. Compos.* 14, 430 (1993).
125. Itoi, M., Yamada, Y., and Pipes, R. B. *Polym. Compos.* 13, 15 (1992).
126. Drzal, L. T., Rich, M. J., and Lloyd, P. *J. Adhes.* 16, 1 (1983).
127. Fitzer, E. and Weiss, R. *Carbon* 25, 455 (1987).
128. Drzal, L. T. and Madhukar, M., *J. Mater. Sci.* 28, 569 (1993).
129. Ehrburger, P. and Donnet, J. B., *Philos. Trans. R. Soc. London* A294, 495 (1980).
130. Ehrburger, P., Hergue, J. J., and Donnet, J. B., *Proc. 5th Intern. Conf. Carbon Graphite*, Vol. I, Soc. Chem. Ind., London (1978).

132. Fitzer, E., Jaeger, H., and Weiss, R., *16th Bienn. Conf. Carbon*, 1983, extended abstracts and program, p. 471.
133. Peng, J. C. and Buttry, D., summited to *"Carbon."*
134. Peng, J. C. and Buttry, D., summited to *" J. Electrochem."*
135. Le Berre, A., Delacroix, A., *Bull. Soc. Chim. Fr.,* 33, 640 (1973).
136. Wudl, F., Li, Q., Hirsch, A., *Angew. Chem. Int. Ed. Engl.* 30, 1309 (1991).

4
SURFACE PROPERTIES OF CARBON FIBERS

Tong Kuan Wang, Jean-Baptiste Donnet, and Jimmy C. M. Peng
Ecole Nationale Supérieure de Chimie and Université de Haute-Alsace, Mulhouse, France
Serge Rebouillat
DuPont de Nemours International S. A., Geneva, Switzerland

4.1 INTRODUCTION

In the past decade advances have occurred very rapidly in the field of composites. Carbon fibers with excellent mechanical properties can be produced, but the efficient translation of these outstanding mechanical properties into useful composite form through a matrix has not been achieved. The composites thus obtained have poor interfacial shear strength. This has been attributed to poor adhesion or weak bonding between carbon fiber surface and matrix molecules. The degree of adhesion between carbon fibers and binders depends considerably on the surface properties of the carbon fibers.

A complete characterization of carbon fiber surface is absolutely essential for the manufacturing of carbon fiber composites. This requires detailed knowledge of the amount and nature of the atoms or groups present on the carbon fiber surface. It is also necessary to have knowledge of the surface area, microporous structure, surface energy of carbon fiber, and the variation between different surface treatments or the adsorption of different gaseous and liquid species. Several techniques have been applied effectively to characterize the surface properties of carbon fiber, (e.g., voltammetry, infrared spectroscopy, x-ray photoelectron spectroscopy, scanning tunneling microscopy, and atomic force microscopy). All of these methods will be discussed in this chapter.

4.2 SURFACE MICROSTRUCTURE OF CARBON FIBERS

Carbon fibers are manufactured from various precursor materials, such as rayon, polyacrylonitrile (PAN), and pitches. The extent and nature of the porous structure and the surface area will, therefore, be largely dependent on the nature and morphology of the precursor material and the history of its manufacturing processes. Surface area is an important parameter that influences the interaction of a carbon fiber with the matrix materials and hence its behavior in a composite. The adhesion between matrix molecules and fibers is primarily affected by the surface properties of fiber, (e.g., its total surface area available for contacting with matrix molecules, its active surface area for forming chemical bonds with matrix functional groups, and the type and size of pores present in the internal surfaces) which could enhance the mechanical interlocking between fibers and matrixes [1]. It is important to have a better understanding of the surface properties of carbon fibers so they can be used more efficiently to improve the fiber/matrix adhesion and enhance the overall mechanical properties of composites.

4.2.1 Surface Area of Carbon Fiber

The surface area of carbon fiber is obtained by the adsorption of nitrogen or krypton at 77°K using the Brunauer-Emmett-Teller (BET) equation [2]. Carbon fibers of known weight are put inside a flask under high vacuum. A known volume of inert gas is expanded from a reservoir into the flask that contains the fibers. The inert gas is then adsorbed onto carbon fibers and a decrease of gas pressure is observed. Several measurements are repeated at different values of flask pressure. The volume of gas adsorbed by the fibers can thus be calculated by using the BET equation [2]. Adsorption depends upon the nature of the precursor material, the temperature of fiber manufacturing process, and the types or degree of surface treatments given to carbon fibers. The carbon fibers manufactured at low temperature have less organized structures, with high ratios of micropores, and therefore have high surface areas. The increase in the fiber manufacturing temperature anneals out most of the pores and reduces porosity open to the surface, resulting in a decrease in surface area. Bobka and Lowell [3], and Pallozzi [4] observed a decrease of two orders of magnitude in surface area on graphitization of rayon carbon fibers. The difference of surface area before and after high temperature treatment (graphitization) is shown in Table 1.

The surface area of carbon fiber increases notably after the surface treatments. The extent of area increase depends on the type and temperature of the treatments. The influence of oxidation with nitric acid on surface area has been carbon fiber increased considerably. Drzal [8,9] measured the surface area of PAN-based carbon fibers before and after air oxidation and observed an increase of about 10% after this treatment. This increase in surface area has been attributed to the creation of micropores and the formation of surface cracks. Bansal and Chhabra [10] studied the effect of several oxidative treatments on the surface area of PAN-based carbon fibers. The results shown in Table 2 indicate that each oxidation treatment causes an increase. This has been attributed to the formation of etch pits on the basal planes of carbon fibers. Nakao and his coworkers [11] measured the surface area of PAN-based carbon fibers before and after the electrochemical treatments in both phosphoric acid and ammonium carbonate electrolytes. The BET surface area was calculated approximately 0.3 m^2/g and they found these electrochemical treatments had little effect on the surface area.

Table 1 Surface Area of Carbon Fibers Before and After Graphitization

Carbon fiber	Surface area(m^2/g)	
	Before graphitization	After graphitization
Cellulose	260	3
VyB (Union Carbide)	340	1–2

Source: from ref. 3.

Table 2 Surface Area of PAN-Based Carbon Fiber (1000°C) After Various Oxidative Treatments

Treatment	Surface area (m^2/g)	
	N_2 adsorption at 77°F	CO_2 adsorption at 273K
Untreated	8	11
Treated with 60% HNO_3	38	45
Treated with 70% HNO_3	35	38
Treated with hydrogen peroxide	64	74
Treated with potassium persulfate	28	36

Source: from ref. 10.

Mimeault and Mckee [12] examined the influence of heat treatment in a vacuum on the surface area of PAN-based Modmer II carbon fibers and observed that the surface area increased gradually with increase in the temperature of evacuation. The value after evacuation at 1000°C, corresponded to a fourfold increase in surface area. However, in the case of Modmer I carbon fiber, the surface area increased a small extent and reached a constant value at about 600°C. The difference in the behavior of the two types of carbon fibers was affected by the history of their manufacturing processes. Modmer II carbon fibers, which had been carbonized at a lower temperature, undergo desorption of residual precursor decomposition products from the pores in the carbon fiber. Turner and Dietz [13] observed that the carbon vapor deposition on the surface of PAN carbon fibers also results in an increased surface area. The surface area was almost doubled when the deposition was carried to a weight increase of about 60%.

Tsutsumi [14] obtained linear adsorption isotherms of nitrogen, carbon dioxide, and water vapor in the low-pressure region for PAN-based carbon fibers carbonized between 800 and 2000°C and calculated the BET surface areas. The nitrogen BET area decreased as the carbonization temperature increased, reaching a maximum value at 1250°C. At a higher carbonization temperature, the surface area decreased and attained a value close to that of graphite at 2000°C HTT (Table 3). The increase in surface area at lower HTT values was attributed to the development of a porous structure, and the decrease at higher temperatures is caused by chains cross-linking. The abnormally high value of water surface area for carbon fibers carbonized at 800 and 1000°C is due to the specific adsorption of water vapor on the oxygen-containing hydrophilic centers.

Ismail [15] examined the dependence of active surface area (ASA) on the level of oxidation treatment. Figure 1 shows that as the burning-off (BO) increases from 0 to about 1%, the ASA abruptly increases by a factor of five from 0.029 to 0.15 m^2/g. ASA slightly increases when the samples are oxidized above 1%. Thus, between 1 and 85% BO, the ASA is linearly related to BO. To the same level of BO, the fibers exposed to air at 873°K have smaller ASA than the corresponding ones activated at lower temperatures (773°K). Possible reason is that at 773°K, the oxidizing agents penetrate deeper below the fiber surface as a result of larger number of internal active sites, and considerably enhance the ASA of the oxidized samples. It is also noted that the samples treated in oxygen plasma exhibit a line located between the above two extremes. It is probably due to the short residence time of the excited species, the probability of oxygen atoms penetrating deeply below the surface is small and thus pits are not formed.

Hoffman [31] patented a process to create new active sites on the fiber surface which dramatically increased both the active and the total carbon fiber surface area with a total weight loss < 1% during the process. The author deposited a single coating of oxides or metals, such as Pt, Ni, Ag, or other transition metals, and their mixtures onto carbon fiber surface, that are capable of catalyzing carbon gasification. The carbon fiber samples were stirred in a silver diamine solution for 24 hours at room temperature. Once the silver is on the surface, the fibers are heated at a rate 20°C/min to a temperature (about 500°C) at which Hoffman described the metal as moving on the surface. In other words, the deposited silver becomes mobile, and catalytic channeling or pitting starts equal to about the half the bulk melting point of the metal. Gasification is terminated when the desired weight-loss is reached. The fibers were then cooled and the silver was removed in $1N$ nitric acid at 50°C. After each surface treatment, the total surface area was measured by krypton adsorption at −195°C, and the active surface area (ASA) was measured by oxygen chemisorption at 300°C. The increase of ASA by the fiber weight loss during the process is shown in Figure 2. The author claimed that the silver catalytic oxidation was the most efficient method in increasing the ASA to a value twice as much as that obtained by the other process at the same weight loss. Meanwhile, the catalytic silver oxidation was also the most efficient technique in augmenting the total surface area of the fiber while keeping the weight loss less than 2%.

Table 3 Surface Areas of Fibers Carbonized at Various Temperatures

Fiber	$S(N_2)$ (m²/g)	$S(H_2O)$ (m²/g)	$S(CO_2)$ (m²/g)	$S(H_2O)/S(N_2)$
HF-800	0.2	201	—	1005
HF-1000	0.2	174	165	870
HF-1250	39	73	28	2
HF-1300	10	59	8	6
HF-1600	7	—	—	—
HF-2000	3	—	—	—
Graphite	3	—	2	—

Source: from ref. 14.

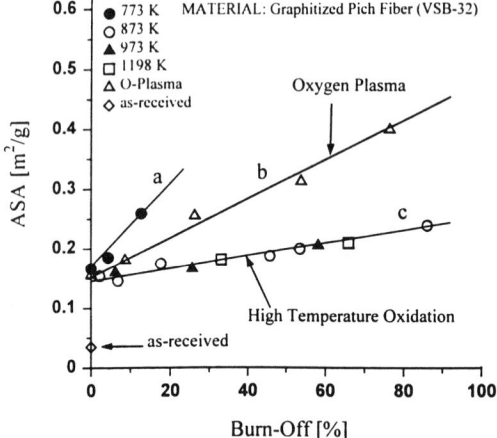

Figure 1 Relation between active surface area (ASA) and burn-off after different treatments: a: high temperature oxidation at 773°K. b: oxygen plasma. c: high temperature oxidation at 873°K, 973°K, and 1198°K (sample: VSB-32 fiber) [15]. (Reprinted with permission from Elsevier Science, Ltd.)

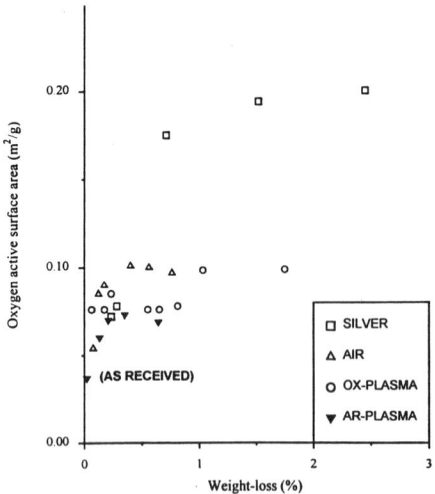

Figure 2 The increase of active surface area by the carbon fiber weight loss during different surface treatment processes [31]. (Reprinted with permission.)

4.2.2 Microporous Structures of Carbon Fibers

The presence of pores in carbon fibers has been intensively studied by the small-angle x-ray scattering method, for example, Perret and Ruland [16,17], Fourdeux et al. [18], Johnson and Tyson [19], Badami et al. [20], Peterlik et al. [21], and several other investigators [22–24]. It has been shown that carbon fibers are made up of graphite basal planes in the form of microfibrils with imperfect stacking. This imperfect stacking gives rise to empty spaces between the microfibrils, where pores or voids are generated. Although their population varies, the micropores are present in all types of carbon fibers. Generally, carbon fibers manufactured at low temperature have small but numerous pores which are distributed throughout the bulk material. However, pores become larger but less numerous when the carbon fiber is made at higher temperatures. The pores are usually needle-shaped and vary in diameter between 1 and 2 *nm*.

The development of voids in the fibers and in the matrix of a carbon/carbon composite due to different temperature treatments (1800, 2100, and 2400°C) was studied by Peterlik et al. [21] using small-angle x-ray scattering (SAXA). It was observed that the heat treatment resulted in a small increase of the pore size, and a more preferred orientation in fiber length direction. The matrix, however, exhibits a pronounced increase of pore size. Large pores develop in the material as a consequence of the stress-induced graphitization of the matrix. By using x-ray diffraction, Young [32] studied the microstructure of a series of mesophase pitch-based carbon fibers. Figure 3 presents the wide-angle x-ray diffraction patterns of P25, P75, and P120 fibers and shows well-defined (002) peak at ca. $2\theta = 26°$. There is a tail on the low-angle side for the low modulus fiber P25 that indicates certain amount of disorder in the structure and the (002) peak is asymmetric. On the other hand, in the case of P75 and P120 fibers, the presence of three-dimensional crystal structure is more evidenced by their sharp peaks. The crystal sizes along the *a*- and *c*-directions, La and Lc, and (002) *d*-spacing are listed in Table 4. The *d*-spacing of high-modulus fibers P120 and C700 is close to the ideal value for graphite of 0.335 *nm*, indicating the two fibers are both highly graphitized and have an ordered structure. In contrast, the values of the *d*-spacing increases with decreasing fiber modulus from P120 to P25, revealing a decreasing degree of graphitization and turbostractic structure. The authors found that the mechanical properties of the fibers, especially the tensile strength and the Young's modulus of the fibers, are strongly related to the fiber microstructure. Void content on the mechanical properties of carbon fiber-reinforced composites were examined [29,30]. Ghiorse [29] indicated that in a manufacturing setting, a 2% void content increase will cause an approximately 20% decrease in both interlaminar shear strength and flexural strength, accompanied by an approximate 10% drop in flexural modulus. He concluded that void content has a major consideration in composite materials' quality control, regardless of the loading mode.

The porous structure of carbon fibers has also been investigated by density measurements and by studying the adsorption of different gases and vapors. By measuring the adsorption of nitrogen, Didchenko [25] compared the surface area and pore size distribution of several kinds of carbon fibers produced at different temperatures. He observed that the type II carbon fiber (low temperature produced) showed a much larger total pore volume and surface area, whereas the high-temperature fibers had a much larger mean pore diameter. Similarly, Ismail [15] studied the microporosity of carbon fibers by measuring the CO_2 adsorption at 298°K. The carbonized polyacrylonitrile (PAN)-based and graphitized rayon-based carbon fibers as well as pitch-based fibers have small surface area and insignificant open porosity. By contrast, at a lower temperature healing, most of the carbonized rayon-based fabrics have considerable micropore volumes.

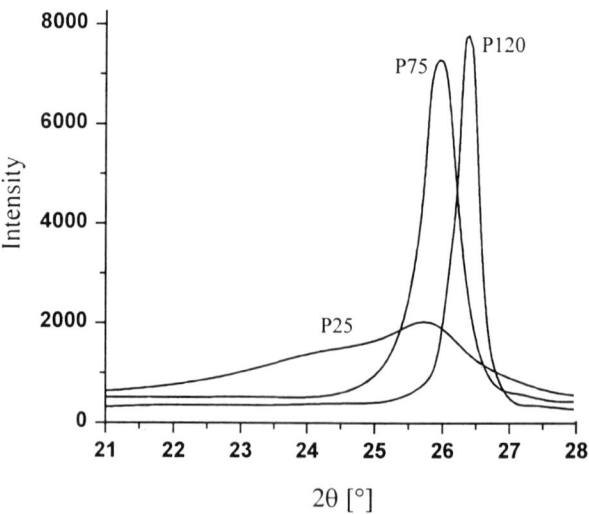

Figure 3 WAXD patterns of fibers P25, P75, and P120 in the 2θ range of 21–28° [32]. (Reprinted with permission from Chapman and Hall.)

Table 4 Structural Parameters Determined by X-Ray Diffraction and TEM

Fiber	Diameter (μm)	d-spacing (nm)	L_a (nm)	L_c (nm)	θ_h (°)
P25	11.5	0.344	5.7	3.2	19.2
P55	10.6	0.342	11.0	16.2	11.6
P75	10.7	0.341	11.4	19.6	10.1
P100	10.5	0.339	43.0	29.1	6.4
P120	11.1	0.337	45.7	37.2	4.8
S1	11.3	0.351	6.0	3.2	14.3
C700	10.2	0.337	45.7	37.2	5.0

Source: from ref. 32. (Reprinted with permission from Chapman and Hall.)

Connolly [26] investigated the degree of porosity in polymeric matrix composite materials by an IR thermogravity technique. This technique consists of heating one side of the sample with a high-power laser and observing the thermal response through a very sensitive infrared camera on the other face as the heat diffuses through the part. The temperature distribution is affected by the presence of internal defects, such as voids and cavities, and these signals will be detected by the infrared camera as brighter or darker regions on the specimen surface. Composite materials normally have low to intermediate thermal conductivities that produce a slow thermal response. Thus, this technique works well and the signals are easy to capture using standard infrared cameras. Experiments were conducted on carbon-epoxy samples with a range of known porosities and the thermal responses were measured. A good correlation is obtained between the diffusivity and the porosities for the sample tested. The results can be explained in terms of the influence of porosity on the thermal diffusivity, since small pores present a higher thermal resistance and retard the conduction of heat flow through the parts. It is known that the surface microstructure changed after a certain degree of surface treatment on the carbon fiber.

The change in microporosity of phenolic resin-based and isotropic pitch-based carbon fibers after activated in steam (a H_2O/N_2 ratio equal to 0.66 for 30min at 1170°K) and CO_2 (pure, 1170°K and 1270°K) was compared by Ryu et al. [27]. Micropore volume was determined from the adsorption isotherms of nitrogen at 77°K. Their results indicate that steam activation led to a more pronounced shift from micropores to meso and macropores than activation by CO_2.

Economy et al. [28] used scanning tunneling microscopy to compare the micropore size and shape of carbon fibers. A regular fiber with BET surface area of 5 m^2/g (ACF05) showed little or no microporosity while an activated carbon fiber with BET surface area of 1400 m^2/g (ACF15) exhibted a pronounced microporosity. Figure 4a is a 30 *nm* * 30 *nm* image of the cross-section of ACF15. The authors observed that there is a certain amount of microporosity and the micropores are oval in shape indicated by the darker regions. The width distribution in pore sizes is ranged from 0.4 *nm* to 1.6 *nm*. On the other hand, for the ACF05 fiber with little microporosity the surface is relatively smooth and uniform as seen in Figure 4b.

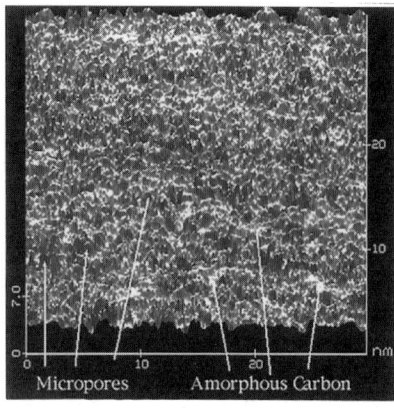

(a)

(b)

Figure 4 STM image of an activated carbon fiber, (a) ACF15, (b) ACF05 [28]. (Original photos from the authors and reprinted with permission from Elsevier Science, Ltd.)

The microstructure of a two-dimensional mesophase pitch fiber, fabricated with phenolic resin char plus chemical vapor infiltrated (CVI) matrix carbon-carbon composite was examined by scanning electron microscopy (SEM) and transmission electron microscopy (TEM) [33]. The C-C composites were fabricated from two-dimensionally woven carbon fibers, and then densified a few times by CVI using methane as precursor to modify the density, microstructure and properties of the composite. The final product was composed of approximately 50% (by volume) fiber, 35% resin char, and 15% CVI carbon. Figure 5 is a secondary electron SEM micrograph illustrating a fracture surface structure of a fiber bundle. As seen, most of the fibers are separated from their surrounding matrix. The authors indicated the weak fiber-matrix bonding could be advantageous from the point of view of tension or oxidation in assisting in the formation of a wear rubbish layer. However, depending on the amount of tension or oxidation, the authors claimed such interfacial separation can degrade the mechanical properties by facilitating fiber pullout [34] as well as oxidation resistance by stimulating oxidation in the interior of the composite. Figure 6 and 7 are the TEM images of transverse and longitudinal structures of an individual fiber. The bright field (BF) image (Figure 6a), and the dark field (DF) image of the same area (Figure 6b), reveal that a microheterogeneous structure exists within a single fiber. The numerous tiny "dots" appear to be the smaller structural unit in a single fiber, that are also identified as long, straight fibrils shown in the longitudinal section TEM micrographs. In the direction of transverse sections, the basal planes in a fiber are randomly oriented evidenced by the diffraction patterns in Figure 6c. In contrast, the basal planes are primarily parallel to the fiber axis in the longitudinal sections, illustrated by the two strong (002) arcs in Figure 7c. For a high modulus mesophase pitch fiber may be resulted from such a strong (002) preferred orientation along the fiber axis. Ju and Murdie [33] suggested that the appearance of the sharp, although weak, higher-index diffraction rings indicate a fairly good three-dimensional order within individual fibrils.

Figure 5 Secondary electron SEM micrograph showing a rough fracture surface [33]. (Original photos from authors and reprinted with permission from Elsevier Science, S.A.)

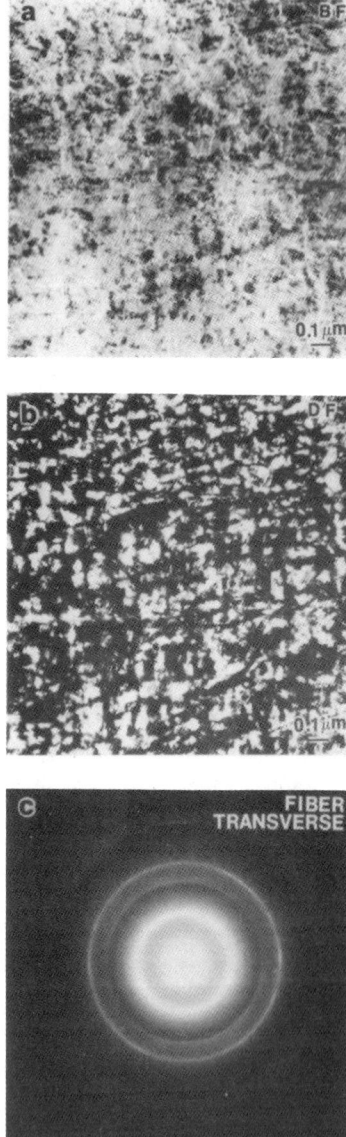

Figure 6 TEM micrographs (bright field (a), dark field (b)), and selected area (~ 1 μm) diffraction pattern (c) of a transverse fiber [33]. (Original photos from authors and reprinted with permission from Elsevier Science, S.A.)

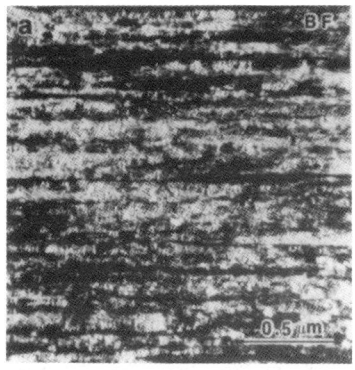
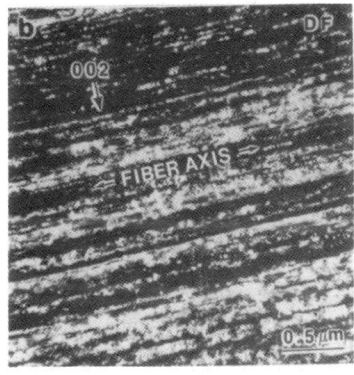
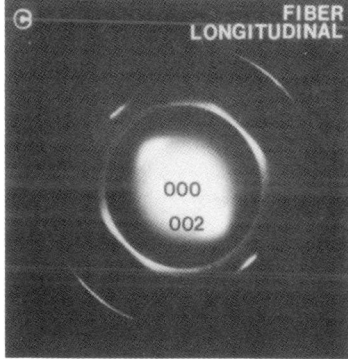

Figure 7 TEM micrographs (bright field (a), dark field (b)), and selected area (~ 1 μm) diffraction pattern (c) of a longitudinal fiber [33]. (Original photos from authors and reprinted with permission from Elsevier Science, S.A.)

4.3 SURFACE ENERGY OF CARBON FIBERS

It is well known that the mechanical properties of fiber-reinforced composites are governed mainly by the quality of the interface between the components concerned, and good wettability of the fiber surface by the matrix is necessary [35]. The wettability is determined by the surface energy (dispersive and nondispersive) of the carbon fiber, which is influenced by both the physical and chemical properties of the surface. Thus, the investigation of carbon fiber surface energy is of great significance. Several methods have been used to determine the surface energy of carbon fibers, including one liquid wetting [42], two liquid wetting [36], calorimetry [9], and inverse gas chromatography (IGC) [37]. Among them, the IGC technique has experienced great success because of its simplicity and precision, as well as its flexibility of measurement. In the IGC, carbon fibers to be characterized are used as the stationary phase, and a solute with well known properties is injected as a probe. Several thermodynamic quantities, including surface energy, can thus be derived from the retention volume of the probe passing through the column filled with the carbon fibers.

4.3.1 Inverse Gas Chromatography Method

In the IGC experiment, helium was used as a carrier with a flow rate between 20 and 30 ml/min, determined by the theoretical plate number [38]. A typical stainless steel column with a length of 60 cm and a diameter of 4.4 mm was filled with approximately 5–6 g of carbon fibers. The determination of the thermodynamic parameters of probe adsorption at infinite dilution has been extensively described in the literature [37,39,40]. The free energy of adsorption, ΔG_A, and the enthalpy of adsorption, ΔH_A, can be calculated from the experimental results of the IGC net retention volume, V_N:

$$\Delta G_A = -RT \ln V_N + C \tag{1}$$

$$\Delta H_A = -R \cdot d(\ln V_N)/d(1/T) \tag{2}$$

where R is the gas constant, T is the temperature in Kelvin, and C is a constant depending on the reference state [37].

The dispersive component of surface energy, γ_s^d, is calculated by the following equation. The free energy of adsorption per mole of methylene group, ΔG_{CH2}, is derived from the slope of the linear plot of ΔG_A of a series of n-alkanes vs. their carbon atom numbers.

$$\gamma_S^d = \frac{\Delta G_{CH2}^2}{4N^2 \cdot \sigma_{CH2}^2 \cdot \gamma_{CH2}} \quad (3)$$

where σ_{CH2} and γ_{CH2} are the group area (0.063 nm^2), and surface energy of methylene, respectively [41,42]. N is Avogadro's number.

The specific interaction, I^{sp}, can be calculated from the difference between the adsorption energies of specific probes and n-alkanes (real or hypothetical) having the same molecular area [39,40], i.e.,

$$-\Delta G_A = N \cdot \sigma_a \cdot \gamma_a^d + N \cdot \sigma_a \cdot \gamma_\alpha^{\sigma\pi} \quad (4)$$

where γ_a^d and γ_a^{sp} represent the adsorption energies per unit surface area resulting from the dispersive and specific interactions, respectively. The molecular area of the probe, σ_a, may vary with adsorbents and measurement conditions. However, for the comparison on an unified scale, the σ_a can be calculated from liquid density and molecular weight by using the spherical model for linear long-chain hydrocarbons and their derivatives [41].

At finite concentration, the adsorption isotherms were obtained by the method of "Elution by Characteristic Points," more generally called the one peak method [43], since the maxima of the chromatographic peaks fall over the diffusion side of the peak at elevated concentration (Figure 8). The quantity injected was controlled to correspond to the submonolayer region, *i.e.*, not beyond the point B in Figure 8. The partial pressure, P, of the adsorbate and the adsorbed quantity, Γ, were obtained from the equations:

$$P = RT \cdot Q \cdot H / (F' \cdot S) \quad (5)$$

$$\Gamma = Q \cdot S_{ads} / (A \cdot S) \quad (6)$$

where

Q = quantity of adsorbate injected (mole);
F' = corrected flow rate of the carrier;
A = total volume of the stationary phase;
H = height of chromatographic peak;
S = area of chromatographic peak; and
S_{ads} = adsorption area of chromatographic peak.

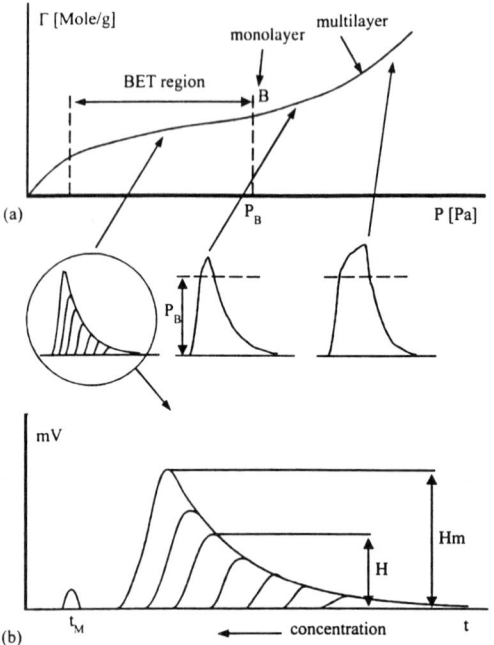

Figure 8 (a) General adsorption isotherm of Type II, and (b) chromatographic peaks corresponding to the isotherm points in (a) [50]. (Reprinted with permission from Elsevier Science, Ltd.)

The adsorption energy distribution function of sites, $\chi(E)$, which is defined as the change in the number of sites with the adsorption energy, E, has been calculated from the partial pressure dependence of the retention volume [44–46]. According to Hobson [47]:

$$\chi(E) = -\left(\frac{P}{RT}\right)^2 \cdot \frac{\partial V_N(P, T)}{\partial P} \qquad (7)$$

with

$$E = -RT \, Ln \, (P/K) \qquad (8)$$

where K is the pre-exponential factor, which is related to the molecular weight of the probe, M, and the experimental temperature, T, can be expressed as K $(mmHg) = 1.76 * 10^4 \, (MT)^{1/2}$. Several expressions for the constant K have been where K is the pre-exponential factor, which is related to the molecular weight of the probe, M, and the experimental temperature, T, can be expressed as K $(mmHg) = 1.76 * 10^4 \, (MT)^{1/2}$. Several expressions for the constant K have been reported in the literature [48]. In this regard, the adsorption energy, E, calculated by Eq. 8 can only be considered as relative. However, as one can see, the K value does not change anything but the position of the distribution function with regard to the energetic axis.

Donnet [49] studied the surface energetics of several pitch-based carbon fibers which were manufactured at different graphitization temperatues. One type of pitch fiber which was further electrochemically oxidized at four different levels was also examined. He found that surface untreated fibers exhibit a relatively high *London's* dispersive component and electron donor (base) character. On the other hand, the fibers which were treated by anodic oxidation show that the increase in the current density of the treatment increases the electron acceptor (acid) character. Donnet and his co-workers [50] discovered that the surface of carbon fibers is energetically heterogeneous with the predominance of graphitic basal plane sites. Fibers A, B, C, D, and E, were submitted to the HTTs ranging from 1500 to 2500°C through equal increments of 250°C. Fibers E1, E2, E3, and E4 were prepared by oxidizing Fiber E electrochemically (anodically) at current densities of 0.2, 0.6, 2, and 6 A/m^2, respectively. As shown in Table 5, the elevation of the final heat treatment temperatures (HTT) reduces both the dispersive component of surface energy γ_s^d, which is related to the high energy sites such as surface defects and highly polarizable heteroatoms, and also the surface acidity which is governed by the

Table 5 γ_s^d and its Temperature Dependence

Sample	γ_s^d (mJ/m²)	$-d\gamma_s^d / dT$ (mJ/m²/°C)
Fiber A	136.1 ± 2.4	0.37
Fiber B	127.2 ± 2.0	0.21
Fiber C	121.9 ± 4.1	0.20
Fiber D	101.1 ± 2.2	0.19
Fiber E	99.8 ± 1.8	0.15
Fiber E1	116.8 ± 8.1	0.29
Fiber E2	113.8 ± 3.4	0.13
Fiber E3	110.4 ± 4.8	0.15
Fiber E4	108.7 ± 2.5	0.27

[a]Values extrapolated at 20°C with 90% confidence.

surface oxygenated functional groups. The small increase of the γ_s^d after anodic oxidation was observed probably due to the creation and/or the reappearance of the high energy sites. A high current density of oxidation results in the recoverage of these high energy sites by involving the oxygen-containing groups, which possess generally smaller polarizabilities and, thus, reduced γ_s^d. It was found that the high energy sites and surface acidity play a dominant role in reinforcing ability of the carbon fibers with organic matrix.

The I^{sp} values determined from the reference line of n-alkanes are given in Table 6. As expected, the general tendency is that the augmentation of HTT diminishes the I^{sp}, while the anodic oxidation process increases the I^{sp}. A strong specific interactions of the probe CH_3CN suggests the existence of the hydrogen bond, thus, existence of functional groups containing the hydrogen, such as O-H at the surface of carbon fibers. Therefore, as shown, the drop of I^{sp} at higher HTT indicates the diminution of amounts of these oxygen-containing groups, which is in good agreement with the ESCA analysis (O_{1s}/C_{1s} atomic ratio) as shown in Table 7. Figures 9 represents the distribution of energy sites with regard to adsorption energy of n-octane on the carbon fibers to different HTTs. One can see that in the region preceding the monolayer the distribution function of sites is characterized principally by a peak situated between 44 and 45 kJ/mol of adsorption energy. It conforms the predominant role of the sites of graphitic basal planes and by a long-diffused tail toward high adsorption energies, demonstrating the surface heterogeneity of carbon fibers. It is interesting to point out that the number of sites adsorbed on Fiber A is generally inferior to that of fibers B, C, and D, but superior in the region of high energies (see the insets on both figures) where the order of magnitude of the number of energetic sites becomes identical to that measured by IGC at infinite dilution. This is not surprising since the infinite dilution is the extreme case of the finite concentration. The authors indicated that the distribution function does not vary sensitively with the temperature of measurement.

Tsutsumi and Ban [51] observed similar results in analyzing different modified carbon fibers by means of IGC. The γ_s^d value obtained from PAN-based fiber carbonized at 1280°C is 67 mJ/m^2, and the γ_s^d value remained almost the same (68 mJ/m^2) after the fibers further oxidized in a flow of O_2/N_2 gas. However, they also found the γ_s^d value dropped to 56 mJ/m^2 when the same fibers were further electrolytically oxidized in HNO_3. Cuesta and Bradley [52] reported in Table 8 that the O/C ratios for the materials used, and the dispersive component of the surface free energy of untreated and anodically or plasma oxidized fibers using V_n data for n-alkanes at 60°C. The DA/AN (donor and acceptor) numbers of the modified fibers were then calculated using Papirer's approach [53]. The results showed that the untreated fiber surface is slightly acidic with a low affinity for interaction with polar species. Anodic oxidation leads to an increase of surface acidity without much change to basic character.

On the other hand, the plasma oxidation results in enhancing base character while the acidity is similar to that produced by the anodic treatment. Bernardo et al. [54] have used IGC to study the changes in the surface energetics of two types of carbon fibers, high-strength PAN-based carbon fibers (HT sample), and ultrahigh-modulus, pitch-based carbon fibers (UHM sample), after an oxygen plasma treatment (excitation frequency, 2.45 *GHz*; power, 75 *W*; oxygen pressure, 1 *mbar*; treatment time, 3 minutes). As shown in Table 9, a strong increase in γ_s^d following the oxygen plasma treatment of both carbon fibers can be observed. The interfacial shear strength (τ) in composites prepared by embedding carbon monofilaments in a polycarbonate matrix increased from 10.3 MPa (UHM) to 33.0 MPa (UHM-O), and from 20.9 MPa (HT) to 26.5 MPa (HT-O). They concluded that the surface chemistry as established from IGC results does not account alone for the interfacial behavior of surface-treated carbon fibers.

Table 6 Specific Interactions, I^{sp}, in kJ/mol, and Acidity/Basicity Index, K_i

					Fiber				
	Probe[a]	A	B	C	D	E	E1	E2	E3	E4
I^{sp}	CH$_3$CN	18.03	12.36	/	/	/	15.27	16.03	18.04	19.03
	THF	11.90	9.41	8.54	7.68	/	15.46	15.92	17.29	18.70
	CHCl$_3$	1.43	9.00	7.87	7.60	/	9.22	/	9.12	/
	EtAc	8.25	5.67	4.92	4.66	/	12.18	12.31	14.44	14.88
	1-luoro-C6	3.96	1.93	1.59	1.52	1.44	2.23	2.14	2.53	2.57
K_i	CHCl$_3$/THF	−1.47	−0.41	−0.67	−0.07	/	−6.23	/	−8.18	/
	F-C6/EtAc	−4.29	−3.74	−3.33	−3.14	/	−9.95	−10.18	−11.92	−12.31
	ILLSS[b] (Kg/mm^2)	9.3	6.9	5.3	3.4	3.1	7.6	8.7	/	/

[a]THF = tetra-hydrofuran; EtAc = ethyl acetate; 1-F-C6 = 1-fluoro-hexane.
Source: from ref. 50. (Reprinted with permission from Elsevier Science, Ltd.)

Table 7 Analyses of Carbon Fibers Studied

Sample	Elemental analysis				ESCA	BET surface area
	%C	%O	%N	%S	%(O_{1s}/C_{1s})	(m²/g)
Fiber A	8.96	0.10	0.44	0.17	8.2	0.43
Fiber B	8.89	11	0.36	0.19	4.6	0.51
Fiber C	9.74	0.10	0.18	0.21	2.4	0.49
Fiber D	9.93	0.10	0.12	0.17	2.0	0.48
Fiber E	9.55	0.10	<0.10	0.11	2.1	0.41
Fiber E1					4.5	0.47
Fiber E2					5.5	0.45
Fiber E3					6.8	0.42
Fiber E4					10.2	0.42

Source: from ref. 50. (Reprinted with permission from Elsevier Science, Ltd.)

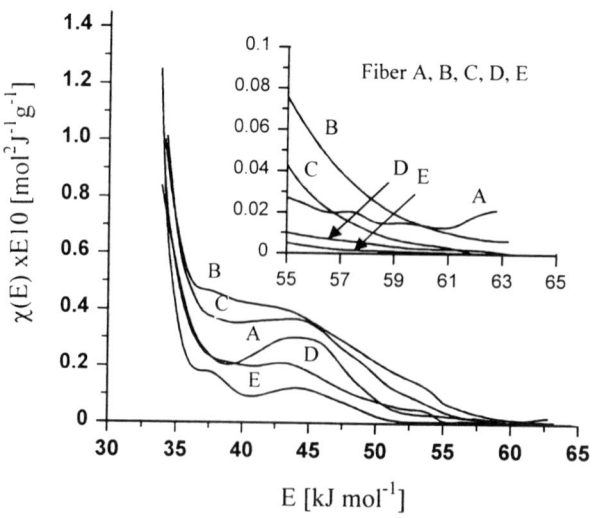

Figure 9 Adsorption energy distribution of sites probed by *n*-octane at 60°C on the surface of carbon fibers submitted to different HTTs [50]. (Reprinted with permission from Elsevier Science, Ltd.)

Table 8 Energetic Parameters and Oxygen Levels

	Untreated	Anodically oxidized	Plasma oxidized
O/C ratio	0.05	0.12	0.13
γ_s^d (mJm^{-2})	117	94	112
AN (arbitrary units)	0.21	0.60	0.60
DN (arbitrary units)	0.06	0.10	0.30

Source: from ref. 52. (Reprinted with permission.)

Table 9 Dispersive Component (γ_s^d) of the Surface Tension of Fresh and Plasma-Treated Carbon Fibers

	γ_s^D (mJm^{-2})			
T (K)	HT	UHM	HT-O	UHM-O
303.15	103.2	120.6	125.4	130.6
313.15	101.5	114.2	128.5	125.5
323.5	99.2	111.0	121.4	124.8
333.5	97.3	109.7	120.0	120.7
343.5	96.3	106.6	122.6	118.5
353.5			113.9	109.0

Source: from ref. 54. (Reprinted with permission.)

4.3.2 Contact Angle Measurement

The surface energy of solids and its two components, the polar and dispersive, could also be determined by measuring the contact angles of a series of liquids with known surface energy components. The contact angles are determined on the flat surface by using either a goniometric or a gravimetric method. However, for high-energy solids such as carbon fibers almost all the liquids spread simultaneously on the surface, hindering the measurement of the contact angle. In the case of very fine monofilaments such as carbon fibers, it is very difficult to measure accurately the contact angle directly via the usual methods which are adopted from the plane surface method. Therefore, Kaelble et al. [61] applied a Wilhelmy balance method to determine the contact angle of carbon fibers.

Table 8 Energetic Parameters and Oxygen Levels

	Untreated	Anodically oxidized	Plasma oxidized
O/C ratio	0.05	0.12	0.13
γ_s^d (mJm^{-2})	117	94	112
AN (arbitrary units)	0.21	0.60	0.60
DN (arbitrary units)	0.06	0.10	0.30

Source: from ref. 52. (Reprinted with permission.)

Table 9 Dispersive Component (γ_s^d) of the Surface Tension of Fresh and Plasma-Treated Carbon Fibers

	γ_s^D (mJm^{-2})			
T (K)	HT	UHM	HT-O	UHM-O
303.15	103.2	120.6	125.4	130.6
313.15	101.5	114.2	128.5	125.5
323.5	99.2	111.0	121.4	124.8
333.5	97.3	109.7	120.0	120.7
343.5	96.3	106.6	122.6	118.5
353.5			113.9	109.0

Source: from ref. 54. (Reprinted with permission.)

4.3.2 Contact Angle Measurement

The surface energy of solids and its two components, the polar and dispersive, could also be determined by measuring the contact angles of a series of liquids with known surface energy components. The contact angles are determined on the flat surface by using either a goniometric or a gravimetric method. However, for high-energy solids such as carbon fibers almost all the liquids spread simultaneously on the surface, hindering the measurement of the contact angle. In the case of very fine monofilaments such as carbon fibers, it is very difficult to measure accurately the contact angle directly via the usual methods which are adopted from the plane surface method. Therefore, Kaelble et al. [61] applied a Wilhelmy balance method to determine the contact angle of carbon fibers.

threefold accompanied by a triple of surface oxygen content and a double of surface nitrogen content after surface treatment of carbon fibers.

4.4 SURFACE FUNCTIONAL GROUPS ON CARBON FIBERS

Carbon fibers from various manufacturing processes have a surface composition consisting of appreciable amounts of oxygen and traces of nitrogen and hydrogen. The oxygen in carbon fibers may be derived either from the starting material that becomes part of the chemical structure of carbon fiber as a result of imperfect carbonization, or becomes chemically bonded to the surface during activation or subsequent surface treatment. All surface treatments tend to modify the interfacial region of carbon fiber composite by increasing the surface area, thus increasing the number of contact points, or by increasing the reactivity of the region by forming surface functional groups which provide a chemical type of bonding with the resin matrix. Although both surface area and surface functionality have been invoked to explain improvement in mechanical properties, it is difficult to assess quantitatively the relative importance of each parameter because both have been found to be enhanced by almost all the oxidative treatments.

4.4.1 Oxygen-Containing Functional Groups

The surface chemistry of carbon fibers is mainly contributed by oxygen-containing functional groups. The present knowledge is based on the studies by Boehm [94,95] and other workers [96,97] in the 1960s and 1970s. The oxygen-containing groups on the surface of carbon fibers according to Boehm et al. [94] is represented in Figure 10 [98]. The surface structure is composed of neutral (quinone-like structures), acidic (hydroxylic and carboxylic groups), and basic surface groups (pyrone-like structures). These acidic surface groups vary in the acidity from strongly acidic carboxylic groups to weakly hydroxylic groups. In the past, many studies have been involved in investigating acid-base neutralizations and thermal decomposition of the surface compounds into CO_2 and CO [67–70]. Emphasis has also been focused on more direct analysis of the surface oxide layer by studying specific chemical reactions [71,72], as well as by employing infrared spectroscopy [73,74], internal reflection IR spectroscopy [75,76], and polarographic techniques [77,78]. These analytical methods have yielded a good deal of information regarding functional groups, such as carboxyl, phenol, lactone, and carbonyl, present on carbon fiber surface.

Donnet and others [79–81] have carried out extensive works on the formation and estimation of oxygen functional groups on type I and type II acrylic carbon fibers under different oxidation conditions. The treatment with nitric acid resulted in the chemisorption of appreciable amounts of oxygen in type II carbon fibers, but only small amounts in type I fibers. The oxygen was present in the form of acidic surface groups that were measured by titration with decinormal sodium hydroxide solution. ESCA studies indicated the presence of carboxylic and phenolic groups. The treatment of type I carbon fibers with Hummer's reagent (mixture of $NaNO_3$, $KMnO_4$, and H_2SO_4) created an appreciably larger amount of surface functionality as shown in Table 10. It can be seen that even a moderate treatment of 3 minutes at 35°C increases the surface oxygen groups about 70-fold. Elemental analysis and surface group determination of carbon fibers oxidized anodically in nitric acid and sodium hydroxide solution showed the oxygen content increased appreciably only in the case of treatment in nitric acid. Analysis of the oxide layer carried out by a differential neutralization technique showed the presence of two types of acidic groups with varying acidic strength. Strong acidic groups titrated with sodium hydroxide were suggested to be carboxylic and phenolic, whereas weak acidic groups titrated with sodium ethoxide were designated as hydroxylic and carboxylic.

Table 10 Development of Surface Functionality on Treatment of Type I PAN Carbon Fibers with Hummer's Reagent

Time of treatment (min)	Temperature of treatment (°C)	Surface functionality (—COOH + —OH) (µEq/g)
0	35°	3
3	35°	200
3	40°	270
10	40°	400
40	40°	835

Source: from ref. 79.

Surface Properties of Carbon Fibers 255

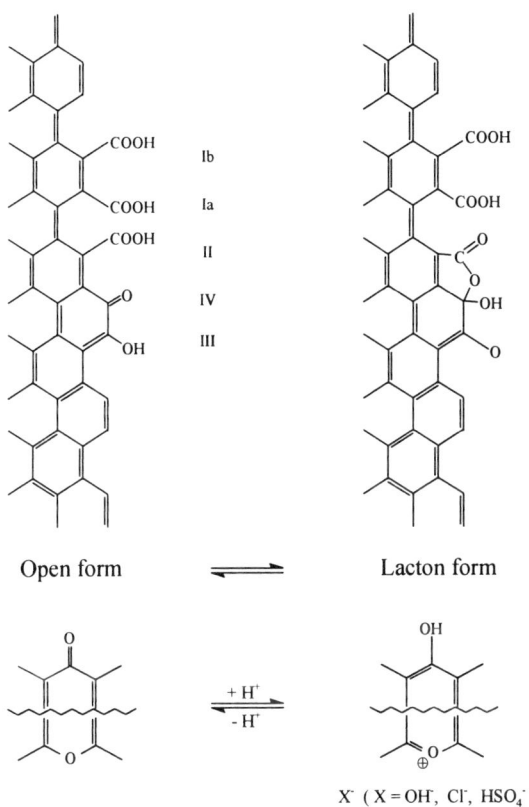

Figure 10 Possible oxygen-containing groups at carbon fiber surface [94].

Bansal et al. [82] oxidized PAN-based carbon fibers with boiling nitric acid, hydrogen peroxide, and potassium persulfate. It was observed that each oxidation treatment resulted in the chemisorption of appreciable amounts of oxygen, which was evolved as CO_2, CO, and water vapor on evacuation shown in Table 11. Treatment with a mild oxidizing agent such as hydrogen peroxide results in the highest increase in oxygen content, while treatment with 70% nitric acid results only in appreciable burn-off of fiber material without showing high chemisorption of oxygen. Most of the chemisorbed oxygen was evolved as CO_2 on high-temperature evacuation and only a small part was evolved as CO and water vapor. The evolution of CO_2 began between 300 and 400°C and was completed at about 700°C, whereas that of CO started at 600°C and was completed only at about 1000°C. This suggests that the oxygen in carbon is present in the form of different functional groups with varying thermal stability. The acidic character of the surface groups was determined by neutralization with sodium hydroxide solution. The surface groups were found to be partly acidic and partly neutral toward sodium hydroxide. The groups that are neutralized by sodium hydroxide are suggested to be carboxylic and phenolic, whereas the neutral groups may be quinonic and carboxylic. The carboxylic groups evolve CO_2 on evacuation, whereas the other groups evolve CO.

Table 11 Amount of Burn-Off During Oxidation Treatments and Amounts of Oxygen Evolved on Evacuating Untreated and Treated PAN Carbon Fibers at 1200°C

Fiber sample	Burn-off (%)	O_2 evolved (g/100g)			H_2 evolved (g/100 g)		
		CO_2	CO	H_2O	H_2O	H_2	Total
Beslon carbon fiber							
Untreated	Nil	3.80	0.30	0.18	0.02	2.22	2.24
Fiber treated with:							
60% nitric acid	9.1	7.60	0.40	1.80	0.20	1.25	1.45
70% nitric acid	27.5	6.80	0.60	0.98	0.11	0.95	1.06
Hydrogen peroxide	13.6	9.60	0.70	3.40	0.20	1.40	1.60
Potassium persulfate	6.8	4.90	0.22	3.40	0.38	Nil	0.38
Toray carbon fiber							
Untreated	Nil	3.70	0.00	0.44	0.05	1.40	1.45
Fiber treated with							
60% nitric acid	8.1	6.50	0.30	1.41	0.17	1.08	1.25
70% nitric acid	20.1	5.50	0.40	1.15	0.13	1.20	1.33
Hydrogen peroxide	9.1	7.80	0.40	1.95	0.22	1.11	1.33
Potassium persulfate	5.0	5.80	0.10	1.87	0.21	1.00	1.21
Ozonized oxygen	2.5	4.80	0.00	1.50	0.17	0.90	1.07

Source: from ref. 82.

The acidity of carbon fibers before and after an electrolytic surface treatment has been investigated by XPS [85,86,88,89] and wetting studies [86]. For analysis of the surface chemistry of the fibers, Hoffman et al. [86] employed x-ray photoelectron spectroscopy to identify and semi-quantitatively determine the functional surface groups. The XPS results of Celion fibers with different surface treatments are summarized in Table 12. A commercial-oxidation treatment (ou) Celion fibers exhibit an equal increase of the surface concentrations of all oxygen-containing functional groups in comparison with the unoxidized fibers (uu). Treating the Celion fiber (uu) in pure oxygen at 400°C for 30 minutes resulted in the highest surface concentrations of hydroxylic and ether groups, while the increase of carboxylic and lactone groups was not significant. It was suspected that the high treatment temperature of 400°C, at which these carboxylic and lactone surface groups are not stable, was responsible. On the other hand, a treatment of the Celion fibers (uu) with ozone in oxygen (0.75% O_3) at 100°C for 1 minute was found to be the most effective means of producing carboxylic groups. A comparison of the wetting behavior among three types of commercially oxidized fibers is presented in Figure 11 [86]. With increasing pH values, all fibers show a stepwise increase in the work of adhesion by a two-steps processes. The values of the work of adhesion at both extremes pH 1 and at pH 11 are very similar for all three fibers. However, in the region between pH 1 and 11 the increase pattern of three fibers is quite different. By varying the pH value, specific acid-base complexes with relevant surface groups could be formed on the carbon fiber surface. The formation of such a complex at a specific pH value results in a decrease of the contact angle (or an increase in the work of adhesion). It was demonstrated that the highest steps in the work of adhesion represent the strongest complexes formed by carboxylic groups at the carbon fiber surface [87]. A similar study of surface acidity on Amoco T300 fibers after mild electrochemical oxidation was investigated by Gustafson et al. [89]. The authors selected diazomethane, an effective reagent to methylate phenolics and carboxylics functional groups on the fibers and subsequently to form ethers and esters, to investigate the selectivity of the selective-neutralization procedure [90]. Esters are easily hydrolyzed in dilute alkali or acid, while ethers are stable in these reagents. After methylation and subsequent hydrolysis, only carboxylics will remain acidic and, therefore, titratable. It was proved from the results that the diazomethane treatment has formed methyl esters with surface carboxylic groups.

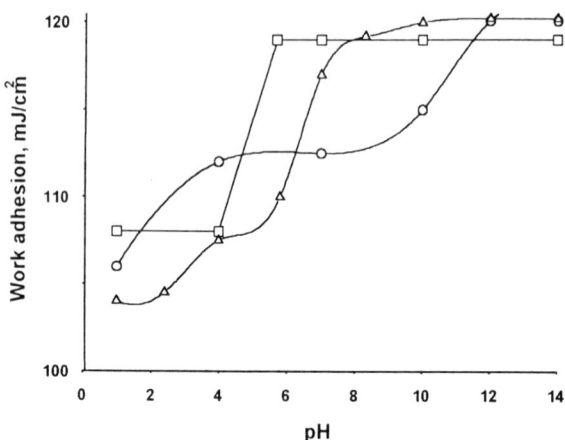

Figure 11 The work of adhesion of different types of commercially oxidized fibers as a function of the pH value of the aqueous test liquids: (□) Celion (ou), (○) Tenax (ou), and (△) AS4 (ou) [86]. (Reprinted with permission from Chapman and Hall.)

Table 12 Surface Compositions Expressed, as a Percentage of Carbon, of Differently Treated Carbon Fibers Analyzed by Fitting of C_{1s} Photoelectron Peak

		Surface composition (% carbon)			
		-C-C-	-C-O-R, -C-NR$_2$	-C=O	-COOR
Fiber	Treatment	284.7 eV	286.1 eV	287.6 eV	288.6 eV
Celion (uu)	None	76.9	18.4	3.4	4.4
Celion (ou)	None	65.9	23.8	3.8	6.5
Celion (uu)	O$_2$/400°C/0.5h	61.5	27.9	6.9	3.7
Celion (uu)	0.75% O$_3$/100°C/60S	65.8	18.3	6.3	9.7

Source: from ref. 86. (Reprinted with permission from Chapman and Hall.)

4.4.2 Other Types of Functional Groups

Several research groups [83,84,91] tried to functionalize carbon fiber surface in order to improve the adhesion of composites. A two-step functionalization of carbon fiber using diazide derivative, p-(azidosulfonyl) benzoyl azide was studied by Palma and Ibarra [83]. The first stage comprises surface oxidation of the fibers, which causes an increase of the number of superficial COOH groups. During the second stage the oxidized fibers reacted with the diazide derivative via the isocyanate group, obtained by means of the Curtius transposition of the carbonyl azide group. The functionalization onto oxidized fiber was evidenced both by the appearance of an exothermic peak around 190°C through DSC analysis due to the decomposition of the azide grafted onto the carbon fiber, and a progressive decrease of SO_2N_3 group content in the reagent solution. The degree of attainable functionalization is a function of COOH group concentration, generated during fiber oxidation, and the ratio of the functionalization agent with regard to the carboxyl groups formed. Désarmot et al. [91] introduced a difunctional chemical coupling agent shown in Figure 12 to improve the adhesion of carbon fiber composites. The first function is an unsaturated compound (methacrylic, styrenic) which is able to copolymerize with the acrylic resin, while the other end is an isocyanate, which is able to react with the hydroxyl groups present on the carbon fiber surface. Four types of difunctional compounds shown in Table 13 were synthesized and used as coupling agents. In addition, a mixture of catalysts (dibutyltindilaurate and diazabicyclooctane) has been added to the hydroxyl-isocyanate reaction between the sizing agent and the fiber. Composites were fabricated by filament winding with the acrylic resin (30% w/w) through the conventional dipping process after these sizing agents were cured onto the fiber surface. The use of carbon fibers sized with compounds 1, 2, 3, and 4 resulted in a substantial improvement of the fiber/matrix adhesion as presented in Table 14, in which the matrix has been cured by electron irradiation. They obtained values of the interlaminar shear strength and shear debonding stress are at least 40 % higher than those for unsized fibers (IM6 and IM7) or epoxy-sized fibers (IM7G). The best interlaminar shear strength results have been obtained with compounds 3 and 4 (about 100% increase). This may be explained by the better mobility and accessibility of the isocyanate group in compound 3 or 4 linking to six methylenic groups for the immobilized processes, compared to the steric hindrance of the isocyanate group in compound 1 in the ortho position of the aromatic ring.

Table 13 Difunctional Unsaturated and Isocyanate Compounds Synthesized

Diisocyanates	Unsaturated-hydroxy compounds	Products
2,4-Toluylenediisocyanate (CH₃-C₆H₃(NCO)₂)	HOCH₂CH₂OC(=O)C(OCH₃)=CH₂ Hydroxyethylmethacrylate (HEMA)	1
OCN—C₆H₄—CH₂—C₆H₄—NCO Methylenediphenyldiisocyanate	HOCH₂CH₂OC(=O)C(OCH₃)=CH₂ Hydroxyethylmethacrylate (HEMA)	2
OCN—(CH₂)₆—NCO Hexamethylenediisocyanate	HOCH₂CH₂OC(=O)C(OCH₃)=CH₂ Hydroxyethylmethacrylate (HEMA)	3
OCN—(CH₂)₆—NCO Hexamethylenediisocyanate	CH₂=CH—C₆H₄—CH₂OH Hydroxymethylstyrene	4

Source: from ref. 92. (Reprinted with permission from Elsevier Science, Ltd.)

Peng and Buttry [84] have demonstrated that both thermal reactions and electrochemical oxidation of amines at carbon fibers allow the covalent bonding of these molecules directly to the fiber surface, presumably via nucleophilic attack of the amine at electrophilic C=C sites at the surface and subsequent formation of C-N bonds between the surface and the amine. Tonen HMU pitch-based carbon fibers were thermally reacted with 1,6-diaminohexane, and then an electrochemical method was developed to assay the yield of amines immobilized at the surface. This method uses the electrostatic binding of $Fe(CN)_6^{-3}$ to the protonated, cationically charged interfacial amine groups as a way of "marking" these cationic groups. Then, an electrochemical measurement is used to determine the number of interfacial metal complexes, from which the

Surface Properties of Carbon Fibers

number of amine groups can be calculated. Figure 13A presents a set of cyclic voltammetry data typical of the response of amine-derivatized carbon fibers at a variety of concentrations of the metal complex in solution, and Figure 13B is the results of control experiment for the bare (untreated) carbon fiber. The curves in Figure 13A clearly show the accumulation of excess $Fe(CN)_6^{-3}$ at the surface as its concentration in solution is raised, as evidenced by the considerably larger peak current for its reduction at the derivatized fiber (at ca. 0.22 V) than at the bare fiber but not so much as for the reduction process. Two possible reasons were explained by the authors, could be either due to ion pairing between $Fe(CN)_6^{-4}$ at the cations from the supporting electrolyte [92], or the spatial distribution of the cationic amine groups hinders adsorption of the more highly charged species. Quantitative measurement of the relevant reductive charge was accomplished by measuring the area under the peak and subtracting the area under the matching curve for the control experiment. These data are shown as a reciprocal Langmuir plot [93] in Figure 14. The plot shows that the behavior of the "adsorbing" $Fe(CN)_6^{-3}$ species is reasonably well approximated by the Langmuir equation. A linear least squares fit through the data gives a saturation surface coverage for $Fe(CN)_6^{-3}$ of $3.3 * 10^{-11}$ mol cm^{-2}. Assuming a 3:2 ratio for amine groups to $Fe(CN)_6^{-3}$, this gives approximately $5 * 10^{-11}$ mol cm^{-2} for the amine coverage.

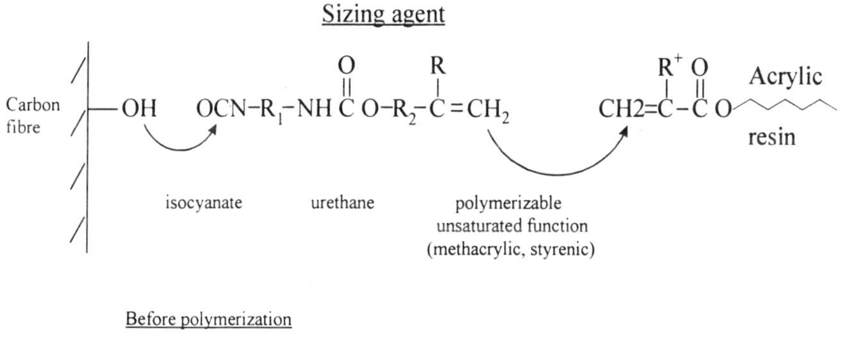

Figure 12 Nature of the sizing agent: difunctional chemical coupling agent [92]. (Reprinted with permission from Elsevier Science, Ltd.)

Table 14 Influence of Sizing Agents on Composite Mechanical Properties

Fiber	Sizing agent	Matrix	Interlaminar shear strength (MPa)	Shear debonding strength (MPa)	Process of polymerization
IM6	–	M_1	30 ± 3	40 ± 3	electron bean
IM7	–	M_2	26 ± 2	49 ± 3	electron bean
IM6	G	M_1	25 ± 3	42 ± 3	electron bean
IM7	G	M_2	32 ± 2		electron bean
IM7	G	epoxide	62 ± 3		thermal
IM6	1	M_1	40 ± 3	63 ± 3	electron bean
IM6	2	M_1		70–90 ± 3	electron bean
IM7	3	M_2	50 ± 2	113 ± 4	electron bean
IM7	3		49 ± 2		thermal
IM7	4	M_2	49 ± 3		electron bean

Source: from ref. 92. (Reprinted with permission from Elsevier Science, Ltd.)

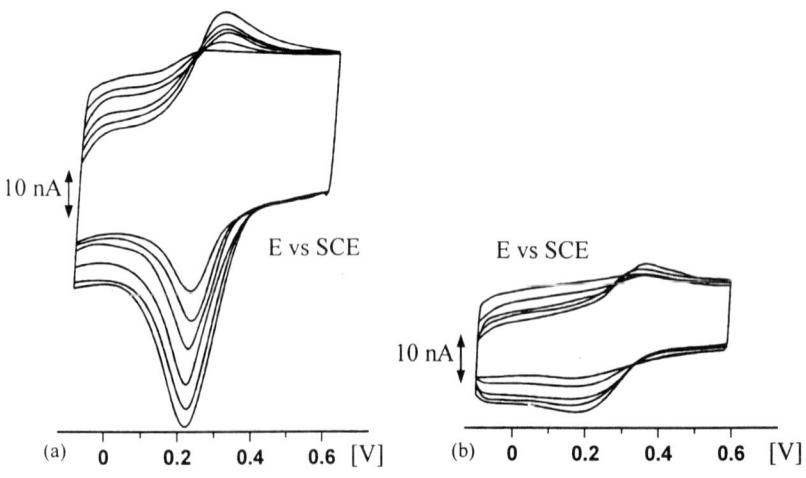

Figure 13 Cyclic voltammograms of a Tonen HMU pitch-based carbon fiber in solutions of various concentrations of $Fe(CN)_6^{-3}$, (a) 1,6-diaminohexane-treated fiber, concentration of $Fe(CN)_6^{-3}$: 0.5 μM, 1 μM, 1.5 μM, 2 μM, 2.5 μM, (b) untreated fiber. Geometrical surface area of carbon fiber electrode: $1.3 * 10^{-2}$ cm^2. Scan rate: 100 mV/s [84].

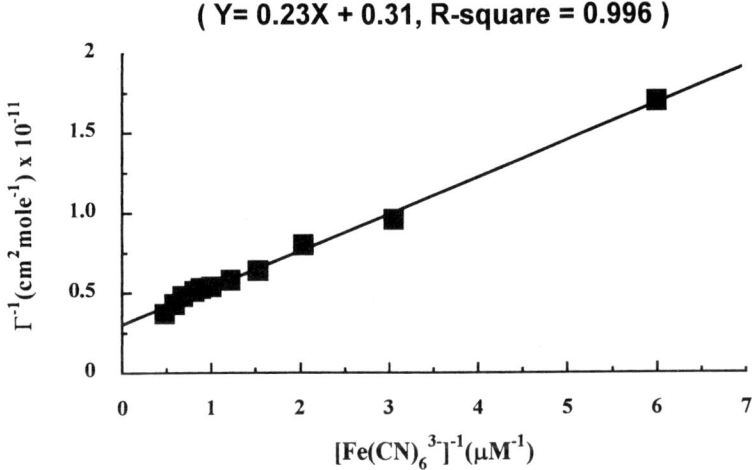

Figure 14 Reciprocal Langmuir isotherm of $Fe(CN)_6^{-3}$ adsorbed on 1,6-diaminohexane-derivatized HMU carbon fiber electrode. Electrolyte: 0.1 M $HCl/NaCl_{(aq)}$ (pH = 1.25) [84].

4.5 ADSORPTION CHARACTERISTICS OF CARBON FIBERS

The amount of activated carbon used in water treatment facilities to control organic pollutants shows a steady increase because of more strict environmental standards. Protection of human beings against hostile environmental conditions is an important aspect of the adsorption process. The production of highly effective fibrous carbon adsorbents with smaller diameter, excluding or minimizing external and intradiffusional resistance to mass transfer, and there exhibiting high sorption rates, is a challenging task for researchers in the science and technology of carbon fibers (see also Chapter 7). These carbon adsorbents can be converted into a wide variety of textile forms and nonwoven materials, with low hydrodynamic resistance, to be used in thin layers for treatment of high gas flows. These materials will increase efficiency and permit far greater flexibility and simplification in the design of sorption processes for environmental pollution control. Carbon fibers could also be potential molecular sieve materials, as they have limited pore diameter. Adsorption characteristics of activated carbon are determined by its pore structure (magnitude and distribution of pore volume), and surface functionality (kind and quantity of surface-bound heteroatomic functional groups). Surface treatment can affect adsorption of organic substances through the modification of pore volume distribution and surface chemistry. Thus efforts have been under way for almost a decade to prepare cheaper activated carbon fibers with an uniform pore-size distribution of high surface area.

4.5.1 Organic Compounds Adsorption

Rand and Robinson [99] characterized PAN carbon fiber surfaces by studying the adsorption of n-butylamine, n-butyric acid, n-butanol, and 1,2-epoxypropane from n-heptane solutions using flow microcalorimetry. The adsorption of n-butanol and 1,2-epoxypropane was reversible and relatively unaffected by surface treatment of the fiber. However, the adsorption of n-butylamine was very strong, partially irreversible, and increased markedly by surface treatments. There was no adsorption of n-butyric acid observed on carbon fibers. The area of exothermic peaks of irreversibly adsorbed n-butylamine, which was considered to be a semiquantitative measurement of the surface acidity, increased by the degree of surface treatment. On the other hand, the external area of the carbon fiber was almost unchanged, which suggested the fraction of the surface acidic groups had increased.

Brooks et al. [7] studied the adsorption behavior of rayon carbon fibers using spectrophotometry and atomic absorption analysis before and after oxidation with air and nitric acid in several cationic solutions, such as sodium hydroxide, methylene blue chloride, and lithium hydroxide, and anionic solutions such as chloroplatinic acid and metanil yellow. The oxidation with nitric acid or air resulted in a remarkable increase (up to 10 times) in the adsorption of cationic species but decreased the adsorption of anionic species by a factor of one-sixth to one-third, which is summarized in Table 15. There is, however, a large variation in the adsorption of different solutes which can not be attributed to their molecular dimensions because methylene blue, the largest cation, shows the largest adsorption capacity after an 8-hour nitric acid treatment whereas lithium, the smallest ion, shows the largest adsorption capacity after the 12-hour oxidation. The adsorption of metanil yellow is favored over that of methylene blue in untreated and air oxidized carbon fibers, but this trend is reversed for the nitric acid-treated fibers. The increase in the adsorption of cationic solutes has been attributed to the creation of acidic functional groups such as carboxyl and hydroxyl during oxidation treatment. The decrease in the adsorption of anionic solutes on oxidation is caused by the loss of surface carbonyl groups, which provide hydrogen bonding to anionic solutes.

Table 15 Adsorption of Cationic and Anionic Solutes from Aqueous Solutions on HMG50 Carbon Fibers After Various Treatments

Fiber treatment	Amount of cation adsorbed (mol/m²)			Amount of anion adsorbed (mol/m²)	
	Li^+	Na^+	Methyl chloride	$PtCl_6^-$	Metanil yellow
Untreated	0.120	0.080	0.117	0.605	0.717
Oxidized in air	0.173	0.138	0.171	0.229	0.347
Oxidized in HNO_3 for 4 hr	-	0.599	0.738	0.260	0.266
Oxidized in HNO_3 for 8 hr	0.204	0.227	1.320	0.255	0.585
Oxidized in HNO_3 for 12 hr	2.10	1.380	1.090	0.117	0.420

Source: from ref. 7. (Reprinted with permission from Elsevier Science, Ltd.)

Tsutsumi [100] studied the adsorption of nitrogen, carbon dioxide, and water vapor of PAN-based carbon fibers carbonized between 800 and 2000°C. The adsorption isotherms shown in Figure 15 of nitrogen at 77°K and of water vapor at room temperature were type II of the BET classification. The adsorption of both nitrogen and water vapor decreased with increased carbonization temperature. This was attributed to the gradual removal of oxygen-containing hydrophilic surface groups from the carbon fiber surface. The adsorption isotherms were linear in the low-pressure region and were used to calculate the surface area using the BET equation. The nitrogen BET area increased with increased temperature, attaining a maximum value at 1250°C (represented by the curve HF-1300) and showing a decrease thereafter at higher carbonization temperatures, and attained a value close to that of graphite at 2000°C HTT. The increase in surface area at lower HTT values has been attributed to the development of a porous structure, and the decrease at higher temperatures to increase cross-linking. The abnormally high value of water surface area for carbon fibers carbonized at 800 and 1000°C (curves HF-800 and HF-1000) is due to the specific adsorption of water vapors on the oxygenated hydrophilic centers.

Liquid-phase adsorption of organic compounds by granular activated carbon (GAC) and activated carbon fibers (ACFs) was studied by Lin and Hsu [101]. Me_2CO, phenol and THF were used as model adsorbates. The results show that adsorption of organic compounds by GAC and ACF depends on the Brunauer-Emmett-Teller (BET) surface area of the adsorbate. The authors discovered that the adsorption characteristics of GAC and ACFs differ rather significantly. ACFs are considerably better than GAC in terms of the adsorption capacity of organic compounds, the time to reach equilibrium adsorption, and the time to complete desorption. Adsorbed by selected organic compounds, GACs are characterized well by the Langmuir isotherm, while the ACF adsorption can be described adequately by either the Langmuir or the Freundlich isotherm. They also did the column adsorption tests and the results indicated that the exhausted ACFs can be regenerated by staticly in situ thermal desorption at 150°C, but not as effective for the GACs under the same regeneration conditions.

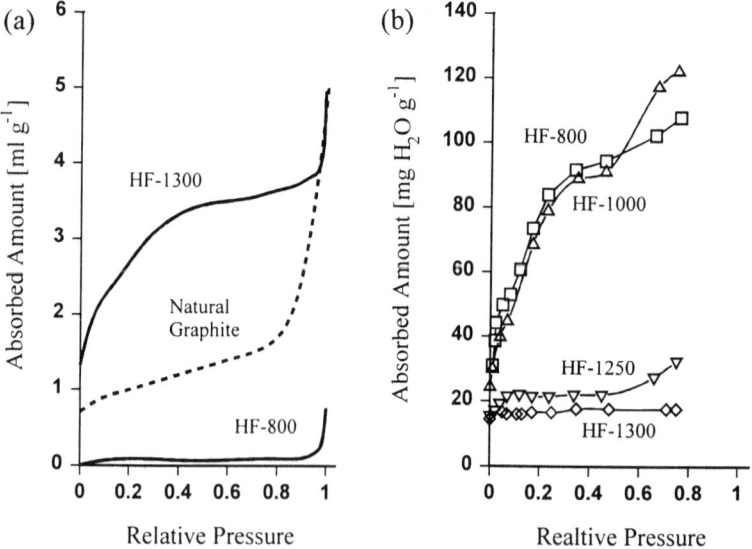

Figure 15 Adsorption isotherms of (a) nitrogen and (b) water vapor on carbon fibers carbonized at different temperatures [100]. (Reprinted with permission).

4.5.2 Moisture Adsorption

The moisture adsorption [102–105] is the first consideration to determine the suitability of carbon fiber composite materials for structural applications for continuous immersion in seawater. Bradley and Grant [102] studied the moisture adsorption of different types of fiber composites when soaked at ambient temperature in simulated seawater with 20.7 MPa (3000 psi) hydrostatic pressure. The laminated composites are composed of graphite and glass fibers with thermosetting resins which are mixed from tetraglycidlyl-4,4'-diaminodiphenylmethane with diaminodiphenyl sulfone crosslinking agent, fluorene-based resins, and vinyl ester resins. The matrix of the composites adsorbs moisture with saturation occurring at 0.6%–2% of the matrix weight of additional moisture over approximate 1% present after fabrication. Pure water absorption resulted in a slightly higher saturation level than that obtained with simulated seawater. With the exception of the graphite/vinyl ester composites, the degradation in transverse tensile strength and interfacial shear strength due to moisture absorption was found to vary from 0%–22%. The observed correlation in the decrease in interfacial shear strength due to moisture absorption with decreases in transverse tensile strength supports the hypothesis that the moisture-induced degradation is associated with a decrease in the interfacial strength rather than the degradation of matrix mechanical properties. In situ fracture observations with a scanning electron microscope further support this hypothesis.

The moisture absorption behavior in T300/934 graphite/epoxy composites was investigated by Zhou and Lucas [104]. Specimens were immersed in distilled water at different temperature conditions (45, 60, 75, and 90°C) for more than 8000 h. Figure 16 [104] present some of the optical micrographs of polished specimens exposed at different environment conditions. There are no visible cracks in the dry specimen, whereas the 75°C and 90°C specimens clearly exhibit multiple cracks after being immersed in water for 4300 h. The profiles of the moisture ratio for graphite/epoxy material exposed at different temperatures are shown in Figure 17 [104]. Solid lines represent a theoretical weight gain by the assumption of obeying Fickian diffusion [106], and the four symbols represent experimental data. Apparently, the experimental data of the lower temperature curves (45 and 60°C) agree better with the Fickian diffusion curves. However, differences between the theoretical and the experimental data profiles are more profound when time increases. In other words, the non-Fickian behavior is inversely related to the exposure temperature. The authors concluded that the anomalous (non-Fickian) behavior in the composite resulted from chemical modification and physical damage to the epoxy resin. As known, cracks, voids, and surface peeling occur with increasing exposure temperature. Thus, cracks can retain water which contribute to absorption behavior higher than the theoretical Fickian diffusion curve (seen by the 75 and 90°C experimental dots higher than their theorical lines). In contrast, surface peeling and dissolution contribute to reduction in the specimen weight and, consequently, the weight change profile data falls below the theoretical Fickian diffusion curve. They concluded that cracking and mass loss are the two main factors in the weight change profile. It depends on which mechanism that dominates the experimental data, real profile either higher or lower than the theoretical Fickian curve is possible.

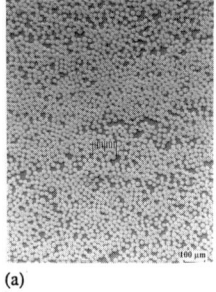
(a)

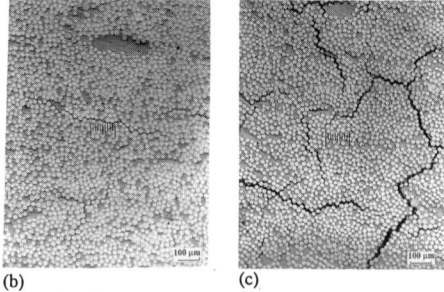
(b)   (c)

Figure 16 Optical micrographs of T300/934 graphite/epoxy composite before and after immersion in water for 4300 h: (a) dry specimen, no crack can be seen: (b) 75°C: and (c) 90°C specimens. Both of the 75 and 90°C specimens have visible cracks [104]. (Original photos from the authors and reprinted with permission from Elsevier Science, Ltd).

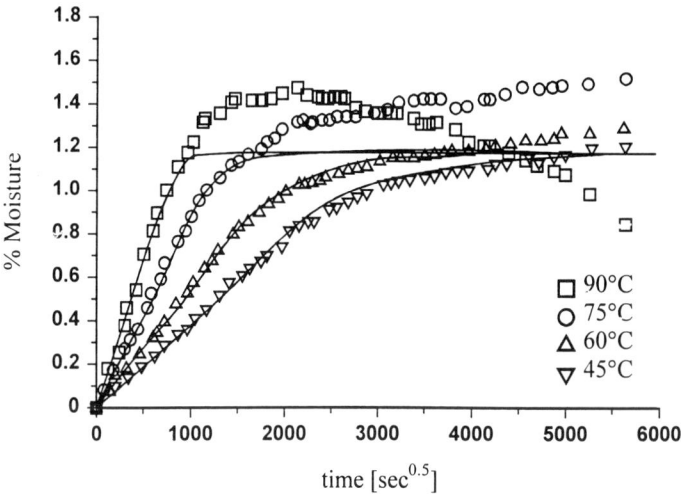

Figure 17 The weight change of T300/934 graphite/epoxy composite immersed in distilled water at different temperatures. Solid lines represent theoretical Fickian diffusion and the symbols are the experimental data at different exposure temperatures [104]. (Reprinted with permission from Elsevier Science, Ltd.)

In order to determine the content of primary adsorption centers, created on the fiber surface after the oxidative treatment, the adsorption values of water vapor were measured [107] at 298°K by the gravimetric method. It was believed that the oxidized carbon fibers possess oxygen-containing groups on their surfaces which form polar (more active) adsorption centers. The polar centers can be regarded as the primary active sites for adsorption of water molecules either as single molecules or perhaps with some cluster formation [108]. The results in Table 16 show that the oxidized fibers (1, weak; 2,3,4, strong, in order of increasing degree of oxidation) contain a greater number of primary adsorption centers than the epoxy-treated fibers (marked as CE), and the adsorption (a-value) increases with the degree of oxidative treatment. By the XPS results [107], the oxidized fibers exhibit a higher content of carbonyl and especially carboxyl functional groups which preferentially act as primary sites for water vapor adsorption, as determined by adsorption measurements. Active sites present on the fiber surface are potentially important in the formation of chemical bonding with epoxy-matrix molecules [109]. Carboxyl and ester groups were proposed to be more important for fiber/matrix adhesion [110], because these functional groups are able to form chemical bonds at the interface

Table 16 Adsorption Values of Water Vapor at a Relative Pressure of 0.25, the Contact Angle Data for Carbon Fibers with Different Surface Treatment and ILSS of the Corresponding Composites

Fiber	a(molkg^{-1})	Epoxide				Glycerol	ILSS (MPa)
		α_{meas} (deg)	θ (deg)	CV (%)[a]	S.D.[b]	α_{means} (deg)	
CO-1	0.1161	56.5	62.8	3.95	2.13	57.5	51
CO-2	0.2349	52.5	57.3	3.31	2.77	49.5	58
CO-3	0.2895	48.0	47.1	6.85	3.26	46.0	69
CO-4	0.2999	47.7	42.0	7.50	4.50	45.0	69
CE	0.1432	55.3	59.5	3.66	1.96	44.0	56
CU	—	62.8	71.5	2.90	2.10	59.8	37

[a]CV, coefficient of variation.
[b]S.D., standard deviation.
Source: from ref. 107. (Reprinted with permission from Chapman and Hall.)

with matrix molecules. Therefore, the authors suggested [107] that the higher ILSS values of composites shown in Table 16, observed from oxidized (CO) fibers, should be primarily related to a higher degree of oxygen-containing groups at the CO fiber surface, and secondly the reaction between the carboxylic acid groups with the epoxy-end groups.

4.6 SURFACE CHARACTERIZATION METHODS

Since the 1970s a tremendous amount of research works has been carried out using the newly developed instrumentation such as Fourier Transform Infrared Spectroscopy (FTIR), X-ray Photoelectron Spectroscopy (XPS), Auger Electron Spectroscopy (AES), Scanning Tunneling Microscopy (STM), and Atomic Force Microscopy (AFM). The examination of carbon fiber surface by these methods has undoubtedly led to a better understanding of the processing and functioning of this material as a filler. Among them, STM and AFM adapted to the investigation of carbon fiber surface properties, have been shown as powerful analytical tools to study the surface properties of carbon fiber at the atomic scale.

4.6.1 Infrared Spectroscopy (IR)

The surface structure of carbon fiber is still not well understood due in part to the technical limitation of the spectroscopic instrument used. XPS is very sensitive to the elemental detection of surface species, but it is not sensitive enough to minor chemical structural differences which are important in understanding the adhesion mechanism of carbon fiber to the resin matrix. Even though Fourier transform infrared (FTIR) spectroscopy has been successfully adopted to study Kevlar fiber surface [111] and glass fiber composite interface [112], the severe scattering phenomena and the high absorptivity of carbon fibers make this analysis a difficult task. However, reflectance measurements, such as diffuse reflectance (DR) technique and attenuated total reflection spectroscopy (ATR), offer an alternative nondestructive method with a high sensitivity for the surface study.

The spinning and stabilization steps involved in the carbon fiber manufacturing processes have been analyzed by Fourier transformed IR spectroscopy [113]. The results obtained show that the starting material, a petroleum pitch precursor with a low amount of oxygen groups, is mainly constituted by highly substituted aromatic groups that do not form condensed aromatic structures. A chemical transformation of the pitch does not happen in the spinning step, however, a molecular redistribution and ordering is produced as deduced from the irreversibility of successive pitch melting processes. The oxidative stabilization starts at temperatures lower than the melting temperature being mainly restricted to the surface of fiber. At higher temperatures, a large amount of oxygen arrives in the bulk of the fiber producing a significant amount of carbonyl, carboxyl, ether, and alcohol groups. In the final step of the stabilization (T ≈ 573°K) the presence of an excess amount of anhydride groups and almost complete consumption of the aliphatic hydrogen are clearly observed.

4.6.1.1 FTIR photoacoustic spectroscopy

Yang [114,117] also used FTIR photoacoustic spectroscopy (PTIF/PAS) to evaluate the degree of oxidation in the near surface of the stabilized fiber, and to study the distribution of the oxidation products between the surface and the bulk of the precursor carbon fiber during the stabilization process. A photoacoustic cell is used as a detector in a photoacoustic infrared spectrometer. Infrared radiation modulated in an interferometer impinges upon the sample. The infrared radiation absorbed by the different vibration modes in the sample is converted to heat, which propagates onto the surface of sample and causes pressure variation in the cell, thus generating photoacoustic signals [118]. The author found that the distribution of the oxidative products is homogeneous during the first 70 minutes of the stabilization process. However, the near surface of the carbon fiber is completely oxidized, whereas the bulk has a lower degree of oxidation, in the last 10 minutes of the process. It was also found that inferior mechanical properties of the fiber may be caused by an incomplete oxidation in the near surface of a precursor carbon fiber. An increase in the rate of the stabilization process temperature reduces the degree of oxidation in both the near surface and the bulk of the precursor carbon fiber. In this research, FTIR/PAS has demonstrated the ability to differentiate the near surface properties of a carbon fiber from its bulk. The authors used the same technique to study the carbonization processes of the petroleum pitch precursor carbon fiber [115]. Various saturated and unsaturated aliphatic hydrocarbon groups, substituted aromatic ring structures, as well as small amounts of carbonyl and hydroxyl were detected in the green precursor fiber. Dehydration and elimination of carbonyls and hydrocarbons in the stabilized carbon fiber were detected in the carbonization process.

4.6.1.2 Diffuse reflectance IR

The diffuse reflectance technique has been proved to be extremely sensitive and useful in a number of applications [127–129]. The principal advantage of diffuse reflectance for the study of composites is the fact that it is essentially a nondestructive method. The infrared beam is directed onto the sample surface, which could be inhomogeneous and rough, and a portion of the back-scattered radiation is captured and analyzed. No physical contact with the surface required for DR is essentially an advantage, because ATR is designed with an intimate contact between the sample surface and the prism. Therefore, it is usually limited to samples which are either deformable or else have a very smooth surface. Cole and Casella [116] used FTIR spectroscopy to study the thermal degradation that occurs in composite materials consisting of PEEK reinforced with carbon fibers. Spectra were measured by diffuse reflectance for samples of prepreg and molded laminate aged at 400–485°C for different periods of time in both air and nitrogen atmospheres. Despite of some spectral distortion arising from front-surface reflection, useful information could be obtained in regard to the reactions occurred under different conditions. their activation energy of the reaction is similar to that observed for neat PEEK film. They found the degradation in an inert atmosphere involved a pyrolytic-type mechanism which produced a new CO species absorbing at 1711 cm^{-1}, probably a fluorenone-type structure; the activation energy for this process was 240 kJ mol^{-1}. On the other hand, the same species was produced at a faster rate in an oxidizing atmosphere, but with a lower activation energy of 200 kJ mol^{-1}. In addition, there was a reaction detected which required the participation of oxygen and produced a species absorbing at 1739 cm^{-1}, probably ester group; the activation energy for this reaction was 130 kJ mol^{-1}. Some samples were also analyzed by photoacoustic spectroscopy. Within the limits of the technique, no significant difference could be noticed between the degradation behavior of PEEK-carbon composites and that of neat PEEK films.

Xue et al. [121] pretreated carbon fibers with copper 2-mercaptobenzimidazolate (Cu^+MBI^-) and copper benztriazolate (Cu^+BAT^-) inner complexes, and observed an enhancement of the interfacial interaction between the treated carbon fiber and the epoxy resin. The DRIR spectrum of the Cu^+MBI^- inner complex film covering the fiber surface is shown in Figure 18A. In comparison with the IR spectrum of bulk MBIH in the solid state (Figure 18B), the decrease in the intensity of the N-H stretching vibration absorption in the region 3300-2300 cm^{-1}, the C=S stretching absorption near 1180 cm^{-1}, and the N-C=S vibration absorption at 920 cm^{-1} indicate deprotonation of some of the amino groups and opening of the C=S bond. The tensile strength data for CF-reinforced epoxy composites are presented in Table 17. Although the tensile strength of these three kinds of CF/epoxy composites was reduced to varying degrees after immersion in water or salt water at 25°C, the treated CFB or CFM-reinforced epoxy composites possess a better resistance to water and salt water than does the composite reinforced with untreated CF.

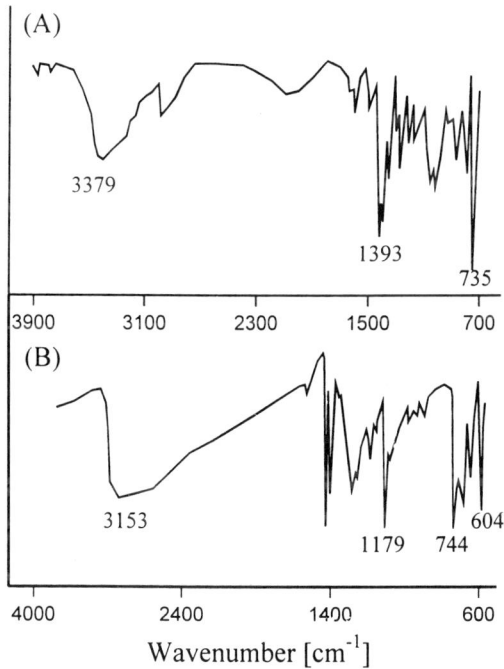

Figure 18 (A) DRIR spectrum of the MBIH-Cu inner complex film on the CF surface; (B) IR spectrum of neat MBIH [121] (Reprinted with permission).

Table 17 Tensile Strength for Carbon Fiber-Reinforced Epoxy Composites

Sample	Average tensile strength (MPa)		
	Stored in air for 10 days	Immersed in water for 10 days	Immersed in salt water for 10 days
CF/Epon 828	96.4	66.7	52.1
CFB/Epon 828	104.2	72.9	67.0
CFM/Epon 828	118.7	85.8	100.8

CF: Carbon fiber; CFB: BTAH-Cu complex-treated CF; CFM: MBIH-Cu complex-treated CF.
Source: from ref. 121. (Reprinted with permission.)

4.6.1.3 Fourier transform infrared attenuated total reflection spectroscopy

Fourier transform infrared attenuated total reflection spectroscopy (FTIR-ATR) has been employed to study the effects on the carbon fiber surface functionalities after the surface treatments [119,120,122,123]. This method has been shown to be an useful technique in the characterization of chemical functionalities and orientation of molecules at surfaces or in thin films [124]. A criterion for using ATR on samples is the refractive indices of the optical element and the sample, n_1 and n_2, must obey the condition $n_1 > n_2$ in order to obtain total reflection of the incident light. Also, the incident angle has to be larger than the critical angle θ ($sin\ \theta = n_2/n_1$). These conditions can be satisfied by using a germanium element due to its highest refractive index of the commonly used ATR materials ($n_{ZnSe} = 2.4$, $n_{KRS-5} = 2.4$, $n_{Ge} = 4.0$), resulting in total reflection of the incident light even for a sample such as graphite, with a refractive index of 3.3 [125]. The ATR configuration maximizes interaction of the IR source radiation with the sample surface, rather than its bulk, which leads to a relative enhancement in the contribution of species at the surface to the spectrum.

Peng and Buttry [119] have recently proposed a new type of reaction which has relevance to carbon fiber-epoxy interfacial adhesion. This is the reaction between amines, acting as nucleophiles, and the vinylic C=C bonds on the carbon fiber surface, acting as electrophiles. Several types of amines, including didecylamine and 1,6-diaminohexane, were thermally reacted with Tonen HMU pitch-based carbon fibers at a neat condition. The ATR spectra as shown in Figure 19, there are no C-H stretching peaks detected on the untreated pitch-based carbon fiber specimens in curve B. However, the asymmetrical (υ_{as}, CH_2)

and symmetrical (υ_s, CH_2) C-H stretches [126] of diaminohexane, near 2930 cm^{-1} and 2852 cm^{-1}, are observed clearly in curve A. These data are suggestive of the covalent immobilization of the amine-derived adducts at the fiber surface following the reaction. However, they do not provide information on the type of linkage to the surface. Attempts to observe the C-N stretch of the amine-derived adducts were unsuccessful due to interference from residual atmospheric moisture in the spectrometer. Ishida et al. [120] oxidized carbon fibers by reacting the samples in 70% nitric acid for different periods of time and used ATR as a tool to examine the surface compositions difference after the treatments. As shown in Figure 20, the difference spectra (subtracted from the spectra of germanium and unoxidized sample) amplify the effect of chemical or physical treatments on a sample. After the oxidation, three main peaks arose proportionally with the treatment times at approximately 1200 cm^{-1} (C-O stretching and O-H bending modes) 1580 cm^{-1} (quinone groups), and 1720 cm^{-1} (carboxyl groups). The presence of the peaks around 1460 cm^{-1} and 1377 cm^{-1} was indicated to be an undersubtraction of the Nujol (used to increase the optical contact between the fibers and the ATR element).

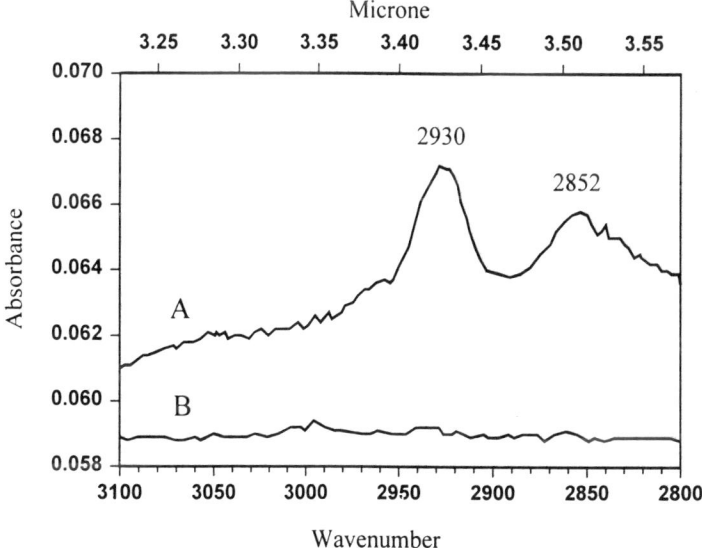

Figure 19 FTIR-ATR spectra of CF's (A) after reaction with 1,6-diaminohexane, (B) untreated [119].

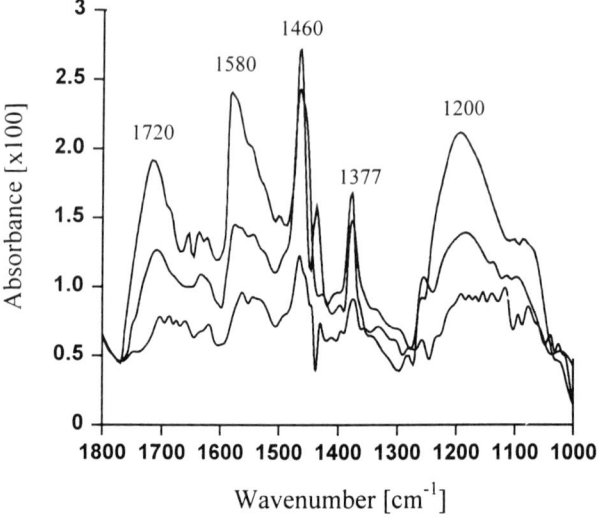

Figure 20 Spectra of graphitized carbon cloth oxidized for 25 h (top spectrum), 15 h (middle spectrum), and 5 h (bottom spectrum) [120]. (Reprinted with permission from Elsevier Science Ltd.)

4.6.2 X-Ray Photoelectron Spectroscopy

X-ray photoelectron spectroscopy (XPS) is an effective and powerful technique to investigate the surface chemistry of carbon fibers. This spectroscopic method has been employed extensively by many researchers [64,110,130–134] to study the surface oxygen-contained functionalities and elemental composition before and after surface treatments. Surface analysis by XPS is conducted by bombarding the carbon fiber specimen with monoenergetic x-ray photons. The binding energy of the core electron is calculated by subtracting the measured energy of the electron from the known energy of the x-ray photon. The basic equation of XPS is

$$E_k = h\nu - E_B$$

where E_k is the kinetic energy of the photoelectrons, $h\nu$ the incident photon energy and E_B the electron binding energy. The variation in binding energy of the electron with the environment of the atom gives rise to different peaks in the spectrum, with varying intensity for the same atom depending on the nature of

Surface Properties of Carbon Fibers

specific since the mean free path of these electrons is very small. Bascom [135] reported that the sampling depth for an XPS measurement on carbon fibers is 10–15 *nm* (i.e., about 10–20 atomic layers from the surface). Table 18 lists the binding energies for carbon, oxygen, and nitrogen that are commonly seen on carbon fiber surfaces.

4.6.2.1 XPS study on acidic solution oxidized carbon fibers
Sutherland et al. [64] examined the surface effect of anodic oxidation on high and low modulus PAN-based carbon fibers by using a high-energy resolution XPS. Data for low modulus oxidized fibers indicate marked increases in surface oxygen compared to untreated fibers. These increased species are C — O, C = O, and COOH. C — N bonding is also detected. An increase in the polar contribution of the fiber surface free energy are reflected in these surface oxygen containing functionalities. The compositions of the four types of fiber surfaces studied are shown in Table 19. For the HM untreated fibers, the measured levels of oxygen and nitrogen are relatively low (3.2% and 0.3%, respectively). In other words, for high-modulus carbon fibers, the effects of oxidation are less evident, indicating a more graphitic, less reactive surface. This result is consistent with the highly ordered basal planes of this type of high modulus fibers under a high temperature heat-treatment process. They concluded that low-modulus fiber has a less ordered structure with a high degree of edge areas of basal plane exposed to environment than the high-modulus fiber surfaces. Such graphitic planes are of low chemical reactivity for high-modulus fiber, and therefore the effect of the electrochemical oxidation was observed as only small changes in the surface levels of oxygen and nitrogen functionalities (0.8% and 2% increase, respectively). In contrast, the oxidation effect is much more profound on the low modulus fiber, in which oxygen and nitrogen functionalities were both observed with about 4.5% enhancement after treatment. XPS investigation of ultrahigh-

Table 18 Binding Energies of Elements on Carbon Fiber Surfaces

Element (bond type)	Binding energy (eV)
Carbon (C — C)	285.0
Carbon (R — C — O)	286.5
Carbon (R — C = O)	288.0
Carbon (R — COOH)	289.0
Oxygen (bonded to carbon)	532.0–534.0
Nitrogen (bonded to carbon)	399.0–401.0

Source: from ref. 136.

vacuum storage effects on carbon fiber surfaces was reported by Desimoni et al. [131]. Surface contaminants can be desorbed during storage in the case of fibers contaminated before insertion in to the UHV chamber. When uncontaminated carbon fibers are inserted in to the chamber layers of hydrocarbons can be gradually built up, probably coming from the UHV system. The contaminated species can be conveniently monitored in situ, in the analytical chamber of an XPS spectrometer. The higher the level of contaminants and/or the content of oxidized species on the fibers' surface, the more evident the influence of the UHV storage time on XPS spectra. Their findings suggest that preliminary tests on time-dependent modifications of the surface status of carbon fibers under UHV conditions can provide a more reliable XPS data.

4.6.2.2 XPS study on electrocemical-treated carbon fibers
Sherwood and coworkers [134, 137–141] have used XPS extensively to study the effects of electrochemical oxidation in different electrolytes, such as H_2SO_4, HNO_3, H_3PO_4, NaOH, and a number of ammonium salts on PAN- and pitch-based carbon fibers. Each treatment resulted in a substantial oxidation of the carbon fiber surface. For example, the authors [141] examined the difference in the effectiveness of two methods of electrochemistry, potentiostatic and galvanostatic oxidation, onto DuPont pitch-based E-120 fibers in a nitric acid solution. They found that potentiostatic treatment generated reproducible surface oxidation only after longer duration, while galvanostatic treatment produced uniform and reproducible oxidation after short and long durations.

Table 19 Carbon Fiber Surface Composition, Surface Free Energy, and Fiber Diameter

		High modulus	High modulus oxidized	Low modulus	Low modulus oxidized
Atm.%	C	96.5	93.6	92.0	83.0
	O	3.2	4.1	4.7	9.2
	N	0.3	2.3	3.3	7.8
γ (mJm^{-2})	γ_s^p	3.9	14.1	7.1	28.6
	γ_s^d	50.3	50.8	48.2	49.9
	γ_s^{d+p}	54.2	64.9	55.3	78.5
Diameter (μm)		6.1	6.5	7.0	6.2

Source: from ref. 64. (Reprinted with permission from Elsevier Science, Ltd.)

Sherwood et al. [134] treated E-120 fibers electrochemically in an ammonium bicarbonate solution and obtained core level and valence band XPS spectra. Both oxygen- and nitrogen-containing functionalities increase after the treatment; while short time oxidation produced mainly hydroxide (C-OH) groups and few carboxyl (COOH) or ester (COOR) species, but carbonyl (C=O) type of functional groups were generated predominantly for a long duration treatment.

4.6.2.3 XPS study on plasma-treated carbon fibers
Unsized, Hercules PAN-based AU4 type II (tensile strength 3.98 GPa, tensile modulus 241 GPa) carbon fibers were conducted a NO plasma treatment at a power level of 25 Watts for different treatment times by Smiley and Delgass [142]. Figure 21 represent a series of XPS spectra for NO plasma treatment from 2 to 60 minutes. For the C(1s) spectra, there is a main peak at 284.5 eV with a shoulder at higher binding energy which are non-graphitic (oxygen-containing) carbons. The shoulder at higher energy side could be fitted to three peaks of oxygen containing species at approximately 285.8, 286.7, and 288.6 eV, assigned to alcohol/ether, carbonyl, and carboxyl groups, respectively. A significant increase in the shoulder intensity of the C(1s) spectra even only after 2 minutes of NO plasma treatment. This higher binding energy carbon increases in intensity with prolonged exposure, especially the 288.6 eV peak (carboxyl groups) increases from 8.6% after 2-minute treatment to 13.8% after 60-minute treatments. As shown, O_{1s} spectra consist of a featureless broad peak centered at approximately 532.2 eV; it is difficult to further analyze the should components of each spectrum. Therefore, the authors attempted primarily to calculate the overall oxygen amount change on the surface, not tried to interpret the chemical state information from the O_{1s} spectra. The amount of oxygen increases dramatically from 3% to 23% after 2 minute treatments, and then augments slowly with longer duration oxidations to a final value of 33%. For the N_{1s} spectra, the main peak is interpreted corresponding to nitrogen without oxygen nearby (such as amines) 398.2 eV, with a shoulder of nitrogen with an oxygen close but not directly bound (such as amides) at 400.4 eV. An additional higher binding energy peak at 405.5 eV is also observed which can be assigned to $-NO_2$ species [143]. The intensity of $-NO_2$ peak increases gradually up to 5 minutes and then slowly decreases. As the 405.5 eV peak intensity decreases, there is accompanied with an increase in the two lower binding energy peaks at 400.4 and 398.2 eV. Especially the 398.2 peak was increased in area by a factor of 3 between 5 and 60 minutes. Suggested by the authors this might give evidence that nitrates further react to form species such as amides, amines and nitriles. Delgass [132] also investigated changes in surface chemical functionality of PAN-based carbon fibers exposed to low-temperature, low-power oxygen plasmas. Unsized Cellion 6000 (type II) carbon fibers were treated in oxygen

plasmas for 2–60 min at 25 W. An increase in functionalities from alcohol (ether) to carbonyl and carboxyl species was caused by increasing treatment time, but the total amount of oxidized carbon near surface remained constant.

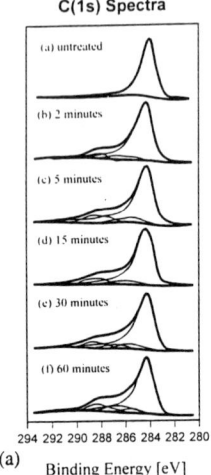

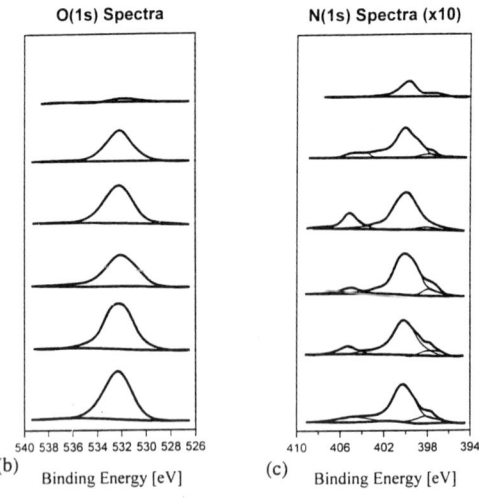

Figure 21 XPS C(1s), N(1s), and O(1s) spectra of AU4 carbon fibers (a) untreated and treated in a 25 Watt NO plasma for (b) 2 minutes, (c) 5 minutes, (d) 15 minutes, (e) 30 minutes, and (f) 60 minutes [142]. (Reprinted with permission from Materials Research Society.)

4.6.3 Scanning Tunneling Microscopy

Scanning tunneling microscopy provides a very powerful way to study the surfaces [144], because of its capacity of atomic resolution. Other practical advantages of this technique include the simplicity of sample preparation, operation in the air, and surface morphology observation is non-destructive. The STM has been successfully applied in the study of carbon fibers [145–151]. This revolutionary new type of microscope is operated by the principle of quantum mechanical tunneling of electrons [152,153] under the influence of a small bias voltage (–25 mV for carbon fibers) between a conducting material and an extremely sharp metallic tip separated by only a few tenths of a nanometer. The STM image can be visualized as a map of the tunneling current as a function of the position, obtained by scanning the surface in X and Y-axis. The tunneling current intensity depends upon the distance between tip and surface as well as upon the electronic density of the observed atoms. Practically two different modes can be operated for STM. In the constant current mode, voltage is applied to the Z piezoelectric translator to move the tip up and down in the Z direction in order to keep the tunneling current constant. When the density of states is homogeneous, a tridimensional image of the solid surface is obtained. This mode is suitable for studying low enlarged images in a range of 75,000 * 75,000 nm^2 to 1000 * 1000 nm^2. On the other hand, constant height mode consists in moving the tip in a quasi-constant height along the Z-axis in order to measure the tunneling current value for each X, Y point. This mode has a higher resolution in the Z direction and is suitable to study the surface at the atomic scale.

Scanning probe microscopy was chosen to study the carbon fiber surface because it is the only technique available that allows one to observe the surface down to the atomic scale. In the past, the scanning electron microscope technique has been successfully adopted to study the carbon fiber surface morphology down to the micron scale. However, the enlargement available with the SEM is not sufficient enough to detect the small changes of fiber surface morphology after the surface treatment. Another technique, transmission electron microscope, has been employed by researchers [145,154,155] extensively to examine the carbon fiber bulk structure on the atomic scale (see Chapter 2). Even the edge-side of graphitic layers can be clearly seen down the atomic resolution, it is not possible to view the actual surface morphology in real space by this technique. After the invention of the STM by Binnig and Rohrer [144] in 1982, with the first commercial STM instrument available in 1987 it is now possible to non-destructively study the fiber surface morphology at the nanometer scale.

Donnet et al. [147–149,178,179] have reported a series of STM study on high modulus, high strength, and activated carbon fibers made from different precursors and under various conditions. The results have clearly shown the difference in surface features and the structural discrepancies between fibers from different precursors [147]. At large scales, some common surface features are always present on a series of petroleum mesophase pitch-based carbon fibers. They are characterized by the ribbons, the smaller stripe-form crystallite stackings, and the discontinuous boundaries (Figure 22). For coal mesophase pitch-based carbon fiber, the ribbons are more like twisted strands (Figure 23). These ribbons have a width of about 200 *nm*. They are further constituted of the smaller stripe-form stackings with the diameter of about 20 *nm* aligned with an angle between 90° and 45° to the fiber axis. Whereas for PAN-based carbon fiber (M40), the ribbons are not discerned but the surface displays rougher rock-like stacks (Figure 24a).

The difference in the surface features between PAN- and pitch-based carbon fibers is evident. Two points were underlined by the authors. First, the ribbon structure along the fiber axis, which can be readily detected on the surfaces of the mesophase pitch-based carbon fibers, were not found on the M40 fiber surface. Secondly, M40 fiber has an oriented rock-like stack surface. The size of these rocks is about 30–60 *nm* by 80–160 *nm*. On each rock, the needle-shaped microvoids with a diameter of 2–3 *nm* are easily observed, and two of them were pointed on Figure 24b. Remarkable results were obtained at atomic resolution scale. At least three kinds of graphitic organizations were found on two types of mesophase pitch-based carbon fibers: the nodular microtexture (Figure 25a) which covers the greater part of the surface examined, the step-like microtexture (Figure 25b) in which the graphitic layers are superimposed, and the deformed graphene (Figure 25c). In the case of M40 fiber, only the latter two textures were observed, not the nodular one.

Figure 22 STM image of Fiber E, 500 nm * 500 nm, topview mode [147]. (Original photos from the authors and reprinted with permission from Elsevier Science, Ltd.)

Similar organization patterns were also observed on other types of PAN-based carbon fibers, T-300 and Courtaulds-XAU by Donnet [148], T650/42 by Hoffman [146], T-300 by Brown [151]. The authors concluded that the various surface microtextures may be the characteristics of the wedge and layer disclinations also observed during the mesophase formation and subsequent heat treatment as discussed by Zimmer and White [157]. Suggesting that these are prominent features in ex-mesophase pitch fibers, as evidenced by fracture behavior [158] and TEM observation [156,159], Zimmer and White believed these features could be retained with only minor modifications throughout graphitizing heat treatments up to 3000°C.

The surface micro-structure of several carbon fibers with different mesophase contents (Mc, < 10, 80, and 100%) and heat treatment temperatures was studied by STM [178]. Both Mc and HTT have a great influence on the mechanical properties of the carbon fibers. The tensile strength decreases with HTT in the case of low Mc (< 10%) while it increases at a higher Mc (80%). The surface structure of the carbonized low Mc (< 10%) sample, 1-01 CF, shows (Figure 26a) that only a small portion of area is occupied by the ordered regions. However, when HTT is raised to the graphitization temperature (2300°C), the surface becomes more crystallized (Figure 26b). On the other hand, for the fiber with a higher mesophase content (sample 4-31, Mc = 80%), the surface organization difference compared with low mesophase sample is notable. At the carbonization temperature, the crystallization has been rather well developed at the fiber surface (Figure 27a) as compared to Figure 26a. The authors suggested that the mesophase structure may favor the carbon organization during the heat treatment and thus improve the mechanical properties of the fiber—the tensile strength of 4-31 CF fiber is double the value of 1-01 CF. The further graphitization gives more locally organized crystalline structure as shown in Figure 27b where several structure patterns can be observed: trigonal spot pattern as often seen on a perfect graphite in AB stacking, hexagonal network where all the carbon atoms can be seen in a perfect graphene layer, and other patterns.

The surface of a mesophase pitch-based carbon fiber followed by anodic oxidation as well as air plasma oxidation has been examined by STM [149]. They found the anodic oxidation (0.6 and 6 A/m^2 in current density were applied in this work) does not change the surface aspects on a microscopic scale, nor at the nanometric scale. This is in a good agreement with the results of the adsorption measurement and the mechanical testing, both values were not affected by the anodic oxidation treatment. A similar result has also been observed by Hoffman et al. [145] on a commercially treated carbon fiber. On the other hand, Figure 28 shows that the air plasma (under the pressure of 0.5 *torr* and the power of 75 *W* for 10 minutes) gives rise to a certain surface roughness. This is evidenced by the etching spots that can be clearly seen on the nanometric image. The air plasma seems to have "burnt" a part of the surface, the ribbons and the small stripe-form stackings become less evident compared with the original fibers. Hoffman et al. [145,146,160] have reported a series of STM images after various carbon fiber surface treatments. Amoco P-55 pitch based carbon fiber was subjected to a treatment in air at 823°K. As shown in Figure 29a, this treatment caused pitting which is localized in a band parallel to the fiber axis and is not uniformly arranged across the surface,. Hoffman also exposed the fibers to an argon plasma until 0.66% of the fiber weight was removed. Pits of less than 35 *nm* in depth are clearly observed spread across the surface (Fig 29b).

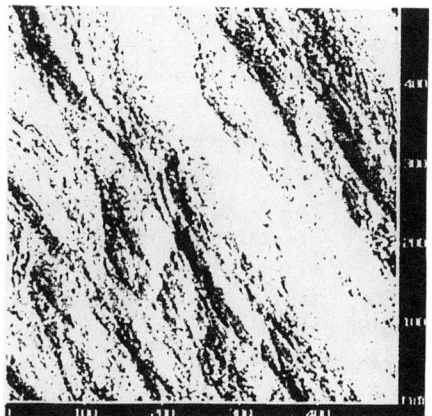

Figure 23 STM image of Fiber G, 500 nm * 500 nm, topview mode [147]. (Original photos from the authors and reprinted with permission from Elsevier Science, Ltd.)

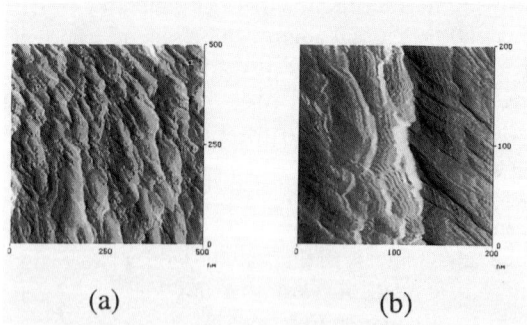

Figure 24 STM topview images of Fiber M40, (a) 500 nm * 500 nm, (b) 200 nm * 200 nm; the arrows indicate microvoids [147]. (Original photos from the authors and reprinted with permission from Elsevier Science, Ltd.)

Surface Properties of Carbon Fibers 285

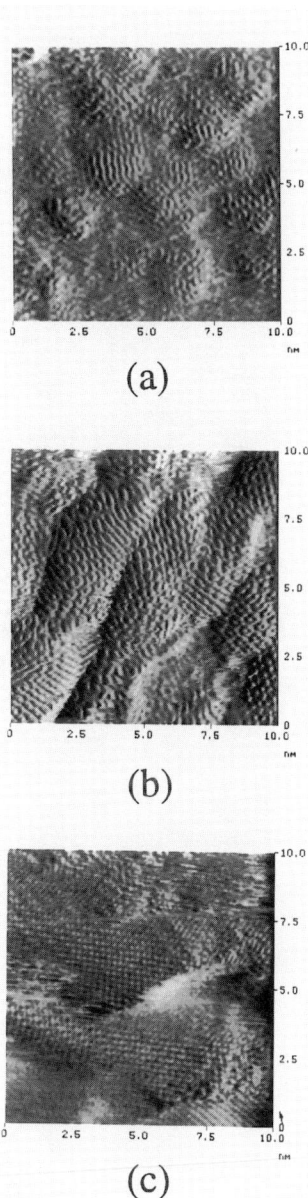

Figure 25 Three kinds of graphitic organizations observed on Fiber A (a) nodular texture, (b) step-like texture, (c) graphene texture [147]. (Original photos from the authors and reprinted with permission from Elsevier Science, Ltd.)

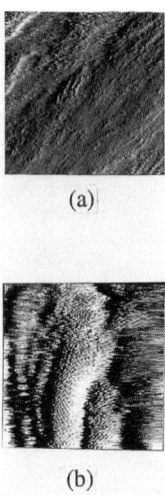

Figure 26 STM images of low mesophase content carbon fiber 1-01, 15 *15 nm^2, (a) carbonized (1000°C), (b) graphitized (2300°C) [178,179] (Original photos from the authors and reprinted with permission.)

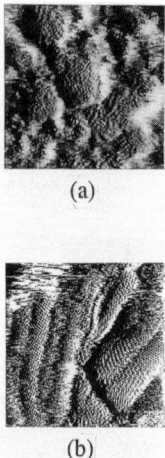

Figure 27 STM images of high mesophase content carbon fiber 4-31, 15 *15 nm^2, (a) carbonized (1000°C), (b) graphitized (2300°C) [178,179]. (Original photos from the authors and reprinted with permission.)

Surface Properties of Carbon Fibers

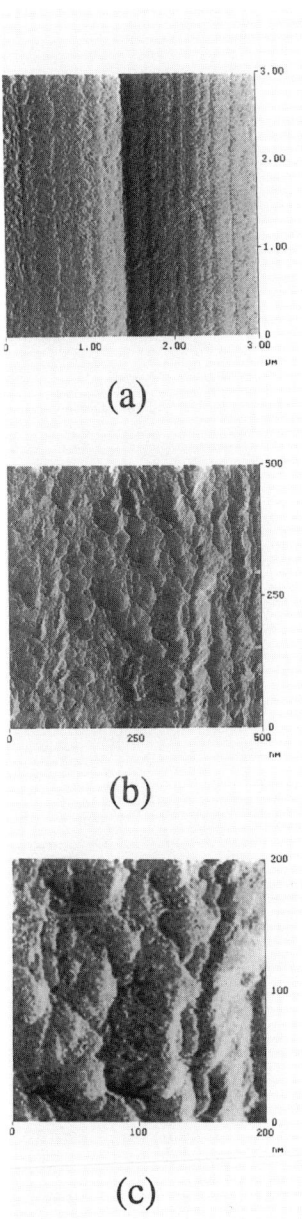

Figure 28 STM top view images of the air plasma oxidized Fiber E5: (a) 3000 nm * 3000 nm, two filaments, (b) 500 nm * 500 nm, (c) 200 nm * 200 nm [149]. (Original photos from authors and reprinted with permission from Elsevier Science, Ltd.)

(a)

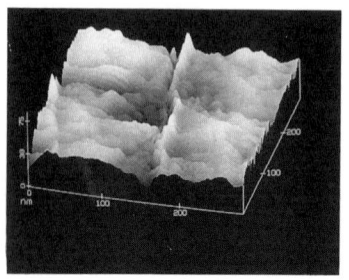

(b)

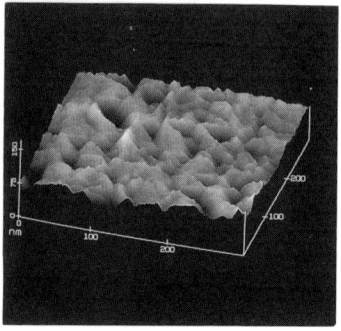

Figure 29 Localized pitting of P-55 carbon fiber surface resulting from (a) oxidation in air at 823 K to 0.76% weight loss, (b) treated in argon plasma to 0.66% weight loss [146]. (Reprinted with permission from Elsevier Science, Ltd.)

4.6.4 Auger Spectroscopy

An alternative method similar as XPS for chemical surface analysis is Auger electron spectroscopy (AES), which involves bombardment of the surface with a beam of electrons in the energy range 1–5 keV (1.6–8.0 $* 10^{-16}$ J), causing vacancies to be created in the core level of the surface. These vacancies represent excited ions in the surface region which may undergo deexcitation and the creation of so-called Auger electrons. Each Auger electron has an energy characteristic of a certain element. Analysis of all the back-scattered Auger electrons in the energy range 0–1 keV (0–1.6 $* 10^{-16}$ J) gives a complete picture of the elemental composition of the outmost atomic layer. The surface sensitivity originates from the fact that the escape depth of electrons is between

Surface Properties of Carbon Fibers

0.5 and 1.5 nm (1–2 nm for XPS). Hopfgarten [165,166] measured the atomic composition of external surfaces of type I, II, and III carbon fibers by Auger electron spectroscopy. The concentration gradients, determined by simultaneous ion etching of the fibers, indicated that the oxygen content of Torayca type I and type II fibers increased with depth up to 30 nm. However, the surface of type III carbon fibers contained more oxygen. It is due to the larger surface area and greater surface roughness of the type III fiber.

4.6.4.1 Scanning Auger microscopy

Scanning Auger microscopy (SAM) is used to study the fiber/matrix interface of composites [161]. This novel application of Auger spectroscopy enables further understanding of the mechanism of failure in composites. Auger spectroscopy can be used to detect the presence of thin polymeric layers on the carbon fibers if a suitable matrix-specific element is chosen to form the scanning Auger image. A high volume fraction of conducting carbon fibers makes Auger analysis possible in the case of unidirectional continuous fiber composites, although the failure surface is predominantly composed of polymer residues. For short-fiber composites, the technique is more difficult considering the low volume fraction of fibers, but Auger spectroscopy enables the identification of microfailure mechanism and of the effect of the fiber surface treatment on the failure mode. SAM was used to study the distribution of oxygen on pitch- and PAN-based carbon fibers surface after different types of chemical oxidative treatments [162,163]. The content of oxygen varied widely from region to region, not uniform along the fiber length. The variation was proportional to the surface treatments. The distribution patterns were observed in various forms such as edge aggregation, wide strips, and striations. The authors proposed that the fiber structure and precursors are related to the oxygen distribution patterns.

Similar results were obtained by Vaidyanathan et al. [167] who conducted low temperature plasma treatment experiments on AS4 and AU4 type fibers in the presence of air, argon, and nitrogen to study their effects on the resulting fiber-matrix adhesion in the composites fabricated with the treated fibers. They concluded that factors such as the flow rate of the gas, power supply, and plasma treatment time play an important role in controlling the amount of oxygen on the fiber surfaces. The AES results of experiments on the AS4 (oxidized-treatment) fibers are given in Table 20 where the time of treatment was varied while the power and the flow rate of gas were kept constant. Similar results were obtained for AU4 (untreated) fibers. It was observed for both fibers that the oxygen and nitrogen contents increase proportionally with the treatment time to approximately 10 minutes, but fall off when the treatment time is increased. Therefore, it was concluded that the optimum treatment time needed to introduce the maximum amount of oxygen and nitrogen on AS4 and AU4 fiber surfaces in a plasma reactor of the size used was approximately 10

Table 20 AES Results and Air Plasma Treatment Conditions for AS4 Fibers With Variable Time of Treatment and Power

Fiber: AS4	Gas: air		Purge gas: air	
Variable: time of treatment (min)	Flow rate: 150 cc - /min		Power: 200 watts	
	Oxygen %	Nitrogen %	Carbon %	Sulphur %
0.0	3.1	4.4	91.9	0.6
5.0	4.6	4.2	91.2	0.0
10.0	13.0	9.0	78.0	0.0
15.0	9.5	5.3	85.2	0.0
20.0	6.1	3.0	91.0	0.0
30.0	5.7	3.0	91.2	0.0
Variable: power (watts)	Flow rate: 150 cc - /min		Time: 10 mins	
	Oxygen %	Nitrogen %	Carbon %	Sulphur %
0.0	3.1	4.4	91.9	0.6
100	5.6	2.8	91.5	0.0
150	5.5	2.9	91.6	0.0
200	12.8	9.1	78.1	0.0
250	5.5	3.0	91.5	0.0
300	6.7	3.3	90.0	0.0
400	8.4	4.6	87.0	0.0

Source: from ref. 167. (Reprinted with permission.)

minutes. Seen in Table 20b, while keeps the gas flow rate (150 ml/min) and treatment time (10 min) constant, the oxygen and nitrogen contents are affected by the variable power supply. One can obtain the highest level of oxygen- and nitrogen-containing groups at the power supply of 200 Watts.

4.6.4.2 X-ray-excited Auger electron spectroscopy
Desimoni et al. [164] have recently developed a proper methodology, X-ray-excited Auger electron spectroscopy (XAES) to investigate the effects of high-temperature annealing and long-term storage on the chemical composition of the fiber surface under ultrahigh vacuum conditions. Both the effects of temperatures (25–60°C), and times (0.5–20 min) of treatment were evaluated by contacting an oxidation treatments on fibers using potassium permanganate in concentrated (98%) sulphuric acid. As a result, the concentrations of sulphur and magnanese increased proportional to all variables (temperature, time, permangnate concentration), but they were always much lower than the oxygen

Surface Properties of Carbon Fibers

is suitable for investigating the surface and layer composition of coated carbon fibers. As known, carbon fibers for metal matrix composites need to be coated with a barrier layer to avoid interface reactions. In Than and his coworkers' [168] study, an electron gun with a beam diameter lower than 1μm was employed for the AES measurements, after a continuous CVD process of depositing SiC onto carbon fibers. In combination with Ar^+ ion etching (3 kV, 200 $nA\ cm^{-2}$) depth profiles were recorded. Figure 30 gives the concentration profile of a SiC coated carbon fiber. The barrier layer mainly consists of C and Si elements in a nearly stoichiometric ratio. The oxygen was also detected (shown in the detail spectrum) which could be caused by either the sample slightly oxidized or contaminated. The chlorine element is originated from part of the reaction products of the CVD process. The authors indicated that the stoichiometry of these barrier layers can be varied by changing the composition of the CVD atmosphere. AES depth profiling on SiC coated single fiber shows a nearly oxygen-free layer of Si and C in the atomic ratio.

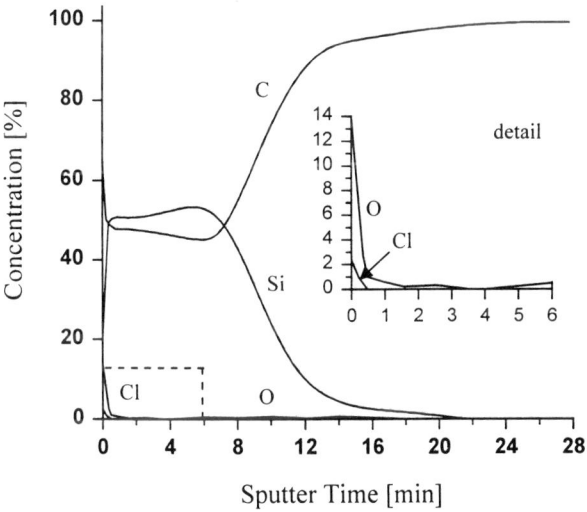

Figure 30 AES depth profile on a SiC coated carbon fiber obtained by single fiber line measurement [168]. (Reprinted with permission from Elsevier Science, Ltd.)

4.6.5 Other Instrumental Methods

4.6.5.1 Secondary ion mass spectroscopy

High-resolution mass spectrometry was successfully employed in combination with gas chromatography [175,176] to identify a number of components in the matrices of carbon fiber composites. Recently [169], time-of-flight secondary ion mass spectroscopy (TOF SIMS) was used to study a two-step surface processing of AS4 carbon fiber: extraction of sizing in CH_2Cl_2 and functionalization with trimellitic anhydride. The primary ions were generated from a Ga ion source (15 kV, 400 pA). The ion gun was pulsed at a frequency of 5 kHz with a pulse width of 4 ns. The secondary ions were accelerated to ± 3 keV and then deflected by three electrostatic analyzers in order to compensate the initial energy distribution of ions with the same masses. The analyzed surface area was a 50 µm * 50 µm square and the mass range scanned was from 0 to 5000 *amu*. As carbon fibers are sufficiently conducting, no charge compensation was needed in the measurements.

The positive TOF SIMS spectra obtained with AS_4 after CH_2Cl_2 extraction and functionalization with trimellitic anhydride are displayed in Figures 31a and b, respectively. Compared with the respective spectra obtained with AS4 before extraction, several differences can be observed. First, the peaks characteristic of the silicone, i.e. $m/z = 28, 73, 147$, etc. in the positive spectrum decrease after CH_2Cl_2 extraction. Second, the intensities of the nitrogen-containing peaks observed in AS4, i.e., NH^{4+} (18), $C_2H_8N^+$ (46), $C_4H_{12}N^+$ (74), $C_5H_{12}N^+$ (86), decrease slightly after extraction. Taking into account all the results obtained with AS4 before and after CH_2Cl_2 extraction, the authors suggested the sizing present on AS4 fibers is composed of at least four different compounds: polydimethylsiloxane (PDMS), dialkyl phthalates, glyceryl monostearate, and phenolic antioxidants. The results demonstrate that CH_2Cl_2 extraction is sufficient for the elimination of PDMS but is not enough for the other three organic compounds, as these molecules are still detected on the CF surface. Nevertheless, the extraction seems to "clean up" the surface of CF as the TOF SIMS results show. Similarly, the negative TOF SIMS spectra obtained with CF after CH_2Cl_2 extraction and functionalization with the chloride of trimellitic anhydride are presented in Figures 32a and b, respectively. Several differences can be noted between the functionalized spectra and the respective spectra of CF after extraction. In the positive spectrum, three new peaks, $m/z = 106, 174$, and 204, appear after the functionalization of trimellitic anhydride. In the negative spectrum, two new peaks appear at $m/z = 146$ and 190, and the intensity of CNO^- (42) increases substantially. The presence of these peaks were explained by two different reactions, one occurs between the anhydride group and the amine functionality, and the other reaction occurs between the chloride group and the amine group on the CF surface.

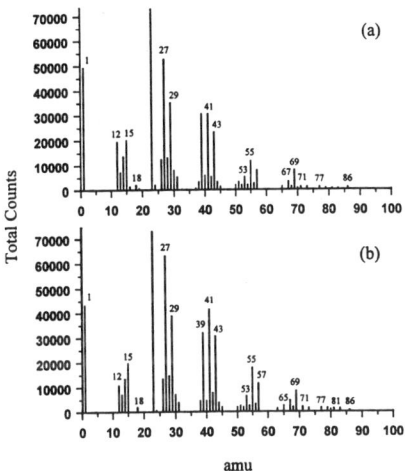

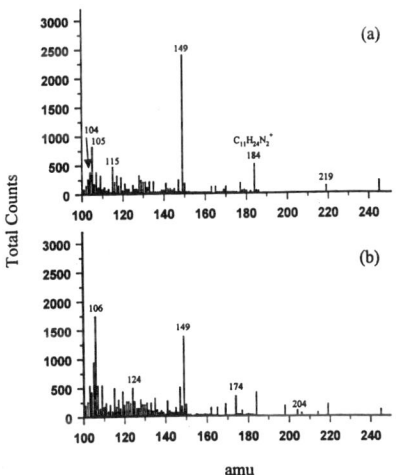

Figure 31 Positive TOF SIMS spectra obtained with (a) AS4 6k after CH$_2$Cl$_2$ extraction, and (b) AS4 6k after CH$_2$Cl$_2$ extraction, and functionalization with trimellitic anhydride [169]. (Reprinted with permission.)

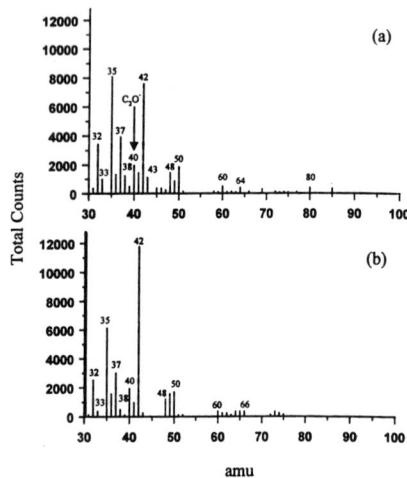

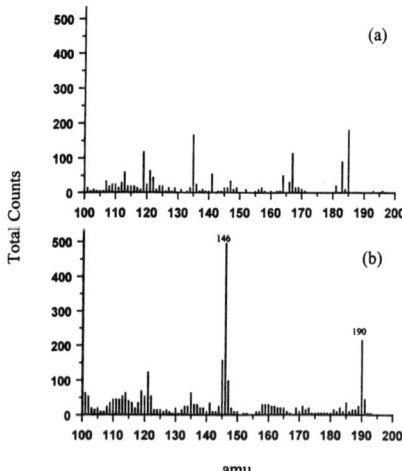

Figure 32 Negative TOF SIMS spectra obtained with (a) AS4 6k after CH_2Cl_2 extraction, and (b) AS4 6k after CH_2Cl_2 extraction and functionalization with trimellitic anhydride [169]. (Reprinted with permission.)

Surface Properties of Carbon Fibers

Verdu et al. [173] also used SIMS technique to study the role of hydrogen in the graphitizing process and mechanical properties of vapor-grown carbon fibers. The ions gun produces O^{-2} ions having 12 keV of energy (the oxygen ions do not destroy the surface under examination with such low energy) with a current flow about 100 nA. The sputtered ions were analyzed in a quadruple mass analyzer. Two types of thick and thin VGCFs, were examined in as-grown condition and after annealing at 510°C for four hours under argon in atmospheric pressure. It is clear that the hydrogen content is higher in thin than in thick-grown VGCF as shown in Figure 33. A small portion of hydrogen was found under the form of CH^+ for both samples, but ions such as CH_2^+ and CH_3^+ do not exist. After annealing at high temperature thin fibers keep more H^+ than thick samples, a difference that is maintained up to the temperature where structural reordering starts. The authors also analyzed a commercial PAN-based fiber (Figure 34) and observed a similar H^+ content compared with the thin VGCF. A notable feature is that the CH_2^+ ion exists in the PAN-based fibers, although it is not seen in the VGCF. VGCF has a duplex structure consisting of an inner core and an outer coating [174]. The inner core is made up by a three-trunk structure, and the outer core is formed of a carbonaceous material with poor structure order. The authors proved that the hydrogen content of the core is greater than in the coating, and the strength of this material is proportional to its content of hydrogen.

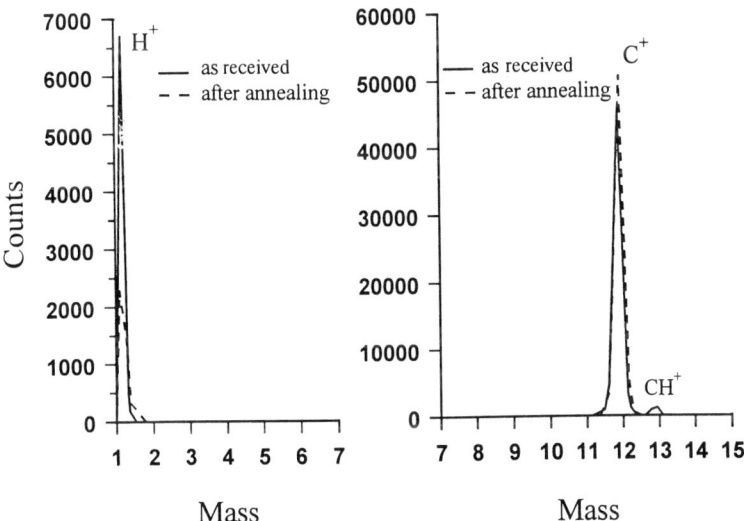

Figure 33a Spectrum of ion mass of VGCF obtained by SIMS; thin VGCF [173]. (Reprinted with permission from Elsevier Science, Ltd.)

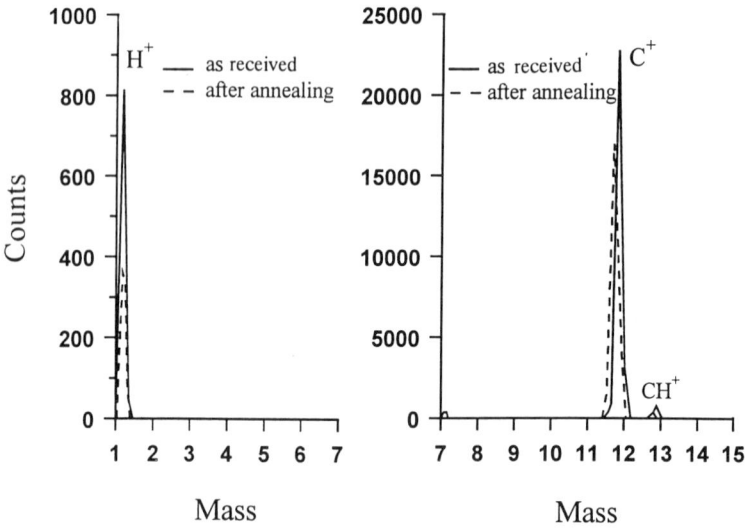

Figure 33b Spectrum of ion mass of VGCF obtained by SIMS; thick VGCF [173]. (Reprinted with permission from Elsevier Science, Ltd.)

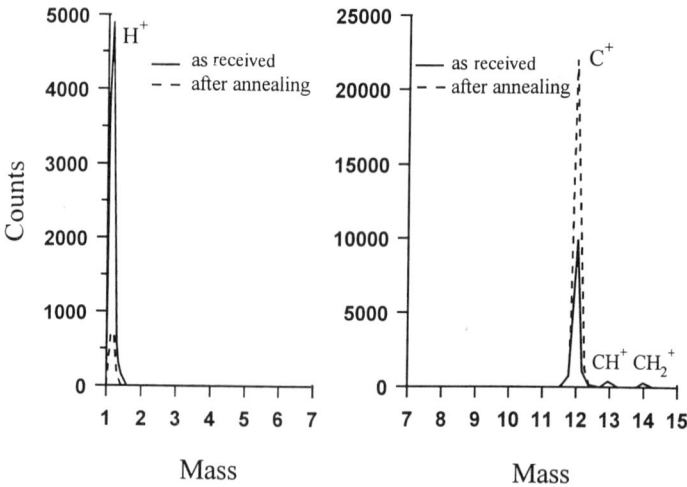

Figure 34 Spectrum of ion mass of Tenax $^@$ ex-PAN fibers obtained by SIMS; a) non thick; b) thick; and c) the three morphological ranges [173]. (Reprinted with permission from Elsevier Science, Ltd.).

4.6.5.2 Ultrasonic spectroscopy

Ultrasonic spectroscopy was employed as a nondestructive method for the evaluation of composite materials [170,171]. In recent studies of the propagation of longitudinal ultrasonic waves impinging at normal incidence on such materials [171,172], the finding of absorption frequencies in the transmitted energy spectrum for crossed-ply composite samples has been reported. Crossed-ply of carbon fiber/epoxy composites is composed of periodic multilayer structures, thus the geometrical dispersion phenomena will influence the propagation of longitudinal waves along a direction normal to the ply boundaries. If the acoustic wavelength is of the same magnitude as the superlayer (constituted of one ply and one residual epoxy layer) thickness, the wave interaction is generated. This absorption phenomenon is clearly related to the material structure and advantage may be taken of its characteristics (central frequency, bandwidth, and relative depth of the absorption dip in the spectrum) to characterize the material and identify the presence of defects or porosities.

There is some useful information obtained by this technique which is best described in Figure 35. There are no notable interactions when experiments are conducted at a low frequency (5 MHz) because the acoustic wavelength (about 600 μm) propagating in the composite is much larger than the ply thickness. However, when the frequency is increased to 10 MHz, the wavelength decreases and approaches the ply thickness, the time echogram is then strongly modified. An oscillating signal is observed due to the interaction between the front and back surface echoes. The back surface echo is more and more severely distorted as the frequency close to a critical 10.8 MHz as shown in Figure 35a. This frequency is independent of the number of plies, and the fiber orientation such as the material anisotropy is also insensitive. The absorption dip from the back surface echo shown in Figure 35b, corresponds to a maximum in the spectrum of the signal [171].

This observation suggests that when the wavelength is near the ply thickness the internal interference occurs during wave propagation. The reflections responsible for these effects occur at the ply boundaries in very thin residual epoxy layers. The absorption frequency is determined by the thickness of the superlayer which is composed of one ply and one epoxy layer. The echogram of Figure 35c was obtained by using a high frequency (100 MHz) ultrasonic transducer driven by a short electrical pulse. The back surface echo is not visible due to the ultrasonic attenuation in the composite at this frequency, but the reflections on the very first ply are clearly detected: these signals are well separated in time, without possible overlap or phase interference.

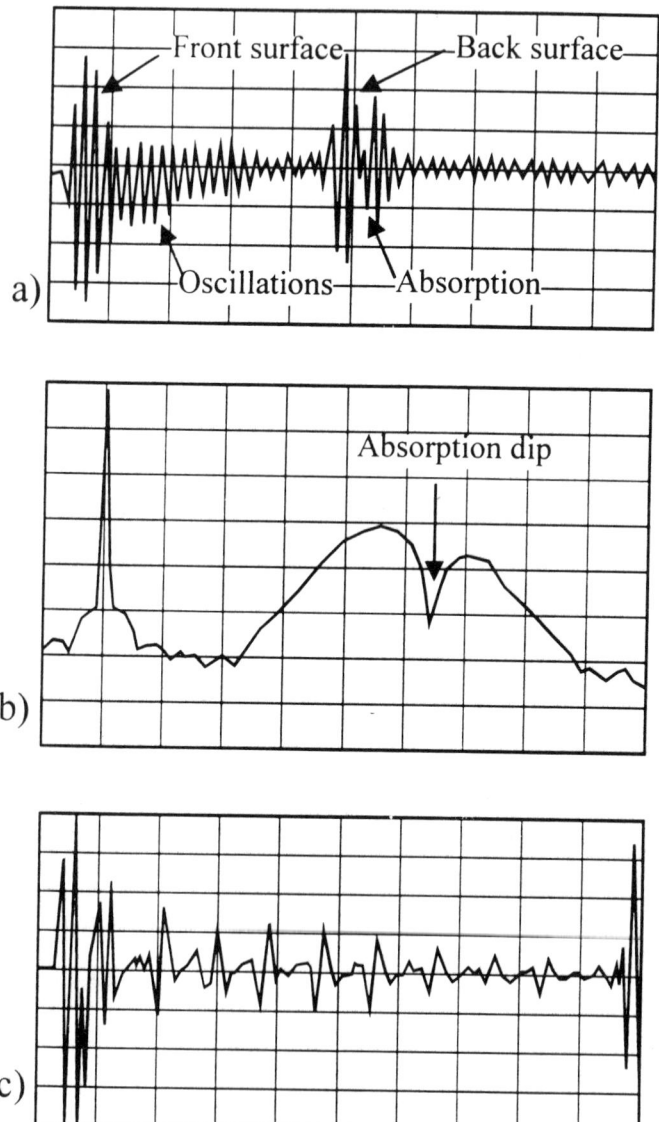

Figure 35 Results from A—scan and ultrasonic spectroscopy, (a) time echogram obtained at 10.8 MHz—vertical scale: 40 mV/div, horizontal scale: 500 ns/div, (b) spectrum of the back face echo—horizontal scale: 2 MHz/div, (c) high frequency echogram at 100 MHz—vertical scale: 10 mV/div, horizontal scale: 100 ns/div [170]. (Reprinted with permission from the publisher of Acoustics Letters.)

Surface Properties of Carbon Fibers 299

4.6.5.3 Atomic force microspectropy

Carbon fibers are coated with various materials, such as sizing compounds at the end of manufacturing process, or the oxidation protection coatings. The atomic force microscopy (AFM) is an ideal technique to study the carbon fiber coatings because the nonconductivity of these organic compounds. Hoffman and his co-workers [145,146] deposited 0.5 equivalent (total surface) monolayers of silica onto Amoco pitch-based carbon fibers and studied the coatings by AFM. Figure 36 is a 400 nm * 400 nm AFM image of the silica coated fiber, on which sharp ridges of silica are clearly seen arranged at a certain angle to the fiber axis.

AFM was used by Smiley and Delgass [132] to investigate changes in topography on PAN-based carbon fibers exposed to low-temperature and low-powered oxygen plasmas. Cellion 6000 carbon fibers (type II, unsized) were treated in oxygen plasmas at a power of 25 W for 2–60 minutes. AFM images clearly indicate that surface morphology changes after oxygen plasma treatments for 2 and 15 minutes. Figure 37a is a 1 μm * 1 μm tilted perspective view AFM image of untreated Cellion 6000 fibers, and the corresponding images of treated fibers in an oxygen plasma for 2 and 15 min are shown in Figures 37b and c, respectively. It appears that the fiber surface goes through an intensive roughening after short plasma treatment (2 min), followed by an overall smoothing as treatment time is increased (15 min).

Figure 38 a–c show enlarged images along the grooves labeled A, B and C in Figures 37a–c. The groove walls are relatively straight and smooth on the untreated fibers. However, the 2 min treated sample shows remarkably different surface features. The grooves contain holes (dark areas) with diameters in the range of 50 nm, which are ring-shaped and 10–20 nm deep. As shown by the gray scale ranging from 0–1.5 nm, overall roughness is gradually reduced to values eventually lower than that of the untreated fibers after 15 min treated in oxygen plasma. Root-mean-square roughness (RMS), average roughness, and maximum peak-to-valley distances (PTV) values calculated from AFM images in Figure 37a–c are given in Table 21 and those figures quantitatively support this observation. Similar results were obtained by Mayer et al. [177] who thermally treated Toray T300 carbon fibers by heating at 400, 500, or 600°C in air for 15 minutes. They found by the AFM measurements a slight increase in fiber surface roughness (Figure 39) following thermal treatment as well as an obvious change of the surface morphology from parallel grooves at 400°C to a more heterogeneous hill-like structure at 600°C (Figure 40).

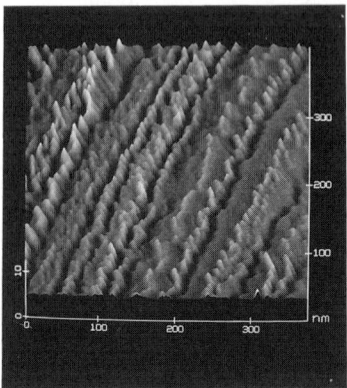

Figure 36 P-120 fiber surface after deposition of 0.5 equivalent monolayers of silica: the micrograph shows preferential deposition, which is evident by the ridges [180]. (Reprinted with permission from Elsevier Science, Ltd.)

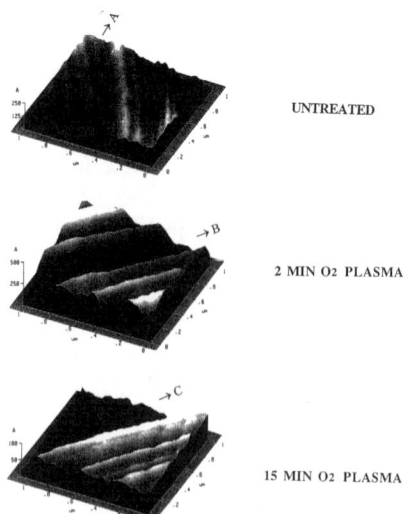

Figure 37 AFM tilted view images of a Cellion 6000 carbon fiber, (A) untreated, treated in an oxygen plasma (B) 2 min, (C) 15 min [132]. (Reprinted with permission from Chapman and Hall.)

Surface Properties of Carbon Fibers 301

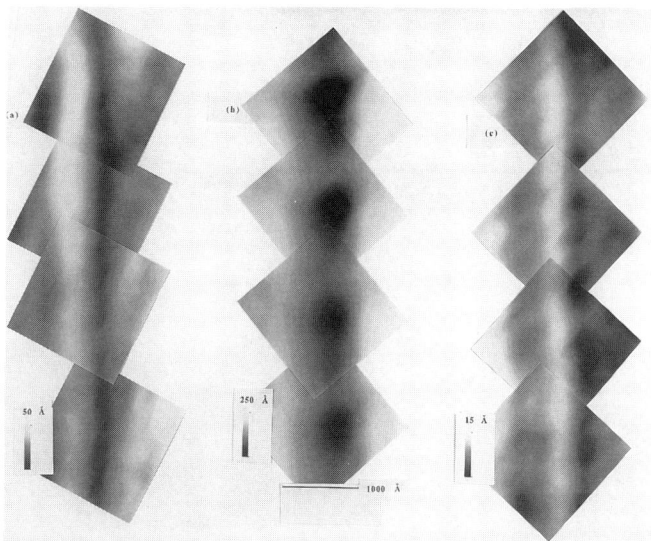

Figure 38 Enlarged AFM images along grooves for (a) untreated, (b) 2 min treated, and (c) 15 min treated Cellion 6000 carbon fibers [132]. (Reprinted with permission from Chapman and Hall.)

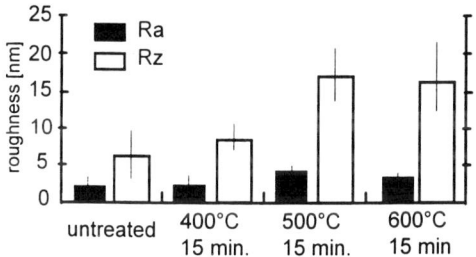

Figure 39 Roughness of T300 fibers after 15 min thermal treatment in air at 400, 500 and 600°C. Data are calculated from AFM measurements (see Figure 40). R_a is the arithmetic mean value of roughness and R_z is the mean value of the 10 maximum values measured [177]. (Reprinted with permission from Elsevier Science, Ltd.)

Table 21 AFM Roughness for Untreated and Oxygen-Plasma Treated Fibers

Treatment time (min)	RMS (nm)	AVG (nm)	PTV (nm)
0	3.8	3.0	22.2
2	9.0	7.3	50.9
15	2.0	1.8	8.3

Source: from ref. 132. (Reprinted with permission from Chapman and Hall.)

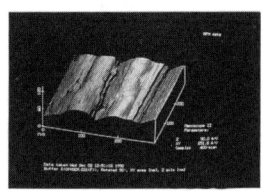
(a)

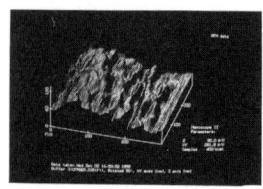
(b)

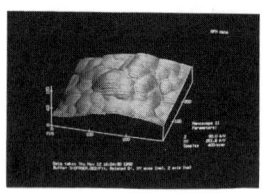
(c)

Figure 40 AFM pictures of thermally oxidized T300 fibers illustrating the change in surface morphology after 15 min treatment at: (a) 400°C; (b) 500°C; (c) 600°C. [177] (Original photos from the authors and reprinted with permission from Elsevier Science, Ltd.)

NOTE

1. Note added in proofs: A recent paper was published on PAN fiber studied by STM, see D. Shi et al., Cailiao Yanjiu Xuebo 11, 305–308 (1997) (in Chinese).

REFERENCES

1. Lehmann, S., Robinson, R., and Tse, M. K. *31st International SAMPE Symposium*, p. 291 (1986).
2. Robinson, R. *SAMPLE J.* November/December, 20 (1985).
3. Bobka, R. J. and Lowell, L. P. Integrated research on carbon fiber composite materials, *Technical report AFML-TR-66-310*, Part 1, Air Force Materials Laboratory (October 1960), pp. 145–152.
4. Pallozzi, A. *A. Soc. Plast. Eng. J.* 8, (February 1966).
5. Herrick, J. W., Grüber, P. E., Jr., and Mansur, F. T. Surface treatments for fibrous carbon reinforcements, *Technical report AFML-TR-66-178*, Part 1, Air Force Materials Laboratory (July 1966).
6. Goan, J. C., and Prosen, S. P. In Interface in Composites, A*STM Special Tech. Publ.* No. 452, p. 3.
7. Brooks, C. S., Gold, G. S., and Scola, D. A. *Carbon* 12, 609 (1974).
8. Drzal, L. T. *Carbon* 15, 129, (1977).
9. Drzal, L. T., Mercher, J. A., and Hall, D. L. *Carbon* 17, 375 (1979).
10. Bansal, R. C., and Chhabra, *P. Indian J. Chem.*, 20A, 449 (1981).
11. Nakao, F., Takenaka, Y., and Asai, H. *Composites* 23, 365 (1992).
12. Mimeault, V. J., and Mckee, D. W. *Nature* 224, 793 (1969).
13. Turner, N. H., and Dietz, V. R. *Carbon* 11, 256 (1973).
14. Tsutsumi, K. *Toyahashi Carbon Conference*, Toyohashi University, Toyohashi, Japan, 1984.
15. Ismail, M. K. I. *Carbon* 29, 777 (1991).
16. Perret, R., and Ruland, W. *J. Appl. Crystallogr.* 3, 525 (1970).
17. Perret, R., and Ruland, W. *J. Appl. Crystallogr.* 2, 209 (1969).
18. Fourdeux, A., Perret, R., and Ruland W. *Carbon Fibers-Their Composites and Applications*, The Plastics Institute, London (1971), p. 57.
19. Johnson, D. J., and Tyson, C. N. *Br. J. Appl. Phys.* D3, 526 (1969).
20. Badami, D. V., Joiner, J. C., and Jones, G. A. *Nature* 215, 386 (1967).
21. Peterlik, H., Fratzl, P., and Kromp, K. *Carbon* 32, 939 (1994).
22. Johnson, D. J. *Carbon Fibers-Their Composites and Applications*, The Plastics Institute, London (1971), p. 52.
23. Donnet, J. B., and Ehrburger, P. *Carbon* 15, 143 (1973).
24. Bacon, R., and Silvaggi, A. F. *Carbon* 9, 321 (1971).

25. Didchenko, R. Carbon and graphite surface properties relevant to fiber reinforced composites, *Technical report AFML-TR-68-45*, Air Force Materials Laboratory (February 1968).
26. Connolly, M. P. *J. Reinf. Plast. Compos.* 11, 1367 (1992).
27. Ryu, S. K., Jin, H., Gondy, D., Pusset, N., Ehrburger, P. *Carbon* 31, 841 (1993).
28. Economy, J., Daley, M., Hippo, E. J., and Tandon, D. *Carbon* 33, 344 (1995).
29. Ghiorse, S. R. *SAMPE Q.* 24, 54 (1993).
30. Bowles, K. J., Kenneth J., Frimpong, S. *J. Compos. Mater.* 26, 1487 (1992).
31. Hoffman, W. P. *US Patent 5,271,917*, December 1993.
32. Huang, Y., Young, R. J., *J. Mater. Sci.* 29, 4027 (1994).
33. Ju, C. P., Murdie, N. *Mater. Chem. Phys.* 34, 244 (1993).
34. Jortner, J., *Carbon*, 24, 603 (1986).
35. Matsui, J. *Critical Reviews in Surface Chemistry*, Vol. 1, 71 (1990).
36. Schultz, J., Lavielle, L., and Simon, H. *in Science and New Applications of Carbon Fibers*, Proc. Int. Symp. Toyohashi University of Technology, 125 (1984).
37. Lloyd D. R., Ward, T. C., Schreiber, H. P., and Pizana, C. C., *Inverse Gas Chromatography-Characterization of polymers and other materials*, ACS Symp. Series 391, Washington, DC (1989).
38. Conder J. R., and Young, C. L. *Physico-Chemical Measurement by Gas Chromatography*, p. 81, Wiley, New York (1979).
39. Wang, M. J., Wolff, S., and Donnet, J. B. *Rubber Chem.Technol.* 64, 559 (1991).
40. Wang, M. J., Wolff, S., and Donnet, J. B. *Rubber Chem. Technol.* 64, 714 (1991).
41. Donnet, J. B., Qin, R. Y., and Wang M. J. *J. Colloid Interface Sci.* 153, 572 (1992).
42. Donnet, J. B., and Qin, R. Y. *J. Colloid Interface Sci.* 154, 434 (1992).
43. Conder J. R., and Young, C. L. *Physico-Chemical Measurement by Gas Chromatography*, p. 387, Wiley, New York (1979).
44. Cooper, W. T., and Hayes, J. M. *J. Chromatography* 314, 111 (1984).
45. Wang, M. J., Wolff, S., and Donnet, J. B. *Kautsch. Gummi Kunstst.* 45, 11 (1992).
46. Wang, M. J., and Wolff, S. *Rubber Chem. Technol.* 65, 890 (1992).
47. Hobson, J. P., *Canad. J. Physics* 43, 1941 (1965).
48. Rudzinski, W., and Everett, D. H. *Adsorption of Gases on Heterogeneous Surfaces*, Academic Press, London (1992).
49. Donnet, J. B., and Park, S. J. *Carbon* 29, 955 (1991).
50. Qin, R. Y., and Donnet, J. B. *Carbon* 32, 165 (1994).

51. Tsutsumi, K., and Ban, K. *Proc. IVth Int. Conf. on Fundamentals of Adsorption*, 1992, p. 679.
52. Cuesta, A., and Bradley, R. H. *Proceedings of the European Carbon Conference, Carbon 96*, Newcastle, UK, 1996, p. 622.
53. Papirer, E. in Composite Interface (Ishida, H., and Koenig, J. L., eds.), Elsevier, 1986, p. 203.
54. Montes-Moran, M. A., Paiva, M. C., Martinez-Alonso, A., Tascon, J. M. D., Bernardo, C. A. *Proceedings of the European Carbon Conference, Carbon 96*, Newcastle, UK, 1996, p. 640.
55. Fowkes, F. M. *J. Adhesion Sci. Technol.* 1, 7 (1987).
56. Fowkes, F. M., in *Acid-Base Interactions: Relevance to adhesion Science and Technology* (Mittal, K. L., and Anderson, Jr., H. R., Eds), 1991, VSP, Zeist, The Netherlands, p. 93.
57. Dilsiz, N., Erinç, N. K., Bayramli, E., and Akovali, G. *Carbon* 33, 853 (1995).
58. Ogawa, T., and Ikeda, M. *J. Adhesion* 43, 69 (1993).
59. Schultz, J., Caseneuve, C., Shanahan, M. E. R., and Donnet, J. B. *J. Adhesion* 12, 221 (1981).
60. Schultz, J., Tsutsumi, K., and Donnet, J. B. *J. Colloid Interface Sci.* 59, 272 (1977).
61. Kaelble, D. H., Dynes, P. J., Maus, L. *J. Adhesion* 6, 239 (1974).
62. Donnet, J. B., Brendle, M., Dhami, T. L., and Bahl, O. P., *Carbon* 24, 757 (1986).
63. Donnet, J. B., Dhami, T. L., Dong, S., Brendle, M. J. *J. Phy.D : Appl. Phys.* 20, 269 (1987).
64. Bradley, R. H., Ling X., and Sutherland, I. *Carbon* 31, 1115 (1993).
65. Hammer, G. E., and Drzal, L. T. *Appl. Surf. Sci.* 4, 340 (1980).
66. Hodge, D. J., Middlemiss, B. A., and Peacock, J. A. *Interfaces in Composites* (Pantano, C. J., and Chen, E. J. H., eds.), Materials Research Society Symposium Proceedings, Vol. 170, Materials Research Society, Pittsburgh, PA, 1989, p. 327.
67. Puri, B. R., and Bansal, R. C., *Carbon* 1, 457 (1964).
68. Barton, S. S., Gillespie, D., and Harrison, B. H., *Carbon* 11, 649 (1973).
69. Donnet, J. B., Lahaye, J., and Benardin, J. *Bull. Soc. Chim. Fr.* 5, 1790 (1969).
70. Boehm, H. P., *Adv. Catal.* 16, 179 (1966).
71. Puri, B. R., in *Chemistry and Physics of Carbon*, vol. 6 (edited by P. L. Walker, Jr.), Marcel Dekker, New York (1970), p. 108.
72. Donnet, J. B., and Voet, A., *Carbon Black*, Marcel Dekker,New York (1976).
73. Donnet, J. B., Hueber, H., Perol, N., and Jaeger, J. *J. Chim. Phys.* 60, 426 (1963).

74. Smith, R. N., Young, D. A., and Smith R. A. *Trans. Faraday Soc.* 62, 2280 (1966).
75. Mattson, J. S., Mark, H. B. *J. Colloid Interfaces Sci.* 31, 131 (1969).
76. Mattson, J. S., Mark, H. B., and Weber, W. *J. Anal. Chem.* 41, 355 (1969).
77. Donnet, J. B., and Henrich, G. *Bull. Soc. Chim. Fr.* 1609 (1960).
78. Hallum, J. V., and Drushel, H. B. *J. Phys. Chem.* 62, 110 (1958).
79. Donnet, J. B. *R.G.C.P.* 50, 51 (1973).
80. Ehrburger, P., and Donnet, J. B. *Philos. Trans. R. Soc.* London A 294, 495 (1980).
81. Ehrburger, P., Herque, J. J., and Donnet, J. B., *Proc. 4th Intern. Conf. Carbon Graphite,* Soc. Chem. Ind., London (1976), p. 201.
82. Bansal, R. C., Chhabra, P., and Puri, B. R. *Indian J. Chem.* 19A, 1149 (1980).
83. Palma, E., Ibarra, L. *Angew. Makromol. Chem.* 220, 111 (1994).
84. Peng, J., Buttry, D., summited to *J. ElectroChem.*.
85. Baillie, C. A., Watts, J. F., Castle, J. E. *J. Mater. Chem.* 2, 939 (1992).
86. Krekel, G., Hüttinger, K. J., Hoffman, W. P. *J. Mater. Sci.* 29, 3461 (1994).
87. Hüttinger, K. J., Höhmann-Wien, S., and Seiferling, M. *Carbon* 29, 449 (1991).
88. Baillie, C. A., Watts, J. F., Castle, J. E., Bader, M. G. *Comp. Sci. Tech.* 48, 97 (1993).
89. Chun, B. W., Davis, R., He, Q., Gustafson, R. R. *Carbon* 30, 177 (1992).
90. Boehm, H. P., Diehl, E., Heck, W., and Sappok, R. *Angew. Chem. Int. Ed.* 3, 669 (1964).
91. Thomas, Y., Parisi, J. P., Boutevin, B., Beziers, D., Chataignier, E., and Désarmot, G. *Comp. Sci. Tech.* 52, 299 (1994).
92. Goldstein, E. L., Van De Mark, M. R. *Electrochimica Acta* 27, 1079 (1982).
93. Stumm, W. *Chemistry of the Solid-Water Interface*, John Wiley & Sons Inc., New York (1992), p. 90.
94. Boehm, H. P., Diehl, E., Heck, W., and Sappok, R. *Angew. Chem.* 76, 742 (1964).
95. Boehm, H. P., Kolloid Z. Z. *Polym.* 227, 17 (1968).
96. Donnet, J. B., and Bansal, R. C., *Carbon Fiber*, 2nd edition, Marcel Dekker, New York (1989).
97. Puri, B. R. *ibid.* 6, 191 (1971).
98. Krekel, G., Hüttinger, K. J., Hoffman, W. P., Silver, D. S. *J. Mater. Sci.* 29, 2968 (1994).
99. Rand, B., Robinson, R. *Carbon* 15, 311 (1977).

100. Tsutsumi, K. *Toyahashi Carbon Conf.*, Toyohashi University, Toyohashi, Japan, 1984.
101. Lin, S. H., Hsu, F. M. *Ind. Eng. Chem. Res.* 34, 2110 (1995).
102. Bradley, W. L., Grant, T. S. *J. Mater. Sci.* 30, 5537 (1995).
103. Jeziorowski, H. *Materialpruefung* 36, 463 (1995).
104. Zhou, J., Lucas, J. P. *Comp. Sci. and Tech.* 53, 57 (1995).
105. Morii, T., Tanimoto, T. *Comp. Sci. and Tech.* 49, 209 (1993).
106. Lee, M. C., Peppas, N. A. *J. Appl. Poly. Sci.* 47, 1349 (1993).
107. Bogoeva-Gaceva, G., Burevski, D., Dekanski, A., and Janevski, A. *J. Mater. Sci.* 30, 3543 (1995).
108. Burevski, D., Poceva, J., and Brezovska, S., *Croat. Chem. Acta* 63, 67 (1990).
109. Jones, C., *Compos. Sci. Technol.* 41, 275 (1991).
110. Nakayama, Y., Soeda, F., and Ishitani, A. *Carbon* 28, 21 (1990).
111. Jang, J., Ishida, H., and Pluedmann,E. P. *Proc. 41st Ann. Conf. Reinf. Plastic* Div. Ed., SPI, 2-C, 1986.
112. Chatzi, E., Ishida, H., and Koenig, J. L. *Composite Interfaces*, (Edited by H. Ishida and J. L. Koenig), Elsevier Sci. Pub., New York (1986).
113 Alcaniz Monge J., Cazorla Amoros, D., Linares Solano, A. Oya, A. *An. Quim.* 90, 201 (1994).
114. Simms, J. R., Yang, C. Q. *Carbon* 32, 621 (1994).
115. Yang, C. Q., Simms, J. R. *Carbon* 31, 451 (1993).
116. Cole, K. C., Casella, I. G. *Polymer* 34, 740 (1993).
117. Yang, C. Q., Simms, J. R., *Fuel* 74, 543 (1995).
118. Rosenwaig, A., Photoacoustics and photoacoustic Spectroscopy, p .94, Wiley, New York (1980).
119. Peng, J. C., and Buttry, D. A., summitted to *Carbon*.
120. Sellitti, C., Koenig, J. L., Ishida, H. *Carbon* 28, 221 (1990).
121. Lu, Y., Xue, G., Wu, F., anf Wang, X. *J. Adhesion Sci. Technol.* 10, 47 (1996).
122. Cole, K. C., Pilon, A., Hechler, J. J., Chouliotis, A., and Overbury, K. C. *Applied Spectroscopy* 5, 761 (1988).
123. Wang, S., Garton, A. *J. Appl. Polym. Sci.* 45, 1743 (1992).
124. Garton, A. *Infrared Spectroscopy of Polymer Blends, Composites and Surfaces*; Hanser Publishers, New York (1992).
125. Foster, P. J., and Howarth, C. R. *Carbon* 6, 719 (1968).
126. Fateley, W. G., Colthup, N. B., Grasselli, J. G. *The Handbook of Infrared and Raman Characteristic Frequencies of Organic Molecules*, Academic Press Inc., London, 1991.
127. Griffiths, P. R., and Fuller, M. P. *Adv. Infrared Raman Spectrosc.* 9, 63 (1982).
128. Chalmers, J. M., Mackenzie, M. W. *Appl. Spectrosc.* 39, 634 (1985).

129. Cole, K. C., Noël, D., and Hechler, J. *J. Polymer Comp.* 9, 395 (1988).
130. Bradley, R. H., Ling, X., Sutherland, I., and Beamson, G. *Carbon* 32: 185 (1994).
131. Desimoni, E., Casella, G. I., Salvi, A. M., Cataldi, T. R. I., and Morone, A. *Carbon* 30: 527 (1992).
132. Smiley, R. J., and Delgass, W. N., *J. Mater. Sci.* 28: 3601 (1993).
133. Bhardwaj, A., and Bhardwaj, I. S. *J. Appl. Polym. Sci.* 51: 2015 (1994).
134. Xie, Y., Wang, T., Franklin, O., and Sherwood, P. M. A. *Appl. Spec.* 46: 645 (1992).
135. Bascom, W. D. *NASA Contractor Report 178306*, Contract NAS1-17918, August, 1987.
136. Devilbiss, T. A., and Wrightman, J. P. *"Carbon Fiber Surface Treatments for Improved Adhesion to Thermoplastic Polymers"*, Final Report to NASA-Langley Research Center, Grant No. NAG-1-343, September 1987.
137. Xie, Y., and Sherwood, P. M. A. *Appl. Spec.* 44: 797 (1990).
138. Xie, Y., and Sherwood, P. M. A. *Chem. Mater.* 2: 239 (1990).
139. Xie, Y., and Sherwood, P. M. A. *Appl. Spec.* 44: 1621 (1990).
140. Xie, Y., and Sherwood, P. M. A. *Chem. Mater.* 3: 164 (1991).
141. Xie, Y., and Sherwood, P. M. A. *Appl. Spec.* 45: 1158 (1991).
142. Smiley, R. J., Delgass, W. N. *Mat. Res. Soc. Symp. Proc.* 318, 361 (1994).
143. Wagner, C. D., Riggs, W. M., Davis, L. E., Moulder, J. F., and Muilenberg, G. E., *Handbook of X-ray Photoelectron Spectroscopy*, Perkin-Elmer, Mn, 1979
144. Binnig, G., Rohrer, H., Gerber, C., and Weibel, E., *Phys. Rev. Lett.* 49, 57 (1982).
145. Hoffman, W. P., Hurley, W. C., Owens, T. W., and Phan, H. T., *J. Mater. Sci.* 26, 4545 (1991).
146. Hoffman, W. P., *Carbon* 30, 315 (1992).
147. Donnet, J. B., and Qin, R. Y., *Carbon* 30, 787 (1992).
148. Donnet, J. B., and Qin, R. Y., *Carbon* 31, 7 (1993).
149. Donnet, J. B., and Qin, R. Y., *Carbon* 32, 323 (1994).
150. Effler, L. J., Fellers, J. F., and Annis, B. K. *Carbon* 30, 631 (1992).
151. Brown, N. M. D., and You, H. X. *Surf. Sci.* 237, 273 (1990).
152. Quate, C. F., *Physics Today* 26 (1986).
153. Hansma, P. K., and Tersoff, J., *J. Appl. Phys.* 61, R1. 15 Jan (1987).
154. Johnson, D. J., *Chem. Ind.* 18, 692 (1982).
155. Johnson, D. J., *in Chemistry and Physics of Carbon* (edited by P. Thrower) Vol. 20, p. 1, Marcel Dekker, New York (1987).
156. Guigon, M., and Oberlin, A. *Compos. Sci. Technol.* 27, 1 (1986).

157. Zimmer, J. E., and White, J. L. *Advances in Liquid Crystals* 5, 157 (1982).
158. Dresselhaus, M. S., Dresselhaus, G., Sugihara, K., Spain, I. L., and Goldberg, H. A., *Graphite Fibers and filaments,* chapter 3/4, Springer-Verlag, Berlin (1988).
159. Bourrat, X., Roche, E. J., and Lavin, J. G. *Carbon* 28, 435 (1990).
160. Hoffman, W. P. *Comp. Inter.* 1, 15 (1993).
161. Cazeneuve, C., castle, J. E., Watts, J. F. C. -R. *Journ. Natl. Compos. 7th* 1990, 1–10.
162. Yip, P. W., Lin, S. S., *Interfaces in Composites*, Pantano, C. G., and Chen, E. J. H., Editors, Materials Research Society Symposium Proceedings, Vol. 170, Materials Research Society, Pittsburgh, PA, p. 339, 1990.
163. Lin, S. S., *J. Vac. Sci. Tech.* A, 8, 2412 (1990).
164. Desimoni, E., Salvi, A. M., Casella, I. G., Damiano, D., *Sur. Interface Anal.* 20, 909 (1993).
165. Hopfgraten, F., *13th Bienn. Conf. Carbon*, Irvine, Calif., 1977, extended abstracts, p. 228.
166. Hopfgarten, F., *Fiber Sci. & Tech.* 11, 67 (1978).
167. Vaidyanathan, N. P., Kabadi, V. N., Vaidyanathan, R., Sadler, R. *Adhesion* 48, 1 (1995).
168. Than, E., Hofmann, A., Leonhardt, G. *Vacuum* 43, 485 (1992).
169. Weng, L. T., Poleunis, C., Bertrand, P., Carlier, V., Sclavons, M., Franquinet, P., and Legras, R. *J. Adhesion Sci. Technol.*, 9, 859 (1995).
170. Ourak, M. Ouaftouh, M., Rouvaen, J. M., and Nongaillard, B., *Acoust. Lett.* 17, 239 (1994).
171. Ourak, M., Nongaillard, B., Rouvaen, J. M., and Ouaftouh, M., *N.D.T. Internat'l*, 24, 21 (1991).
172. Ourak, M., Ouaftouh, M., Nongaillard, B., and Rouvaen, J. M., *Review of progress in NDE*, Plenum Press, New York, 12B, 1217 (1993).
173. Madronero, A., and Verdu, M. *Carbon* 33, 247 (1995).
174. Daumit, G. P. *Carbon* 27, 759 (1989).
175. Bradna, P., Zima, J., *J. Anal. Appl. Pyrolysis* 24, 75 (1992).
176. Bradna, P., Zima, J., *J. Anal. Appl. Pyrolysis* 21, 207 (1991).
177. Mayer, J., Giorgetta, S., Koch, B., Wintermantel, E., Patscheider, J., and Spescha, G. *Composites* 25, 763 (1994).
178. Donnet, J. B., Wang, T. K., and Shen, Z. M., *Carbon* 34, 1413 (1996).
179. Donnet, J. B., Wang, T. K., and Shen, Z. M., *J. Mater. Sci.* 31, 6621 (1996).
180. Krekel, G., Huttinger, K. J., Hoffman, W. P., and Silver, D. S., *J. Mater. Sci.* 29, 2968 (1994).

5
MECHANICAL PROPERTIES OF CARBON FIBERS [1]

L. H. Peebles
Annandale, Virginia

Yuri G. Yanovsky
Russian Academy of Sciences, Moscow, Russia

Anatoly G. Sirota, Valery V. Bogdanov, and P. M. Levit
St. Petersburg State Technical University, St. Petersburg, Russia

5.1 TENSILE PROPERTIES AND TEST METHODS

The tensile properties of carbon fibers are usually determined by one of two methods: the single filament tensile test [2] and the composite strand tensile test [3]. In the single fiber test, a filament is carefully separated from a yarn or tow of carbon fiber, then mounted on a cardboard form as shown in Figure 1. Care must be taken to ensure that the filament is aligned axially with the tensile direction. Prior to test, the sides of the cardboard mount are cut, leaving only the filament between the grips. As it is not reasonable to attach an extensometer to the filaments, the compliance of the testing instrument must be known to obtain accurate strain measurements. This test is labor intensive, requiring many samples for statistical analysis. The composite strand tensile test is more suitable for rapid and accurate measures of filament tensile properties. A yarn or tow of fiber is impregnated with a resin, usually an epoxy, to produce a rigid test specimen capable of sustaining uniform loading of the individual filaments. The elongation-to-break of the cured resin must be larger than that of the fiber. Depending on the breaking stress of the composite, end tabs of a thermoplastic resin may be required to prevent slippage within the grips. An extensometer is used to accurately determine the amount of strain. Information on the tensile and compressive properties of commercial carbon fibers are provided in Table 1. Table 2 provides additional information on the mechanical properties of some commercial fibers.

Table 1 Selected Properties of Manufactured PAN-Type Carbon Fibers. Citations are for Young's Modulus (YM), Tensile Strength (TS), Elongation (ε_b), Compressive Strength (CS), Measured Compressive Failure Strain ($\varepsilon_{c,m}$), Measured Compressive Modulus, ($E_{c,m}$)

Manufacturer	Fiber designation [4]	d [μm]	ρ [g/cm³]	TS [GPa]	YM [GPa]	ε_b [%]	CS [GPa]	Note	Method	Ref.	$\varepsilon_{c,m}$ [%]	$E_{c,m}$ [GPa]
Amoco	T-50	6.5	1.81	2.90	390	0.7	1.61	c	composite	75		
							2.31	c	sgl.beam	44	0.65	
							4.09	c	beam	51		
							0.579	m	recoil	51		
	T-40	5.1	1.81	5.65	290	1.8	2.76	c	composite	75		
							1.88	c	sgl.beam	44	0.73	
							1.59	m	recoil	51		
	T-650/35	6.8	1.77	4.55	241	1.8	≤ .80	c	sgl.beam	44	≤ 0.15	
							7.7	c	piezo	47	3.2	
	T-300	7	1.76	3.45	231	1.4	2.88	c	composite	75		
							2.8	c	composite	74	1.4	
							1.38	m	recoil	51		
							2.2	m	recoil	70		
							3.53	c	loop	45		
							4.4	c	loop	71		
							4.45	c	loop	73		
							3.7	c	loop	74	1.6	
							2.06	c	crit.len.	45		

Mechanical Properties of Carbon Fibers

Source	Fiber										
BASF [5]	Celion GY-70	8.4	1.90	1.86	517	0.36	1.05	c	composite	51	
[5]	Celion G30-500	7	1.78	3.79	234	1.62	0.413	m	recoil	51	
							8.3	c	piezo	47	3.6
							6.2	c	min-comp	68	2.65
Grafil Inc. [58]	Grafil 34-700	6.9		1.8	4.5		2.76	c	composite	72	
							1.66	c	raman	58	0.80
	Gr. IM 43-750	5.0			5.5		>2.2	c	raman	76	>0.65
[58]	Grafil HM-st	6.9					0.8	m	recoil	84	
				3.2	390		>2.0	c	raman	76	>0.55
[58]	Grafil XA	7.0					1.58	c	raman	58	0.45
				3.5	230		1.39	c	raman	58	0.9
							>2.0	c	raman	76	>0.9
Hercules	Magnamite-HMS4	7	1.80	2.34	345	0.8	1.66	c	raman	58	0.50
	Magnamite-IM6	5.4	1.74	5.1	303	1.7	2.8	c	composite	90	
							7.1	c	min-comp	68	2.35
	Magnamite-IM7	5	1.8	5.3	303	1.8	8.8	c	piezo	47	2.9
							6.39	c	min-comp	68	2.11

(Table continues)

Table 1 Continued

Manufacturer	Fiber designation [4]	d [μm]	ρ [g/cm³]	TS [GPa]	YM [GPa]	ε_b [%]	CS [GPa]	Note	Method	Ref.	$\varepsilon_{c,m}$ [%]	$E_{c,m}$ [GPa]
Hercules	Magnamite-IM8	5	1.8	5.3	303	1.6	3.22	c	composite	75		
	Magnamite-AS1	8	1.8	3.1	228	1.3	5.9	c	min-comp	68	2.59	
	Magnamite-AS4	8	1.79	4.0	221	1.6	2.69	c	composite	75		
							1.44	m	recoil	50		
							1.41	m	recoil	51		
							7.03	c	crit.len.	51		
							8.5	c	piezo	47	3.7	
							5.02	c	min-comp	68	2.27	
Toho Rayon	Besfight-HTA	7	1.77	3.72	235	1.6	2.8	c	composite	74		
							1.9	m	recoil	74		
Toray Industries	Torayca-M30						3.7	c	loop	74	1.6	
	Torayca-M40	6.5	1.81	2.74	392	0.6	6.71	c	loop	73		
							1.6	c	composite	74	0.4	
							1.2	m	recoil	74		
							2.8	c	loop	71		
							2.3	c	loop	74	0.6	
							0.78	c	crit.len.	45		
	Torayca M40J	5	1.77	4.41	377	1.2	2.33	c	composite	75		
							3.41	c	loop	73		
	Torayca M46	6.5	1.88	2.55	451	0.6	1.4	c	composite	74		
Toray							1.2	m	recoil	74		
							1.98	c	loop	73		

Industries										
Torayca-M50J	5	1.88	3.92	465	0.8	4.0	c	crit.len.	71	
Torayca-M60J	4.7	1.94	3.92	588	0.7	1.67	c	composite	75	
Torayca-T300	7	1.75	3.53	230	1.5	2.8	c	composite	74	1.4
						3.7	c	loop	74	1.6
[a]						2.55	m	mcompr	77	3
						5.4	c	min-comp	68	2.36
Torayca-T700S	7	1.82	4.8	230	2.1	6.1	c	min-comp	68	2.66
Torayca-T800H	5	1.81	5.49	294	1,9	7.86	c	loop	73	
						2.74	c	loop	45	
Torayca-T1000	5		4.8	294	2.4	2.2	m	recoil	70	
						2.74	c	loop	45	
						9.7	c	piezo	47	3.3

(Table continues)

Table 1 Continued

Manufacturer	Fiber designation [4]	d [μm]	ρ [g/cm³]	TS [GPa]	YM [GPa]	ε_b [%]	CS [GPa]	Note	Method	Ref.	$\varepsilon_{c,m}$ [%]	$E_{c,m}$ [GPa]
Selected properties of manufactured MP-type carbon fibers												
Amoco	Thornel P-120	10	2.18	2.37	827	0.3	0.45	c	composite	75		
							0.3	m	recoil	70		
							0.4	m	recoil	78		
	Thornel P-100	10	2.15	2.37	758	0.3	0.48	c	composite	69		
							0.4	m	recoil	84		
							0.5	m	recoil	78		
	Thornel P-75S	10	2.0	1.9	520	0.4	0.2	m	mcomprn.	54		
							0.69	c	composite	75		
							~1.65	c	beam	49		
							2.69	c	beam	51		
							0.5	m	recoil	84		
							0.558	m	recoil	51		
							0.8	m	recoil	78		
	Thornel P-55S	10	2.0	1.9	380	0.5	0.4	m	mcomprn.			54
							1.29	c	min-comp	68	0.25	
							0.85	c	composite	75		
							0.5	m	recoil	84		

Amoco						0.4	m	recoil	50
						0.655	m	recoil	51
						0.8	m	recoil	78
	Thornel P-25	11	1.9	1.4	160	0.9	m	mcomprn.	54
						1.15	c	composite	75
						0.875	m	recoil	51
						0.5	m	recoil	84
						>1.4	m	recoil	78
						0.5	m	recoil	78
DuPont de	Fiber G-E120	9.2	2.14	3.3	827	0.48	m	recoil	78
Nemours	Fiber G-E105	9.3	2.14	3.1	717	0.5	c	composite	75
						0.74			
						0.8	m	recoil	78
	Fiber G-E75	9.3	2.14	3.1	524	0.57	c	composite	75
						0.81			
						1.1	m	recoil	78
	Fiber G-E55	9.35	2.10	3.2	393	0.75	m	recoil	78
	Fiber G-E35	9.4	2.04	2.9	262	1.0	c	composite	75
						1.26			
						1.6	m	recoil	78
Mitsubishi	Dialead-K135					0.82	c	composite	75
Nippon	NT-20	10.0		2.08	201	1.2	c	composite	74

(Table continues)

Table 1 Continued

Manufacturer	Fiber designation [4]	d [μm]	ρ [g/cm³]	TS [GPa]	YM [GPa]	ε_b [%]	CS [GPa]	Note	Method	Ref.	$\varepsilon_{c,m}$ [%]	$E_{c,m}$ [GPa]
Steel [74]							1.8	m	recoil	74		
							2.6	c	loop	74	1.3	
[a]	NT-40	9.5		3.5	400		1.35	m	mcompr	77	5	
							1.1	c	composite	74	0.8	
							1.9	m	recoil	74		
							2.0	c	loop	74	0.5	
[a]	NT-60	9.4		3.0	595		0.74	m	mcompr	77	3	
							0.8	c	composite	74	0.2	
							0.7	m	recoil	74		
							1.8	c	loop	74	0.3	
[a]							0.44	m	mcompr	77	2	

Note: [a] Identification is based on the tensile properties of coded sample examined in the reference; M = measured; c = computed; sgl.beam = single filament in a beam; crit.len = critical length under compression; mcompr. = microcompression; min-comp = minicomposite.

Sources: from refs. 4, 5, and manufacturer's data.

Table 2 Additional Selected Properties of Commercial Fibers: Transverse Compressive Strength (TCS); Transverse Strain-to-Break (Tεb); Shear Modulus (G); damping factor (δ); Transverse Compressive Modulus (TCM)

Manufacturer	Fiber designation	TCS [MPa]	Tεb [%]	Ref.	G [GPa]	δ	Ref.	TCM [GPa]	Ref.
PAN-type fibers									
Hercules	Magnamite-AS4				17.0	1.10	80		
Toray	Torayca M-30				1.42		79	3.2	79
Industries Torayca T-40					14.0	2.00	80		
					17.0	1.96	81		
					3.52		79	4.0	79
	Torayca T-40J				17.5	1.98	81		
	Torayca T-50				14.0	1.50	80		
					15.0	1.36	81		
	Torayca-T300	556	26	77a	15.0	1.30	80	3.21	79
[a]	NT-60	54.9	5	77					
MP-type fibers									
Amoco	P-25	600		82					
Amoco	P-55S	300		82	6.6	0.85	80		
	P-75	200		82	7.0	0.81	81		
					8.0	0.85	80		
					9.0	0.88	81		
	P-100	130		82	4.7	1.00	80		
					5.0	1.04	81		

(Table Continues)

Table 2 Continued

Manufacturer	Fiber designation	TCS [MPa]	Tε$_b$ [%]	Ref.	G [GPa]	δ	Ref.	TCM [GPa]	Ref.
DuPont de Nemours	E-35	80		82	8.0	0.70	80		
					8.5	0.73	81		
	E-75				5.7	1.00	80		
					6.0	1.01	81		
	E-105				5.0	1.00	80		
					5.5	1.04	81		
Nippon Steel [a]	NT-20	160	8	77					
[a]	NT-40	94.3	10	77					
	NT-60	54.9	5	77					

Notes: [a] see Table 1.

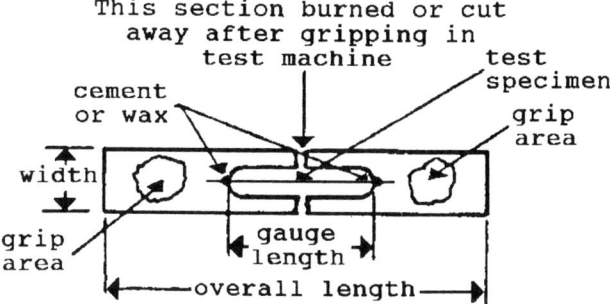

Figure 1 Schematic of cardboard fixture for single filament tensile testing [2]. (Copyright © ASTM, reprinted with permission.)

5.2 TENSILE PROPERTY VARIATIONS

There is a great variation in the tensile strengths of filaments both along and between given yarns within a spool of carbon fibers as well as variation between spools and lot numbers. The distribution of strengths can be described by a number of formulas, one of the most common is the Weibull distribution given by

$$F = 1 - \exp\left\{-\left(\frac{L}{L_0}\right)\left[\frac{\sigma - \sigma_u}{\sigma_0}\right]^m\right\} \quad (1)$$

where L is the gauge length; L_0 is a standard gauge length; m the Weibull modulus; σ, the tensile strength; σ_0, a scaling parameter; σ_u, an arbitrary parameter which many times is set equal to zero; and F, the cumulative probability of failure [6].

The value of F in Equation (1) is determined from the experimental tensile strength measurements by first ranking the data in ascending order then assigning a cumulative probability to each ranked data point. The cumulative probability for a data point of rank R falls somewhere in the interval

$$\frac{R-1}{N} \leq F \leq \frac{R}{N} \tag{2}$$

where N is the total number of data points. One way of assigning a value to F is the formula

$$F = \frac{R}{N+1} \tag{3}$$

For a large number of data points, the precise definition of F in Equation (3) becomes unimportant. Values of m and σ_0 can be determined from the straight line obtained by rewriting Equation (1) in the form

$$\ln\ln\left(\frac{1}{1-F}\right) = m\ln[\sigma - \sigma_u] + m\ln\left[\frac{(L/L_o)^{1/m}}{\sigma_0}\right] \tag{4}$$

The mean strength of the fiber is given by [6]

$$\overline{\sigma} = \sigma_u + \sigma_0 \left(\frac{L}{L_0}\right)^{1/m} \Gamma\left(1 + \frac{1}{m}\right) \tag{5}$$

where Γ is the complete gamma function.

In many cases, the data are not linear, relative to Equation (1). Several different methods have been proposed to account for the non-linearity: Beetz [6] used two Weibull functions with mixing fractions of x and $(1-x)$ while Own et al. [7] chose to use a bimodal log-normal density distribution. The log-normal density distribution is given by

$$W = \frac{1}{(2\pi)^{1/2}\sigma_1\sigma} \exp\left(-\frac{[\ln(\sigma) - \mu]^2}{2\sigma_1^2}\right) \tag{6}$$

where σ_1 is the log-normal standard deviation and μ is the log-normal mean tensile strength. The parameters x, μ_i and σ_i in the bimodal Weibull distribution and the parameters x, σ_{li}, and μ_i for the bimodal log-normal density distribution were determined by the maximum likelihood method (see Olsson [8] or Draper and Smith [9]). Own et al. [7] concluded that the higher-strength distribution was controlled by internal flaws while the lower-strength distribution was controlled by surface flaws, Figure 2, based on the observation that electrodeposition of titanium di(dioctyl pyrophosphate) oxyacetate (TDPA) lowered the weight fraction of the lower-strength distribution, presumably by healing of the surface flaws through the deposition of a protective layer of the titanium derivative.

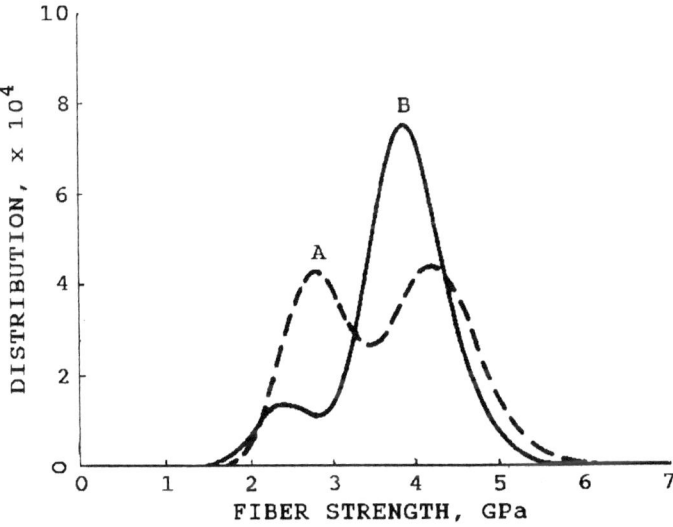

Figure 2 Estimated distributions of fiber strength for untreated Fortifil fiber (A) and electrocoated fiber (B) (Reprinted with permission from Own et al. [7], Copyright © Chapman and Hall, UK.)

5.3 GAUGE LENGTH VARIATIONS

The tensile strength of the fiber also depends strongly on the gauge length. In order to determine the shear strength between fiber and matrix through use of the Kelly-Tyson equation [10]

$$\tau = \left(\frac{\sigma_f}{2}\right)\left(\frac{d}{l_c}\right) \tag{7}$$

where τ is the interfacial shear strength, σ_f is the fiber tensile strength at the critical length l_c, and d is the fiber diameter.

It is necessary to know the tensile strength of fibers at very short gauge lengths. To obtain the tensile strength at short lengths, Beetz [11] provides a consistent method of obtaining fiber strength through use of multi-modal Weibull distributions, matrix algebra, and the maximum likelihood method. His analysis indicates that a linear extrapolation of log tensile strength versus log gauge length underestimates the tensile strength at short gauge lengths. On the other hand, Asloun et al. [12] examined the Weibull probability of failure for two sets of fibers at various gauge lengths. Figure 3 shows that in some cases a bimodal distribution is indicated but not for all gauge lengths. Thus if the bimodal distribution is due to the different contributions of surface and internal flaws, there should be a steady progression of the bimodal distribution as the gauge length is varied. Furthermore, the parameters m and σ_0 are not independent of gauge length. They find that a linear extrapolation of log tensile strength versus log gauge length is appropriate, at least down to 3 mm (critical lengths can be one-tenth this amount). Own et al. [7] find that the log normal distribution parameters for uncoated fibers depend on gauge length as shown in Table 3. For the coated samples, the fraction of surface flaw failure decreases as one would expect; the log-normal parameters, however, are independent of gauge length for the coated fibers.

Waterbury and Drzal [13] have developed a method of measuring the tensile strength at very short gauge lengths by use of the single fiber fragmentation test. Here it is necessary to ensure that the fiber is not under a residual strain owing to sample preparation. In their case, the fiber was embedded in resin through solvent casting rather than chemical curing with the assumption that solvent evaporation does not result in a residual strain. The results obtained thereby (Figure 4) agreed with results obtained by conventional testing methods. The in situ method described here does provide tensile strength measurements at the critical length level used in Equation (7). Examination of gauge length variations through (1) demonstrates that the equation should not be used to predict σ_f [14].

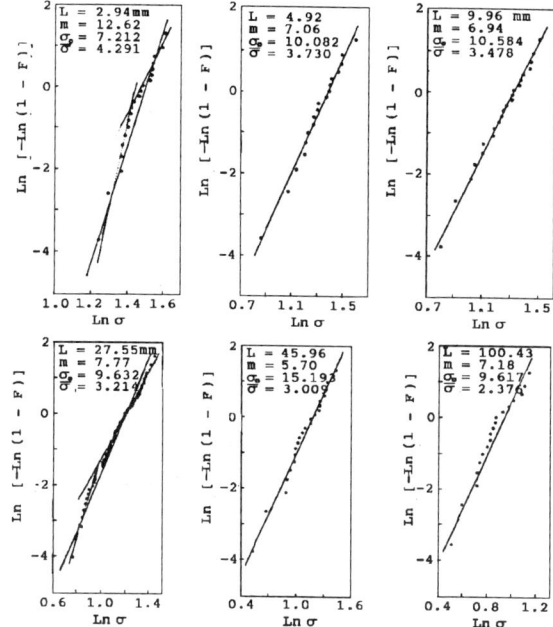

Figure 3 Two parameter Weibull failure probability plot for an untreated high-strength PAN-based carbon fiber for different gauge lengths. Note scale changes in the ln σ axis. (Reprinted with permission from Asloun et al. [12], Copyright © Chapman and Hall, UK.)

Table 3 Gauge Length Dependence of the Fiber Strength Parameters for the Bimodal Log-Normal Distribution

Strength parameters	Uncoated fibers	Coated fibers
Surface flaw fraction, x	Independent of L = 0.48	$1 - AL^{-b}$
Log normal mean, σ_i	$a_i - b_i[\log L]$	Independent of L
Log-normal deviation, σ_{0i}	Independent of L	Independent of L

(Reprinted with permission from Own et al. [7], Copyright © Chapman and Hall, UK.)

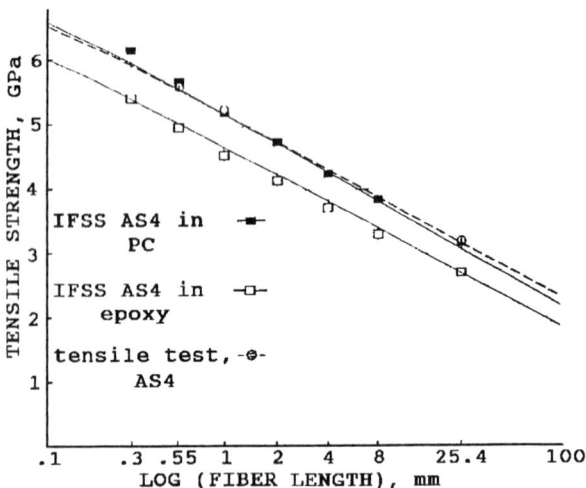

Figure 4 Fiber tensile strength versus log (gauge length) for the traditional tensile test and for single fiber strength test performed in polycarbonate and epoxy matrices. The lines indicate a least squares fit to the experimental data. The strain-free solvent deposited polycarbonate specimens show close agreement with the standard tensile test in both slope and intercept. The epoxy matrix results agree in slope but are displaced owing to pre-tensioned fibers. (Waterbury and Drzal [13], Copyright © ASTM, reprinted with permission.)

5.4 FINE STRUCTURE OF HIGH STRENGTH PAN-TYPE FIBERS

The mechanical properties of carbon fibers are controlled by their fine structures. The basic structural unit of carbon fibers consists of a stack of turbostratic layers with the dimensions shown in Figure 5. The basic structural unit can split, twist, fold, and join other basic structural units to form microdomains, which can also split, twist, fold, join, etc. within a carbon fiber. The fine structure is not a homogeneous, monolithic carbon but rather a somewhat chaotic collection of basic structural units formed into microdomains interspersed with pores. It is not possible within a fiber of 10 μm diameter to show all the structures possible, ranging in size from 1 nm to the fiber diameter, hence the proportions in that figure necessarily are out of scale. The dark sections containing aromatic rings represent the basic structural units with an L_a dimension of about 1 nm on each edge. (see the discussion of the interpretation of L_a and L_c in Chapter 2).

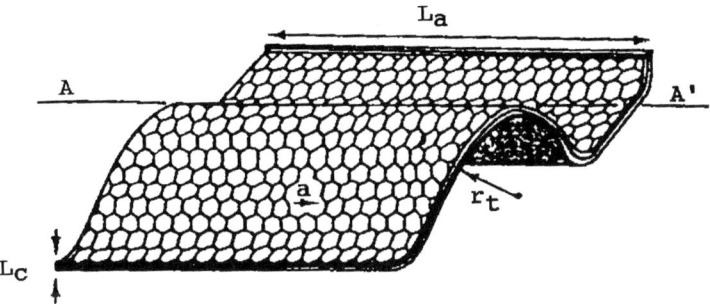

Figure 5 The basic structural unit of carbon fibers. A-A': fiber direction; a: a defect; L_a, L_c: dimensions of the basic structural unit; r_t: radius of curvature in the transverse direction. (Reprinted from Guigon et al. [19] with permission from Elsevier Science, Ltd., Kidlington, UK.)

5.5 MECHANISMS OF TENSILE FAILURE

Much has been written on the effect of flaws on the strength distribution in carbon fibers. The improved tensile strength in recent years is probably a result of higher quality control of production by reducing particulate material and reducing the presence of bubbles created by gas evolution. Indeed, Bennett et al. [15] fractured PAN-based carbon fibers in glycerol so that the broken ends could be examined by SEM and TEM. They found that the internal and surface flaws which initiated failure showed evidence of large misoriented crystallites in the walls of the flaws. Internal flaws (bubbles, holes) that did not initiate failure had walls containing crystallites oriented mainly parallel to fiber axis. They concluded that it is the presence of misoriented crystals that initiate tensile failure. The Reynolds and Sharp [16] mechanism of failure is sketched in Figure 6. The misoriented crystallite in (a) links two crystallites parallel to the fiber axis. In (b) tensile stress exerted parallel to the fiber axis causes layer plane rupture in the direction $L_{a\perp}$ and a crack develops along $L_{a\perp}$ and $L_{c\perp}$. Further application of stress causes failure of the misoriented crystal which propagates across the fiber. The tensile strength of PAN-based carbon fibers is greater than that of the mesophase pitch-based (MP-based) carbon fibers at comparable Young's modulus values, compare Figure 7 [17] and Table 1.

Endo [18] finds that MP-based fibers with high strength have a more turbostratic structure than the ultra high modulus fibers. Further, the fine structure of the strong MP-based fibers was quite similar to that of high modulus carbon fibers produced from PAN (Chapter 2, Figures 25 and 26). The folded structure shown is suggested to result in a lower graphitizability and to provide a more tortuous path for crack propagation. In their examination of high-modulus PAN-based fibers, Guigon et al. [19] found a linear relation between tensile strength and the ratio of radius of curvature in the transverse direction to the axial dimension of the basic structural unit, $\bar{r}_t / \bar{L}_a$, Figure 8. A similar correlation was found for high strength PAN-based carbon fibers, but it was not possible to precisely determine $\bar{r}_t$, therefore a new parameter Δ_t, based on subjective ranking of TEM micrographs, was used [20,21].

One of the unique properties of carbon fibers and carbon-carbon composites is that the tensile strength does not decrease with increasing temperatures. Tanabe et al. [22] measured the tensile strength, Young's modulus and Weibull modulus as a function of temperature up to 1300°C. These values increased very slightly with increasing temperature but within the experimental error range of a zero slope.

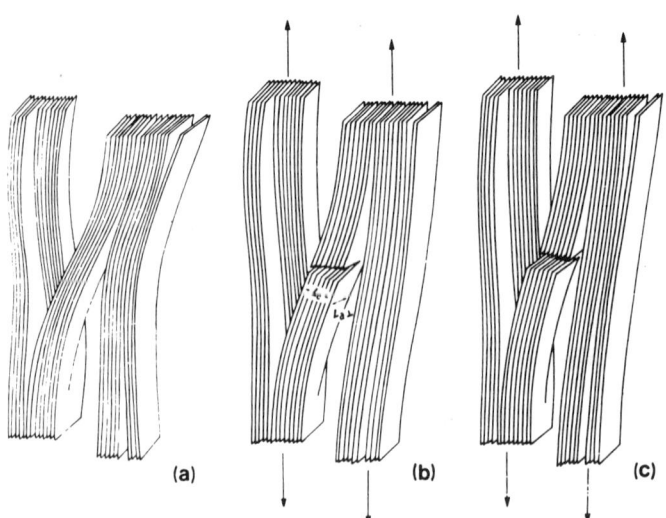

Figure 6 Sketch of the mechanism of tensile failure in a misoriented crystal: (a) the misoriented crystal, (b) crack initiation, (c) crack propagation leading to fiber failure (Reprinted with permission from Bennett and Johnson [15], Figure 3, Copyright © Chapman and Hall, UK.)

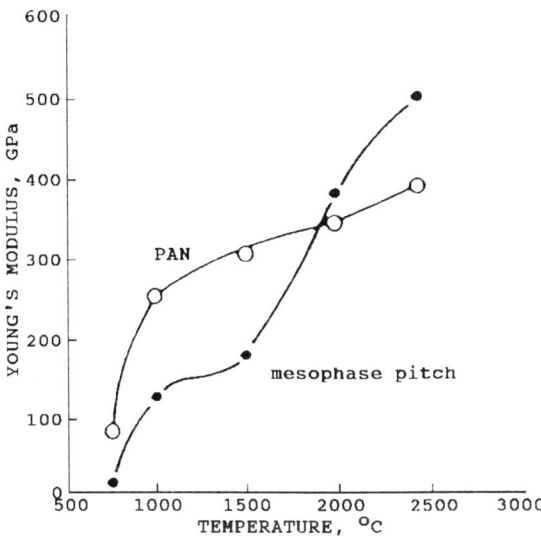

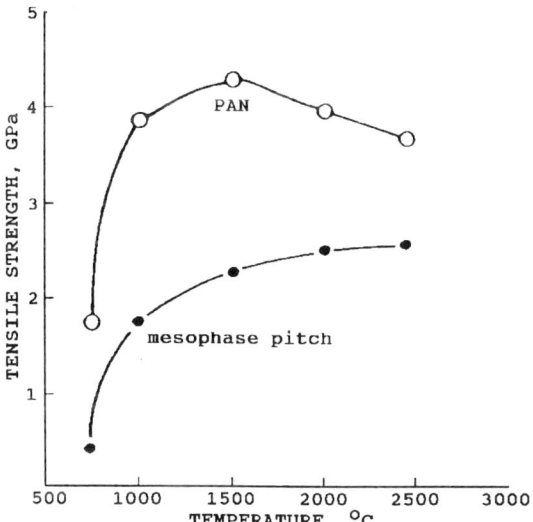

Figure 7 Tensile strength and Young's modulus for a PAN-based carbon fiber and a mesophase pitch-based carbon fiber versus final heat treatment temperature. (Reprinted from Matsumoto [17] with permission from Pergamon Press, Ltd., Oxford, England.)

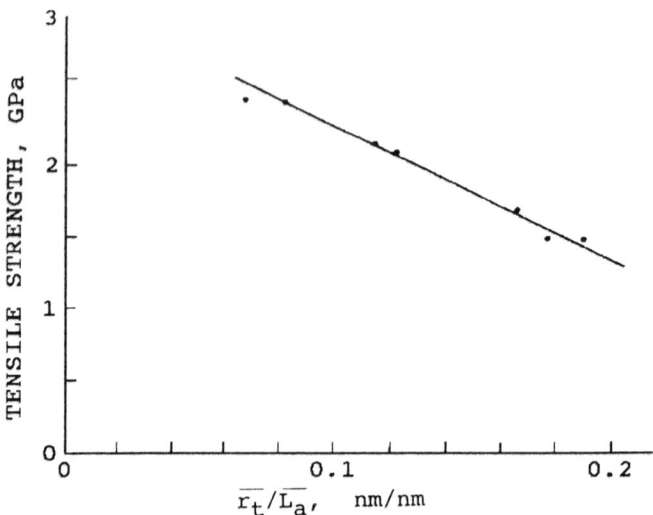

Figure 8 Correlation of tensile strength with average fine structure parameters identified in Figure 5 for high modulus PAN-type carbon fibers. (Reprinted from Guigon et al. [19] with permission from Elsevier Science, Ltd., Kidlington, UK.)

5.6 NON-LINEAR ELASTICITY AND STRUCTURE-MODULUS RELATIONS

Carbon fibers are considered to be linear elastic by many investigators because of an apparent linear stress-strain plot at the usual rates of elongation. Beetz [23], however, cites a number of references indicating non-linear behavior, namely the increase in ultrasonically measured Young's modulus with increasing load [24], an increase in the dynamic Young's modulus with increasing static strain [25], the increase in torsional modulus with increasing tensile stress [26,27], and the decrease in Young's modulus with increased bending strain [28]. Kowalski [29] investigated the nonlinearity of stress-strain data for single filaments, the strand method and unidirectional composites. The data are fitted by the empirical equation

$$\sigma = E_0 \varepsilon^2 + F\varepsilon^2 \tag{8}$$

where E_0 is the initial tangent modulus and F is a measure of the degree of nonlinearity, by least squares analysis under the condition of zero stress at zero strain. The results are given in Table 4 for two PAN-based carbon fibers. Note that the modulus for single filaments is always lower than that of the strand and composite data and further that the coefficients of variation are smaller for the latter two. The strand data agree quite well with the composite data. The lower filament modulus may be due to inadequately taking the instrument compliance into account.

Table 4 Nonlinearity in Stress Strain as Analyzed by Equation (8) for Amoco T-40 and T-700 PAN-based Carbon Fibers

	T-700	T-40
Single filaments		
N	25	25
E_0, GPa	235	259
CV	5.5%	4.2%
F, GPa	1380	1765
CV	27	28
E_{sec}	245	271
Strands		
N	19	13
E_0, GPa	243	267
CV	1.3%	1.4%
F, GPa	1760	2115
CV	5.9%	9.5%
E_{sec}	256	281
Composites		
N	12	32
E_0, GPa	237	263
CV	1.0%	0.9%
F, GPa	2255	2495
CV	2.9%	2.6%
E_{sec}	253	280

Note: E_{sec} is a secant modulus taken between 0.1% and 0.6% strain.
Source: Kowalski, from ref. 29. (Copyright © ASTM, reprinted with permission.)

The interpretation for the increasing modulus with increasing strain is a gradual straightening of the crystalline regions along the fiber axis. This interpretation is consistent with the correlation between preferred orientation and Young's modulus, Figure 9 [30], for a variety of rayon-based carbon fibers. The Young's modulus of PAN-based carbon fibers correlates well with the average value of the crystallite diameter, Figure 10 [19].

The preferred orientation parameter, Z, the full width at half the maximum intensity of the 002 x-ray reflection, obtained in the azimuthal scan and measured in degrees, is an empirical, easily assessed parameter. A more accurate measure of the orientation of crystallites is given by Equation (9)

$$\langle \cos^2 \phi \rangle = \frac{\int_0^{\pi/2} \rho(\phi) \cos^2 \phi \sin^2 \phi \, d\phi}{\int_0^{\pi/2} \rho(\phi) \sin \phi \, d\phi} \tag{9}$$

where ϕ is the azimuthal angle of a reflection arc, usually the 002 arc for carbon fibers, and $\rho(\phi)$ is the intensity of scattering at angle ϕ. Northolt et al. [31] examined the series model of Ruland [32] and by eliminating second order terms arrived at the equation

$$\frac{1}{E} = \frac{1}{e_1} + \frac{\langle \cos^2 \phi \rangle}{g} \tag{10}$$

where E is the Young's modulus in the axial direction, e_1 is the modulus in the direction normal to the c-axis of the scattering crystal, and g is the modulus for shear between planes oriented normal to the c-axis.

All of the data given in [31] are shown in Figure 11, the values of e_1 and g are given in Table 5; those without correlation coefficients are reported by Northolt et al. The values of e_1 and g calculated for the PAN-based fibers are approximately 611 and 39 GPa respectively, those for MP fibers are 776 and 15 GPa while those for rayon-based fibers are 900 and 14 GPa. These values agree reasonably with those reported for graphite crystals, $690 < e_1 < 1020$ GPa, $0.1 < g < 5$ GPa. The shear modulus for graphite crystals would be expected to be lower than the crystals in carbon fibers because the turbostratic layers in the latter are non-planar. The g values for rayon and MP-carbon fibers are lower than those for PAN-based fibers because the former have a more graphitic nature. Part but not all of the scatter in Figure 11 may be due to the combination of crystals of different perfection in the transition from the turbostratic to the graphitic structure.

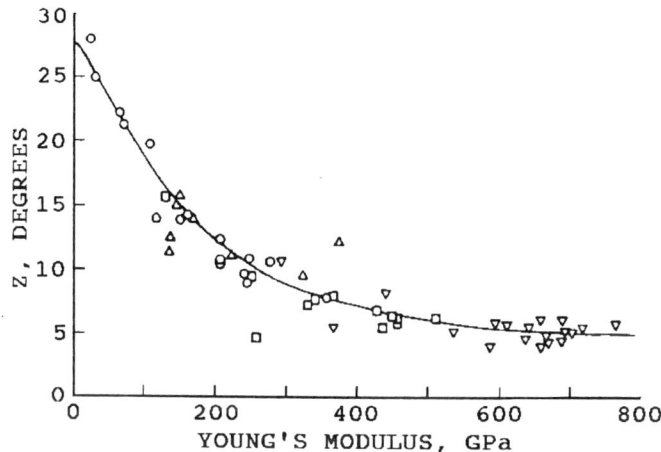

Figure 9 The preferred orientation, half width at half maximum intensity of the 002 arc, as a function of Young's modulus for a variety of rayon carbon fibers (Bacon and Schalamon [30], Copyright © 1969, reprinted with permission from John Wiley and Sons, Inc.)

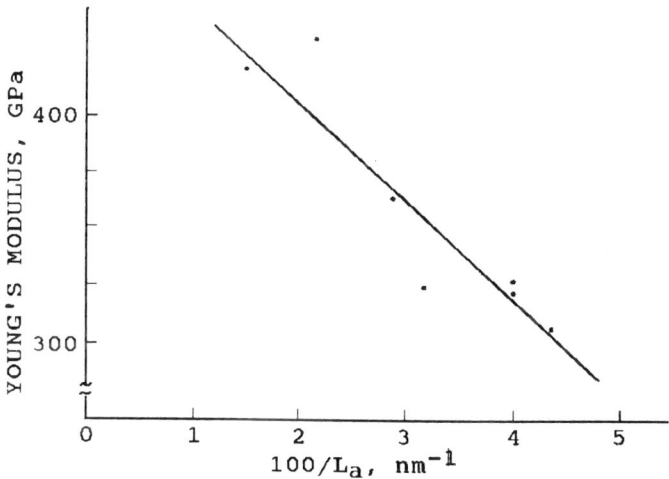

Figure 10 Correlation between Young's modulus and reciprocal crystallite dimension for PAN-based carbon fibers. (Reprinted from Guigon et al. [19] with permission from Elsevier Science, Ltd., Kidlington, UK.)

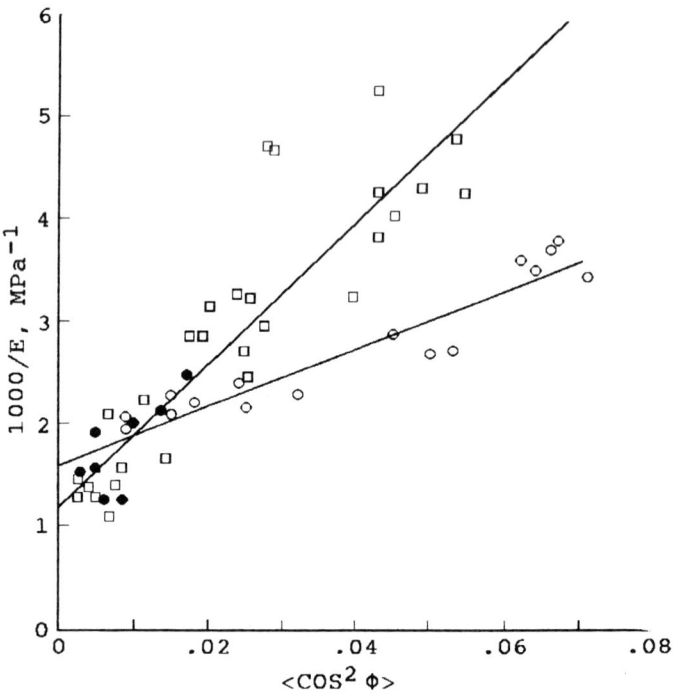

Figure 11 Correlation between 1000/(Young's modulus) and orientation defined by $<\cos^2 f>$ for PAN (O), MP (●), and Rayon (□) carbon fibers (from [1], after Northolt et al. [31])

Table 5 Values of Crystal Young's Modulus, e_1, Perpendicular to the C-Axis, Shear Modulus, Based on Equation (10) and the Correlation Coefficient R^2

Fiber type	Number of Samples	e_1 [GPa]	G [GPa]	R^2
PAN	4	820 ± 90	34 ± 7	—
	17	611	39	0.90
MP	3	870 ± 70	12 ± 2	—
	8	776	15	0.88
Rayon	45	930 ± 320	14.1 ± 0.8	—
	45	900	14.3	0.82

Values with ± reported by Northolt.
Source: from Northolt et al., ref. 31.

The calculation of strength and modulus of carbon fibers requires measurement of the filament diameter. Both optical microscopic and laser diffraction methods have been used. (See Li and Tietz [33] for a discussion of the laser method). Chen and Diefendorf [34] find that optical methods tend to be skewed to smaller diameters and show a far higher standard deviation when compared to laser diffraction methods. Further, the laser method is easier to use as there is much less sample preparation and microscope manipulation. The previously reported variation between fiber modulus and fiber diameter [35] was probably the result of errors in diameter measurements as the new results on fibers of the same vintage did not show a significant variation of modulus with diameter [34].

The modulus within a carbon filament can vary across the fiber radius. Electrochemical etching of the fibers can evenly remove the outer layers, in contrast to oxidation of the fiber which produces pits. The etched fibers are then subjected to mechanical property measurement. The measured and estimated modulus of HMS fiber is shown in Figure 12 as a function of fiber radius. The variation in modulus across the fiber implies that other properties also vary across the fiber diameter.

An experimental fiber with a breaking strain of ~2% was prepared from the PAN precursor used to prepare Torayca T-300 carbon fiber [36], and compared to the T-300 fiber that had undergone a proprietary etching to increase the tensile strength. The increase in tensile strength of the etched T-300 could not be attributed to the removal of surface defects as the etched sample had the same gauge length dependence as the unetched T-300. The increase in tensile strength was attributed to the removal of microvoids concentrated at the outer surface of the fiber. Why microvoids were not considered surface defects was not explained. The etching procedure also showed an increase of tensile strength and a decrease in modulus as a function of fiber weight loss (Figure 13) i.e., across the filament diameter.

The axial and transverse values of tensile (E_a and E_t) and shear moduli (G_a and G_t), Poisson's ratio, and the coefficient of thermal expansion for a series of Amoco fibers were calculated from unidirectional composite properties (Table 6) [37]. The values of E_a and α_t differ from those reported in Table 1, the modulus values being somewhat lower, because the method of Wagoner and Bacon is a resonance measurement measured under zero strain where E_a is always lower due to nonlinear effects. The G values of Table 5 for MP fibers agree rather well with the G_a values for MP fibers. The G values for PAN-type fibers agree reasonably with the high strength varieties of MP-type fibers which were treated at lower temperatures and exhibit less perfection of the crystallites. The difference between the two transverse moduli is because the data in Table 5 is based on x-ray diffraction measurements of orientation while that of Table 6 is based on dynamic mechanical property data.

The Poisson's ratio on three different high strength PAN-type carbon fibers was measured with the laser diffraction technique, Krucinska [38]. The average value is in the range of 0.26–0.28 with an accuracy of 5.7%. For a dog-bone-shaped fiber, the Poisson's ratio is anisotropic around the fiber perimeter. The Poisson's ratio for Torayca T-300 and Amoco P-55 fibers are 0.22 and 0.10, respectively, as measured by photographic techniques [39] which are different from those computed from dynamic mechanical measurements on composites (Table 6).

Table 6 Computed Properties of Some Amoco Carbon Fibers Based on Composite Measurements

Fiber	E_a [GPa]	G_a [GPa]	ν_a	E_t [GPa]	G_t [GPa]	α_a [ppm]	α_t [ppm]
P-55-H	305	15.6	0.32	10.8	3.0	−1.37	12.1
P-55-L	300	12.9	0.29	10.7	3.2	−1.37	11.6
P-75-H	450	13.3	0.24	8.8	2.5	−1.49	12.2
P-75-L	449	12.7	0.23	8.8	2.5	−1.44	12.7
P-100-H	770	23.6	0.30	7.1	2.1	−1.48	9.4
P-100-L	775	20.6	0.22	6.8	2.0	−1.47	9.5
T-50-H	358	18.1	0.13	10.3	3.3	−1.28	6.6
T-50-L	354	16.4	0.15	10.2	3.3	−1.17	6.8
T-300X	341	19.0	0.13	10.3	3.5	−1.20	7.0
T-300	204	22.2	0.26	14.7	5.0	−0.67	8.9
T-300	205	23.4	0.27	14.4	4.9	−0.67	8.8
T-650	243	23.1	0.29	13.8	5.0	−0.84	7.8

Notes: H = high resin volume fraction; L = low resin volume fraction; T-300X = heat treated T-300; two samples of T-300; the higher modulus PAN-type fibers heat treated at a higher temperature relative to the lower modulus PAN-type fibers. E_a, E_t, Young's modulus in the axial and transverse direction; G, shear modulus; n, Poisson's ratio; a, coefficient of thermal expansion.
Source: from Wagoner and Bacon, ref. 37.

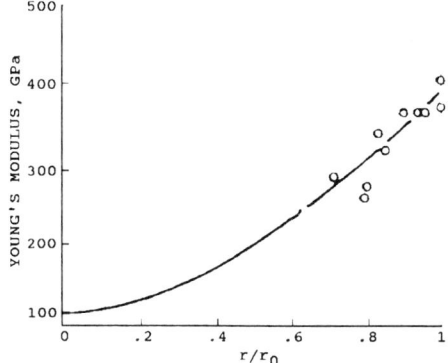

Figure 12 The measured average modulus distribution across the radius of etched HMS fibers (Chen and Diefendorf [34]).

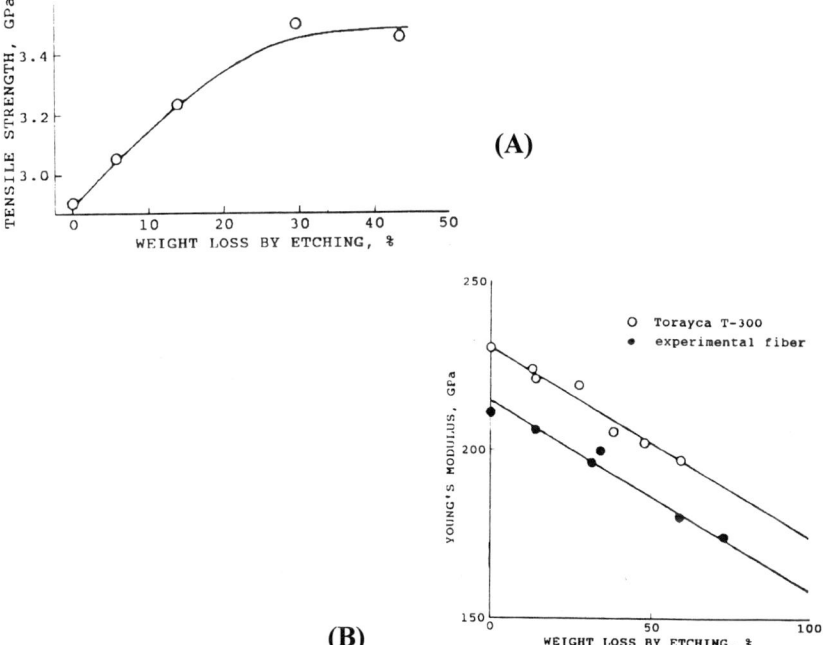

Figure 13 The variation of (A) tensile strength of Torayca T-300 and (B) Young's modulus of T-300 and an experimental fiber, both figures as a function of weight loss due to etching and variation of properties across fiber radius. (Morita et al. [36], Copyright © Pion Limited, London).

5.7 HIGH TEMPERATURE CREEP

At very high temperatures, carbon fibers will undergo creep without necking according to the equation

$$\dot{\varepsilon} = A'\left(\frac{\sigma}{\sigma_f}\right)^n \exp\left(\frac{-Q}{RT}\right) \tag{11}$$

where $\dot{\varepsilon}$ is the rate of extension; σ, the applied stress; σ_f, tensile strength at room temperature; n, the stress exponent; R, the gas constant; and Q, the apparent activation energy. The combined average values of Q for composite and matrix-free strands of Hercules HM was 1082 ± 92 kJ/mole while that for E-130 was 1058 ± 57 kJ/mole. The average value for n was 7.4 for HM and 1.4 for E-130 while the average values of A were 5.37 ± 10.3 E25 and 1.26 ± 0.49 E19, respectively [40]. The mechanism of creep is thought to be due to diffusion of carbon in the bulk and the moving element is the crystallite.

5.8 COMPRESSIVE PROPERTY TEST METHODS

The compressive strength of carbon fibers is about half that of the tensile strength, compare the entries in Table 1. A major reason for improving the compressive strength and a reduction in compressive variability is that composites are subject to flexing, which with low compressive properties require a relatively thicker composite with an accompanying increase in weight. There is a significant problem with determining the compressive strength of carbon fibers because of the assumptions required to make the calculations as will be detailed below. Indeed, the mechanism of compressive failure remains uncertain.

An early compressive strength test method is based on the Sinclair loop test [41], Figure 14 [42] where W is the tension on the fiber, d, the fiber diameter, D and ϕ are the dimensions of the loop. The filament is constrained between two glass cover slides and lubricated with a light oil. If the filament is linear elastic, then the following equations apply

$$\sigma = \frac{60WD}{\pi d^3} \qquad \varepsilon = \frac{16d}{15D} \qquad E = \frac{\sigma}{\varepsilon} \tag{12}$$

and

$$\frac{\phi}{D} = 1.335 \tag{13}$$

Most carbon fibers do not obey (13) but have a larger value resulting in the fiber first yielding to form a "hinge" prior to failure. Other problems include: fibers do not always fail on the compression side, they may fail away from the position of maximum stress, many fibers are non-round, and there may be an effect of strain gradients across the fiber diameter. Ng et al. [43] show large divergences from Equation (13) for the Amoco P-series of MP carbon fibers while T-300 remains linear elastic to near the failure point.

The compressive failure of single filaments embedded in a transparent resin was determined by optical microscopic examination as the resin block was subjected to a bending strain [44]. These authors made the assumptions that the compressive modulus is equal to the tensile modulus and that the modulus is invariant with strain. They find that the apparent axial compressive strengths do not vary much with modulus up to moderate orientation. At higher modulus values there is an inverse relation between compressive strength and tensile modulus.

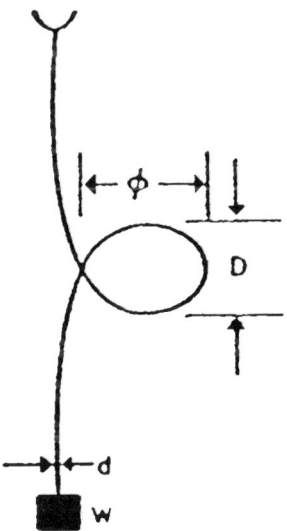

Figure 14 The Sinclair loop test for single filaments (Williams et al. [42], reprinted with permission from the American Institute of Physics).

Use has been made of the Kelly-Tyson relation, Equation (7), to determine the compressive strength under the reasonable assumption that the shear strength between fiber and matrix is independent of whether the composite is under tensile or compressive strains. Ohsawa et al. [45] rearrange this equation into the form

$$(\sigma_f)_{comp} = (l_c)_{comp} \left[\frac{(\sigma_f)_{ten}}{(l_c)_{ten}} \right] \tag{14}$$

Through use of the equation

$$P_T = \frac{\alpha_m E_m \Delta T}{1 + v_m} \tag{15}$$

where P_T is the thermal shrinkage stress; α_m, the thermal expansion coefficient of the matrix; E, the Young's modulus; ΔT, the difference between test and either the cure temperature or the glass transition temperature, whichever is the lower; and v_m, the Poisson's ratio of the matrix.

They found a linear relation between the estimated compressive strength and the thermal shrinkage stress caused by cooling the composite from the molding temperature, Figure 15. A similar result was found by Waterbury and Drzal for tension tests [13]. Boll et al. [46] and DeTeresa [47] both find that the critical lengths are shorter in compression relative to those determined in tension. One would expect a lower compressive strength based on Equation (14), but also that the distribution of lengths is significantly narrower in compression. As electrical continuity is maintained in the compressed sample, some stress can be transferred through the fractured ends [47]. Thus the critical length in compression may not be a true measure of compressive strength.

An alternate approach to the single fiber composite is to mount the fiber perpendicular to the tensile direction within the matrix block. The strain in the perpendicular direction, y, is given by

$$\varepsilon_y = \frac{\sigma_y}{E_y} - v_m \left(\frac{\sigma_x}{E_x} \right) \tag{16}$$

Mechanical Properties of Carbon Fibers

where σ_x is the stress on the composite in the tensile direction. Taking $\varepsilon_c = \varepsilon_y$ and $\sigma_y = 0$, where ε_c is the fiber strain in compression, v_m the Poisson's ratio of the matrix, $E_x = E_m$ the Young's modulus of the matrix and assuming a perfect fiber matrix interface so that $\varepsilon_c = \sigma_c/E_c$ and that the fiber moduli in tension and compression are equal, the equation reduces to

$$\sigma_c = -\left(\frac{v_m E}{E_m}\right)\sigma_x \tag{17}$$

The technique has been applied to two experimental fibers, the results are reported in Table 7 [48].

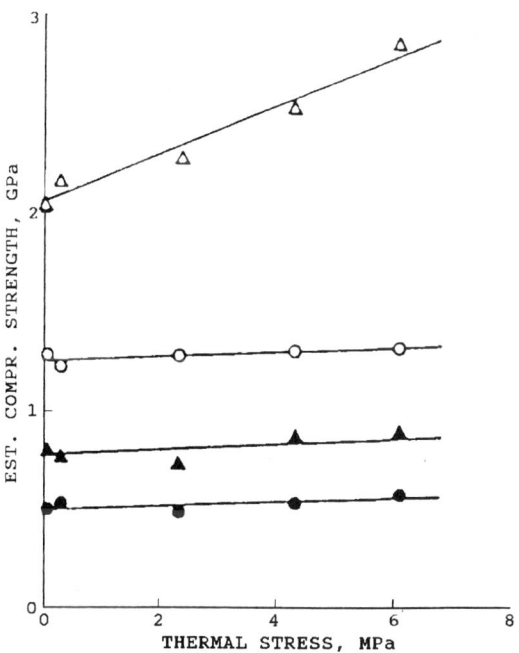

Figure 15 Relation between estimated compressive strength and thermal shrinkage stress: (O) pitch-based carbonized fiber Tonen HTX; (●) pitch-based graphitized fiber Tonen HMX; (Δ) PAN-based carbonized fiber, Torayca T-300; (▲) PAN-based graphitized fiber, Torayca M-40 (Ohsawa et al. [45], Copyright © 1990, reprinted with permission from John Wiley and Sons, Inc.).

Table 7 The Compressive Strengths of Two Experimental Fibers Determined by the Transverse Tension Single Fiber Composite Technique Along with the Reported Tensile Properties Under the Assumption that the Compressive Modulus is Equal to the Tensile Modulus

Fiber	E_t [GPa]	σ_c [GPa]	σ_t [GPa]
PAN	230	0.89	2.9
MP	260	1.31	2.0

Source: from ref. 1 after Wagner et al., from ref. 48.

The compressive strength can be calculated from the amount of strain observed at initial kink-band formation and the assumptions that the sample is linear elastic both in tension and compression and that the compressive strength is less than the tensile strength. DeTeresa et al. [49] devised a rig that can be used to obtain several different measures of σ_c on the same filament through the use of successively smaller diameter rods, Figure 16. The filament is bonded to a transparent beam with a thin coating of an acrylic spray cement and evidence of failure observed with an optical microscope. The strain at coordinate d is given by

$$\varepsilon(d) = \frac{3tv}{2L^2}\left(1 - \frac{d}{L}\right) \tag{18}$$

where the dimensions are given in Figure 16. If failure occurs prior to the first optical manifestation or if there is a residual load on the fiber due to mounting, the calculated results may be biased [49].

If the compressive strength is less than the tensile strength, then the recoil test can be used to provide an estimate of the compressive strength. Here, a series of single fibers are subjected to a static tensile load, then quickly broken with scissors [50], an electric spark [51], or driven coplanar scalpel blades [52] (the first method may impose shear or bending modes during fiber scission [51]). The immediate release of tension initiates a tensile wave back along the two fiber fragments which is then reflected at the grips into a compressional wave. Following the test, the fragments are examined for compressive failure. The compressive strength is then bounded between tensile loads for compressive failure and for no compressive failure. If the fiber fails by buckling instead of compressive failure or if flexural waves are generated during recoil, inconsequentially low results may be obtained [47]. If the fiber is coated with a thin, viscous, non-structural coating, an increase in recoil compressive strength is observed [53].

The compressive properties of very short fibers have been determined through the use of a microcompression apparatus that uses a piezoelectric element to apply force to the sample [54]. The gauge length of the fiber must be carefully chosen to avoid buckling failure, ensure against pure compression failure, and to avoid the effects of the clamped ends. Thus the length L must be within the bounds given by

$$L > 2r\left(\frac{E}{E_t}\right)^{1/2}$$

$$L \approx 20r \tag{19}$$

$$L < \pi r\left(\frac{E}{\sigma_{cr}}\right)^{1/2}$$

where r is the fiber radius; E_t, the transverse modulus, and σ_{cr} an estimated value for the critical buckling stress. At the small gauge lengths involved, the apparent compressive modulus needs to be corrected for instrument compliance. An equation relating compressive stress σ_c and the apparent modulus E_{app} is given by

$$\sigma_c = E_{app}\frac{(l_f + l_m)}{L} \tag{20}$$

where l_f is the deformation of the fiber and l_m is the deformation contributed by the testing instrument and the adhesive used to hold the fiber in place. Rearranging this equation yields

$$\frac{1}{E_{app}} = \frac{1}{E_f} + \left(\frac{l_m}{\sigma}\right)\left(\frac{1}{L}\right) \tag{21}$$

where E_f is the corrected fiber modulus [55]. This instrument can provide stress-strain curves in both tension and compression. The engineering compressive stress—compressive strain diagram for Amoco P-75 fiber is given in Figure 17. Note the non-linearity of stress vs. strain, the yield point at –0.2% strain, and the non-catastrophic failure at higher strain levels. All of the fibers tested exhibited similar non-linearities. The MP P-series of fibers may have a bimodal distribution of compressive strengths, however only a limited number of filaments (7–9) were examined for each fiber type [54].

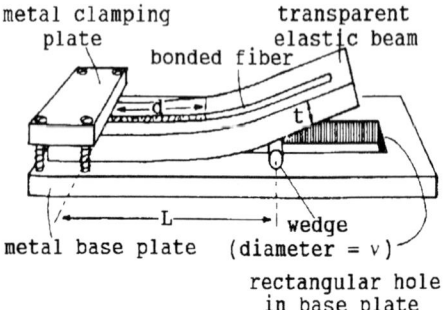

Figure 16 Schematic of an apparatus for determination of critical compressive strain for kink band formation in single fibers (Reprinted with permission from DeTeresa et al. [49], Copyright © Chapman and Hall, UK).

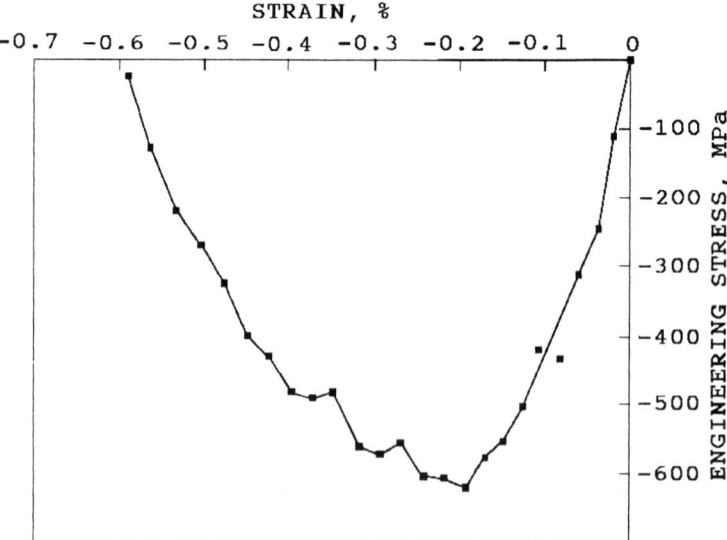

Figure 17 Engineering compressive stress—strain diagram for Amoco P-75 fiber. Note that the coordinate origin is at the upper right hand-corner. (Macturk et al. [54], reprinted by permission of the publishers, Butterworth Heinemann Ltd, Copyright ©).

Mechanical Properties of Carbon Fibers

The Raman spectra of carbon fibers has two bands, the E_{2g} band at 1575 cm^{-1}, also seen in graphite samples with large crystals, and a A_{1g} band at 1355 cm^{-1} attributed to a particle size effect, [56]. The intensity of the A_{1g} band, relative to E_{2g} decreases with increasing perfection of the crystalline regions [57]. The position of both bands shifts linearly with stress and therefore provides a means of determining the failure strain under either tension or compression. The frequency shift depends strongly on the fiber type (varying crystalline perfection at the fiber surface) and on whether strain is in tension or compression [58]. Thus, these authors indicate that compressive modulus is less than the tensile modulus for three out of the four fibers tested, based on the equation

$$E_c = \left(\frac{S_c}{S_t}\right) E \tag{22}$$

where S_c is the slope of the frequency shift-stress correlation under compression, and S_t that in tension. The frequency shift is also dependent on the temperature of the fiber due to heating by the laser beam. Also variations can occur from fiber to fiber and along an individual fiber. These effects can be minimized by use of a low power laser [59,60] or by Raman microline focus spectroscopy [61]. Because of the heat-transfer differences between bare filaments and composites, this technique may not be applicable for comparison between the two.

The piezoresistivity of carbon fibers has also been used to measure the compressive failure strain of carbon fibers, [47]. The procedure is to mount a fiber on a block of resin as shown in Figure 18. There is a non-linear response of piezoresistivity in both tension and compression, Figure 19. The compressive strength is calculated from the compressive failure strain and the tensile modulus.

The compressive strength of fibers can also be calculated from unidirectional composite measurements. Some standard methods include ASTM D-695 [62], ASTM D-3410 [63], and ASTM D-5467 [64]. The compressive strength of the fiber is just the compressive strength of the composite divided by the volume fraction of fiber and the number of fibers. The major problem is the support of the composite and construction of the gauge length. Odom and Adams [65] examined five different ways of constructing the tabs used to hold the composite in the instrument and define the gauge length. They found five different modes of failure, some or all of which appear in a given test configuration, and the composite strengths obtained from some of the tab designs were significantly larger than that of other designs. Indeed, by careful machining of the gauge length section, Curtis et al. [66] found a > 50%

improvement in composite compressive strength over a Royal Aircraft Establishment study. The stiffness of the interphase, that region between the fiber and bulk matrix, also has an influence on composite compression properties [67]. Therefore, the compressive strength of a composite may be more a system property rather than a material property.

Prandy and Hahn [68] prepared mini-composites by embedding a single strand of fibers in an epoxy resin then machining the composite into a dogbone shape prior to compressive testing. Some results are presented in Table 1. It is obvious from the number of tests described above that there is no standard method of measuring the compressive strength of carbon fibers. All of the tests depend on certain assumptions, many of them assume that the compressive modulus equals the tensile modulus, which may or may not be warranted. It is a reasonable assumption based on composite data as shown in Figure 20 [69].

A comparison of the compressive strengths determined by these various methods is given in Table 1. References in Table 1 that are not cited in text include Johnson and Park [70], Johnson [71], Vezie and Adams [72], Morita et al. [73], Furuyama et al. [74] and Kumar et al. [75]], Melanitis et al. [76], Shinohara et al. [77], and Hayes et al. [78]. The high values cited for compressive strength determined by piezoresistivity method is due to the experimentally determined large value of failure strain. Similar high values are obtained by the minicomposite technique [68]. DeTeresa [47] concludes that the compressive strength of composites is not limited by the compressive strength of the fiber and that a large potential exists to improve composite properties.

The transverse compression modulus of round single filaments can be calculated from

$$U = \left(\frac{4F_2}{\pi}\right)\left(\frac{1}{E_T} - \frac{v^2}{E}\right)\left(0.19 + \sinh^{-1}\frac{r}{b}\right) \quad (23)$$

where

$$b^2 = \left(\frac{4F_2 r}{\pi}\right)\left(\frac{1}{E_T} - \frac{v^2}{E}\right) \quad (24)$$

where U is the diameter change under a compression force/unit length F_2, r is the fiber radius, C the transverse compression modulus and E the Young's modulus. For highly oriented fibers, one can assume $1/E \ll 1/E_T$, ν is small so the term ν^2/E can be neglected in both equations [79].

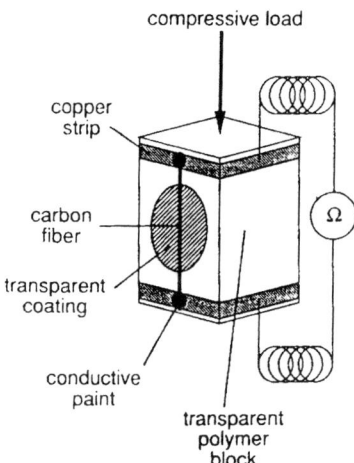

Figure 18 Schematic of a single filament compression specimen using piezoresistance to measure strain. (Reprinted with permission from DeTeresa [47], Pergamon Press, Ltd., Oxford, England.)

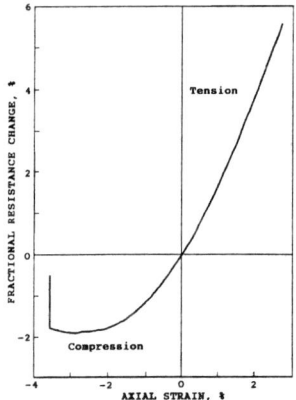

Figure 19 Piezoresistive behavior of T1000 fiber in tension and compression. (Reprinted with permission from DeTeresa [47], Pergamon Press, Ltd., Oxford, England.)

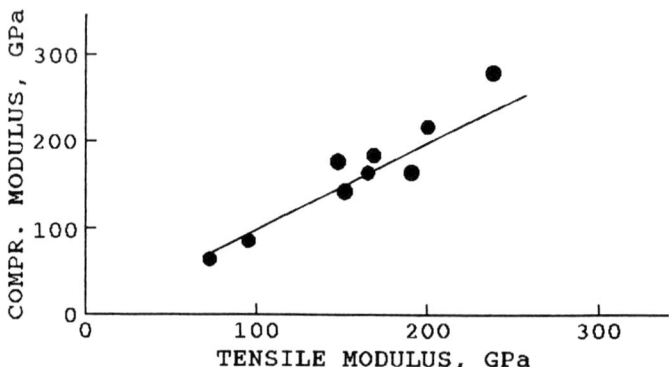

Figure 20 Correlation between composite compressive modulus and composite Young's modulus for different carbon fibers. The line has unit slope. (from [1], after Kumar et al. [69].)

5.9 TORSIONAL PROPERTIES

The torsional or shear modulus and the damping factor of carbon fibers are measured by the torsional pendulum method under vacuum [80]. Fibers under compression can fail by a shear mechanism, see discussion below, hence in many cases shear strength and compressive strength are synonymous.

5.10 MECHANISMS OF COMPRESSIVE FAILURE

The mechanism of failure of carbon fibers under compressive stress is still under debate. Hawthorne and Teghtsoonian [44] suggest that fiber compressive fracture changes from a shear to a microbuckling or kinking mode with increasing fiber anisotropy. Ohsawa et al. [45] investigated the compressive strength as a function of temperature and found that at 20°C, a high compressive strength Tonen MP carbon fiber failed without buckling, but buckling occurs at 140°C, certainly the result of the lower modulus of the matrix which means less support of the fiber at 140°C. Boll et al. [46] find for isolated PAN-type carbon fibers (AS4, IM6, IM7) that failure invariably occurred by transverse shear with no evidence of microbuckling. Ewins and Potter [83] investigated the compressive strength of carbon fiber/epoxy composites as a function of temperature and reported a distinct change from shear failure to microbuckling as the temperature approached the glass transition temperature of the matrix. At low temperatures, failure of the composite is dependent on the compressive strength of the fibers and the mechanism is one of shear across both fibers and matrix on a plane of near-maximum shear stress.

Dobb et al. [84] examined the broken fragments after the recoil test in the SEM. They find kink bands in PAN-type carbon fibers and shear failure in high modulus pitch-based carbon fibers. The failure mechanism of PAN-based and P-25 fibers is a result of crack initiation on the tensile side of a bent fiber which propagates to a kink band on the compression side thereby forming a step. This mechanism is attributed to the effect of buckling instability coupled with the high strain due to bending as recoil takes place. Photomicrographs of failure and a schematic of the two failure mechanisms are given in Figure 21.

Hahn and Sohi [85], on the other hand, studied unidirectional composites, made of a single impregnated yarn, prepared from T-300, T-700, and P-75. The first two fibers failed by microbuckling while the MP fiber failed in shear. The presence of microfibrils has been blamed for the low compressive strength [42,44] as microfibrils are observed in SEM photomicrographs following compressive failure. However, microfibrils exist in all rayon- and PAN-type carbon fibers and probably exist in all MP-type fibers [27]. To increase the compressive strength of carbon fibers it is necessary to a) increase the tensile strength for PAN-type fibers; b) decrease the tensile strength in MP-type fibers [84]; c) decrease the Young's modulus, Figure 22 [86]; d) increase the d (002) spacing [87,88]; e) decrease the crystal size as measured by L_c [87,88]; and f) increase the microvoid content, see Figure 23 [88]. These observations lead to the conclusion that it is the fine structure within the microfibrils that controls compressive failure mechanisms. Ion implantation of Torayca PAN- and MP-based carbon fibers with boron ions (B^+) at ambient temperature under high vacuum results in lowering the perfection and size of crystallites along the c-axis in the surface region [89]. The effect is to increase the compressive strength, the torsional modulus, and for the PAN-type fibers examined, the tensile strength. Therefore, structures containing a very tortuous crack path are required within the microfibrils. Thus, there is a virtue in disorder, as far as compressive strength is concerned; the problem is how to obtain the requisite disorder.

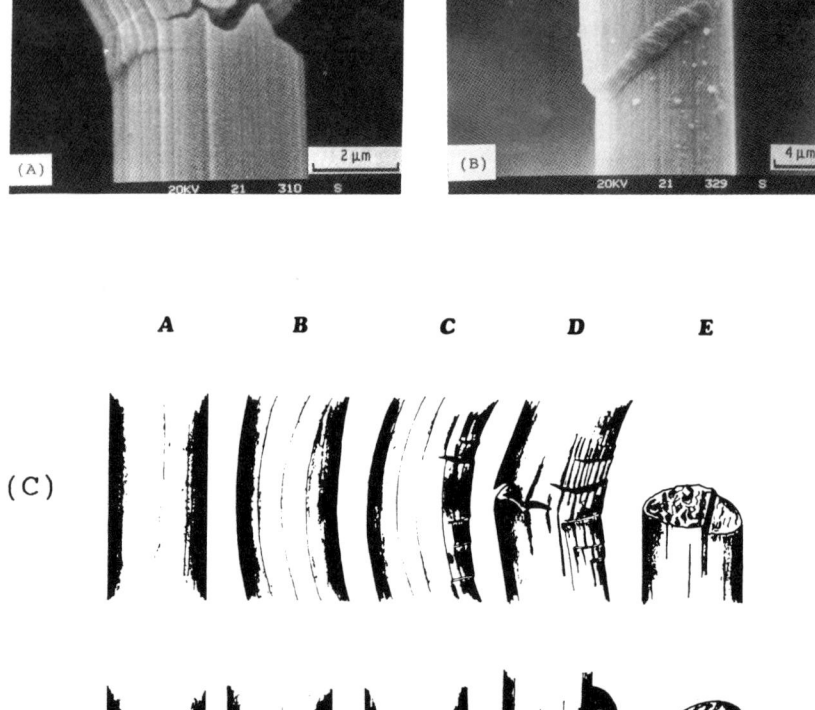

Figure 21 (A): Kink bands in PAN-based carbon fibers after recoil compression under high deformation, (B): shear bands in high-modulus MP-based carbon fibers after moderate deformation, (C): mechanism of compressive failure of PAN-based and P-25 carbon fibers and (D) that for high-modulus MP-based carbon fibers. (Reprinted with permission from Dobb et al. [84], Figures 8b, 9a, and 10, Copyright © Chapman and Hall, UK.)

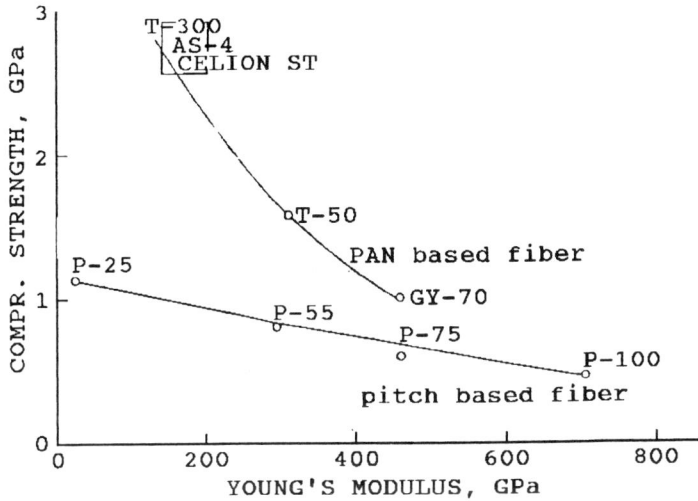

Figure 22 Compressive strength as a function of Young's modulus for PAN- and MP-based carbon fibers (Kumar [86], reprinted with permission of the Society for the Advancement of Material and Processing Engineering).

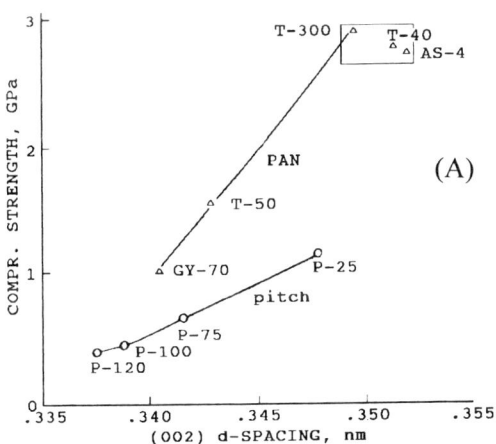

Figure 23 The compressive strength of carbon fibers as a function of A) d(002) spacing (uncorrected for crystallite size effect); B) Crystal size; and C) void content (Kumar and Helminiak [88], reprinted with permission of the Society for the Advancement of Material and Process Engineering).

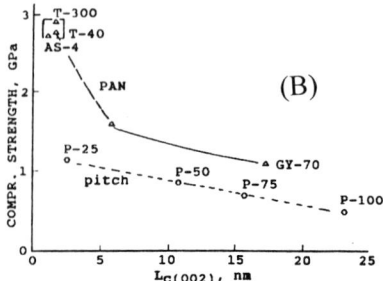

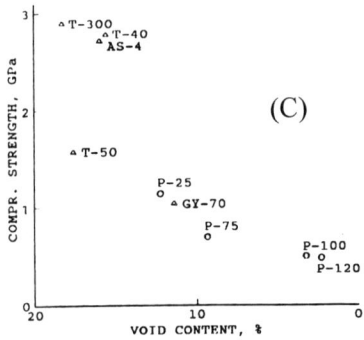

Figure 23 (B) (C) Continued.

The microvoid content in Figure 23c is determined from

$$\text{Void Content (\%)} = \left[1 - \frac{\rho_f d_f}{\rho_g d_g}\right] *100 \qquad (25)$$

where ρ_f and d_f are the density and d(002) spacings of the fiber, respectively, while $\rho_g = 2.26$ g/cm^3, and $d_g = 0.3354$ nm for graphite crystals [90].

The torsional modulus of carbon fibers also depends on the fine structure, an increasing disorder in the structure should lead to higher torsional moduli. A good correlation exists between compressive strength, probably measured by the recoil method, and torsional modulus for a series of six MP- and four PAN-based carbon fibers [91]. The correlation coefficient for the combined fiber data, R^2, is 0.83. The correlation coefficients for the separate MP- and PAN-based fibers are lower, 0.38 and 0.18, respectively, due in part to the scatter in compressive strength data and the paucity of samples with widely different torsional moduli within each group.

In addition to the effect of matrix modulus on the compressive strength of composites there is also an effect of fiber-matrix adhesion. Drzal and Madhukar [92] studied unidirectional composites of Hercules AU-4 (untreated), AS-4 (commercially surface treated) and AS-4C (surface treated and coated) fibers in an epoxy matrix consisting of DGEBA and m-PDA. The adhesion between fiber and matrix increases in the progression AU-4, AS-4, AS-4C. The failure mode of fibers in the composite depends strongly on the adhesion between fiber and matrix as depicted in Figure 24. With low adhesion the filaments delaminate from the matrix and the fibers undergo columnar buckling; at intermediate adhesion strength, the fibers undergo microbuckling along the line of maximum shear stress; and at high adhesion, fiber compressive failure occurs in several planes owing to the strong lateral support to the fiber columns. The fibers, therefore, can be compressively loaded to their maximum capacity.

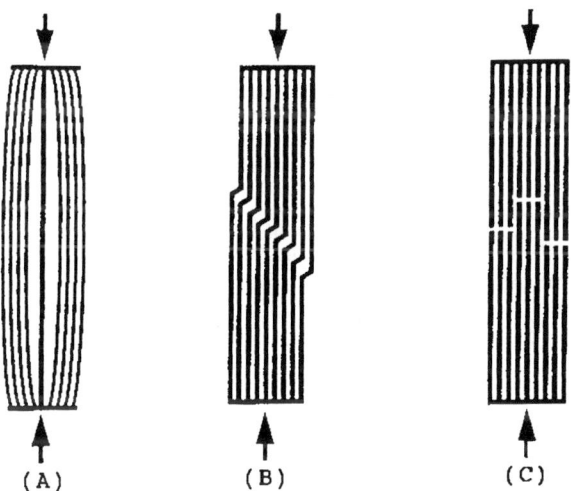

Figure 24 Schematic diagrams of the major failure modes detected for three epoxy composites which differ only in the surface treatment of the fiber which led to low, intermediate and high fiber/matrix adhesion. These failure modes are (A) low adhesion: global delamination buckling; (B) intermediate adhesion: local fiber microbuckling; (C) high adhesion: fiber compressive failure. (Reprinted with permission from Drzal and Madhukar [92], Figure 21, Copyright © Chapman and Hall, UK.)

5.11 INFLUENCE OF PROCESS VARIABLES ON MECHANICAL PROPERTIES

All carbon fiber types, rayon, PAN, and MP, are prepared by essentially the same process. A precursor is selected–viscose for rayon-type fibers, a copolymer for PAN-type fibers, and a specially prepared pitch for MP-type fibers. The precursor is spun into highly oriented precursor fibers. The spinning process controls the density and orientation of ordered material that generally forms a gradient of properties across the fiber diameter. The fiber is then stabilized, usually in air or other oxidizing media, at temperatures between 200 and 300°C. This is an essential step to render the fiber ready for high temperature heat treatment, to maintain the orientation imparted during the fiber formation step, and keep the fiber from fusing. The stabilization process is slow, requiring tens to hundreds of minutes depending upon the process details. Following stabilization, the fiber is subjected either to a one-step or a two-step process, depending on the precursor type, cost considerations and the desired final properties. In the two-step process the fiber is first carbonized at temperatures somewhere between 1000 and 1500°C, depending on the process which eliminates most but not all of the non-carbon atoms. For some fibers, this is the final step. The fiber can then be subjected to high temperature heat treatment, sometimes under high stress, to temperatures in the range 2200–800°C. In general, if the final heat treatment temperature is in the range of 1500°C, a high strength fiber results; if the final treatment temperature is 2200°C or above, a high modulus fiber results.

There are many publications on how one variable influences fiber properties while holding most of the others invariant. While interesting, these results provide only a restricted insight on how to improve fiber properties. The exact conditions of carbon fiber formation are carefully guarded by the fiber manufacturers. As mentioned above, for tensile high strength fibers it is necessary to minimize flaws; for high compressive strength it is necessary to produce a fine structure that results in a tortuous path for crack propagation; for high modulus fibers it is necessary to have highly oriented fibers. A significant difference between PAN-type fibers and MP-type fibers is the influence of the final processing temperature on tensile strength and tensile modulus (Figure 7). Similar plots have been presented by other authors which differ in detail because different fibers and different processing details were used, but the general trend is similar: PAN-type fibers have a maximum in tensile strength at around 1500°C, MP-type fibers generally have a lower tensile strength than PAN-type fibers but have a higher modulus at the higher processing temperatures. The compressive strength as determined by the recoil method on an experimental PAN-type fiber also has a maximum at around 1500°C [93].

5.12 MECHANICAL PROPERTIES OF COMPOSITE MATERIALS ON A BASIS OF POLYMERS AND CARBON FIBERS[1]

The development of engineering technology stimulates creation of new composite materials (CM), capable of long operation in complex conditions; under action of high temperatures, large and diverse mechanical loads, chemically active environment, and radiations, etc. A special place among CM is taken with materials on a basis of thermoplastics and thermoset polymers, reinforced with carbon fibers (CF). It is explained by successful application of such polymer types, in air and space technology, ground and water transport, and chemical industry.

The future of CM development on a polymer basis and CF, is largely dependent upon the success and improvement of these materials properties; durability and modulus of CF. At the earlier stage of research, the increase of physical and mechanical parameters of carbon plastics (CP) was achieved by the application of more high-strength and high-modulus CF. However, soon there was clear absence of proportional conformity between increase of CF mechanical properties and the properties of CP. The durability of high-strength CF was found to affect CP durability by less than 30% and this circumstance has stimulated researches on the improvement of CF surface reactivity and to search for new binding approaches and modification of the matrix to enhance CF adhesion.

In the early works [96-98], attention has been made to the characteristics of CF surface (contents and nature of carbon complexes with oxygen, moisture etc.) and their influence on the properties of polymeric CM. There is a wide range of CF, precursors from polyacrylnitrile and rayon fibers, synthetic resins, and also from petroleum and coal tars. The structure and properties of CF essentially depend on the type of initial raw material and process of technology. To characterize the structure and define the properties of CF and CM, it is necessary to refer to a grade of orientation and perfection of structure, and also to the total volume of pores and their distribution of dimensions, grain of surface, contents of defects, and contents of chemical groups [99,100].

Depending on their mechanical properties the following kinds of CF are categorized by: high-strength (1.8–3.4 GPa) and superhigh-strength (about 4.5 GPa); low, medium, high and superhigh-modulus (accordingly up to 100, 150–300, > 350, > 450 GPa). The mechanical properties of CF and CM essentially depend on the scale factor; sizes of CF. For example, a decrease of CF diameter from 7 microns to 4 microns results in notable increase of its durability and shear stress in epoxy-carbon-plastic [101]. Efficiency of CF reinforcement is to a great extent defined by the length of CF [102,103]. When the length of fibers is close to their diameter, the transfer of mechanical stresses from the matrix to the filler is weak. Thus, the increase of durability is insignificant. Moreover, the introduction in polymer of short-sized fibers causes nonuniformity of stresses in the matrix and reduction of its durability. With reduction of length of a discrete fiber until the critical size is reached, and below this critical size polymeric matrix cannot transfer a load to the fiber, leading to destruction of the fiber/matrix interface. The maximum durability of CM filled with discrete CF is reached when the length of the fiber exceeds its critical length by a factor of 5 to 10 times.

It is necessary to also take into account, that in the manufacture of molding products there is some destruction of fibers and final product appears filled with fibers of different lengths; therefore strength property of CM samples are usually affected. The reduction of length in the molding process is especially probable for CF having a large modulus of elasticity, for which there is a characteristic fragility and crispness. Thus, the fabrication of CM based on discrete CF is preferably in the use of high-strength CF with a small modulus of elasticity, allowing to make CM with improved strength characteristics. Another factor, CF orientation, is equally important for enhancing CM mechanical properties. The orientation of fibers in the direction of applied deformation promotes transfer of stresses on fibers and, accordingly, durability is improved.

The major factor affecting the mechanical properties of CM is the adhesion between the polymeric matrix and CF. The degree of adhesion depends on the properties of the polymeric matrix and CF, and also on the mode of formation of adhesive linking them. Insufficient adhesion of the CF-matrix phases results in limitation of durability of CM. The enhancement of adhesion is usually reached by the processing of CF surface [99,104]. However, too much of an increase in adhesion leads to a brittle composite and a diminution of its resistance to shock loads [103]. The CM properties depending on many factors such as the structure and properties of CF and polymeric binding, this opens many possibilities for deriving new CM with a given complex of properties by means of modification of binding between CF and polymers. The future approaches could be the creation of multicomponent matrixes, special preparation of CF for reaching of adhesive optimum with a given matrix, and hybrid filling.

5.13 CARBON FIBER COMPOSITES WITH THERMOSET MATRIXES

The most widely used thermoset polymers in processing are the epoxy oligomers. The advantage to their use is a significant value of adhesion to CF, which insures high elastic and strength properties of CM. Minor shrinkage by curing and stabilizing both in alkaline and acid media are the main features of these materials. Different thermoset matrixes are selected for different uses, for example, the silicon matrix is applied to deriving materials with increased thermostability, and phenolformaldehyde resins are used for preparation of CM with thermo- and acid stability.

The mechanical properties of carbon plastics with thermoset matrixes depend on the character of the CF surface. Oxidation of the CF surface by different methods allows the growth of epoxy CM durability and diminution of shock resistance. When the CF content is up to 70%, durability and resistance to shock loads will increase [100,104,105]. Reinforced CF polyethers, silicoorganic and phenolformaldehyde resins lead to similar materials on a basis of epoxy binding with high durability parameters [100,106,107]. The reason for such a difference is first, on the epoxy binding which ensures higher adhesion between the polymeric matrix and CF in CM. Secondly, in the curing process other bindings usually form volatile substances, which will form pores in CM as concentrators of stresses.

Important parameters of thermoset polymers reinforced carbon fiber composites are their high fatigue resistance durability. After alternating the load of epoxy-CM, the tension compression at 10^7 cycles, it retains 80% of durability [105]; in the same conditions glass-plastic, titanium and aluminium retain only 30, 52, and 55% of durability.

Durability of thermoset polymers CMs depend also on the porosity [108,109]. Pores, being concentrators of stresses, sharply reduce the durability of CM. Increase of the porosity of polymeric matrix from 0.02 up to 0.12% results in a diminution of shock resistance of CM by a factor of three. The significant influence of porosity on elastic and strength properties of CM is one reason for primary application of a low porosity epoxy resin as a matrix of in the CM. On the other hand, the porosity of CF surface promotes the CM durability as the pores increases the surface of contact between the matrix and CF, and creates the possibility of penetration of the matrix into the CF surface, ensuring an increase of the adhesion between the matrix and CF.

Thermostability of thermoset polymers based CM is limited by the thermostability of adhesive binding. Manufacturing temperatures of carbon fiber composites with resin compounds of polyethers-, epoxy-, phenolformaldehyde-, polyimide-, and silicon-bindings are 95, 175, 250, 250, 350°C accordingly. The thermal stability is increased in accordance with the increase of polymer matrix glass temperature, thus durability of CM interface is maintained. Therefore, for preservation of adhesive durability of the polymer-CF at high temperatures, it is necessary to use a thermoresistant binding [110–112]. The adhesion of CF with a matrix is affected by the temperature-time dependence of interphase durability [113]. Durability of the interphase layer by shear at temperatures from 25 to 200°C and velocity of load increase was examined. It was found that the activation energy by shear increased with increase of concentration of carboxyl groups on a previously treated CF surface.

The major condition of deriving high-strength and thermoresistant materials on a basis of thermoset polymers and CF consists of a sufficient adhesion between the matrix and CF surface. Adhesion between a CF surface and matrix can be reached by modifying both parts of this system. The recent tendency is in modifying thermoset matrixes to improve CM properties- durability, resistance to shock loads, and thermostability. The balance between durability, deformation and thermoresistance can be reached, for example, by a combination of thermoset and thermoplastic polymers [114]. Thus, it is possible to use advantages of individual components of the mixture, such as thermostability of products from thermoset materials, and higher shock resistance of thermoplastic, at proper processing conditions.

Utilization of an epoxy matrix which contains high elastic thermoset microparticles can lead the CM to a good thermoresistance combined with resistance to shock loads [115]. Carbon plastics with better durability are obtained by using a matrix from the copolymers of aromatic diisocyanate and low molecular rubbers with various reactive pendant isocyanate groups [116]. In accordance with a diminution of reactivity of rubber in relation to aromatic diisocyanate the sizes of heterophase inclusions are significantly increased. The greatest energy of destruction are obtained with copolymers with high dispersive heterophase structure and polymodal distribution. The strength characteristic of carbonplastics modified by rubber polyisocyanurates are determined by rubber concentration and reach a maxima at the contents of rubber from 2 up to 8%.

One approach to modify thermoset matrixes for carbon plastics was studied [117]. The authors observed an abnormal alteration of melt viscosity, modulus of elasticity, durability [118,119] in a mixture of high molecular linear polymers. These anomalies usually appear in the field of transition from monophase to biphase structure of a mixture owing to thermodynamic incompatibility of components. Selective interaction of components with carbon fiber surface results in a reinforcement of heterogeneity [119]. These concepts were adopted at processing and investigations of CMs on basis of phenolformaldehyde resin with the addition of oligomers containing free isocyanate groups [117]. The introduction of an oligomeric additives in binding reduces the surface tension and internal stresses in cured binding, and improves wetting of CF as a result of formation on CF boundary of a damping structure, promoting a relaxation of internal stresses. Introduction in the binding small components of the oligomeric modifier allows to lower CF glass temperature. Such modifying results also in notable increase of durability on break and modulus of elasticity. In a word, the introduction in the system small quantities of thermodynamically incompatible oligomers will result in a biphase structure, in which particles of the modifier play a role as a polymeric filler. Such an "elastomeric" filler increases possibilities to a relaxation and decreases the internal stresses in binding. This formation of a gradient of concentration of the component on interphase boundary of the CF-binding, results in improving of durability and viscoelastic properties of CM.

5.14 CARBON FIBER COMPOSITES WITH THERMOPLASTIC MATRIXES

Different thermoplastic matrixes have been used: polyethers, polyamides, polystyrenes, polycarbonate, polytetrafluoroethylene, copolymers of tetrafluorineethylene with ethylene, with vinylidenfluoride [120]. These thermoplastics materials are characterized by their high durability and modulus of elasticity, wearproof, electrical conductivity and antistatic properties, thermal conductivity, reduced thermal shrinking during molding, and a wide interval of working temperatures. Cooling of amorpho-crystalline thermoplastics melts (polyolefins, polyamides etc.) can influence on kinetics of crystallization and promote crystalline formations [121] with the formation of a small spherulitic structure [122,123]. A small scatter on spherulitic sizes results in reduced internal stresses, and it is especially essential for creating a structure of an interphase matrix between CF and polymer.

The durability limitations of thermoplastic reinforced carbon fiber composites are connected, first of all, with weak adhesion of thermoplastics to CF. To a certain extent, the insufficient adhesive interaction of a fiber to a matrix can be overcome by means of preliminary CF processing, in particular, by oxidation treatment. It was shown [124] that the CM compatibility with a polymeric matrix can be reached with thermoplastic with the same strength properties than those of thermoset polymers.

This compatibility is especially possible with amorpho-crystalline polymers. By introducing non-crystallized polar oligomers or polymeric components in the amorpho-crystalline polymer, it is possible to change the properties of a thin surface layer [123]. The non-crystallized additive is displaced from crystallized areas of the melt sample to the least dense zones during the molding process. Central parts of amorphous areas (nucleus) and surface layers are formed in contact with a rigid surface. As a result the concentration of the polar non-crystallized component in these zones appear much greater than in the polymer bulk. This effect was shown in an example of a low density polyethylene with addition of an epoxy resin previously oxidized with phosphorized polyethylene. Increasing the of concentration of the component in a thin polymer surface layer (less than 0.4 microns), and as a corollary increase of moistening by polar liquids of adhesive ability, was observed. The wetting angle by water of plate-sample from a polyethylene mixture with the component of phosphorize polyethylene is smaller in the slow cooling procedure after pressing than in case of fast cooling. This fact is probably connected with a diffusive character (migration) of the non-crystallized polar component in a surface layer in the process of sample molding [125].

One of the conditions of migration of the polar non-crystallized components in a thin surface layer of the amorpho-crystalline polymer is a low density and order of this layer as shown for products from polyethylene [126]. Investigation of the transversal sections of the isotactic polypropylene's sample after extrusion has revealed [127] that surface layer, as against other layers having mainly the spherulitic structure, has a character of a "quasimelt floating" due to separate ordered formations. The smaller packing density of a polymer in a surface layer generated in contact of a rigid surface is explained by a discordance of a space disposition of the rigid surface and the polymer [128].

Introduction of an amorpho-crystalline polymeric matrix of non-crystallized components with various functional groups and concentration of these components in surface layers, binding with previously oxidized CF, it is possible to create conditions for formation of various boundary links between a thermoplastic matrix and CF. The resulting interaction between functional groups created after oxidation of CF (hydroperoxide, carboxyl, and other groups) and thermoplastic additives make it possible for the formation of various links, including covalent, ionic, coordination, dipole-dipole, and hydrogen bonds. Varying components of the interphase interaction could possibly influence the composite properties. Overcoming CF incompatibility with polymers by use of amorpho-crystalline polymers features leads to delocalizing of stresses in CM due to increase of adhesion between phases. This especially appears at large velocities of applied loads, in particular, at shock loading when the structure of the polymeric matrix also promotes delocalizing of stresses. It has been proposed [129,130] that the durability of amorpho-crystalline polymers is connected to an amount of through interfacial chains which keep the load distributed in the amorphous macromolecule zones. Under load there is transfer of stresses from one crystallite to another by shear deformation. The increase of concentration of interfacial chains, as well as crystalline formations leads to an increase of the polymer ability to delocalize stresses and increase the durability in static and dynamic conditions of loading [123].

The amount of interfacial chains can be enlarged by curing of a polymeric matrix, in particular, by radiation or chemical methods. If the rate of radiation or chemical curing is more than the rate of destruction (polyethylene, polyamide, polyvynylidenfluoride etc.), the intermolecular links will mainly be formed at the amorphous intercrystalline sites. These links are formed through the macromolecular ends by closing loops and bend through interfacial chains (the probability of sharing during curing process of flatten, and tense chains is rather small because of lack of mobility of free-radical centers formed on such chains, the velocity of their recombination with formation of covalent links is significant). Consequently, there is a complicated system of correlation between the crystallites, including different length, orientation, configuration, connection with other chains, and different ability to bear a load. Examples of polyethylene, polyvynylidenfluoride were shown [123,131–133] where modifications of the number of through passage chains keeping a load in relation to a total number of macromolecules in crystallite's section (n/N) in process of radiation or chemical curing of the polymers, caused an equivalent modification of durability as the one obtained by an accoustic method at −180°C [134].

Application of these effects to produce thermoplastic CM filled with CF, leads to a diminution of incompatibility of components and a different type of link formed between the fiber and the matrix and resulting in an increase of adhesion in CM. As a result one can observe greater ability of the material to delocalization of stresses and, therefore, an increase of mechanical properties. In addition, a radiation treatment results in a ten times increase of thermostability of cured amorpho-crystalline thermoplastics, that should enlarge the CM working temperatures interval. Another remark is that for polyethylene based-CM, a sharp increase of stability to cracking is observed under operation of long and alternating loads in the curing process [123]. Such processes in the field of modifying of polyethylene, polypropylene, and other thermoplastic by means of elastomeric additives, can provide significant experience to produce CM with remarkable strength and deformation parameters [123,135,136].

Thus, possible directions in the creation of thermoplastic CM reinforced with CF is represented by complex modification: etching CF surface to increase porosity, and inducing functional groups at the surface layers. On the other hand, selection and modifying the polymeric matrix for maintenance of its chemical and physical interaction with CF and for maintenance of an optimum combination of mechanical, thermophysical and other properties are beneficial. By such approach, alongside with the reinforcement of thermoplastics by CF filling (hybrid filling etc.), it will allow to optimize a complex of CM properties and enhance the transfer of stresses from the matrix to the fiber [137].

5.15 COMPOSITE MATERIALS ON A BASIS OF AN ELASTOMERIC MATRIX AND CARBON FIBERS

The existing compositions on a basis of elastomers with carbon fillers can be divided into two large groups: (1) materials in which the polymeric matrix consists only of one polymer, and (2) materials in which the polymeric matrix is a mixture of polymers. From a practical point of view; the second group of materials presents the greater interest for many industries. For example, use of elastomeric mixtures in the tire industry allows for the reduction in cost of products, and facilitates the possibility of producing tires from less complicated construction, and improving the composition's properties. Carbon black is an unique component in production of tires and other rubber-technical products, adding to an elastomer compositions the "reinforcement" factor, leading to magnification of hardness, module of elasticity, resistance to tear and wear. This phenomenon of rubber reinforcement has been investigated for a long time [138], and this process is the result of several chemical and physical interactions.

The preparation of elastomeric matrixes/CM filled of short fibers essentially differ from the usual fillers of elastomers by the form-ratio of length and diameter [140]. In the case of short fibers the stress, applied to the matrix, is partially transmitted on a short fiber under condition of high adhesive strength of linkage or high degree of friction on boundary of fiber-matrix. Thus it is necessary to take into account the fact that the stresses in short fibers will be always less than in continuous CM at identical deformation, as the short fibers ends are loaded always less than their middle. Thus, only under condition of significant adhesion between a fiber and matrix, it is possible to improve the CM properties.

The existing methods of orientation of short fibers are however not effective [141]. It is known that the fibers which do not lie on a plane parallel to the shear plane are severely damaged. At the same time it is necessary to take into account the fact that a number of products (such as tires) work in conditions of tension-compression, and the distribution of a carbon fiber optimum for conditions of extension will be unsatisfactory for compression. In the case of CM with polymeric basis composed not by one but by two rubbers, the problem becomes much more complicated. For creation of such CM it is necessary to reconsider some positions of the theory of mixing [142,143], considering this process not only as reaching general homogeneity, but also as a process of creation of oriented structures. In addition, the prognostic is still not very favorable to have a positive contribution of the mixing process [144].

NOTE

1. Note added in proofs: A review paper entitled "Single Fibre Fragmentation Test for Assessing Adhesion in Fibre Reinforced Composite," has recently been published (83 references). See D. Tripathi and F. R. Jones, *J. Mater. Sci.* 33, 1–16 (1998).

REFERENCES

1. Portions of this chapter (Section 5.1–5.10) are taken from Peebles Jr., L. H., *Carbon Fibers—Formation, Structure and Properties*, © CRC Press, Inc. Boca Ratan, FL, 1995, Chapter 5, with permission.
2. Standard test method for tensile strength and Young's modulus for high-modulus single-filament materials, *ASTM Standard D 3379–75* (Reapproved 1989).
3. Tensile properties of continuous filament carbon and graphite yarns, strands, rovings, and tows, *ASTM D* 4018–4081.
4. Starr, T., *Carbon and High Performance Fibers Directory and Handbook*, 6th edition, Chapman & Hall, 1995.
5. Lovell, D. R., *Carbon and High Performance Fibers Directory*, 5th Edition, Chapman & Hall, 1991.
6. Beetz, Jr., C. P., *Fiber Science and Technology 16*, 45 (1982).
7. Own, S. H., Subramanian, R. V., and Saunders, S. C., *J. Materials Sci. 21*, 3912 (1986).
8. Olsson, D. M., *J. Qual. Tech. 11*, 153 (1979).
9. Draper, N. R., and Smith, H., *Applied Regression Analysis*, John Wiley & Sons, Inc., New York, 1966, 2nd Edition, 1981, Chapter 10.
10. Kelly, A., and Tyson, W. R., *J. Mech. Phy. Solids 13*, 329 (1965).
11. Beetz, Jr., C. P., *Fiber Science and Technology 16*, 81 (1982).
12. Asloun, E. l. M., Donnet, J.-B., Guilpain, G., Nardin, M., and Schultz, J., *J. Materials Sci. 24*, 3504 (1989).
13. Waterbury, M. C., and Drzal, L. T., *J. Compos. Tech. Res. 13*, 22 (1991).
14. Dai, S.-R., and Piggott, M. R., *Compos. Sci. Tech. 49*, 81 (1993).
15. Bennett, S. C., Johnson, D. J., and Johnson, W., *J. Materials Sci. 18*, 3337 (1983).
16. Reynolds, W. N., and Sharp, J. V., *Carbon 12*, 103 (1974).
17. Matsumoto, T., *Pure Appl. Chem. 57*, 1553 (1985).
18. Endo, M., *J. Materials Sci. 23*, 598 (1988).
19. Guigon, M., Oberlin, A., and Desarmot, G., *Fiber Sci. Tech. 20*, 177 (1984).

20. Guigon, M., Oberlin, A., and Desarmot, G., *Fiber Sci. Tech. 20*, 55 (1984).
21. Guigon, M., and Oberlin, A., *Composite Sci. Tech 27*, 1 (1986).
22. Tanabe, Y., Yasuda, E., Bunsell, A. R., Favry, Y., Inagaki, M., and Sakai, M., *J. Materials Sci. 26*, 1061 (1991).
23. Beetz, Jr., C. P., *Fiber Sci. Tech. 16*, 219 (1982).
24. Curtis, C. J., Milne, J. M., and Reynolds, W. N., *Nature 220*, 1024 (1968).
25. Voet, A., and Morawski, J. C., Extended Abstracts, 12th Biennial Conf. on Carbon, Pittsburgh, Pennsylvania, 1975, pp. 87–88.
26. Henrichsen, R. E., and Fischbach, D. B., Extended Abstracts, 12th Biennial Conf. on Carbon, Pittsburgh, Pennsylvania, 1975, pp. 135–136.
27. Fischbach, D. B., and Srinivasagopalan, S., 5th London Int'l Conf. on Industrial Carbon and Graphite, 1978, pp. 389–397.
28. Jones, W. R., and Johnson, J. W., *Carbon 9*, 645 (1971).
29. Kowalski, I. M., *Composite Materials: Testing and Design*, ASTM STP 972. (J. D. Whitcomb, ed.), ASTM, Philadelphia, 1988, p. 205.
30. Bacon, R., and Schalamon, W. A., *J. Appl. Polym. Sci., Appl. Polym. Symp. 9*, 285 (1969).
31. Northolt, M. G., Veldhuizen, L. H., and Jansen, H., *Carbon 29*, 1267 (1991).
32. Ruland, W., *J. Appl. Polym. Sci., Appl. Polym. Symp. 9*, 283 (1969).
33. Li, C.-T., and Tietz, J. V. V., *J. Mater. Sci. 25*, 4694 (1990).
34. Chen, K. J., and Diefendorf, R. J., *Progress in Science and Engineering of Composites* (T. Hayashi, K. Kawata and S. Umekawa, eds.), ICCM-IV, Tokyo, 1982, p. 97.
35. Jones, B. H., and Duncan, R. G., *J. Mater. Sci. 6*, 289 (1971).
36. Morita, K., Miyachi, H., Kobori K., and Matsbara, I., *High Temperatures-High Pressures 9*, 193 (1977).
37. Wagoner, G., and Bacon, R., Extended Abstracts, 19th Biennial Conf. on Carbon, State College, Pennsylvania 1989 pp. 296–297.
38. Krucinska, I., and Stypka, T., *Composites Sci. Tech. 41*, 1 (1991)
39. Villeneuve, J. F., Naslain, R., Fourmeaux, R., and Sevely, J., *Compos. Sci. Tech. 45*, 89 (1993).
40. Kogure, K., Lavin, J. G., Sines, G., Extended Abstracts, 21st Biennial Conf. on Carbon, Buffalo, NY. 1993, pp. 16–17.
41. Sinclair, D., *J. Appl. Physics 21*, 380 (1970).
42. Williams, W. S., Steffens, D. A., and Bacon, R., *J. Appl. Physics 41*, 4893 (1970).

43. Ng, C. B., Henderson, G. W., Buechler, M., and White, J. L., Extended Abstracts, 16th Biennial Conf. on Carbon, San Diego, California, 1983, pp. 515–516.
44. Hawthorne, H. M., and Teghtsoonian, E., *J. Mater. Sci. 10*, 41 (1975).
45. Ohsawa, T., Miwa, M., Kawade, M., and Tsushima, E., *J. Appl. Polym. Sci. 39*, 1733 (1990).
46. Boll, D. J., Jensen, R. M., Cordner, L., and Bascom, W. D., *J. Compos. Mater. 24*, 208 (1990).
47. DeTeresa, S. J., *Carbon 29*, 397 (1991).
48. Wagner, H. D., Migliaresi, C., Gilbert, A. H., and Marom, G., *J. Mater. Sci. 27*, 4175 (1992).
49. DeTeresa, S. J., Porter, R. S., and Ferris, R. J., *J. Mater. Sci. 23*, 1886 (1988).
50. Allen, S. R., *J. Mater. Sci. 22*, 853 (1987).
51. Wang, C. S., Bai, S. J., and Rice, B. P., *Polym. Mater. Sci. Eng. 61*, 550 (1989).
52. McGarry, F. J., and Moalli, J. E., *Polymer 32*, 1811 (1991).
53. Crasto, A. S., and Kumar, S., Int. SAMPE Symp. Exhib., 35 (1, Adv. Matr.: Challenge Next Decade) (1990) 318–331.
54. Macturk, K. S., Eby, R. K., and Adams, W. W., *Polymer 32*, 1782 (1991).
55. Fawaz, S. A., Palazotto, A. N., and Wang, C. S., *Polymer 33*, 100 (1992).
56. Tuinstra, F., and Koenig, J. L., *J. Compos. Mater. 4*, 492 (1970).
57. Chieu, T. C., Dresselhaus, M. S., and M. Endo, *Phys. Rev. B26*, 5867 (1982).
58. Melanitis, N., and Galiotis, C., *J. Mater. Sci. 25*, 5081 (1990).
59. Everall, N. and Lumsdon, J., *J. Mater. Sci. 26*, 5269 (1991).
60. Everall, N. J., Lumsdon, J., and Christopher, D. J., *Carbon 29*, 133 (1991).
61. Bowden, M., Gardiner, D. J., Southall, J. M., and Gerrard, D. L., *Carbon, 31*, 1057 (1993).
62. Standard test method for compression properties of rigid plastics *ASTM D 695-691*.
63. Standard test method for compressive properties of unidirectional or crossply fiber-resin composites, ASTM Standard D 3410-87.
64. Standard test method for compressive properties of unidirectional polymer matrix composites using a sandwich beam, *ASTM Standard D 5467-5493*.
65. Odom, E. M., and Adams, D. F., *Composites 21*, 289 (1990).
66. Curtis, P. T., Gates, J., and Molyneux, C. G., *Composites 22*, 363 (1991).

67. Lesko, J. J., Swain, R. E., Cartwright, J. M., Chin, J. W., Reifsnider, K. L., Dillard, D. A., and Wightman, J. P., Proc. 16th Annual Meeting of The Adhesion Society, Williamsburg, VA 1993, pp 41–43.
68. Prandy, J. M., and Hahn, H. T., *SAMPE Q.* 22, 47–52 (1991).
69. Kumar, S., Adams, W. W., and Helminiak, T. E., *J. Reinf. Plast. Comp.* 7, 108 (1988).
70. Johnson, D. J., and Park, C. R., Extended Abstracts, 20th Biennial Conf. on Carbon, Santa Barbara, California, 1991, pp. 224–225.
71. Johnson, D. J., *Carbon Fibers, Filaments, and Composites*, (J. L. Figueiredo, C. A. Bernardo, R. T. K. Baker and K. J. Hüttinger, eds.), Kluwer Academic Publishers, Dordrecht, The Netherlands, 1990, p. 119.
72. Vezie, D. L., and Adams, W. W., *J. Mater. Sci. Lett.* 9, 883 (1990).
73. Morita, T., Kitano, A., and Noguchi, K., *Proc. 4th Japan-U.S.Conf. on Composite Materials*, Technomic Press, Inc. Lancaster, PA, 1988, p. 548.
74. Furuyama, M., Higuchi, M., Kubomura, K., Jiang, H., and Kumar, S., *J. Mater. Sci.* 28, 1611 (1993).
75. Kumar, S., Anderson, D. P., and Crasto, A. S., *J. Mater. Sci.* 28, 423 (1992).
76. Melanitis, N., Tetlow, P. L., Galiotis, C., and Smith, S. B., *J. Mater. Sci.* 29, 786 (1994).
77. Shinohara, A. H., Sato, T., Saito, F., Tomioka, T., and Arai, Y., *J. Materials Sci.,* 28, 6611 (1993).
78. Hayes, G. J., Edie, D. D., and Kennedy, J. M., *J. Mater. Sci.* 28, 3247 (1993).
79. Kawabata, S., *Proc. 4th Japan-U.S.Conf. on Composite Materials*, Technomic Press, Inc. Lancaster, PA, 1988, p. 253.
80. Mehta, V. R., and Kumar, S., *J. Materials Sci.,* 29, 3658 (1994).
81. Kumar, S., Mehta, V. R., Anderson, D. P., and Crasto, A. S., *Int. SAMPE Exhib.,* 37, 967 (1992).
82. Kozey, V. V., Jiang, H., Mehta, V. R., and Kumar, S., *J. Mater. Res.* 10, 1044 (1995).
83. Ewins, P. D., and Potter, R. T., *Phil. Trans. Roy. Soc. A294*, 507 (1980).
84. Dobb, M. G., Johnson, D. J., and Park, C. R., *J. Mater. Sci.* 25, 829 (1990).
85. Hahn, H. T., and Sohi, M. M., *Compos. Sci. Tech.* 27, 25 (1986).
86. Kumar, S., *SAMPE Quart.* 20, 3 (1989).
87. Dobb, M. G., Guo, H., Johnson, D. J., and Park, C. R., *Carbon 33*, 1553 (1995).
88. Kumar, S., and Helminiak, T. E., *SAMPE J.* 26, 51 (1990).

89. Matsuhisa, Y., Washiyama, M., Hiramatsu, T., Fujino, H., and Katagiri, G., Extended Abstracts, 20th Biennial Conf. on Carbon, Santa Barbara, California, 1991, pp. 226–227.
90. Anderson, D. P., and Kumar, S., *SPE Antec 36*, 1248 (1990).
91. Mehta, V. R., and Kumar, S., *J. Mater. Sci. 29*, 3658 (1994).
92. Drzal, L. T., and Madhukar, M., *J. Materials Sci. 28*, 569 (1993).
93. Jiang, H., Damodaran, S., Abhiraman, A. S., Desai, P., and Kumar, S., *Mater. Res. Soc. Symp. Proc. 305*, 135 (1993).
94. Budnizky G. M. *Chemical Fibers*, 15, n.5. (1990) (in Russ.).
95. Budnizky G. M. *Chemical Fibers*, 12, n.7. (1982) (in Russ.).
96. Kraus, I. G., *J. Phys. Chem.*, v. 59, p. 343 (1955).
97. Gregg, S. J. The *Surface Chemistry of Solids*, Reinhold Publishing Corp., N4, 1961, p. 14.
98. Pallozzi, A. A. *SPE J.*, v.22, N2, p. 80, (1966).
99. Thermo heat resistant and incombustible fibers. Ed. by A.A.Konkin, Ìoscow, Chemistry, 1978, p. 424 (in Russ.).
100. Guniaev G.M. Structure and properties of polymeric fiber aggregates, Ìoscow, Chemistry, I981, p. 232 (in Russ.).
101. Guniaev, G. M. *Eng.Mat.Design*, v.27, N4, p. 14.(1983).
102. Bader, M. G., Bower W.N., *J. Composites*, 1973, July, p. 150.
103. Berlin, À. À., Volfson S.A, Oshmyn V.G., Enikolopov N.S., Principles of creation of composite polymeric materials. Ìoscow, Chemistry, 1990, p. 240 (in Russ.).
104. Konkin, À. À. Carbon and other heat resistant fiber materials, Ìoscow, Chemistry, 1974, p. 376 (in Russ.).
105. Konkin, À. À., Varshavsky V.Y. *Chemical fibers*, 11, p. 4 (1982) (in Russ.).
106. Semenova, G. P., Pavlov V.V. *Mechanics of Polymers*, 14, p. 585 (1970) (in Russ.).
107. Kaznelson, M. D., Balaev G.A. *Polymeric Materials*, St.Petersburg, Chemistry, 1982, p. 315 (in Russ.).
108. Kobez L. P., Guniaev G. M. *Plastics for Constructional Purpose*, Moscow, Chemistry, 1974, p. 45. (in Russ.).
109. Yakovlev V. M., *Plastics*, 11, p. 53 (1979)(in Russ.).
110. Shul G. S., Gorbatkina Y. A., Shchukina L. A. A *Mechanics of composite materials*, 14, p. 452. (1993) (in Russ.).
111. Gorbatkina Y. A. et al., *Vysokomolekularnye Soedineniya*, À, v.23, 11, p.110. (1981) (in Russ.)
112. Gorbatkina Y. A. Adhesive durability in systems a polymer-fiber, Moscow, p. 192 (1987) (in Russ.).
113. Nakanishi Y., *J. Soc. Mater. Sci., Jap.*, v.42, N476, p. 542. (1993).
114. Nakanishi Y., *Sci. and Ind. (Jap.)*, v.67, N1, p. 32. (1993).

115. Nakanishi Y., *Techno Jàðan*, v.25, N3, p.108. (1992).
116. Pankratov V. A. et al., Vysokomolekularnye Soedineniya, 1993, À-Á, v.35, 17, p. 803. (in Russ.)
117. Lipatov Y. S., Matyushova V. G., Rosovitsky V. F. *Mechanics of Composite Materials*, 1993, ò.29, 14, p. 440. (in Russ.)
118. Lipatov Yu. S., et al., *Rheol.Acta,* 1982, v.21, p.270.
119. Lipatov Yu. S., *Colloid Chemistry of Polymers*, Amsterdam, Elsevier, 1988, p. 450.
120. Molchanov B. I. et al., Composite materials on the basis of carbon fibers and polymeric matrix, Moscow, Proceedings of Inst.Chem.Materials, 1979, p. 12. (in Russ.).
121. Zekina I. G. The tendencies in the field of deriving of high-strength reinforced plastics, Moscow, Proceedings of Inst. Chem. Materials, 1972, p. 63 (in Russ.)
122. Lipatov Y. S., Peryshkin N. G. *Polymer Adhesion,* Ìoscow, Academy of Sciences, 1963, p. 107. (in Russ.).
123. Sirota A. G. Modifying of a structure and properties of polyolephynes, St.Petersburg, Chemistry, 1984, p. 152 (in Russ.).
124. Giltrow J. Ð., Lankaster J. K., *Wear*, v.16, N5, p. 359, (1970).
125. Budtov V. P., Sirota A. G. The reports of USSR Academy of Sciences, 1982, 13, p. 627. (in Russ.).
126. Kagan D. F., Popova L. A. *Mechanics of Polymers*, 14, p. 761(1970). (in Russ.).
127. Fitchmun D.R., Mencik Z., *J. Polym. Sci.*, 1973, v.11, p.951.
128. Malinsky Y.M. Successes of chemistry, 1970, v.39, 18, p.1511. (in Russ.).
129. Peterlin A., *J. Polym.Sci.,* 1965, v.9, p.61.
130. Peterlin A., *J. Polym.Sci.*, 1966, v.15, p.427.
131. Sirota A. G., Verhovetz A. P., Utevsky L.E., Vysokomolekularnye Soedineniya, 1976, p. 661. (in Russ.).
132. Chlyabich P. P., Sirota A. G., Budtov V. P., Vysokomolekularnye Soedineniya, 1990, v. 32, A, p. 1444. (in Russ.).
133. Sirota A. G., Verkhovets A.P., Auslender V.L., *Radiat. Phys. Chem.,* 1995, v. 46, N4-6, p. 999.
134. Utevsky L. E., *Vysokomolekularnye Soedineniya*, 1972, v.14, B, 14, p. 308. (in Russ.).
135. Manson J. A., Sperling L.H., *Polymer Blends and Composites*, Plenum Press, N.Y.–London, 1976, p. 440.
136. Kuleznev V. N. *Mixtures Polymer Blends*, Moscow, Chemistry, 1980, p. 304. (in Russ.).
137. Chun-Hway Hsueh, *Trends in Polymer Sci.*, 1995, v.3, N10, p. 336.

138. *Reinforcement of elastomers*, Ed. D.Kraus, Moscow, Chemistry, 1968, p. 484. (translation from English).
139. Bogdanov W. W., Savvateev S. G., Christoforov E. J., *Plaste und Kautschuk*, 1985, N10, pp. 362–369.
140. *Industrial polymeric composite materials*, Ed. M.Richardson, Moscow, Chemistry, 1980, p. 472 (translation from English).
141. Parratt N.J., *Fibre-Reinforced Materials Technology*, Von Nostrand / Reinhold, London, 1972.
142. Bogdanov V. V., Torner R. V., Reger E. O., *Polymer Mixing*, St.Petersburg, Chemistry, 1979, p. 192. (in Russ.).
143. Bogdanov V. V., Metelkin V. I., Savvateev S. G. *Bases of technology of polymer mixing technology*, St. Petersburg, St.Petersburg State University, 1984, p. 192. (in Russ.).
144. Bogdanov W. W., Metjolkin W. J., Krassowski W. N., *Plaste und Kautschuk*, 1982, N5, pp. 286–290.

6
ELECTRICAL AND THERMAL TRANSPORT IN CARBON FIBERS

Jean-Pierre Issi and B. Nysten*
Université Catholique de Louvain, Louvain-la-Neuve, Belgium

6.1 INTRODUCTION

Carbon fibers are considered by purists to be dirty physical systems because of the large number of lattice defects they usually contain, however, they are interesting entities on which transport properties can be extensively investigated. Their variety of structures, particular geometry, and the fact that some may be intercalated, have lead to interesting observations, that could not be made on the corresponding bulk carbons and graphites. For example, their large length to cross-section ratio made possible the discovery of quantum transport effects on pristine (Section 6.2) and intercalated fibers (Section 6.5.2.2) and the separation of electronic and lattice thermal conductivities (Section 6.5.3.3). In addition to their favorable mechanical properties, their exceptional thermal conductivities offer unique possibilities for heat transfer devices in practical applications.

Transport properties have been reported for polyacrylonitrile-based (PAN), pitch-derived (PDF), vapor grown carbon fibers (VGCF) and, carbon nanotubes. Indeed, carbon nanotubes may be considered a particular kind of carbon fiber. For basic studies, VGCF are the more adequate because they have the highest structural perfection when heat treated at high temperature, their transport properties are much like those of highly oriented pyrolytic graphite (HOPG).

---

*Current affiliation: Unité de Chimie et de Physique des Hauts Polymères.

They display the highest electrical and thermal conductivities with respect to other carbon fibers. Also, they may be intercalated with donor and acceptor species which results in higher electrical conductivities. As is the case for HOPG, their thermal conductivity and thermoelectric power are modified by intercalation. However, for practical applications, their main drawback is that they are obtained in limited lengths.

The transport properties of carbon fibers depend on the in-plane coherence length, which in turn depends on the heat treatment temperature. Some commercial pitch-derived carbon fibers have thermal conductivities close to 1,000 $Wm^{-1}K^{-1}$, which is about twice that of copper. These high thermal conductivities, along with the fact that they may be obtained in a continuous form, make them the ideal candidates for thermal management applications [1–4].

PAN-based fibers are also continuous fibers, but they are generally disordered and their conductivity levels are rather low. One exception is the Celanese GY70, which exhibits a room temperature thermal conductivity of almost 200 $Wm^{-1}K^{-1}$ when it is heat treated at very high temperature [5].

Some carbon nanotubes may have electrical conductivities comparable to VGCF heat treated at high temperature [6–8]. Their thermal conductivities have not been measured, but one may expect for such materials very high values associated to unique mechanical properties.

Pristine HOPG is a semimetal with an equal density of electrons and holes, while some carbons are narrow-gap semiconductors. This means that in any case these materials have very few charge carriers compared to metals. Also, like HOPG, carbons generally have more than one type of charge carrier, which complicates the analysis of electronic transport properties data. Intercalation of HOPG increases the carrier density leading to metallic behavior.

The small density of charge carriers in carbons and graphites, associated with a relatively large lattice in-plane thermal conductivity, leads to a negligible contribution of the charge carriers to this property above the liquid helium temperature range. Heat is thus almost exclusively carried by the lattice vibrations at high temperature.

The particular geometry and the varying defect structure of carbon fibers have allowed the observation of *quantum transport effects* in these materials. These were first observed in intercalated acceptor carbon fibers [9], then on pristine fibers [10,11] and more recently on nanotubes [7]. The observed negative magnetoresistance in pristine carbon fibers, which is one of the signatures of the two-dimensionnal (2D) weak localization, was attributed to the turbostratic layers. In addition to their fundamental interest, these findings confirmed what was already known from the classical behavior (i.e., that magnetoresistance measurements may be used as a tool to characterize carbon fibers at the level of the microstructure). The fact that weak localization effects may be used for this purpose results from the direct relation between the quantum aspects of scattering and the material defect structure.

It was also clearly demonstrated that *thermal conductivity* measurements performed at low temperature allow to determine directly the in-plane coherence length in carbon fibers [12]. They also allow comparisons between shear moduli (C_{44}) and yield information about point defects. It is interesting to note that from the temperature dependence of the lattice thermal conductivity, one may assert that the phonon mean free path for boundary scattering is almost equal to the in-plane coherence length as determined by x-ray diffraction, L_a [12]. This enables use of thermal conductivity measurements as a tool to determine this parameter, especially for high L_a values where x-rays are inadequate. So, such measurements may be a very useful substitute for x-ray analysis for large in-plane coherence length values. One may also deduce from thermal conductivity measurements that the concentration of point defects (i.e., of impurities, intersticial atoms or vacancies) decreases with graphitization. This is in agreement with results obtained by other techniques. As it is the case for electrical resistivity, the interpretation of thermal conductivity data for characterization purposes presents the advantage of an overall view over the entire sample. This is in contrast to microscopic techniques which only probe a tiny portion of the sample.

Oddly enough, though the electrical and thermal conductivities of carbon fibers are generated by different entities, charge carriers for the electrical resistivity and phonons for the thermal conductivity, a direct relation between the two parameters was observed [5,13]. This is related to the fact that both transport properties depend extensively on the structure of the fibers. They both increase with the in-plane coherence length. As a result of the direct relation between these transport coefficients for fibers with the same precursor, one can determine the thermal conductivity once the electrical resistivity is measured.

As with bulk graphites, *intercalation* of some carbon fibers with acceptor or donor species largely increases their electrical conductivity and modifies drastically their thermal conductivity [14]. Some fibrous acceptor graphite intercalation compounds (GIC) exhibit room temperature electrical conductivities comparable to that of the best metallic conductors. However, because of the particular 2D electronic structure of acceptor GICs, a conductivity independent of charge transfer for a given stage is expected if only scattering by in-plane graphitic phonons are considered [14,15]. This limits the conductivity which could be attained in these compounds.

The analysis of the residual electrical resistivity of low stages acceptor GICs leads to the estimation of the size of the large-scale lattice defects [16]. Though the temperature variation of the ideal electrical resistivity is not yet definitely understood, at low temperature it is most probably due to hole-hole interaction in the presence of weak disorder, while electron-phonon scattering dominates at higher temperature. A simple model of a 2D hole gas, introduced by the intercalate through charge transfer and interacting with the phonons and defects of the host layers, gives a good picture of electronic transport in the Boltzmann conductivity range [14,15].

Due to the inherent 2D nature of the electronic structures of acceptor GICs, weak localization effects and electron-electron interactions were found to be particularily interesting in these materials. The possibility of varying the Fermi level of the compound through charge transfer and the defect structure of the host material over wide ranges offers a number of different experimental possibilities for the investigation of these phenomena [14].

Intercalation decreases the total thermal conductivity at high temperature and increases it at low temperature with respect to that of the pristine material [14]. Here again, it was shown that phonons are powerful and direct tools to probe the lattice defects. As for the pristine material, the interpretation of the low temperature lattice thermal conductivity allows the estimation of the size of the large scale defects and the concentration of point defects. An additional contribution due to the intercalate phonons was also evidenced [17].

Observations concerning the in-plane thermoelectric power of fibrous GICs that were made also proved interesting. While at high temperature almost all compounds show the same temperature variation, in the low temperature region the behavior varies according to the stage of the compound. At low temperatures, the diffusion thermoelectric power dominates in non-magnetic compounds. For stages higher than stage-1, a stage-dependent linear function of the temperature is observed for the diffusion thermoelectric power.

Polymer-matrix composites filled with carbon fibers may be electrically conductive provided they contain a percentage of fibers exceeding the percolation threshold, which is relatively low for these fillers because of their geometry. For unidirectional composites with highly conductive continuous pitch-based carbon fibers, room temperature thermal conductivities comparable to that of pure copper are now easily attained [2]. With chopped fibers, thermal conductivities superior to those of metallic alloys may be obtained in composites [3]. Thus, with carbon fibers one may tailor at the macroscopic scale the thermal conductivity of composites to the desired values for practical applications.

Among the numerous advantages of using polymeric materials as heat exchangers are their low specific gravity, their chemical resistance, their ease of processing, and energy saving abilities.

Throughout this review, after general considerations about results obtained on various fibers for the property are considered, we will report on measurements made on a set of well-characterized pitch fibers on which a variety of data were taken, including electrical and thermal transport. This will help compare the different properties obtained on the same samples and show how they are related to the microscopic properties. These fibers (E35, E55, E75, E105, E120, and E130), which were produced by Du Pont, were spun from 100% mesophase pitch and graphitized at different temperatures. Their mechanical properties are summarized in Table 1. Their structural parameters, which were determined by x-ray diffraction, are also presented in Table 1.

There have been comprehensive reviews recently pertained to the transport properties of pristine [18] and intercalated [18,19] carbon fibers. We shall often refer to them for more detailed information and, for an update, we will mainly concentrate on results which were obtained after the mid-eighties.

In the following, we shall discuss the electrical and thermal transport properties of various types of carbon fibers. For each property, after briefly overviewing the mechanisms of conduction, we will discuss the main experimental data obtained. We will first discuss the situation of pristine carbon fibers relative to the electrical resistivity (Section 6.2), thermal conductivity (Section 6.3), and thermoelectric power (Section 6.4). The effect of intercalation on these transport properties will then be discussed (Section 6.5). Then, after reviewing the situation in nanotubes (Section 6.6) and showing how transport measurements may be used to characterize carbon fibers (Section 6.7), we shall briefly consider the situation for carbon fiber composites (Section 6.8).

Table 1 Mechanical Properties of the Pitch-Derived Carbon Fibers Investigated and their Structural Parameters as Determined by X-Ray Diffraction[a]

Fibers	Tensile modulus [GPa]	Tensile strength [GPa]	d_{002} [nm]	Lc [nm]	La [nm]
E35	241	2.8	0.3464	3.2	7.2
E55	378	3.2	0.3430	8.2	16.2
E75	516	3.1	0.3421	10.7	22.4
E120	724	3.3	0.3411	17.3	46.1
E105	827	3.4	0.3409	18.9	51.4
E130	894	3.9	0.3380	24	180.4

[a] d_{002}, the interlayer spacing, L_c, the out-of-plane coherence length, and L_a, the in-plane coherence length parallel to the fiber axis. The interlayer spacing, d_{002}, was determined from the position of the (002) reflection. The out-of-plane coherence length, L_c, was calculated from the analysis of the (002) reflection and the in-plane coherence length, L_a, from the (100) reflection. All measurements were made in symmetrical-transmission geometry with curved-crystal diffracted beam monochromator, using as-produced fibers
Source: from refs. 11,12, and 42.

6.2 ELECTRICAL RESISTIVITY

6.2.1 Introduction

Since the early eighties, the electrical resistivity of metals, ρ, was considered as the sum of the contributions from a temperature independent residual (extrinsic) term, ρ_r, due to lattice defects and a temperature sensitive ideal (intrinsic) term, ρ_i, due to electron-phonon or, to a lesser extent, to electron-electron interactions:

$$\rho = \rho_r + \rho_i \qquad (1)$$

This relation is known as Matthiessen's rule. The interpretation of the electrical resistivity of carbons and graphites and their intercalation compounds, whether bulk or fibrous, was made along these lines. However, it was realized in the beginning of the last decade from experiments performed on metallic thin films that, in the presence of weak disorder, one should consider these two contributions, weak localization resulting from quantum interference effects [20–22] and/or Coulomb interaction effects [23,24]. It was demonstrated that this is also the case for acceptor GICs [9,14,25–27] and later on was proposed to explain the peculiar behavior of pristine carbons [10]. These quantum effects, though they do not generally affect significantly the magnitude of the resistivity, introduce new features in our understanding of low temperature transport effects. So, in addition to the classical ideal (Section 6.2.5) and residual (Section 6.2.2) resistivities of carbon fibers, we must take into account the contributions due to quantum localization and interaction effects (Section 6.2.3). These localization effects were found to confirm the 2D character of conduction in turbostratic carbons [10], carbon nanotubes [7] and acceptor GICs [27].

In contrast to HOPG, four-probe electrical DC measurements may be readily performed on graphite fibers, whether pristine or intercalated, because of the favorable length to cross-section ratios [14,28]. It is thus possible to perform high resolution electrical resistivity measurements even on graphite acceptor compounds when they are in a fibrous form. This allows the investigation of weak localization effects and to separate the ideal resistivity from the residual resistivity in spite of their very low residual resistivity ratio (RRR)—the ratio of the resistance at 300°K to that at 4.2°K [26,29]. Indeed, in acceptor fibrous GICs, the RRR is rather small: roughly less than 5 for vapor deposited carbon fiber (VGCF) compounds [29] and less than 2 for mesophase fiber-based compounds (PDF) [26]. For comparison, the RRR of high purity metals may exceed a few thousands. Because of these low RRR for carbon fibers, unless high resolution measurements starting from the liquid helium range and covering about two decades of temperature are performed, it is difficult to extract the ideal term from the measured total resistivity using Equation (1), particularly in the low temperature range, since ρ_i results from subtracting two large numbers. Also, in carbon fibers, one may simultaneously measure the electrical and thermal conductivities on the same sample (Section 6.5.3.3) and determine the electronic component from the measured thermal conductivity in some temperature ranges using the Wiedemann-Franz relation (Section 6.5.3.3). Thus, measurements on fibers allow to extract some basic information one can not obtain on bulk samples.

The diameters of the fibers are not the same along a given filament and, a fortiori, from one filament to another of the same batch. It is also known that the cross-sections of the samples are not always cylindrical and may present a complex shape. This complicates the experimental procedure, since, in order to calculate the conductivity (or resistivity) from the measured conductance (or resistance), the value of the cross-sectional area is needed. So, one has to measure the cross-section of each fiber investigated along the fiber and determine an average value along the axis. This is a delicate step in the experimental procedure, since it introduces the largest uncertainties in the estimation of the absolute values of the resistivities or conductivities.

6.2.2 Boltzmann Zero-Field Resistivity

Whatever the nature of the conductor considered, the expression for the electrical conductivity for a given group of charge carriers is always given by:

$$\sigma = qN\mu \tag{2}$$

where q is the electronic charge and N is the charge carrier density. Since the mobility, μ, is related to the relaxation time, τ, and the carrier effective mass, m*, by:

$$\mu = \frac{q\tau}{m^*} \tag{3}$$

we may alternatively write:

$$\sigma = \frac{q^2 N\tau}{m^*} \tag{4}$$

Since the carrier mean free path, l, is equal to $v\tau$, the conductivity may also be expressed in the form:

$$\sigma = \frac{q^2 Nl}{m^* v} \tag{5}$$

where v is the velocity of the charge carriers.

The relaxation time is defined as the time elapsed between two collisions, and its inverse $1/\tau$ reflects the probability for a carrier to experience a scattering event. The mean free path, $l = v\tau$, is the distance between two scattering centers. In a naive way, one may imagine the mobility, which is defined as the drift velocity, v_d, of the charge carriers per unit electric field, E:

$$\mu = \frac{v_d}{E} \tag{6}$$

as expressing the ease with which the carriers move in the crystal lattice. Among the three parameters, τ, l, μ, which are all related, the mean free path is the more concerned with the use we want to make here of transport measurements. We will show below that it may be directly related to the size of the large scale defects.

In order to analyze the electrical resistivity results in a consistent manner, a necessary prerequisite is to have a reliable knowledge of the distribution of electrons and holes. This requires to know the electronic band structure for the specific type of material we are investigating. Though the semimetallic band structure of HOPG and the values of the relevant electronic parameters are rather well known, this is not the case for other forms of carbons [18]. So, some reasonable assumptions have to be made relative to these materials in order to analyze the transport data.

Concerning the *residual resistivity*, scattering is determined by the defect structure of the graphitic layers. In carbon fibers, the defect structure may vary widely according to the type of fiber (vapor deposited, pitch-derived, PAN-based), the heat treatment temperature, and the quality of the precursor for a given type of fiber.

Equations (2–5) are for a single type of charge carrier. If there is more than one type of carrier, (i.e., electrons and positive holes), as in pristine graphite fibers, or many bands for electrons or holes as in intercalation compounds for stages higher than 1, the contribution of each type of carrier should be taken into account. In that case, the total electrical conductivity is given by the sum of the partial conductivities, σ_j, of each group of charge carriers, j:

$$\sigma = \sum \sigma_j \tag{7}$$

Applying Matthiessen's rule (1) for each band and taking into account the various scattering events, m, one may write:

$$\rho_j = \sum \rho_{jm} \tag{1'}$$

where m denotes scattering by static defects (m = r) or by charge carriers or phonons (m = i).

Equation (7) shows that the contributions to the conductivity from different carrier groups add, while Equations (1) and (1') show that it is the resistivities due to various scattering mechanisms that add. Matthiessen's rule should be applied for each type of carriers.

6.2.3 2D Localization and Interaction Effects

A variety of quasi-two dimensional (2D) systems exhibit a logarithmic increase of resistivity with decreasing temperature. This behavior, which is more pronounced in the presence of significant defect scattering [30–32], is interpreted in terms of weak electron localization [20–22] and electron-electron Coulomb interaction [23,24].

The two mechanisms which contribute to the correction term that gives rise to the logarithmic increase (relation 6.9) in resistivity are:

- a single-carrier weak localization effect produced by constructive quantum interference between elastically back-scattered partial-carrier-waves [21] and
- charge carrier many-body Coulomb interactions [23,24].

Both effects are enhanced in the presence of weak disorder, or, in other words, by defect scattering. A weak external magnetic field suppresses the phase coherence of the backscattered waves, but does not influence the Coulomb interaction phenomenon. Thus magnetoresistance measurements allow to distinguish between the two effects.

In the weak disorder limit, which is also the condition for transport in the Boltzmann approximation, i.e., when $k_F.l \gg 1$, where k_F is the Fermi wave vector and l the mean free path of the carriers, a correction term, $\delta\sigma^{2D}$, is added to the Boltzmann classical electrical conductivity, σ^{2D}_{Boltz}:

$$\sigma^{2D} = \sigma^{2D}_{Boltz} + \delta\sigma^{2D} \tag{8}$$

with σ^{2D}_{Boltz} as given in relation (4) or (5).

The additional term $\delta\sigma^{2D}$ accounts for localization and interaction effects which both predict similar temperature variation. With $\delta R = -R^2 \delta\sigma^{2D}$, the temperature dependence of the resistance of two dimensional disordered systems, due to weak localization and Coulomb interaction effects, is given by [22,23]:

$$\frac{\delta R(T)}{R(T_0)} = -\frac{q}{2\pi^2 \hbar} \rho_{2D}(\alpha p + \gamma) \ln\left(\frac{T}{T_0}\right) \tag{9}$$

where $\delta R(T) = R(T) - R(T_0)$ and ρ_{2D} is the 2D resistivity at the temperature T_0. The weak localization contribution is expressed by the product αp, while the Coulomb interaction contribution is given by γ, which is a measure of the screening by other charge carriers, taking a value close to 1 when the screening is weak [23,33]. p is the exponent in the temperature dependence, T^p, of the inverse of the inelastic scattering time τ_i. The value of α depends on the effect of spin-dependent processes due to magnetic scattering and spin-orbit coupling. The characteristic scattering times for these two last processes are respectively τ_s and τ_{so} which are both assumed to be temperature independent.

The different scattering times are associated with the physical image that for times smaller than τ_k (k = i,s,so) the phase coherence of interfering complementary waves, which is the source of the weak localization effect, is maintained [21]. Therefore, weak localization occurs only when scattering by static defects is the dominant mechanism, i.e. when $\tau_r \ll \tau_k$, where τ_r is the elastic scattering time due to defect scattering. In the limit where $\tau_i \ll \tau_{so}, \tau_s$, $\alpha = 1$; while for $\tau_s \ll \tau_i, \tau_{so}$, $\alpha = 0$; and if $\tau_{so} \ll \tau_i, \tau_s$, $\alpha = -1/2$.

The weak localization theory for 2D systems [22] predicts that a uniform perpendicular magnetic field, H, introduces a phase shift between interfering partial carrier waves, thus weakening the localization effect when the characteristic magnetic time:

$$\tau_H = \frac{\hbar}{4qDH} \tag{10}$$

becomes comparable or smaller than the dominant scattering time, τ_k. D is the diffusion constant.

One may also express the relaxation time for scattering by magnetic impurities, τ_s, associated to a characteristic magnetic field, H_s in the same way:

$$\tau_s = \frac{\hbar}{4qDH_s} \tag{10'}$$

6.2.4 Zero-Field Resistivity Results

The temperature dependence of the electrical resistivity of various classes of carbon-based materials is very sensitive to their lattice perfection. The higher the structural perfection, the lower the resistivity. Differences observed between samples are more pronounced when the heat treatment temperature (HTT) varies within one class of carbon-based materials than when various classes of samples are heat treated at approximately the same HTT. This is obvious when we compare the results presented in Figure 1–3. Whether we consider bulk (Figure 1) or fibrous (Figure 1–3) materials, the general trend of the electrical resistivity is as follows [18].

Samples of *high structural perfection* exhibit resistivities below a few 10^{-6} Ωm. For these samples, one can describe the resistivity results using the semimetallic graphite band model. At room temperature electrons and holes are scattered by both defects and phonons. Since the carrier mobilities and densities are very sensitive to defects and temperature and have opposite temperature dependences, the exact temperature variation cannot be predicted with confidence. Contrary to metals, where the degenerate electron system is temperature insensitive and only mobility varies with temperature, a semimetal like graphite, due to the small carrier densities is sensitive both to defects and temperature. Partially carbonized samples exhibit resistivities higher than $10^{-4}\Omega$m that generally increase with decreasing temperature. An *intermediate behavior* between these two extremes is represented by curves which depend less on the HTT and does not show significant temperature variations.

The first comprehensive measurements of the temperature dependence of the electrical resistivity of pitch-based carbon fibers were performed by Bright and Singer [34]. They investigated radial and random samples heat treated at various temperatures ranging from 1000°C to 3000°C. They have shown that the magnitude of the resistivity, as well as its temperature variation, depends on the heat treatment temperature (HTT). The same observations were made for VGCF (Figure 3).

We present in Figure 2 the temperature dependence from 2 to 300K of the zero field electrical resistivity of the duPont E-series of six samples of pitch-based carbon fibers heat treated at various temperatures [11]. The thermal conductivities of these fibers are presented in Figure 13 and the thermoelectric powers in Figure 14a. It may be seen that, except for the E35 fibers, all the samples investigated present an increase of their zero-field electrical resistivity with decreasing temperature (negative temperature coefficient) down to the lowest temperature. The same behavior was observed on other carbon fibers [10] and pyrocarbons [35]. Such a behavior could not be explained in terms of the simple two bands (STB) model [36] or by the Bright model [37]. Bayot and co-workers [10] have shown that the weak localization effect may explain this phenomenon. As shown above (§6.2.3), the weak localization generates an additional contribution to the low temperature electrical resistivity which adds to the classical Boltzmann resistivity [10,35]. A magnetic field destroys this extra contribution [10,35] and restores the classical temperature variation predicted by the STB model. This results in an apparent negative magnetoresistance. As discussed previously, weak localization is also destroyed when the carriers are scattered by magnetic impurities or by phonons since the phase coherence is not conserved in these processes. Weak localization effects thus take place at low temperatures where elastic scattering by lattice defects is the main mechanism limiting electrical conduction. They occur whatever the dimensionality of the system but the effect is more pronounced for quasi 2D electronic systems than for 3D systems.

The fact that the temperature dependence of the E35 fibers does not follow the same trend as the other fibers of the series may be explained in the following way. The E35 fibers present an interlayer spacing greater than 0.344 nm (Table 1), which is characteristic of a fully turbostratic structure [18]. The structure of these fibers is thus highly disordered although weak disorder condition is no more important since $k_F \cdot l$ was found to be slightly smaller than 1 [11]. The negative magnetoresistance observed for all the other fibers is indicative of the dominating presence of a turbostratic structure [38–41].

In Table 2 we have also reported the values of the resistivities at 4.2°K, $\rho_{4.2}$, and the "apparent" residual resistivity ratios, RRR = $\rho_{300}/\rho_{4.2}$. In Figure 4 we present the room temperature resistivity of various experimental pitch-based carbon fibers versus in-plane coherence length [42]. We may see that there is a direct correlation between the two parameters, and, as expected, the electrical resistivity decreases with increasing structural perfection.

Table 2 Zero Field Electrical Resistivity of the Fibers at 4.2°K, $\rho_{4.2}$, and Residual Resistivity Ratio (RRR)[a]

Fibers	$\rho_{4.2}$ [$\mu\Omega$m]	RRR	R [Ω]	H_s {10^{-1} Tesala]
E35	11.85	0.93	961	4.58
E55	10.90	0.90	1792	2.50
E75	9.39	0.88	1860	2.08
E120	9.00	0.78	2737	1.61
E105	6.67	0.72	2524	1.21
E130	2.69	0.48	1435	0.61

[a]Sheet resistances, R, and characteristic field for magnetic impurity scattering, H_s, as determined by the analysis of the magnetoresistance data according to the 2D weak localization theory, are also presented.
Source: from ref. 11.

Electrical and Thermal Transport in Carbon Fibers

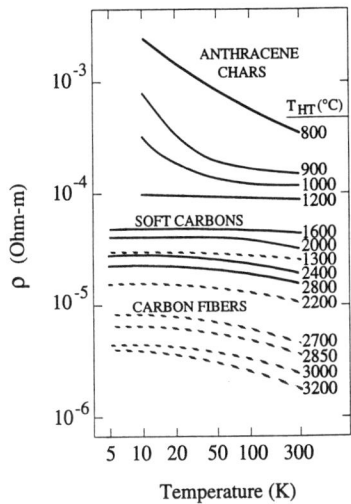

Figure 1 Comparison of the temperature dependence of the electrical resistivity of ex-PAN fibers heat treated at various temperatures with that of some heat treated bulk carbons [18].

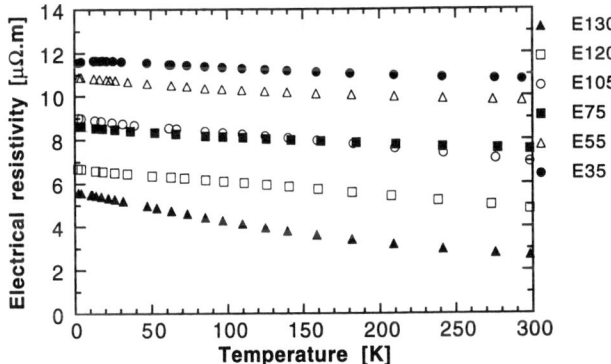

Figure 2 Temperature dependence, from 2 to 300°K, of the zero-field electrical resistivity of the six samples of the E-series of pitch-based carbon fibers heat treated at various temperatures [11]. The values of the liquid helium resistivity and of the residual resistivity ratio are given in Table 2.

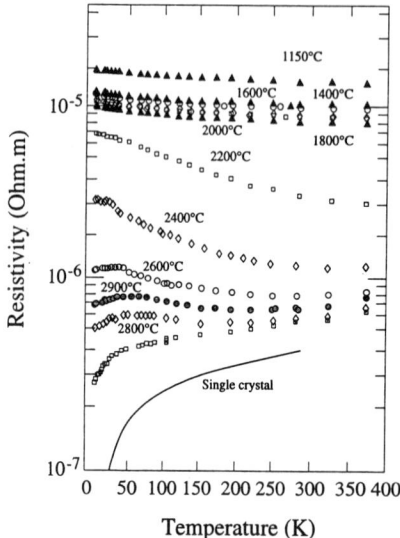

Figure 3 Temperature dependence of the electrical resistivity of benzene-derived carbon fibers heat treated at various temperatures, as indicated in the figure. The data are compared to those relative to single crystal graphite (From J. Heremans, *Carbon*, 23, 431 (1985)).

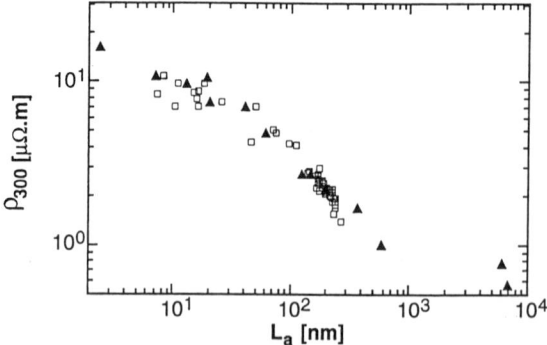

Figure 4 Room temperature resistivity of various experimental pitch-based carbon fibers versus in-plane coherence length, L_a [42]. It may be seen that the electrical resistivity decreases with increasing structural perfection.

6.2.5 Electron-Phonon Interaction

The electron-phonon interaction in GICs has been previously discussed [14,16,43] and it was shown that it was more complex than in 3D metals. We will briefly discuss this particular case below (Section 6.5.2.1). In pristine carbons and graphites the electron-phonon interaction should be given even more specific attention. The effectiveness of a scattering event for a degenerate electron gas depends on the angle that the charge carriers are scattered on the Fermi surface and, for a given scattering angle, on the number of available phonons for scattering. The main difference between a 3D metallic system and the quasi 2D electron and hole systems in semimetallic graphites is that the charge carriers do not interact with the same class of phonons. For a 3D metal, electrons are scattered through large angles by the Debye high energy phonons around and above the Debye temperature, which value for metals is in most cases of the order of room temperature, and through small angles by phonons of energy $\sim k_B T$ well below the Debye temperature. This leads to small relaxation times around room temperature and to large ones at low temperatures. For small Fermi surfaces, like those characterizing semimetals, in order to satisfy energy and momentum conservation requirements, the wave numbers of the phonons interacting with charge carriers are restricted to values smaller than twice the Fermi wave number. As a result, the electrons and holes are scattered through large angles by low energy *subthermal phonons* at all temperatures, except at very low temperatures. The limit is determined by a characteristic temperature [14,16]:

$$\theta^* = \frac{2k_F v_s \hbar}{k_B} \tag{11}$$

Above θ^* the subthermal phonons, which are those allowed to interact with the charge carriers, are of lower frequencies than those which dominate at the temperature considered and which mainly contribute to the lattice thermal conductivity.

For graphite, the velocity of sound in the graphene planes, $v_s \sim 2.1 \times 10^6$ cm/s, is one order of magnitude higher than in metals, while k_F, the Fermi wave vector, is orders of magnitudes smaller than in ordinary metals. The result is that θ^* is much smaller than the Debye temperature of 3D metals and, *a fortiori*, of the in-plane Debye temperature of graphite.

As a result, the electron-phonon interaction should be much weaker in pristine carbons and graphites than in 3D metals at a given temperature because, though scattering is through large angles above θ^*, the number of interacting phonons is very small, since we are well below the in-plane Debye temperature, even at room temperature. Thus, for pristine carbons and graphites, at low and high temperatures, phonon scattering should be much less effective than in 3D metals leading to high relaxation times. This, associated to the fact that the effective masses are smaller than the free electron mass, explains why we observe very high mobilities in these materials (Equation 3). It explains also why, despite the very small densities of electrons and holes, about four orders of magnitudes lower than in metals at room temperature, well graphitized samples display resistivities only two orders of magnitudes higher than in metals (Equation 2).

6.2.6 Magnetoresistance

Owing to the effect of the Lorentz force on the charge carriers, the application of a magnetic field on a conductor carrying an electrical current leads to an increase in resistance. We shall call hereafter this effect *"positive magnetoresistance."* The fractional change in the resistance caused by the application of an external magnetic field is expressed:

$$\frac{\Delta \rho}{\rho_0} = \frac{\rho_H - \rho_0}{\rho_0} \qquad (12)$$

where ρ_H and ρ_o are the electrical resistivities with and without magnetic field respectively. Solid state theory predicts that the classical positive magnetoresistance at low magnetic fields depends essentially on the carrier mobilities. However, *negative magnetoresistances*, (i.e., decreases in resistivities with increasing magnetic fields), have been observed in pregraphitic carbons by Mrozowski and Chaberski [44] This unusual phenomenon has also been observed in other forms of carbons, including poorly graphitized bulk carbons [38,39], PAN-based fibers [45,46], pitch-derived fibers [34] and vapor-grown fibers [40]. We will show below that, as it is the case for the zero-field resistivity (Section 6.2.4), these observations are consistent with the weak localization theory for two dimensional systems that we discussed in Section 6.2.3.

It was initially realized that the amplitude and the sign of the magnetoresistance was closely related to the microstructure of the material [34,38,40,44,45]. Highly graphitized samples present large positive magnetoresistances at all temperatures which were attributed to the high mobility of the charge carriers. When the sample is more disordered, a negative magnetoresistance shows up at low temperature. The amplitude of the negative magnetoresistance and the temperature at which it was observed increased as the relative amount of turbostratic planes increased in the material.

For many years, the most popular model which was invoked to justify the presence of a negative component in the magnetoresistance was the Bright model

Electrical and Thermal Transport in Carbon Fibers 387

[37]. In this model it is assumed that the electronic structure for turbostratic graphite is nearly two-dimensional (2D). The magnetic field induces changes in the electronic density of states which lead to an increase in carrier concentration. A critical analysis of this model is given in the book of Dresselhaus et al. [18].

Bayot et al. [10,35] proposed an alternative explanation to the negative magnetoresistance. This was based on weak localization effects resulting from weak disorder in a 2D electronic system, which were previously observed in intercalated graphite by Piraux and co-workers [9]. Bayot et al. suggested that the 2D weak localization effects observed in pristine carbons occur in the quasi 2D turbostratic phase [10,35].

Nysten and co-workers [11] have measured the magnetoresistance at 4.2°K and 77°K of the six samples of pitch-derived carbon fibers (E35, E55, E75, E105, E120, and E130). Their results are presented in Figure 5. At both temperatures, the magnetoresistance was found to be negative for all the samples investigated. We may see that for a given field, say 1T, the magnitude of the negative magnetoresistance increases with increasing structural perfection from the E35 fibers to the E120 fibers, then it decreases for the E130 fibers. A similar trend was also observed on another set of pitch fibers [34] and on PAN fibers [45].

The results of Nysten and co-workers [11] were also interpreted using the 2D weak localization formalism discussed above and adapted by Bayot et al. [10,35] to the particular case of pregraphitic carbons. The parameters derived from the analysis of Nysten et al. were directly related to the structural parameters as measured by x-ray diffraction. They were also compared to results obtained on other pitch-based fibers [10] and on bulk pyrocarbons [35]. It was shown that the analysis of the magnetoresistance data enables to determine the portion of turbostratic regions in the fibers and also the degree of disorder in their microstructure: boundaries, voids, amorphous regions, and magnetic impurities.

The good agreement between the data obtained on different carbon materials confirmed what was previously suggested, though not fully understood, (i.e., that magnetoresistance measurements could be used as a tool to characterize carbon fibers). So, whatever the mechanism involved, the fact that the magnetoresistance was found to be directly related to the structural perfection in carbons and graphites, lead to the use of magnetoresistance data to gather information about the microscopic parameters in these materials [41,45,46] (Section 6.7.3).

The analysis of the field dependence of the magnetoresistance, which was made using the 2D weak localization theory [10,35], allowed the determination of two essential parameters [11]:

- The sheet resistance, R, which is the resistance of the two-dimensional system as calculated from the classical Boltzmann theory [10,35]. In the case of thin metal films, R, which is expressed

in Ohms, is equal to the 3D electrical resistivity divided by the thickness of the quasi 2D system.
- H_s, the characteristic magnetic field associated with scattering by magnetic impurities (Equation 10'). Scattering by magnetic impurities was found to be the main mechanism destroying the phase coherence at low temperatures [10,35].

The values of these parameters are presented in Table 2.

The sheet resistance in carbons has been shown [35] to be related to the probability of finding turbostratic layers in the material, p^2:

$$\frac{\rho_{4.2}}{R} = \frac{d_{002}}{\lambda p^2} \qquad (13)$$

where λ is a factor introduced in the weak localization theory for inhomogeneous systems containing many sort of defects and irregularities such as cracks, voids, amorphous domains, and extra resistive boundaries [47]. It was assumed that λ is equal to 1 in totally homogeneous systems, while it should be lower than 1 in inhomogeneous systems. This factor takes into account the fact that the resistivity should not solely be attributed to the 2D phase.

p is the probability of finding two neighboring layers in a randomly stacked configuration [48]. It should thus be equal to 1 for a fully turbostratic carbon and to 0 for a perfect single graphite crystal. The probability to have a turbostratic layer is equal to p^2 since it must be randomly stacked with respect to its two neighbours. p, may be evaluated from the analysis of the *(hkl)* reflexions. It has been empirically related by Franklin [48] to the mean interlayer spacing:

$$d_{002} = 0.3354 + 0.0086 p^2 \qquad (14)$$

Since d_{002} does not only depend on the degree of graphitization but also on the presence of interstitial atoms, Equation (14) allows only a rough estimation of p. On the other hand, the sheet resistances was found to follow Equation (13) for pyrocarbons suggesting that the 2D weak localization effect should be ascribed to the turbostratic structure [35].

In order to visualize the $\lambda.p^2$ variation for the E-series samples, we present in Figure 6a the ratio $\rho_{4.2}/R$ versus the interlayer spacing d_{002}. These data are also compared with those obtained on other pitch-derived fibers [10] and on pyrocarbons heat-treated at different temperatures [35]. The scattering in the data was attributed to the uncertainty in the evaluation of the cross-sectional area of the fibers when determining the value of the electrical resistivity from that of the measured resistance.

The minimum values for $\rho_{4.2}/R$ occur for d_{002} around 0.341–0.340 nm corresponding to the strongest observed 2D weak localization effects. Starting at the minimum value for $\rho_{4.2}/R$, when d_{002} decreases, $r_{4.2}/R$ increases as p decreases. This result is not unexpected since the 3D graphitic structure is known to grow at the expense of the quasi-2D turbostratic structure leading to the disappearance of 2D weak localization effects [35]. On the other hand, when d_{002} increases starting from 0.341 nm, the values of $\rho_{4.2}/R$ for the fibers increase. In the same way, it has been observed for PAN fibers [45] and for pitch-derived fibers [34] heat treated at different temperatures that the amplitude of the negative magnetoresistance decreases with decreasing HTT. Concerning the E-series samples, if the assumption that the 2D turbostratic planes are responsible for the 2D weak localization, relation 6.13 should apply as long as the condition $k_F l > 1$ is satisfied. This is the case for all these samples, except for the E35 fibers. As p^2 tends to 1 when d_{002} increases (relation 6.14), the corresponding increase of $\rho_{4.2}/R$ should be ascribed to a decrease of λ. In Figure 6b, we report the values of λ versus d_{002} as calculated from relation 1, p^2 being determined from Equation (14). We see that λ strongly decreases when the interlayer spacing increases above 0.340 nm. This decrease of λ should be attributed to an increase of the concentration of large scale defects (grain boundaries, cracks, and voids) with increasing d_{002}. In addition, the samples become more and more disordered tending to a structure where the condition for weak disorder is no more respected. As a result, the negative magnetoresistance, and with it the 2D weak localization effects, disappear. This should be related to the decrease of the coherence lengths measured by x-ray diffraction (Table 1). The low values of λ found in the E-series fibers compared to those relative to other pitch-derived fibers may be related to the presence of a large number of disclinations in these fibers [49,50] which are responsible for plane foldings, thus hindering the graphitization process. It was also suggested that these plane foldings were responsible for the high tensile strength of these fibers [41].

H_s, the characteristic magnetic field associated with scattering by magnetic impurities [10,35], is the other parameter which may be determined from an analysis based on the 2D weak localization theory. H_s is related to the relaxation time of the charge carriers due to the scattering by magnetic impurities, τ_s, as shown above (Equation 10). Since H_s results from the interaction of the charge carriers with magnetic scatterers, if it is found to play an effective role, it should reveal the presence of magnetic impurities in the fibers. We have plotted in Figure 7 H_s versus d_{002}–0.3354 nm. The abcissa is a measure of the deviation of the samples structure from that of a totally graphitized specimen. When these data are compared to those obtained on other pitch-based fibers [10] and on pyrocarbons [35], the same general trend is observed. The variation of H_s, and thus of τ_s, suggests that the concentration of magnetic impurities decreases with increasing graphitization. This confirms conclusions made on the light of results obtained by means of other techniques. These observations suggest that impurities diffuse between the graphene layers and are removed from the sample as the structure becomes more and more ordered [51]. Similarly, we see in Figure 5 that thermal conductivity measurements indicate that phonon scattering by point defects decreases with increasing order [12,52].

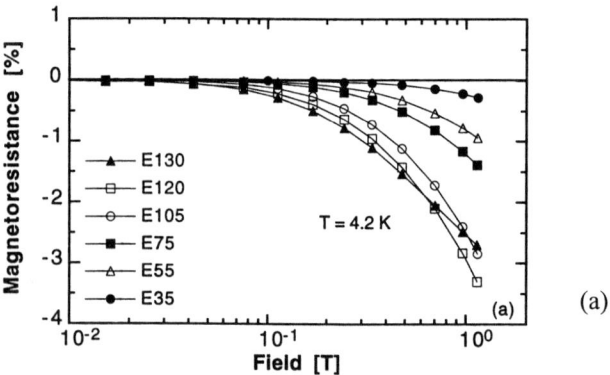

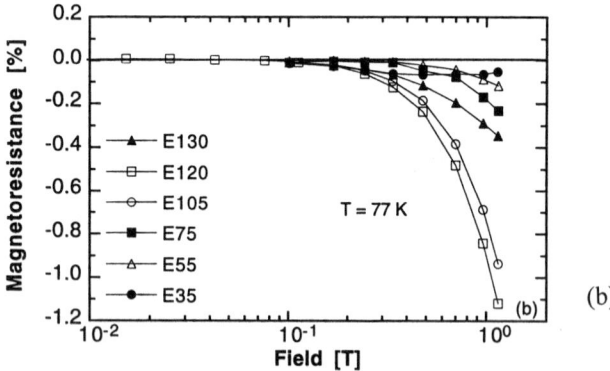

Figure 5 Field dependence of the magnetoresistance of the six samples of the E-series of pitch-derived carbon fibers at 4.2°K (a) and at 77°K (b). At both temperatures, all the samples exhibit negative magnetoresistances.

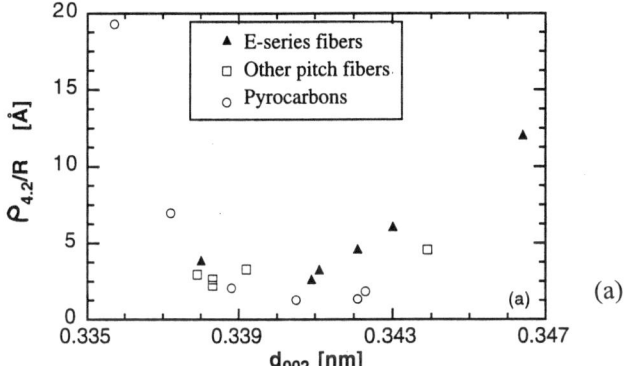

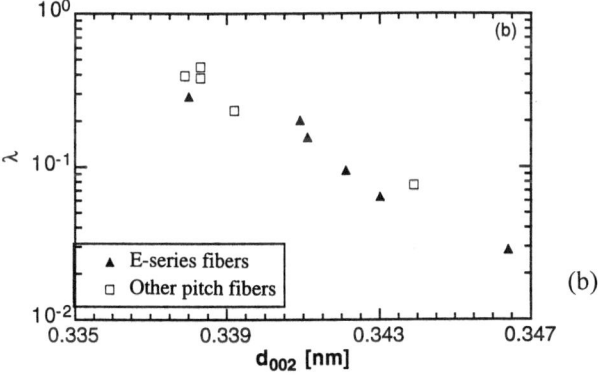

Figure 6 Variation of the ratio $r_{4.2}/R$ (a) and of λ (see text) (b) as a function of the interlayer spacing d_{002} for the six samples of the E-series pitch-derived carbon fibers, compared to the results obtained on other pitch-derived carbon fibers and on pyrocarbons. The scattering in the data is attributed to the uncertainty in the evaluation of the cross sectional area of the fibers when determining the value of the electrical resistivity from that of the measured resistance. The minimum values for $r_{4.2}/R$ occur for d_{002} around 0.341–0.340 nm corresponding to the strongest 2D weak localization effects observed.

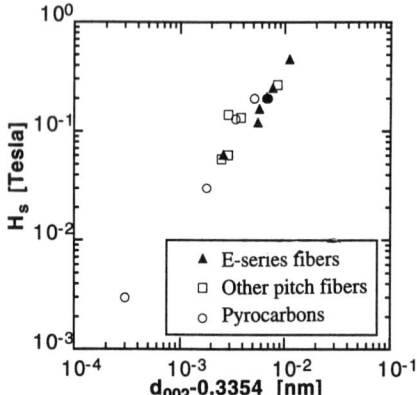

Figure 7 Variation of the parameter H_s (see text) as a function of d_{002} 0.3354 for the six samples of the E-series of pitch-derived carbon fibers, compared to the results obtained on other pitch-based fibers and on pyrocarbons.

6.3 THERMAL CONDUCTIVITY

6.3.1 Introduction

Little attention had been paid to the thermal conductivity of carbons fibers until the beginning of the last decade. This, despite of the fact that it was known that phonons were powerful tools to probe the lattice defects in carbons and graphites and their compounds. It is most likely that the importance of this property as a method of sample characterization had not been given proper credit because of the complexity of thermal conductivity measurements. Indeed, thermal conductivity measurements are time consuming and very delicate to perform [53]. This is particularily true for samples of small cross-sections, as it is the case for carbon fibers [54].

To measure the temperature variation of the fibers thermal conductivity one should use a sample holder specially designed to reduce significantly heat losses. This sample holder, which was designed for measuring samples with very small thermal conductances, is described in detail elsewhere [54,55]. Measurements should be performed below room temperature, say from 3 to 300°K, in a variable temperature liquid helium cryostat. Since, contrary to electrical resistivity, thermal conductivity measurements were found to be reproducible from one fiber to another of the same batch [55], the measurements could be performed on small bundles of fibers in order to increase the sample thermal conductance with respect to that due to the heat losses.

Very often the knowledge about thermal conductivity is limited to that of the Wiedemann-Franz Equation (16). The result of this law is that metals, which are good electrical conductors are also good thermal conductors. Metals have indeed a high electronic thermal conductivity, but the Wiedemann-Franz Equation holds exactly over limited temperature ranges and only for pure metals.

In fact, there are essentially two mechanisms for thermal conduction in solids around and below room temperature [56,57]. The first, the electronic thermal conductivity, κ_E, is due to the charge carriers, while the second, which is due to the quantized lattice vibrations, the phonons, is called the lattice thermal conductivity, κ_L. In electrical insulators heat is exclusively carried by phonons, while in pure metals it is predominantly carried by the charge carriers.

For some materials like highly doped semiconductors, metallic alloys and group V semimetals, the lattice thermal conductivity may be comparable to the electronic contribution in certain temperature ranges [56,57]. This is also the case for GICs (Section 6.5.3) [14]. In that case, the total thermal conductivity is expressed:

$$\kappa = \kappa_E + \kappa_L \tag{15}$$

In the case of magnetic materials, magnons may also contribute to the transport of heat energy, but we shall not discuss this particular case here.

6.3.2 Electronic Thermal Conductivity

The electronic thermal conductivity is caused by the transfer of energy due to the difference in the broadening of the Fermi-Dirac distribution around the Fermi level caused by a temperature gradient. The Wiedemann-Franz law directly relates the electronic thermal conductivity to the electrical conductivity, σ:

$$\kappa_E = LT\sigma \tag{16}$$

When the electrical conductivity and electronic thermal conductivity may both be expressed in terms of the same relaxation time, the Lorenz ratio, L, takes the value of the Lorenz number ($L_o = 2.44 \; 10^{-8} V^2 K^{-2}$) [56,57,59]. This applies for a degenerate free electron system which undergoes elastic collisions. Thus $L = L_o$

for metals in the temperature ranges where these last conditions apply. This is the case in the low temperature residual resistivity range when scattering is dominated by impurities and lattice defects. For 3D metals the Wiedemann-Franz law holds also around and above the Debye temperature when large angle intravalley electron-acoustic phonon interaction is the main scattering mechanism. In these two temperature ranges, Equation (16) allows to compute the electronic thermal conductivity from the measured electrical resistivity.

From elementary kinetic arguments, one may derive the electronic thermal conductivity in terms of electronic parameters [56,57]. One finds that k_E is directly proportional to the electronic specific heat, C_E, to the velocity of the electronic charges at the Fermi energy, v_F, which increases with the carrier density, and to the mean free path of the charge carriers, l_E

$$\kappa_E = C_E v_F l_E \tag{17}$$

In the lowest temperature range, where the electrical resistivity is in the residual range, the corresponding electronic thermal conductivity, κ_E, of pure metals varies linearily with temperature. At higher temperature, κ_E reaches a maximum, which is sharper for samples with less impurities and defects. Then κ_E decreases and levels off at higher temperatures. The electronic thermal conductivity is constant in the temperature range where the electrical resistivity is due to large angle electron-phonon interactions and should vary linearly with temperature. For samples with a high concentration of impurities or lattice defects, the linear increase may be directly followed by a temperature insensitive κ_E without any intermediate thermal conductivity peak [57].

6.3.3 Lattice Thermal Conductivity

For nonmetals there is another mechanism for thermal conduction. This explains why diamond, an electrical insulator, is the best thermal conductor at room temperature. This is also true for graphite, which as a semimetal is not as good an electrical conductor as metals, but is a very good thermal conductor in-plane.

In order to understand how heat is transmitted in crystalline materials, let us consider the case of graphite in-plane, in which we are interested here, and which has the advantage of being a material which exhibits at high temperature a typical lattice conduction behavior, (i.e., the transmission of atomic vibrations). As a further simplification, we shall consider highly oriented pyrolytic graphite (HOPG) and suppose that it is a two-dimensional (2D) system, which is not too far from the real situation around room temperature. The atoms in such a system may be represented by a 2D array of balls and springs and any vibration at one end of the system will be transmitted via the springs to the other end. Since the carbon atoms have small masses and the interatomic covalent forces are strong, one can expect smooth transmission of the vibrational motion and thus good lattice thermal conductivity. Any disruption in the regular arrangement of the atoms, such as defects or atomic vibrations, will cause a disruption in the heat flow, thus giving rise to a scattering which decreases the thermal conductivity.

In order to treat this naive picture in a quantitative way, a system of particles, the phonons, is associated to the crystal vibrational modes. Then a formalism is developed which allows to describe the thermal conductivity in terms of phonon transport. Once the equivalence between the properties of the vibrating and particle systems is established, a simple kinetic picture is then used. The lattice or phonon thermal conductivity κ_g is given by [56,57]:

$$\kappa_g = \frac{1}{3} \sum_s \int_{\omega_0}^{\omega} c_s(\omega) v_s(\omega) l_s(\omega) d\omega \qquad (18)$$

where $c_s(\omega)d\omega$ is the contribution to the specific heat of phonons of polarization, s, and frequency, ω, in the range $d\omega$, $v_s(\omega)$ is the velocity and $l_s(\omega)$ the mean free path. The integral is taken from the lowest frequency, ω_0, to the Debye cut off frequency, ω_D. The phonon mean free path is directly related to the phonon relaxation time, τ, through the relation $l = v\,\tau$.

The discussion of lattice thermal conductivity results is greatly simplified if we consider an isotropic material and use the Debye relation [56,57]:

$$\kappa_g = \frac{1}{3} Cvl \qquad (19)$$

in the dominant phonon mode approximation, i.e., if we use an average phonon frequency which is proportional to the absolute temperature. C is the lattice specific heat per unit volume and v is an average phonon velocity, the velocity of sound. Equation (19) is convenient to discuss qualitatively the thermal conductivity results of isotropic materials. For a given solid, since the specific heats and the phonon velocities are the same, the thermal conductivity is directly proportional to the phonon mean free path.

It was shown that the thermal conductivity of carbon fibers may vary widely, about two orders of magnitude, according to their microstructure [5,12]. Some carbon fibers are among the best heat conductors around room temperature [3,5]. This is not surprising since it is well known that diamond and highly oriented pyrolytic graphite (HOPG) in-plane present heat conductivities exceeding 2000 $Wm^{-1}K^{-1}$ at room temperature (Figures 8 and 9). VGCF carbon fibers heat treated at 3000°C present room temperature thermal conductivities of this order. In these materials, the transport of heat above the liquid helium temperature range is entirely due to the phonons. In the lowest temperature range, the lattice thermal conductivity is mainly limited by phonon-boundary scattering. We will show below that, as is the case for electrical conductivity (Section 6.24), low temperature thermal conductivities are directly related to the in-plane coherence length, L_a. Depending on the precursor and heat treatment temperature L_a will vary, the higher the HTT the larger the in-plane coherence length, up to a certain limit [5]. Furthermore, it will also be shown below that for carbon fibers the phonon mean free path coincides almost exactly with the in-plane coherence length.

When scattering is mainly on the crystallite boundaries, the phonon mean free path should thus be temperature insensitive. Since the velocity of sound is almost temperature insensitive, the temperature dependence of the thermal conductivity should follow that of the specific heat. Thus, the larger the crystallites the higher the thermal conductivity. Well above the maximum, phonon scattering is due to an intrinsic mechanism: phonon-phonon umklapp processes, and the thermal conductivity should thus be the same for different well-ordered graphites.

Around the thermal conductivity maximum, scattering of phonons by point defects (small scale defects) will be the dominating process. The position and the magnitude of the thermal conductivity maximum will thus depend on the competition between the various scattering processes (boundary, point defect, and phonon). So, for different samples of the same material the position and magnitude of the maximum will depend on the point defects and L_a, since phonon-phonon interactions are assumed to be the same. Thus, by measuring the low temperature thermal conductivity, one may gather information about the in-plane coherence length L_a and point defects.

This shows also that by adjusting the microstructure of carbon fibers, one may control their thermal conductivity to a desired value around and below room temperature. The highest value attained so far is for a vapor deposited carbon fiber with a room temperature thermal conductivity around 1400 $Wm^{-1}K^{-1}$ [55]. For pitch-derived fibers, samples of conductivities around 1000 $Wm^{-1}K^{-1}$ [5,12] are now available. Thus, by acting on the microstructure of carbon fibers, thermal conductivities as high as 1400 $Wm^{-1}K^{-1}$ may be reached (see Figures 8–11).

Since in the lowest temperature range the in-plane phonon mean free path, l, is constant, the lattice thermal conductivity for well-ordered graphites follows a T^n law with $n = 2.3$

$$\kappa_L = BlT^n \qquad (20)$$

where B is a constant, which has roughly the same value for all graphitic samples, and l is directly related to the in-plane crystallite size, as will be shown below. The fact that B should not vary significantly from one sample to another may be easily understood, since it depends on the vibrational modes of the crystal lattice, which do not vary from one sample to another provided their defect structure is not too much pronounced. Equation (20) is derived from the Debye Equation (19) [60].

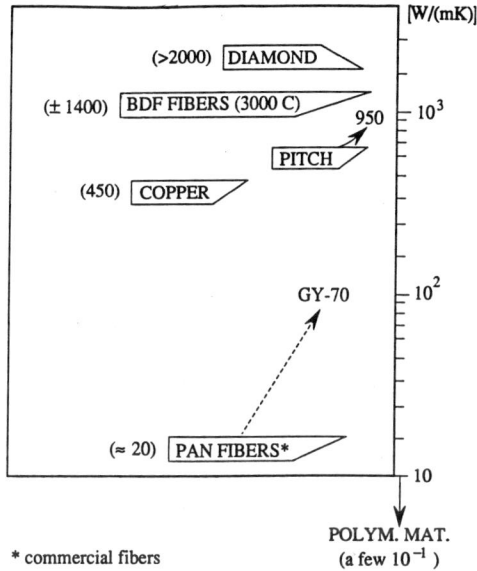

Figure 8 Comparison of the room temperature thermal conductivity of some selected carbon fibers with that of diamond, copper and polymeric materials.

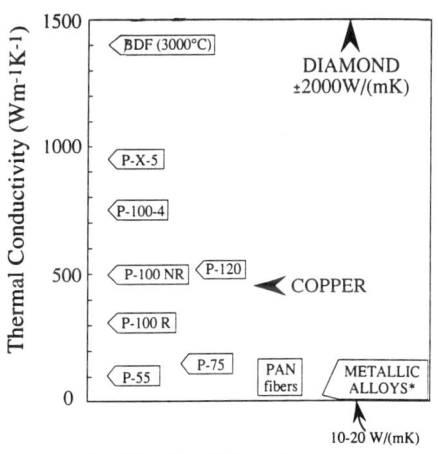

Figure 9 Comparison of the room temperature thermal conductivity of various experimental and commercial carbon fibers from Amoco with that of some typical solid materials.

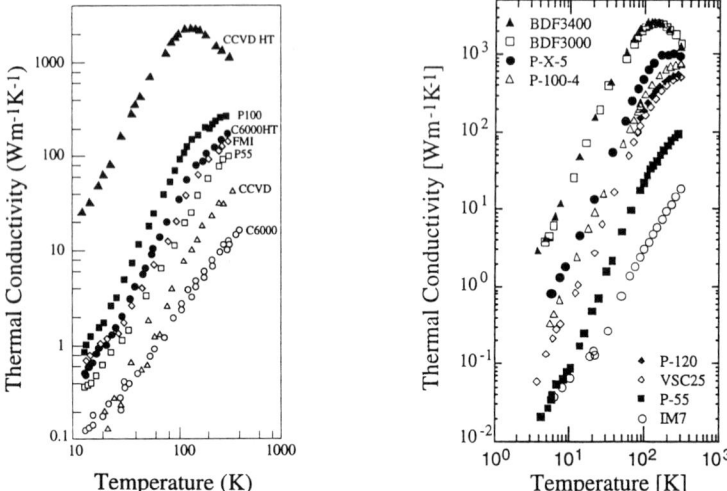

Figure 10 Comparison of the temperature variation of the thermal conductivity of pristine carbon fibers of various origins and precursors. Since scattering below room temperature is mainly on the crystallite boundaries, the phonon mean free path at low temperatures, i.e. below the maximum of the thermal conductivity versus temperature curve is temperature insensitive and mainly determined by the crystallite size. The larger the crystallites the higher the thermal conductivity. Note that some VGCF and PDF of good crystalline perfection show a dielectric maximum below room temperature. For decreasing lattice perfection the maximum is shifted to higher temperatures.

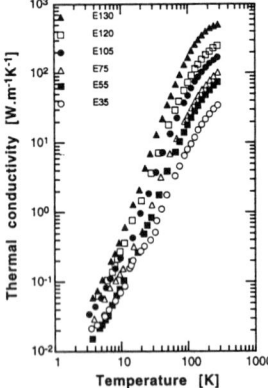

Figure 11 Temperature dependence of the thermal conductivity of the carbon fibers of the E-series, the same as that which electrical resistivity is presented in Figure 2.

6.3.4 Analysis of Results

In Figures 8 and 9 we compare the room temperature thermal conductivity of various carbon fibers to that of other solids. In Table 3 the room temperature thermal conductivity of the fibers of the E-series is presented. These conductivity values range from 35 $Wm^{-1}K^{-1}$ for the E35 fibers up to 525 $Wm^{-1}K^{-1}$ for the E130 fibers. In Table 3 the resistivity at 4.2°K is shown and the coefficient, n, of the temperature dependence of the low temperature thermal conductivity (Equation 20). It is seen that VGCF and pitch-derived carbon fibers have thermal conductivities higher than that of copper when heat treated at sufficiently high temperatures. Also, there are orders of magnitude differences between the thermal conductivity of the highest conducting fiber, a benzene derived carbon fiber heat treated at 3000°C, and that of the lowest ones, PAN-based fibers. We do not present here the thermal conductivity of activated carbon fibers which have thermal conductivities much lower than the PAN fibers presented here [61].

In Figures 10 and 11 we present the temperature variation of the thermal conductivity of various types of fibers. In Figure 10 fibers of various origins are compared, while in Figure 11 the E-series of du Pont fibers heat treated at various temperatures are presented. This last series is the same as that which electrical resistivity is presented in Figure 2. By comparing Figures 2 and 11, note the parallel behavior of electrical and thermal data (see also Table 3).

In order to discuss the thermal conductivity results in terms of the relevant parameters, we present in Table 4, the structural parameters of the fibers of the E-series determined by x-ray diffraction. These parameters are the interlayer spacing, d_{002}, the in-plane coherence length, L_a, and the out-of-plane coherence length, L_c, (i.e., the distance over which planes are parallel to one another). Note that there is a limitation to the x-ray analysis for large coherence lengths, (i.e., those exceeding approximately 100 nm for the values of L_a). For comparison, we have also shown in Table 1 the mechanical properties of the various fibers.

We will now discuss the temperature variation of the thermal conductivity of the six samples of the E-series (Figure 11). The fibers exhibit increasing thermal conductivity values with decreasing d_{002} and increasing tensile modulus. In the temperature range between 10 and 100°K, the fibers present increasing slopes for the temperature dependence from T^2 for the E35 fibers to approximately $T^{2.7}$ for the E130. This corresponds to increasing moduli, C_{33} and C_{44}, from the E35 to the E130. In the less graphitized fibers, planes are weakly coupled and this gives rise to a quasi 2D behavior, (i.e., a T^2 variation) [62]. In better graphitized fibers, the increase of C_{33} and C_{44} leads to an increase of the slope in the temperature dependence [62].

In the lowest temperature range, the thermal conductivity does not follow the simple power law $\kappa \propto T^{2-2.7}$, characteristic of phonon boundary scattering for graphite and apparent at intermediate temperatures. This departure from a simple power law dependence is more pronounced for the higher values of d_{002}. This is due to the presence of an additional electronic contribution to the thermal conductivity whose relative contribution increases when the lattice contribution decreases. Since the Wiedmann-Franz law should be valid even in disordered system [63,64], the electronic thermal conductivity may be computed using Equation (16) with $L = L_o = 2.44 \times 10^{-8} V^2 K^{-2}$. The lattice thermal conductivity may then be calculated by subtracting the electronic contribution from the total measured thermal conductivity.

Using Kelly's theory for the lattice thermal conductivity of graphite [62,65], Nysten et al. [12] have carried out an extensive analysis of the experimental data presented in Figure 11. They have shown that the thermal conductivity measurements performed at low temperature allow to determine directly the magnitude of the in-plane coherence length in carbon fibers. Also they were able to extract information about point defects and to compare between shear moduli (C_{44}).

The temperature variation of the lattice thermal conductivity below the dielectric maximum was fitted using the relations developed by Kelly for the thermal conductivity and the relaxations times [12]. For the compression modulus, C_{33}, it was found that it decreases by 15% from its value for graphite when the interlayer spacing increases to 0.341 nm. It was also expected that C_{33} should be directly related to d_{002} [18]. So, the following dependence for the compression modulus was assumed:

$$C_{33} = 36.5 - 9.777(d_{002} - 0.3354) \qquad (21)$$

where d_{002} is expressed in nm and C_{33} in GPa.

The shear modulus, C_{44}, which influences the slope of the low temperature dependence of the thermal conductivity [62], is also related to the interlayer spacing and to defects [18]. However, this dependence is not so clear as for C_{33}. It was one of the three adjustable parameters. The two other adjustable parameters were the phonon mean free path for boundary scattering, l_B and the constant for point defects scattering, A. For L_c, the value of the out-of-plane coherence length as determined by x-ray diffraction was taken (Table 1).

The results of the parameters adjustment are presented in Table 4, where they are compared to the in-plane coherence length. In Figure 12 we present the variation of both L_a and l_B with d_{002}. These combine the results of the x-ray studies carried out by Nysten et al. [12] with those of their thermal conductivity measurements carried out on samples from the same batches of the E-series.

The most important result is that the phonon mean free path for boundary scattering is found to be almost equal to the in-plane coherence length as determined by x-ray diffraction (Figure 12). This enables us to say that thermal conductivity measurements are suitable for the determination of this parameter. Moreover x-ray determination of L_a are limited to low values, while thermal conductivity measurements have no limit in resolution. So these may prove to be a very useful substitute for x-ray analysis in the determination of large in-plane coherence length values. In fact, the measurement of the thermal conductivity at liquid nitrogen temperature allows to get directly a rough estimation of L_a by comparing the measured value with the thermal conductivity of fibers with known in-plane coherence lengths.

Table 3 Room Temperature Thermal Conductivity, κ_{300}, and Electrical Resistivity at 4.2°K, $\rho_{4.2}$, of the Fibers Investigated[a]

Fibers	κ_{300} [Wm^{-1}K^{-1}]	$\rho_{4.2}$ [$\mu\Omega$m]	n
E35	37	11.6	2.0
E55	75	10.9	2.4
E75	110	8.6	2.4
E120	180	9.0	2.5
E105	265	6.7	2.5
E130	525	5.6	2.6

[a]Also presented is the exponent n of the power law T^n followed by the temperature variation of the thermal conductivity between 10 and 100°K.

Table 4 Results of the Parameters Adjustment on the Lattice Thermal Conductivity

Fibers	$L_{a\|}$ [nm]	$L_{a\perp}$ [nm]	l_B [nm]	A [10^{-23}m^2]	C_{44} [GPa]
E35	7.2	6.5	6.9	2.051	< 0.10
E55	16.2	14.5	12.9	0.782	0.52
E75	22.4	16.7	20.8	0.738	0.41
E120	46.1	26.5	40.4	0.435	0.57
E105	51.4	26.5	61.5	0.146	1.31
E130	± 180.0	46.5	147.9	0.056	2.29

l_B; the phonon mean free path for boundary scattering, A; the constant for point defects scattering, and $C_{44;}$ the shear modulus, compared to the in-plane coherence length, $L_{a\|}$ and $L_{a\perp}$, as determined by x-ray diffraction.

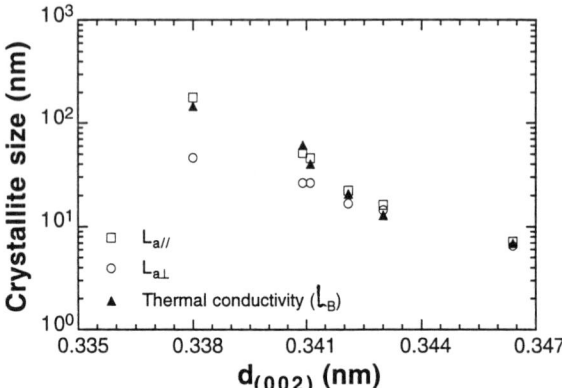

Figure 12 Dependence on the interlayer spacing d_{002} of the in-plane coherence lengths and the phonon mean free path for boundary scattering, lB.

For point defects, the values of A were found to agree with those found by other authors for graphites [66,67]. The parameter, A, on which depends the phonon relaxation time for diffusion by point defects (vacancies, intersticial atoms, and impurities), is directly related to the concentration of point defects [12,42,60,65]. Thus, the decrease of A with decreasing d_{002} may be interpreted as a decrease of the concentration of point defects, (i.e., of impurities or vacancies, with increasing graphitization). This is in agreement with results obtained by other techniques [68] which shows that impurities are removed from the material as the structure is improved. This also corresponds with the magnetoresistance measurements performed on the same fibers (cfr. §6.26), which showed that scattering of the charge carriers by magnetic impurities decreases when the fibers graphitize [11], observation which indicates a decrease of the concentration of these impurities.

Finally, the general decrease of C_{44} with increasing interlayer spacing coincides with the expected dependence due to a decrease of the interlayer bonding [18]. The value found for the E130 fibers agree with those found by other authors on graphite and HOPG by specific heat measurements [69] or neutron spectrometry [70]. Moreover, indirect evaluations of C_{44} by methods such as specific heat or thermal conductivity measurements present the advantage to give the "real" value of the shear modulus, (i.e. the dislocation-free modulus contrary to method such as ultrasonic- or sonic-resonance measurements) [71,72]. Indeed, the determination of C_{44} by these latter methods is influenced by the presence of glissile basal dislocations and provides values which are one or two orders of magnitude lower than those obtained by specific heat or neutron spectrometry. After irradiation with neutrons, these dislocations are pinned and the measured values increase and become comparable to those obtained by other methods.

6.3.5 Relation Between k and ρ

In Figure 13 the correlation of thermal conductivity and electrical resistivity in pitch-derived carbon fibers at 300K is shown [13]. The data are relative to 45 samples among which the E-series and experimental fibers from DuPont and commercial fibers from Amoco are presented.

Though the carriers involved in both phenomena, phonons for the thermal conductivity and electrons and/or holes for the electrical resistivity, are different the correlation is quite good. The data may be fitted to the relation:

$$\kappa = \frac{44.10^4}{\rho + 258} - 295 \qquad (22)$$

where ρ is expressed in 10^{-6} Ωcm and κ in $Wm^{-1}K^{-1}$. So, for the same precursor, once the electrical resistivity is known, the thermal conductivity may be calculated. This may avoid delicate and time consuming thermal conductivity measurements.

In the same way the electrical and thermal conductivities are directly related to Young's modulus in carbon fibers [5].

6.4 THERMOELECTRIC POWER

6.4.1 Introduction

The thermoelectric power or Seebeck coefficient, S, is defined as the potential difference, ΔV, per unit temperature difference, ΔT, across an electrical conductor. In order to measure this coefficient one has to measure ΔT and ΔV on the same spot on the sample.

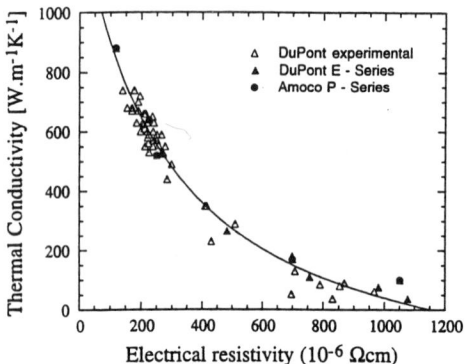

Figure 13 Correlation of thermal conductivity and electrical resistivity in pitch-derived carbon fibers at 300K. The data are relative to 45 samples among which the E-series and experimental fibers from DuPont and commercial fibers from Amoco.

The theory of the thermoelectric power is well established [56] and the effect has been measured on various types of materials. However, even for the simplest cases, there is generally no agreement between theoretical predictions and experimental results. This is true for pristine carbons and graphites where the results are difficult to analyze correctly.

If we exclude magnetic materials, there are essentially two mechanisms for thermoelectric power generation:

- The *diffusion thermoelectric power*, which is due to the spontaneous diffusion of the charge carriers from hot to cold caused by the redistribution of their energies due to the difference in temperature. Charge carriers tend to accumulate at the cold end of the sample giving rise to an electric field. This field acts to counterbalance the stream of diffusing carriers until a steady state is reached. Thus, a measure of the thermoelectric power allows to determine the sign of the dominant charge carriers, since the cold end should be negative for electrons and positive for holes.
- The *phonon drag thermoelectric power* which consists in an anisotropic transfer of momentum from the phonon system to the electron system when the coupling between the two systems is strong. This results in a drag on the charge carriers causing an extra electronic motion with an additional electric field to counterbalance it.

The mechanisms invoked for thermoelectric power generation concern a given group of charge carriers and we shall call the thermoelectric power of this group the partial thermoelectric power, S_j, where j denotes the group of carriers considered. When there is more than one type of carriers, the total thermoelectric power is obtained by considering the different groups of carriers with partial thermoelectric powers that contribute to the total thermoelectric power, S, as emf's in parallel. The total thermoelectric power is thus expressed as:

$$S = \frac{\sum \sigma_j S_j}{\sum \sigma_j} \quad (23)$$

The general expression for the diffusion thermoelectric power for a given group of charge carriers is given by the Mott formula:

$$S_d = \frac{\pi \, k_B^2}{3q} T \left[\frac{\delta \ln \sigma}{\delta \varepsilon} \right]_{\varepsilon_F} \quad (24)$$

where ε_F is the Fermi energy, T is the absolute temperature, σ the electrical conductivity. The derivative is taken at the Fermi level. Equation (24) shows that, for a given scattering mechanism, the diffusion thermoelectric power depends mainly on the Fermi energy.

For a degenerate electron gas the diffusion thermoelectric power depends on the magnitude of the Fermi energy, or, in other words, the carrier density. The smaller the Fermi energy, the higher the relative fraction of the charge carriers concerned with respect to the total Fermi gas [73]. This is why semimetals like graphites should have higher partial diffusion thermoelectric powers than metallic graphite intercalation compounds (GIC). In GICs the diffusion thermoelectric power depends on the charge transfer which rules the magnitude of the Fermi energy.

It is worth noting that, if the relaxation time dependence on energy is isotropic, the partial diffusion thermoelectric power is also isotropic since it depends on the Fermi energy. This is in contrast to the partial phonon drag thermoelectric power, which depends on the highly anisotropic electron-phonon interaction in these compounds.

The partial phonon-drag thermoelectric power of a group of degenerate carriers of density N is given by [74]:

$$S_g = \frac{C_g}{3Nq} \left(\frac{\tau_{p-x}}{\tau_{p-x} + \tau_{p-e}} \right) \quad (25)$$

where C_g is the lattice specific heat at constant volume, q the electronic charge, τ_{p-e} the phonon-electron relaxation time and τ_{p-x} the phonon relaxation time for all other scattering events.

In the simplest case, if we assume that we are at low temperature and only phonon-electron scattering is important, i.e. that $\tau_{p-x} \gg \tau_{p-e}$, Equation (25) becomes:

$$S_g = \frac{C_g}{Nq} \qquad (26)$$

In the latter case the temperature dependence of the phonon-drag thermoelectric power should follow that of the specific heat. It means that, if it is the thermal phonons that drag along the charge carriers, the mean free path of these phonons should be limited by electron scattering. Since the lattice thermal conductivity is expressed by Equation (19), one should then expect a behavior similar to that of the κ_g in that temperature range. This is what was observed in HOPG-Graphite Intercalated C_S (GIC_S)[14].

In semiconductors and semimetals with small Fermi energies like graphite, where the lattice thermal conductivity is usually predominant, the phonons that mainly contribute to the lattice thermal conductivity below the Debye temperature are those which dominate the spectrum at a given temperature. These have the highest energies at the temperature considered. These phonons are called the *thermal phonons*. These phonons do not necessarily exert drag on the charge carriers. Because of energy and momentum conservation requirements, the phonons that drag along the charge carriers are subthermal phonons (see Section 6.2.5) which in semimetals and semiconductors may be of much lower energy than the thermal phonons [14].

In the case of the electrical resistivity, the contribution of normal and umklapp processes add in a way that both lead to increasing scattering of the charge carriers. For the phonon drag thermoelectric power the contribution of normal and umklapp processes have effects opposite in sign [74]. In the case of umklapp phonon-electron interaction the electron is driven after scattering on the side of the Fermi surface opposite to that on which it is driven in a normal interaction. Thus, while for a normal process the charge carrier is dragged along the temperature gradient in one direction, for an umklapp process it is dragged in the opposite direction. This gives rise to thermoelectric powers of opposite signs. Moreover, an additional complication in calculating the thermoelectric power is due to the presence of more than one group of carriers.

6.4.2 Experimental Data

Robson et al. [45,46] measured the thermoelectric power of ex-PAN fibers, but the main data published pertain to pitch-based fibers [42] and VGCF [42,75,76]. Nysten [42] measured the temperature dependence of the thermoelectric power of a large number of pitch-based carbon fibers of various structural perfection. We present in Figure 14a the results relative to the E-series and in Figure 14b those pertaining to fibers with different precursors. In Figure 15 the room temperature thermoelectric power of a large number of pitch-derived carbon fibers as a function of the in-plane coherence length, L_a is shown [42].

It may be seen from Figure 14 that, as is the case for amorphous carbons, the fibers IM7 and E35 have a low thermoelectric power that does not vary significantly with temperature. For highly disordered fibers the room temperature thermoelectric power may even be negative (Figure 15), as is also the case for highly disordered amorphous bulk carbons.

Decreasing disorder leads to a more marked temperature dependence and higher magnitudes for the thermoelectric power. The highest values are observed for fibers with the highest proportion of turbostratic regions, corresponding to the highest values of negative magnetoresistances (E105, E120). Then, the room temperature thermoelectric power decreases progressively with increasing order to reach finally negative values as is the case for HOPG. For the fibers of highest structural perfection, such as E130, P-X-5, and BDF3400, we observe the same behavior displayed by graphites with the typical low temperature peak around 30K.

Nysten analyzed the experimental results in his thesis [42] in the following way. In the case of a two-band model, as for turbostratic graphite, Equation (24) may be written:

$$S = \frac{\pi^2 k_B^2}{3q} T \frac{s+1}{\varepsilon_F} \qquad (27)$$

where s is the scattering parameter, which determines the energy dependence of the relaxation time, $\tau = \tau_0 E^s$. For the case of grain boundary scattering, s is equal to $-1/2$ for a parabolic dispersion relation and to 0 for a linear dispersion.

Using the results obtained by Bayot [77] for pregraphitic carbons, relation (6.27) gives for parabolic bands, a value of 3.1 $\mu V K^{-1}$ at 10K for the thermoelectric power of the majority hole carriers and values varying between 8.7 and 76 $\mu V.K^{-1}$ for the minority holes. Taking into account both contributions by using relation (6.23), one obtains a total thermoelectric power varying between 4.7 à 6.2 $\mu V.K^{-1}$. If we assume linear dispersions instead, the values obtained are multiplied by 2. These values are much higher than those presented in Figure 14 at the same temperature (< 1 μV.K^{-1}). This shows that the thermoelectric power of carbon fibers with a turbostratic structure cannot be interpreted only in terms of the diffusion mechanism described above. One should consider instead that disordered carbon fibers contain more than one phase with various degrees of disorder.

Kaburagi et al. [78] interpreted along these lines the temperature dependence of the thermoelectric power of amorphous carbons. In these materials the electrical conductivity is essentially due to two mechanisms [79]. The first one is a metallic type conduction between extended states in a highly disordered material. The second mechanism is attributed to a variable-range hopping between localized states. The thermoelectric power of the sample should then be interpreted by combining the contribution from these two mechanisms.

For metallic conduction in a highly disordered system, the electrical conductivity is proportional to the square of the density of states at the Fermi level, $\{g(\varepsilon_F)\}^2$ [80]. In that case, Equation (24) reads:

$$S_1 = \frac{2\pi^2 k_B^2}{3q} T \left\{ \frac{d \ln g(\varepsilon_F)}{d\varepsilon} \right\}_1 \qquad (28)$$

where $\{d \ln g(\varepsilon_F)/d\varepsilon\}_1$ is the variation of the density of states at the Fermi level for extended states.

The electrical conductivity due to variable range hopping is given by Mott and Davis [80]:

$$\sigma_{vrh} = B \exp\left[-\left(\frac{T}{T_0}\right)^{1/4} \right] \qquad (29)$$

where B and T_0 are two constants. The thermoelectric power is then given by [80,81]:

$$S_2 = \frac{k_B^2}{2q} (T_0 T)^{1/2} \left\{ \frac{d \ln g(\varepsilon_F)}{d\varepsilon} \right\}_2 \qquad (30)$$

where $\{d\ln g(\varepsilon_F)/d\varepsilon\}_2$ represents the variation of the density of states at the Fermi level for localized states.

The thermoelectric power of a sample containing two regions connected in series, where the diffusion of charge carriers is different is given by [82]:

$$S = \frac{W_1 S_1 + W_2 S_2}{W_1 + S_2} \tag{31}$$

where W_1 and W_2 are the thermal resistances of the two regions considered. Assuming that both regions have almost the same thermal resistance, Kaburagi et al. [78] derived an expression for the total thermoelectric power of the sample:

$$S = aT + bT^{1/2} \tag{32}$$

where:

$$a = \frac{\pi^2 k_B^2}{3q} \left\{ \frac{d \ln g(\varepsilon_F)}{d\varepsilon} \right\}_1 \tag{33}$$

and:

$$b = \frac{k_B^2 (T_0)^{1/2}}{4q} \left\{ \frac{d \ln g(\varepsilon_F)}{d\varepsilon} \right\}_2 \tag{34}$$

For this model to be verified, one should observe a linear relation in a diagram representing $S/T^{1/2}$ as a function of $T^{1/2}$. This is what we observe in Figure 16. By means of a linear regression, one could determine the two coefficients a and b. Nysten [42] calculated these coefficients (Table 5) and concluded that the predominant charge carriers in the extended states are electrons for the more disordered fibers (IM7). With increasing order, the contribution of holes increases. This corresponds to a lowering of the Fermi level in the valence band when the structure tends to a turbostratic one. The inverse occurs for charge carriers in the localized states.

As was the case for the electrical and thermal conductivities, the behavior of the thermoelectric power of carbon fibers as a function of the degree of disorder confirms that already observed in bulk carbons. This may be summarized as follows. The as-grown fibers exhibit a low TEP, characteristic of amorphous carbons. Fibers with turbostratic structures (HTT = 2000°C) present a positive thermoelectric power which increases with increasing temperature. Graphitic fibers behave very much like bulk graphites.

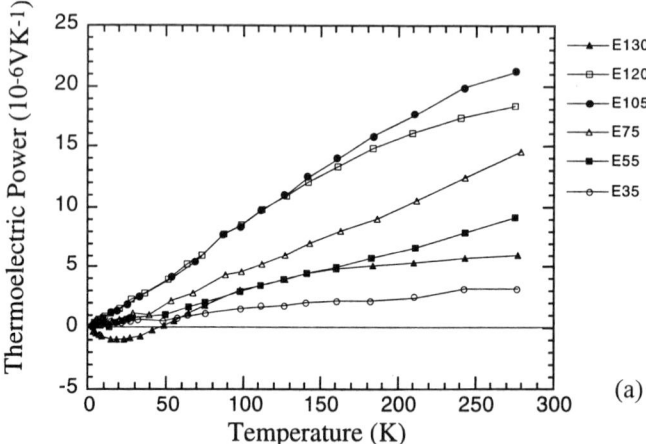

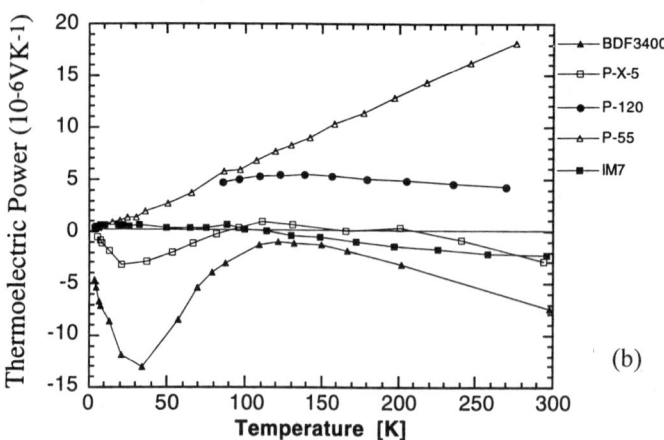

Figure 14 Temperature dependence of the thermoelectric power of different types of carbon fibers with varying structural perfection. Figure 14a is relative to the E-series pitch-based carbon fibers [42], while Figure 14b is relative to other fibers of various origins and precursors [42]. BDF 3400 is a benzene-derived fiber of high structural perfection heat treated at 3400°C and the P-X-5 is an experimental pitch-based fiber from Amoco.

Electrical and Thermal Transport in Carbon Fibers 411

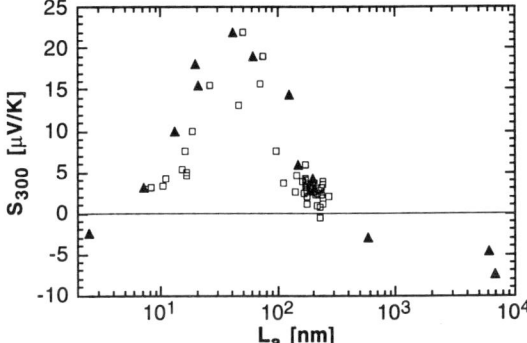

Figure 15 Room temperature thermoelectric power (S_{300}) of a large number of experimental pitch-derived carbon fibers as a function of the in-plane coherence length (L_a) [42].

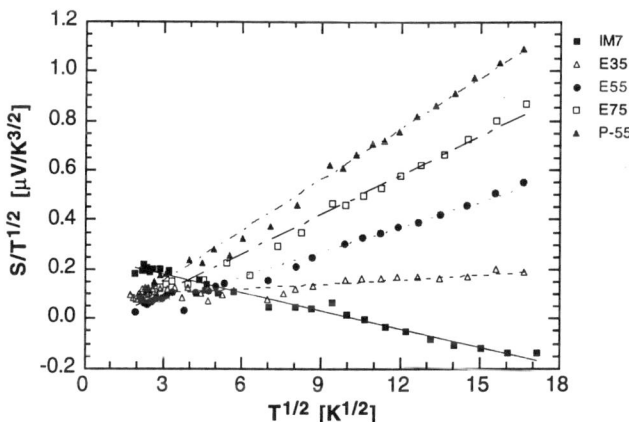

Figure 16 $S/T^{1/2}$ plotted versus $T^{1/2}$ for the most disordered fibers ($k_F l = 1$) measured by Nysten [42] and which temperature variation of the thermoelectric powers are presented in Figure 14. The straight lines are the result of linear regressions made to determine a and b (see text).

Table 5 Coefficients a and b and $\{d \ln g (\varepsilon_F) / d\varepsilon\}_1$ for a Set of Fibers Whose Thermoelectric Powers have been Analyzed (see text)

Fibers	a [μVK^{-2}]	b [$\mu VK^{-3/2}$]	$\{d \ln g (\varepsilon_F) / d\varepsilon\}_1 [eV^{-1}]$
IM7	−0.0244	0.255	1.00
E35	0.0061	0.087	−0.25
E55	0.0334	−0.027	−1.37
E75	0.0525	−0.048	−2.15
P-55	0.0680	−0.047	−2.78

6.5 INTERCALATED FIBERS

6.5.1 Introduction

A large amount of experimental data has been reported in the last decade on the temperature variation of the electrical resistivity and to a lesser extent on the thermal conductivity and thermoelectric power of intercalated carbon fibers of various origins. These have been recently reviewed [14,18].

The main effect of intercalation is an increase in the carrier density resulting from charge transfer which is accompanied by a decrease in the electronic mobility, but generally in a smaller relative amount. The net result of intercalation is thus an increase in electrical conductivity.

We have shown above that for in-plane conduction pristine HOPG and well ordered fibers along with diamond are the best heat conductors around room temperature. This is also the case for some pristine fibers of good crystalline perfection. We will consider here how intercalation can modify this property.

The work on the transport properties of graphite intercalation compounds was initiated by the promise of realizing electrical conductors with conductivities that could reach or even exceed that of copper. Indeed, as a result of charge transfer, one can increase significantly the electrical conductivity of graphitic materials by intercalation of various species [19,83]. In Figure 17 we present the effect of intercalation on the temperature variation of the electrical resistivity of a benzene-derived VGCF (BDF) heat treated at 2900°C [84]. There is a dramatic decrease in resistivity due to intercalation either for a donor (Rb) or an acceptor ($FeCl_3$) intercalate. The resulting intercalation compounds exhibit typical metallic behaviors with room temperature electrical resistivities 3 to 4 times that of pure copper.

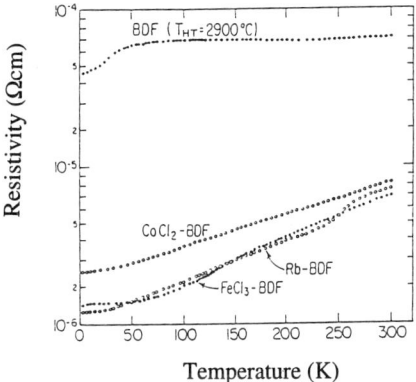

Figure 17 Effect of intercalation by a donor (Rb) and acceptor ($FeCl_3$) intercalate on the temperature variation of the electrical resistivity of a benzene-derived VGCF heat treated at 2900°C [84].

As it is the case for metallic systems, the electrons and holes in acceptor graphite intercalation compounds are described by their Fermi surfaces. Since acceptor GICs are highly anisotropic electronic systems, their charge carriers may be considered as 2D hole gases and their Fermi surfaces consists in circles for stage-1 compounds and in cylinders for higher stage compounds [85]. For donor compounds, the situation is more complicated, since the anisotropy may vary considerably from one type of compound to another, and from one stage to another for a given compound. For this reason, in spite of the important amount of theoretical investigations relative to the band structure of donor compounds, the situation is not as clear as for acceptor compounds with regard to transport properties.

As a result of the anisotropic band structure, the electrical resistivity of acceptor GICs is also highly anisotropic. For some acceptor GICs, the ratio of the in-plane conductivity to that along the c-axis may reach six orders of magnitude at room temperature [83]. This allows to consider the hole system in-plane as a quasi two-dimensional (2D) gas. Since it is partly dependent on the phonon spectrum, which is less anisotropic than the electron distribution, the thermal conductivity is also less anisotropic than the electrical resistivity. On the other hand, the anisotropy of the thermoelectric power is very small compared to that of the other transport properties [14,86].

We shall mainly be concerned here with low-stage acceptor compounds since most of the reported transport data on fibers are relative to these types of compounds. Also, for acceptor GICs, we may rely on a 2D band structure [85] for which quantitative data are available. As long as graphene planes host the carriers, the same 2D model may be applied for all acceptor compounds and the value of the Fermi level is directly related to the magnitude of the charge transfer.

6.5.2 Electrical Resistivity

6.5.2.1 Boltzmann resistivity
In order to interpret the in-plane resistivity data in acceptor GICs, one considers a 2D hole gas in the graphene layers (host layers) originating from the charge transfer from the intercalate. This 2D hole gas is fully described once its Fermi energy is known. This hole gas interacts with phonons and defects in the host layers causing the resistivity. In the lowest temperature range hole-hole interaction may also take place in the presence of weak disorder.

For 3D metals, when the *ideal electrical resistivity* is due to electron-phonon scattering, the Bloch-Grüneisen relation [74] predicts a T^n variation of resistivity with n = 5 at very low temperature with a gradual decrease in n with increasing temperature until n = 1 around and above the Debye temperature. From earlier experimental data on GICs, the temperature dependence of the ideal resistance in GICs was found instead to fit the relation [14,83]:

$$R(T) = BT + CT^2 \qquad (35)$$

where B and C are constants. The values of these constants may vary according to the compound considered.

In acceptor GICs the electrical resistivity depends on the charge transfer from the intercalate, with determines the carrier density, N. The ideal resistivity is related in addition to the phonon spectrum of the host material [16,43], on which depends the electron-phonon relaxation time. The amount of charge transfer will vary according to the nature of the intercalated species and to the stage of the compound. When the host material is not too disordered, the phonon dispersion relations do not vary significantly from one compound to another. Thus, the differences in ideal resistivities for various fibrous GICs samples with different host materials may be entirely ascribed to the Fermi energy as long as the same 2D band model is assumed for all acceptor GICs. Concerning the case of metal chlorides GICs, since their Fermi energies are almost the same for a given stage and do not vary significantly from one stage to the other for lower stages. The ideal resistivities are not expected to be much different from one chloride to another [87–90]. This is what can be observed as presented in Figure 18 for various intercalates and host materials.

It may also be seen from Figure 18 that in order to extract the ideal resistivity from the total measured resistivity high resolution measurements are required. Measurements with resolutions better than one part per 10^5 in the lowest temperature range are needed. The residual resistivity, ρ_r, is first determined by extrapolating to 0K the low temperature experimental values of the total Boltzmann resistivity, ρ. Then, using Mattiessen's rule (1), the temperature dependence of the ideal resistivity, ρ_i is computed.

Though, as will be shown below (Section 6.5.3), hole-hole interaction should be important in the lowest temperature range, the main scattering mechanism for charge carriers, which determines the ideal resistivity at high temperature, is that by acoustical phonons. Let us consider now how this works for GICs.

We have discussed in Section 6.2.5 the electron-phonon interaction in pristine carbons and graphites and concluded that, contrary to 3D metals, the charge carriers interact with subthermal phonons except at very low temperatures. In the case of acceptor GICs also the 2D hole system do not interact with the same class of phonons than in 3D metals around room temperature. For the 2D electronic systems of stages 1 and 2 acceptor GICs, the holes are scattered through decreasing angles with decreasing temperature, by low energy ($\sim k_B T$) thermal phonons below θ^* (see Equation 11). Above θ^*, the holes are scattered by subthermal phonons through large angles. However, since the Fermi surfaces are much larger in GICs than in the pristine material, the interacting phonons at a given temperature in GICs are of higher energy than those in the pristine material. Since higher energy phonons are more numerous at a given temperature than lower energy ones, this should lead to a higher scattering probability and thus to a lower ideal mobility in the GICs than in the pristine material.

Taking the velocity of sound of graphite, $v_s \sim 2.1 \times 10^6$ cm/s in-plane, and for metal chloride GICs, $k_F \sim 2 \times 10^7$ cm^{-1} and using Equation (11) one obtains for θ^* a value around 540K. The Fermi wave vector in acceptor compounds is of a smaller magnitude than in ordinary metals. However, the velocity of sound is one magnitude higher in the graphene planes. The result is that, in GICs, θ^* is accidentally of the same order than in metals. However, in metals θ^* corresponds to θ_D, the Debye temperature, while in acceptor GICs θ^* is roughly equal to $\theta_D/4$.

The electron-phonon interaction is thus weaker in GICs than in 3D metals at a given temperature. Though the scattering angle is not too different in both cases at a given temperature, the number of interacting phonons is smaller in acceptor GICs, since we are well below the in-plane Debye temperature. Thus, for acceptor GICs, at low and high temperatures, scattering should be less effective than in 3D metals, but significantly more effective than in the pristine material.

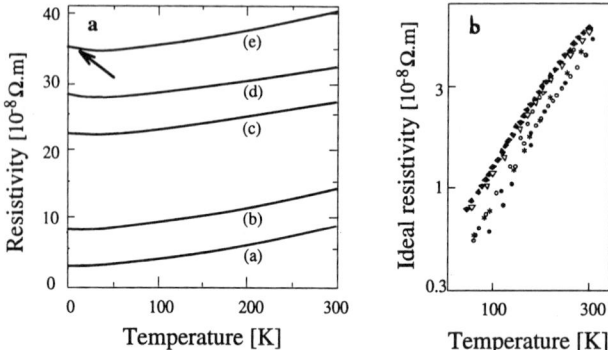

Figure 18 Temperature dependence of the as measured total (a) and ideal (b) electrical resistivities of various low stage fibrous acceptor GICs. The resistivities are given as the equivalent of the 3D electrical resistivity, and not in 2D. The data in Figure 1a are relative to a $CuCl_2$ intercalate for different host materials: a) BDF, n = 2, b) PX5, n = 4, c) VSC25, n = 1–2 (mainly 2), d) P55 (n = 3) and e) P100-4, n = 1–2 (mainly 2). The ideal resistivities are almost the same in spite of the large differences in total resistivities as shown in Figure 1a, (i.e., whatever the stage or host material). Note also that the temperature dependent part of the resistivity is a tiny fraction of the total resistivity in most of the compounds. For the ideal contribution (b) an almost linear variation is observed in the lowest temperature range and an almost T^2 behavior around room temperature.

Results of measurements performed on low stages fibrous acceptor GICs verified relation (6.35) in the temperature range 1.5 < T < 300°K [20,24]. An almost linear variation of the ideal resistivity was observed in the low temperature range and an almost T^2 behavior around room temperature for the various samples investigated (Figure 18). Oddly enough, contrary to what is observed in a 3D metal and predicted by the Bloch-Grüneisen relation, a higher power law is found at higher temperature. This indicates that electron-phonon interactions are probably not the dominant mechanism in the lowest temperature range.

On the other hand, the relaxation time for 2D hole-hole interactions instead is expected to vary linearly at low temperatures in the presence of weak disorder. The experimental findings concerning the low temperature behavior of these compounds suggest indeed that strong hole-hole interactions might be effective at low temperatures (cfr.§6.5.2.2).

Electrical and Thermal Transport in Carbon Fibers 417

The *residual resistivity* of acceptor GICs, is mainly determined by the hole-defect scattering in the graphene layers where conduction takes place. The defects that we consider here are those that were initially in the pristine host material to which were added those possibly introduced during the intercalation process. Scattering from the defects of both origins will combine (Equation 1') provided the mean free paths associated to the two scattering processes are comparable in magnitude. The residual resistivity is then expected to vary depending on the host material and on the intercalate species. As we already stated in Section 6.2.2, in pristine fibers the defect structure varies widely according to the type of fiber, heat treatment temperature and quality of the precursor for a given type of fiber. Thus, carbon fibers provide a large variety of host defect structures in intercalation compounds.

It is worth noting that the intercalation process in acceptor GICs occurs through the Daumas-Hérold [91] mechanism for stages equal and higher than two. The additional 2D defects, which constitute the Daumas-Hérold domain boundaries, introduced during intercalation are thus inherent to the process and cannot be avoided.

6.5.2.2 2-D weak-localization and interaction effects
It was shown that fibrous acceptor GICs are among the best candidates to investigate 2D weak localization and interaction effects [14]. Acceptor GICs, and especially stage-1 compounds which have a circular Fermi surface, are natural 2D electronic systems [14]. This 2D behavior results from the distribution of the charge carriers, which are strongly localized in the graphene planes and which may be considered as quasi free carriers—though weakly localized—only for motion along these planes. This leads to a 2D electronic band structure. It was also emphasized that the possibility of varying the defect structure of the host material over wide ranges in acceptor GICs allows large experimental possibilities for investigating the phenomena of weak localization and electron-electron interactions. Also, the Fermi level may be modified by varying the nature of the intercalate and its concentration.

In low stage acceptor graphite fiber intercalation compounds high resolution low temperature resistivity measurements performed by Piraux and co-workers [25–27] displayed logarithmic increases of the resistivity with decreasing temperature for different host structures and intercalates. In most cases, 0.1% to 1% increases in electrical resistivities were detected over a decade of temperature in the liquid helium range (Figure 19). Except in the case of fluorine intercalation compounds (Figure 20), the increase in resistivity was thus found to be usually small above 1.5°K, the lowest temperature at which measurements were performed. The better the samples are, the smaller was the increase in resistance due to localization (Figure 19). This increase in zero-field resistivity was accompanied by a negative magnetoresistance.

A logarithmic increase in resistivity was displayed by most fibrous acceptor GICs investigated, whatever the host material or intercalate. The temperature at which the mimimum occured in the ρ(T) curve was found to increase with the value of the residual resistivity, ρ_r (Figure 21). This is consistent with what is predicted from the weak localization theory. The most spectacular effects of the localization and Coulomb interaction effects were indeed observed on fiber-based compounds of low crystalline perfection. This is the case for fluorine compounds (Figure 20) where intercalation may lead to significant distortion of the graphene layers [92].

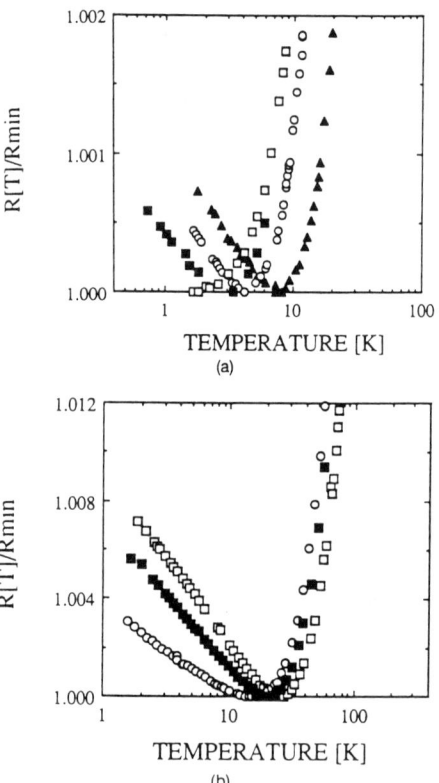

Figure 19 Low temperature dependence of the resistance for low-stage fibrous acceptor GICs, with various hosts and intercalates, showing the logarithmic increase in resistivity with decreasing temperature, charateristic of localization and electron-electron interaction effects. All data are normalized to the minimum value of the resistance [25].

Electrical and Thermal Transport in Carbon Fibers

The case of fibrous fluorine compounds is indeed particularly interesting [92]. By varying the fluorine content one observes a transition from weak to strong localization (Figure 22). Also, at a given temperature, starting from semimetallic graphite with quasi 2D electron and hole gases, increasing the fluorine percentage leads to a metallic 2D hole gas, and ends with a 3D insulator. This is a unique situation where with two constituents, one may achieve such profound modifications [15,92].

Results as those shown in Figure 19 were quantitatively interpreted in the frame of weak localization and electron-electron Coulomb interaction theories for 2D weakly disordered systems [27]. The relation between the magnitude of the effect and the temperature at which the resistance minimum occurs to the residual resistivity of the compound was compared to that observed experimentally (Figure 21). Also, the values of the temperature dependent inelastic scattering time obtained from the temperature and field dependence of the magnetoresistance were found to be in good agreement with those predicted by both theories of electron-electron Coulomb interaction and electron-phonon interaction in 2D disordered systems [27].

Since a uniform perpendicular magnetic field, H, introduces a phase shift between interfering partial-carrier-waves, the magnitude of the weak localization effect should decrease when the characteristic magnetic time, τ_H (Equation 10) becomes comparable or smaller than the dominant scattering time, τ_k. In Equation (10) $D = 1/2(v_F 2\tau_0)$ is the 2D diffusion constant, and $v_F \sim 8 \times 10^5$ m/s in low stage acceptor GICs, is the Fermi velocity. Since the Coulomb interaction is not affected when a low magnetic field is applied [32,93], the measurement of the magnetoresistance at various temperatures allows to estimate the relative contribution to the resistance of both effects at a given temperature. Such an analysis of the results have shown [27] that hole-hole interaction contribute 65 to 85% of the additional resistance to the Boltzmann term and is thus the dominant quantum mechanism at low temperature.

Altshuler et al. [94] derived an expression for the scattering time τ_i due to electron-electron interaction in a 2D electron gas:

$$\left(\frac{1}{\tau_i}\right)_{ee} = \frac{k_B T}{2\varepsilon_F \tau_0} \ln\left(\frac{\varepsilon_F \tau_0}{\hbar}\right) \tag{36}$$

For acceptor GICs, using $\tau_0 = l_0/v_F \cong 2 \times 10^{-14}$ s and $\varepsilon_F \cong 0.8$ eV, a value of $(1/\tau_i)_{ee} \cong 3.8 \times 10^9$ T is obtained. This was found to be in good agreement with the values obtained from the analysis of other experimental data [27].

On the other hand, Takayama [95] derived an expression for the electron-phonon lifetime in the dirty limit:

$$\left(\frac{1}{\tau_i}\right)_{ep} = \frac{2\pi^2 a k_B}{k_F l_0 \hbar \theta_D} T^2 \tag{37}$$

where a is a numerical factor of the order of unity. In the case of acceptor GICs, taking for θ_D the value of $\theta^* = 540°K$, it is found from relation (6.28) that $(1/\tau_i)_{ep}$ ~ $1.5 \times 10^8 T^2$. The relation between this value of τ_i and the one that may be obtained from ideal resistivity measurements is discussed in reference [27].

In brief, weak localization effects have been observed in all intercalated pitch-derived carbon fibers whose electrical resistivity have been measured at low temperature with the required resolution. They were also observed in intercalated vapor deposited fibers and HOPGs of low structural perfection. The compounds investigated were found to verify the weak disorder condition $k_F l \gg 1$. The temperature dependence of the resistivity increase was found consistent with the 2D electronic structure and the results obtained concerning the temperature and magnetic field dependences of the effect were found to fit the theoretical predictions [27].

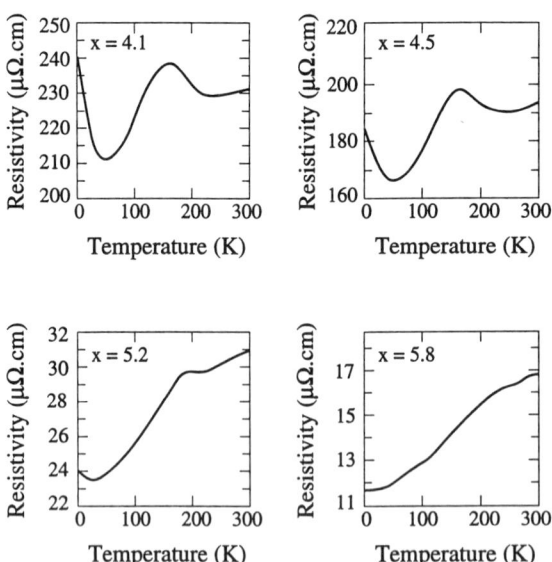

Figure 20 Temperature dependence of the resistivity for four C_xF samples with x = 4.1, 4.5, 5.2, and 5.8 showing weak localization and Coulomb interaction effects [92].

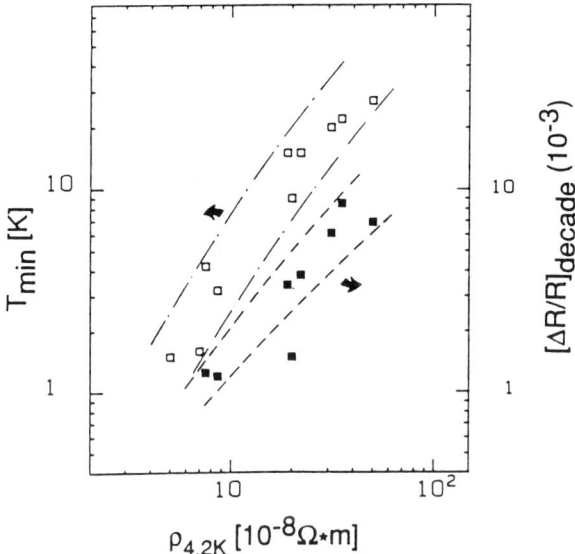

Figure 21 Relation between the value of the residual resistivity $\rho_{4.2K}$ of carbon fibers intercalated with metal chlorides and the temperature at which the resistivity is minimum, T_{min}, (left scale) on one hand, and the logarithmic variation of resistance over a decade of temperature $[\Delta R/R]_{decade}$ (right scale) [25].

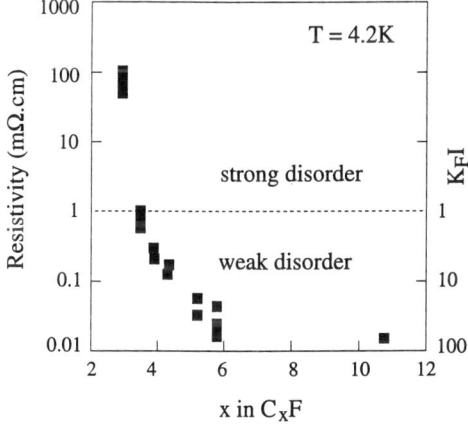

Figure 22 Electrical resistivity at 4.2°K for various fluorine GICs showing the effect of the transition from weak to strong disorder by varying the fluorine content [15].

6.5.2.3 *Electrical conductivity and charge transfer*

Since it was announced that some acceptor GICs may exhibit electrical conductivities higher than that of copper, a question which was often raised was what could be the highest electrical conductivity which could be attained with GICs. A survey of the data in the literature relative to the temperature variation of the electrical resistivities of various metallic chlorides GICs allow to answer this question for these compounds [14,15].

In the 2D Blinowski-Rigaux model [85], N is proportional to k_F^2 and m*, the effective mass varies linearly with k_F [43]. Thus, from Equation (4), the 2D electrical conductivity may be expressed as:

$$\sigma_{2D} \approx k_F \cdot \tau(k_F) \tag{38}$$

and since the relaxation time, $\tau(k_F)$, is expected to decrease with increasing wave number or carrier density:

$$\tau(k_F) \approx k_F^{-a} \tag{39}$$

One may see from Equations (38) and (39) that, if the magnitude of a is larger than 1, the electrical conductivity decreases when the Fermi wave vector or carrier density increases. Since around room temperature one should expect a value of a equal or higher than 1, the conductivity should be constant or decrease with increasing charge transfer in the range where the dispersion relation is linear in k . If only scattering by in-plane graphitic phonons is considered, $a = 1$ [96,97], then $\tau(k_F) \sim k_F^{-1}$, and in this case one should expect a conductivity independent of charge transfer.

From the above considerations and from the results presented in Figure 18 it may be concluded that the room temperature electrical resistivity of metallic chloride GICs could never be less than roughly 5×10^{-6} Ωcm, the value of the intrinsic resistivity at this temperature. In practice, one should add to this value that of the residual resistivity, which depends on defects, some of them being inherent to the intercalation process.

6.5.3 Thermal Conductivity

6.5.3.1 *Introduction*

We have already discussed in Section 6.2 the various mechanisms which contribute to the thermal conductivity of solids and discussed the case of pristine carbons and graphites. Let us consider now the effect of intercalation on the thermal conductivity of fibers.

We have previously shown how intercalation modifies the thermal conductivity of graphites (Figure 23) [14,43,58,98]. In contrast to the pristine material, electronic conduction in GICs may contribute to heat transport well above the liquid helium temperature range [14,43,58,99]. Indeed, one expects that the large increase in the charge carrier concentration in GICs due to charge transfer should give rise to an electronic contribution, especially at low temperatures where the lattice thermal conductivity of the pristine material decreases almost quadratically with temperature. Also, the fact that intercalation introduces lattice defects leads to a decrease of the lattice thermal conductivity around the maximum. The overall effect is a decrease of the total thermal conductivity at high temperature and an increase at low temperature with respect to that of the pristine material [14,43,58,98]. A schematic representation of the effect of intercalation on the thermal conductivity of pristine graphite is presented in Figure 23. [14].

Contrary to the case of intercalated HOPG, where the thermal conductivity of both acceptor [58,100–105] and donor [58,106] compounds have been investigated, for intercalated fibers, only a few acceptor compounds have been studied [92,107,108]. The results obtained with HOPG and fibers as hosts were found to be qualitatively the same. We shall present the results obtained on intercalated fibers below.

6.5.3.2 Electronic and lattice thermal conductivities
From inspection of either Equation (16) or of the kinetic Equation (17), since the charge transfer accompanying the intercalation process leads to a large increase in carrier density, we should expect to observe, as in the case of electrical conductivity, an increase in the electronic thermal conductivity. In most fibrous acceptor GICs the residual resistivity contributes significantly to the total resistivity, even at high temperature. We should also expect in fibrous GICS that the linear increase in the temperature dependence of the electronic thermal conductivity be followed by a temperature insensitive κ_E without any intermediate thermal conductivity peak [57].

Concerning the lattice thermal conductivity, we have seen above that, from the analysis of the lowest temperature region and that around the maximum, κ_{max}, one may get information on lattice defects. Such an analysis was also carried for GICs [58], and the results are discussed below.

In order to analyze quantitatively the low temperature thermal conductivity results of GICs and to determine the various phonon scattering mechanisms, we must take into account all the resistive mechanisms, including those specific to the intercalation process. These include scattering with lattice defects, point defects and boundary scattering, and phonon-phonon scattering. Phonons may experience umklapp scattering and normal processes. Though they are not direct resistive processes, normal processes lead to the creation of higher frequency phonons, which are more apt to undergo umklapp and point defect scattering, which are both resistive processes. Besides, in GICs, one should take into account the additional low frequency phonon modes introduced by intercalation [17].

Large scale lattice defects are those which sizes are much larger than the phonon wavelength, such as crystal boundaries or Daumas-Hérold domains. Scattering by these defects lead to relaxation times which are independent of the phonon frequencies:

$$\tau_B = \frac{L}{v_s} \tag{40}$$

where the boundary scattering length L is almost equal to the distance between such defects.

Point defects are those which sizes are small compared to the phonon wavelength. The relaxation times for this type of scattering vary rapidly with the phonon frequencies ω:

$$\tau_D = D\omega^r \tag{41}$$

where D is a constant and r = 4 for 3D solids [57] and 3 for 2D solids [109].

For phonon normal and umklapp processes, the relaxation times, τ_N and τ_U, respectively, are both frequency and temperature dependent [57].

Callaway [110] expressed the lattice thermal conductivity k_L as a sum of two terms in order to take into account in a quantitative way the resistive and non resistive phonon relaxation mechanisms:

$$\kappa_L = \kappa_1 + \kappa_2 \tag{42}$$

where κ_1 considers all relaxation mechanisms, including normal processes, as resistive mechanisms and κ_2 is a correction term which takes into account the fact that normal processes do not contribute directly to the thermal resistance.

The relaxation frequency τ_R^{-1}, which takes into account all resistive processes, is then:

$$\tau_R^{-1} = \tau_U^{-1} + \tau_B^{-1} + \tau_D^{-1} \tag{43}$$

and the relaxation frequency which takes into account both resistive and normal processes is:

$$\tau_C^{-1} = \tau_N^{-1} + \tau_R^{-1} \tag{44}$$

Using the Callaway formalism [110] and taking into account the pronounced anisotropy of phonon conduction in GICs, an analysis of the thermal conductivity of HOPG-FeCl$_3$ compounds was carried out [58]. This allowed the estimation of the size of the point defects and of the large scale defects in these compounds [58]. The same kind of analysis was made to determine the boundary scattering lengths in fibrous compounds.

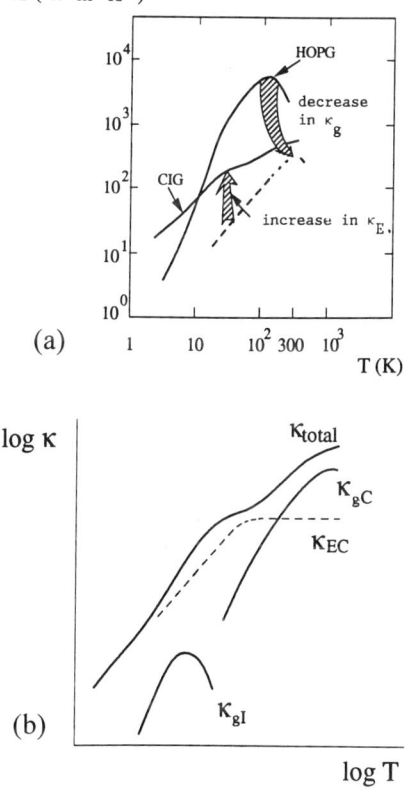

Figure 23 (a) Schematic representation of the effect of intercalation on the thermal conductivity of pristine graphites. Intercalation decreases the lattice thermal conductivity and increases the electronic thermal conductivity. The overall effect is a decrease of the thermal conductivity at high temperature and an increase at low temperature with respect to the pristine material. (b) Schematic representation on a log-log plot of the contribution of the graphene layers; electronic, k_{EC}, lattice, k_{gC}, and due to the intercalate phonons, k_{gI}.

6.5.3.3 Separation of the electronic and lattice contributions
There are two methods for separating the electronic and lattice thermal conductivities when they are of comparable magnitude. The first consists of measuring simultaneously the same sample on the electrical and thermal conductivities. Then, using the Wiedemann-Franz law (Equation 16), one may compute from the measured electrical resistivity the corresponding electronic thermal conductivity. By subtracting the latter from the total measured thermal conductivity (Equation 15), one obtains the lattice thermal conductivity for each temperature. This method is only applicable in the temperature range where the Wiedemann-Franz ratio is equal to L_o, the free electron Lorenz number.

Fibrous acceptor GICs have high residual resistivities which remain a large part of the total resistivity up to relatively high temperatures (Figure 18). Thus the validity of the Wiedemann-Franz law is expected to be maintained up to temperatures higher than those for pure 3D metals. This is particularily true for pitch-derived fibers of poor structural perfection, where the electrical resistivity is very weakly temperature dependent (Figure 18). In that case the electronic thermal conductivity can be directly computed over a large temperature range from the measured electrical resistivity via the Wiedemann-Franz law using the free electron Lorenz number [107,108]. For a P100-4-$CuCl_2$ compound, the electronic contribution may be computed in this way from liquid helium temperature up to room temperature and up to 130°K for a P × 5-$CuCl_2$ compound, where the host material is of higher structural perfection [111]. At these limiting temperatures, the ideal resistivity is less than 20% of the residual resisitivity and this percentage decreases as the temperature is lowered.

Concerning the lattice thermal conductivity, if we compare the results obtained at high temperature on HOPG, benzene and pitch derived carbon fibers, we see that the relative decrease in the lattice thermal conductivity due to intercalation is less pronounced as the lattice perfection of the pristine material decreases. As a result, while in intercalated pitch-based fibers of low quality the lattice contribution dominates the thermal conductivity above liquid nitrogen temperature, both electronic and lattice contributions are comparable in intercalated BDF and HOPG.

So, the fibers present two major advantages with respect to HOPG when we need to separate the two contributions to the thermal conductivity. First, one can measure electrical and thermal conductivities on the same sample, and second the Wiedemann-Franz law applies over much wider temperature ranges.

The second method of separation of κ into κ_E and κ_L is through the application of a large magnetic field which will deplete the electronic thermal conductivity so that $\kappa_E(H)$ may become negligible in comparison with κ_L at high magnetic fields where $\kappa(H)$ saturates so that $\kappa(H) \approx \kappa_L$ and κ_E and κ_L may be separated. This method was used for the case of HOPG intercalation compounds [103].

6.5.3.4 Experimental results

In Figure 24 we present on a log-log plot the temperature dependence of the thermal conductivity of a benzene-derived carbon fiber (BDF) intercalated with $CuCl_2$ [108]. For comparison we have presented on the same figure the temperature dependence of the thermal conductivity of the pristine material heat treated at the same temperature (3,000°C). Starting from the liquid helium temperature range, the thermal conductivity increases with increasing temperature up to nearly 230K. Measurements above this temperature were not reported because the heat losses by radiation in the experimental system were too large compared to the conductance of the sample to allow accurate data to be taken. Using the Wiedemann-Franz relation, the total measured thermal conductivity was separated into its electronic and lattice contributions.

In Figure 25 we present on a log-log plot the temperature dependence of the thermal conductivity of $CuCl_2$ intercalated pitch-based carbon fibers [107]. Here again we have shown on the same figure the data relative to a pristine sample taken from the same batch as that of the host material of the intercalation compound. Above 80°K, the thermal conductivity of the intercalated compound is reduced with respect to that of the pristine material, while the situation is inverted below 80°K. Also, while the functional behavior above 80°K is not too different for the two samples above 80°K, this is not the case below this temperature. In fact, below 80K the behavior for the intercalated compound is qualitatively different than that of the pristine material. In intercalated fibers, the thermal conductivity decreases less rapidly with temperature than the pristine fibers which present a $T^{2.4}$ variation. Thus, below 25°K, we observe a thermal conductivity enhancement for the intercalation compound with respect to the pristine fibers. At still lower temperature, say around 5°K, we observe a steeper decrease, immediately followed by a quasi linear temperature dependence.

Finally, in Figure 26 we present on a log-log plot the temperature dependence of the thermal conductivity from 2.5 to 300°K of a fibrous sample of $C_{4.1}F$ [92]. Contrary to the results obtained on other low stage intercalation compounds, the electronic contribution to the thermal conductivity, calculated from the Wiedemann-Franz relation, is negligible in the entire temperature range investigated. Above 50°K, the temperature dependence of the thermal conductivity is much like the one observed in the pristine material. Piraux et al. [92] have estimated the phonon mean free path in the boundary scattering regime, using Kelly's model [60]. They found for their sample a phonon mean free path around 25 nm at 80°K. This value is about three times smaller than that estimated for the low-stage VGCF-$CuCl_2$ compound discussed above. For the fluorine compound, the mean free path for the holes is more than one order of magnitude smaller than the phonon mean free path, while for the VGCF-$CuCl_2$ compound the two mean free paths were comparable in magnitude [16]. In the lowest temperature range, there is a significant enhancement of the lattice thermal conductivity with respect to that of the pristine fiber. This result is similar to that obtained in the pitch-derived intercalation compound, which was presented above. The enhancement is also a consequence from a modification of the phonon low-frequency modes due to the additional modes of the intercalate which will be discussed in the next paragraph.

Thermal conductivity measurements performed on intercalated graphitic fibers confirmed the existence of a low-temperature extra phonon contribution to the thermal conductivity which is due to the intercalate [17,111]. This extra contribution was previously inferred from the first thermal conductivity results on a stage-2 HOPG-FeCl$_3$ intercalation compound [101]. As was shown in Figure 25, this effect was found to be more pronounced in pitch-based carbon fiber intercalation compounds [17,111]. Further studies have shown a decrease of this extra contribution with increasing stage for low stages [111]. This was justified by the fact that as the compound becomes more dilute there are less extra phonons per unit of volume due to the intercalate. For very dilute compounds, the situation was found to be qualitatively different since the relative contribution of the electronic term as well as the extra lattice term should become negligible with respect to that of the lattice thermal conductivity of the graphene planes and κ_L was only due to the graphite phonons. The above considerations should then only apply for low stages.

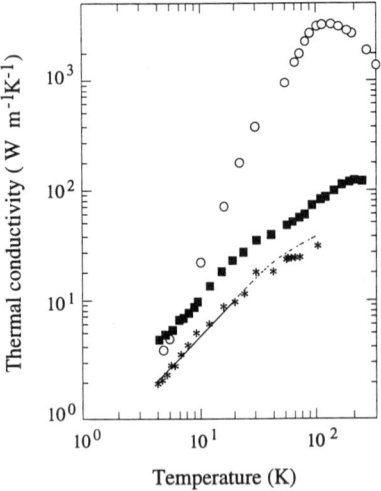

Figure 24 Temperature dependence of the thermal conductivity of CuCl$_2$ intercalated benzene-derived carbon fibers compared to that of pristine fibers (open circles) heat treated at the same temperature (3000°C). The total measured thermal conductivity of the intercalated sample (black circles), is separated into its electronic (curve), and lattice (crosses) contributions.

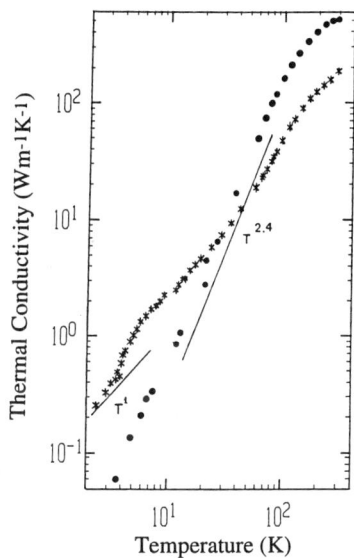

Figure 25 Temperature dependence of the thermal conductivity of $CuCl_2$ intercalated pitch-derived carbon fibers (✱) compared to that of pristine fibers (●) from the same batch.

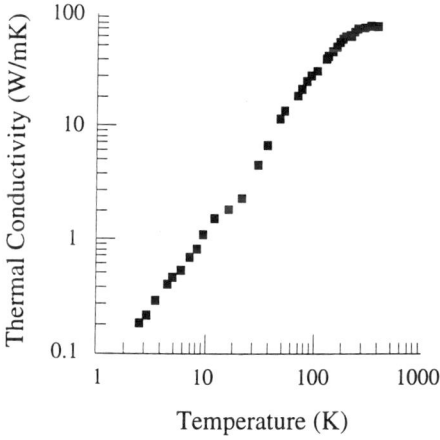

Figure 26 Temperature dependence of the thermal conductivity of a fibrous $C_{4.1}F$ sample.

6.5.4 Thermoelectric Power

In Figure 27 we present the temperature dependence of the thermoelectric power of the benzene-derived carbon fiber (BDF) intercalated with $CuCl_2$ [108], which thermal conductivity is presented in Figure 24 and discussed in Section 6.5.3.4. The thermoelectric power is positive over the entire temperature range, indicating that hole conduction dominates. The temperature dependence is typical of all low-stage GICs whether fibrous or bulk. It starts with a linear dependence in the lowest temperature range, then increases more rapidly, and tends to saturate around 200°K. Similarly, in Figure 28 we present the temperature dependence of the thermoelectric power of the pitch-based carbon fiber intercalated with $CuCl_2$ [107], which thermal conductivity is presented in Figure 25.

In Figure 29 we present the temperature dependence of the thermoelectric power from 2.5 to 300°K of the fibrous sample of $C_{4.1}F$ [92], which thermal conductivity is presented in Figure 26 and discussed in Section 6.6.3.4. Here again, the temperature dependence of the thermoelectric power is also quite similar to that previously observed in other low-stage acceptor GICs. In the lowest temperature range, we first observe an increase in thermoelectric power with temperature, followed by a broad maximum around 200°K. However, the fluorine compound displays a slightly lower thermoelectric power value at the maximum, 16 X $10^{-6} VK^{-1}$ compared to a value ranging from 20 to 30 × $10^{-6} VK^{-1}$ for other stage-2 compounds. In the lowest temperature range, an almost linear temperature variation is observed with a coefficient S/T equal to 3.5 × $10^{-8} VK^{-2}$. From this coefficient the Fermi energy was computed (see below) and was found to be in good agreement with values obtained by means of optical measurements. So, in spite of the relatively high electrical resistivity of the compound (230 × $10^{-6} Wcm$ at 300°K), the carrier density is comparable to that of other low-stage acceptor GICs. This confirms that fluorine compounds have low mobilities resulting from strong defect scattering.

Applying Equation (24) for a stage-1 acceptor compound and using the Blinowski-Rigaux model [85], we find for the diffusion thermoelectric power of holes [100]:

$$S_d = 2.45 \cdot 10^{-8} T(1+s)\varepsilon_F^{-1} \qquad (45)$$

where S_d is expressed in V/K and s is the scattering parameter, which expresses the energy dependence of the relaxation time:

$$\tau = \tau_0 \varepsilon^s \qquad (46)$$

If we assume defect scattering in the low temperature regime, with an energy independent relaxation time, as is the case for boundary scattering (s = 0 in the Blinowski-Rigaux model), and a Fermi energy of 1 eV, one obtains for a stage-1 compound:

$$S_d = 2.45 \cdot 10^{-8} T \tag{47}$$

Taking a Fermi energy $\varepsilon_F = 0.8$ eV and $\gamma_1 = 0.38$ eV, we obtain for the hole bands 1 and 2 [100] of a stage-2 compound a diffusion thermoelectric power:

$$S_{d1} = 2.45 \cdot 10^{-8} T(1.1 + 1.25 s_1) \tag{48}$$

$$S_{d2} = 2.45 \cdot 10^{-8} T(2.0 + 2.38 s_2) \tag{48'}$$

since scattering in these compounds is mainly on large scale defects at low temperature [14,58], one may assume that $s_1 = s_2 = 0$ in the residual resistivity range and thus that $S_{d1} = 2.7 \times 10^{-8}$ T and $S_{d2} = 4.9 \times 10^{-8}$ T. Equations (48) and (48') concern the partial thermoelectric powers, S_j, of one group of charge carriers (Section 6.5.1). For GICs of stages higher than stage-1, the total thermoelectric power is obtained by considering the different groups of carriers and computing the total thermoelectric power by means of Equation (23).

Since it was shown that $\sigma_1 \sim 2\sigma_2$ [100], combining Equations (23), (48), and (48') one finds a value for the total diffusion thermoelectric power of a stage-2 compound, S_d, equal to 3.5×10^{-8} T.

From the low temperature linear temperature dependence of the thermoelectric power of stage-2 compounds at low temperatures, Piraux and co-workers [100] have estimated the Fermi energies of a few compounds. Assuming that $s_j = 0$, they obtained $\varepsilon_F = 0.9$ eV, 0.55 eV and 0.40 eV for stage-2 HOPG-SbCl$_5$, BDF-CuCl$_2$, and P100-4-CuCl$_2$ intercalation compounds respectively. They attributed the lower values obtained for fibers with respect to HOPG to a lack of uniformity in the intercalation of the fibers and/or to a lower charge transfer from the CuCl$_2$ intercalate.

A thorough analysis of the data obtained on the thermoelectric power of various acceptor intercalation compounds lead to the conclusion that there is a dominant phonon drag contribution for low stage acceptor GICs in the high temperature range [14]. This is a unique situation, since such an effect was never observed in a 3D metal, which should be attributed to the particular 2D nature of acceptor GICs.

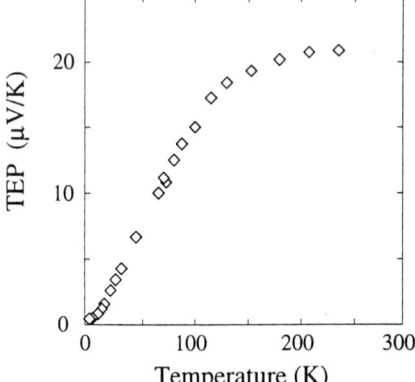

Figure 27 Temperature dependence of the thermoelectric power of CuCl$_2$ intercalated benzene-derived carbon fibers, which thermal conductivity is presented in Figure 24.

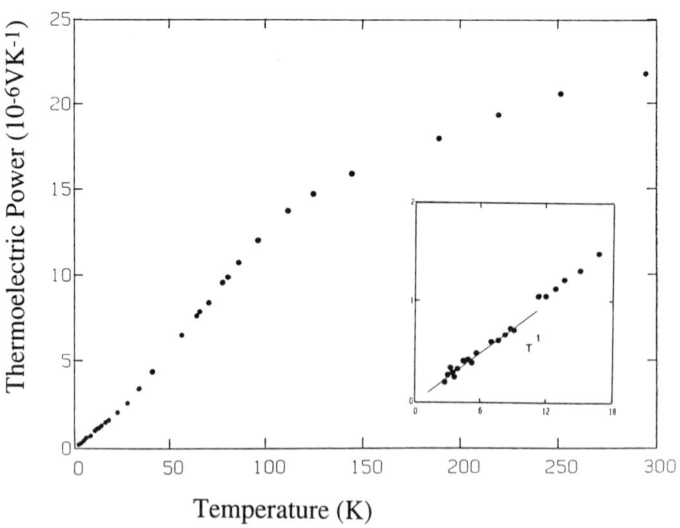

Figure 28 Temperature dependence of the thermoelectric power of CuCl$_2$ intercalated pitch-derived carbon fibers, which thermal conductivity is presented in Figure 25. The inset shows the lowest temperature linear variation.

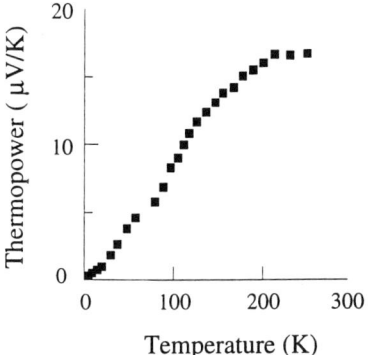

Figure 29 Temperature dependence of the thermoelectric power of the fibrous $C_{4.1}F$ sample which thermal conductivity is presented in Figure 26.

6.6 CARBON NANOTUBES

6.6.1 Introduction

With the discovery by Iijima [112] of carbon nanotubes, a new class carbon materials of reduced dimension has been introduced. The nanotubes consist of one or more seamless cylindrical shells of graphitic sheets. Each cylinder can be visualized as the conformal mapping of a 2D honeycomb lattice onto its surface. The structure of carbon nanotubes has been studied at the atomic scale using high-resolution transmission electron microscopy and scanning probe microscopy. Theoretical calculations predict that the nanotube diameter and helicity strongly influence its electronic properties [113–115]. As a result, a carbon nanotube can either be a metal, a semimetal, or an insulator, depending on the diameter and the degree of helicity.

Carbon nanotubes has been described by Iijima [112,116,117] and others [118,119]. Dresselhaus et al. [120] have recently reviewed their energy band structure calculations performed by a number of authors. The calculated density of states [115] shows singularities which are characteristic of one-dimensional systems. Also, contrary to the case of a rod-shaped quantum wire, for which the energy gap should scale with the inverse square of the diameter, the gap for the nanotubes that are semiconducting scales linearly with the inverse of the tube outer diameter [114,115].

The theoretical predictions concerning the electronic structure of carbon nanotubes are difficult to verify experimentally because of the strong dependence of the predicted properties on their diameter and chirality. This means that measurements should be made on individual, preferably single-wall nanotubes, that have been assigned the right diameter and chiral angle. Furthermore, nanotubes are often produced in bundles, so that obtaining data on single, well-characterized tubes is a very delicate performance. We will discuss below some recent experimental observations relative to the electrical resistivity and magnetoresistance of individual nanotubes or bundles of nanotubes.

6.6.2 Electrical Resistivity and Magnetoresistance

In order to test the theoretical predictions concerning the electronic properties of nanotubes by means of electrical resistivity measurements, one has to solve at least two delicate experimental problems:

- Realize a four-probe measurement on a single nanotube. This means that one has to attach four electrical connnexions on a sample of a few nm diameter and about a μm length. This requires the use of nanolithographic techniques [6,7].
- Characterize this sample with its contacts in order to determine its diameter and helicity.

The fractal-like organization of carbon nanotubes produced by classical carbon arc discharge suggested by Ebbesen et al. [119] lead to conductivity measurements which were performed at various scales. The results provided useful information on the electronic properties at different levels.

Ebbesen and Ajayan [118] measured a conductivity of the order of 10^{-2} Ωcm in the black core bulk material, inferring that the carbon arc deposit contains electrically conducting entities. A subsequent analysis of the temperature dependence of the electrical resistivity of similar bulk materials [121,122] revealed that the resistivities were strongly sample dependent.

Song et al. [123] performed a four-point resistivity measurement on a large bundle of carbon nanotubes of 60 μm diameter and 350 μm distance between the two voltage probes. They interpreted their resistivity, magnetoresistance, and Hall effect results in terms of semimetallic conduction and 2D weak localization like for the case of disordered turbostratic graphite.

Whitesides and Weisbecker [124] developed a technique to estimate the conductivity of single nanotubes by dispersing nanotubes onto lithographically defined gold contacts to realize a "nano-wire" circuit. From this 2-point resistance measurement and, after measuring the diameter of the single nanotubes by non-contact AFM, they estimated the room temperature electrical resistivity along the nanotube axis to be around 10^{-2} Ωcm.

Using submicronic lithographic patterning of gold films with a scanning tunneling microscope to attach two electrical contacts to a *single microbundle*, Langer et al. [6] have measured the electrical resistivity of a single microbundle of 50 nm diameter down to 0.3°K (Figure 30). Also, later on, a single nanotube has been measured, with three electrical contacts, by the same group down to 20 mK [7]. A room temperature electrical resistivity of $\approx 10^{-4}$ Ωcm was estimated for the single nanotube and of $\approx 10^{-3}$ Ωcm for the microbundle (Figure 31).

Above 2°K, the temperature dependence of the zero-field resistivity of the microbundle was found to be governed by the temperature dependence of the carrier densities and well described by the simple two-band (STB) model derived by Klein [36] for electrons n, and hole p, densities in semimetallic graphite:

$$n = C_n k_B T \ln\left[1 + \exp\left(\frac{\varepsilon_F}{k_B T}\right)\right] \tag{49}$$

and

$$p = C_p k_B T \ln\left[1 + \exp\left(\frac{\Delta - \varepsilon_F}{k_B T}\right)\right] \tag{49'}$$

where ε_F is the Fermi energy and Δ is the band overlap. C_n and C_p are the fitting parameters.

A value of $\Delta = 3.7$ meV was obtained for the band overlap, with the Fermi energy right in the middle of the overlap. This value of the overlap has to be compared to that of 40 meV for crystalline graphite. The difference was ascribed to the turbostratic stacking of the adjacent layers which should reduce drastically the interlayer interactions, like in disordered graphite. The smaller overlap implies also that, within the frame of the STB model, the carrier density is 10 times smaller than in graphite.

By applying a magnetic field normal to the tube axis of the microbundle, Langer et al. [6] observed a magnetoresistance (Figure 30) which, in contrast to the case of graphite, remains negative at all fields. The negative magnetoresistance was found consistent with the formation of a Landau level predicted by Ajiki and Ando [125]. This Landau level, which should lie at the crossing of the valence and conduction bands, increases the density of states at the Fermi level and hence lowers the resistance. Moreover, the theory predicts a magnetoresistance which is temperature independent at low temperature and decreasing in amplitude when k_BT becomes larger than the Landau level. This is also what was observed experimentally.

Below 2°K, an unexpected temperature dependence of the resistance and magnetoresistance was observed. As shown in Figure 30, the resistance presents, after an initial increase, a saturation or a broad maximum and an unexplained sharp drop when a magnetic field was applied.

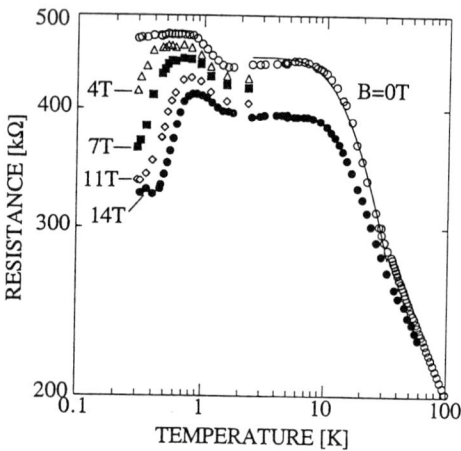

Figure 30 Electrical resistance of a microbundle of nanotubes of approximately 50 nm diameter versus temperature at various magnetic fields. The plain curve is calculated from a fit of Klein's 2D-band model for graphite with an overlap D = 3.7 meV and a Fermi level right in the middle of the overlap.

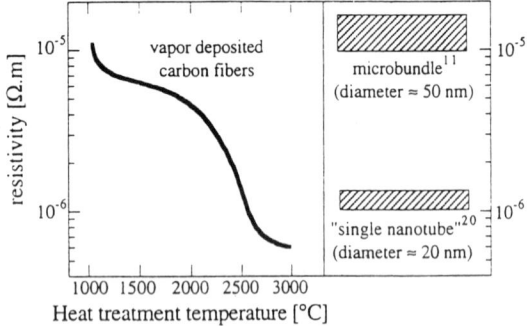

Figure 31 Estimated room-temperature electrical resistivity of carbon nanotubes samples compared to those of VGCF heat treated at different temperatures.

6.6.3 Quantum Transport

Langer et al. [7] performed also electrical resistance measurements on *individual multiwalled carbon nanotubes* down to very low temperatures and with a magnetic field applied perpendicular to the tube axis.

As for the case of the microbundle, the carbon nanotubes were synthesized using the standard carbon arc-discharge technique. Electrical gold contacts have been attached to the carbon nanotubes via local electron beam lithography with a scanning tunneling microscope (STM). The measured single multiwalled carbon nanotubes had a diameter of about 20 nm and a total length of the order of 1 µm.

In Figure 32 we present the temperature dependence of the conductance for one of the carbon nanotubes measured in respectively zero magnetic field, 7 T and 14 T. In zero magnetic field a logarithmic decrease of the conductance at higher temperature, followed by a saturation of the conductance at very low temperature, is observed. The saturation occurs at a critical temperature; $T_c = 0.3°K$ at zero magnetic field that shifts to higher temperatures in the presence of a magnetic field.

In the presence of a magnetic field normal to the tube axis, a significant increase in conductance appears, (i.e., a *positive magnetoconductance* or negative magnetoresistance). Both the temperature and field dependences of the nanotube conductance were interpreted consistently in the frame of the theory for 2D weak localization [7] that we discussed above. However, in the particular case of nanotubes one must take into account that, owing to the very small dimensions of the sample, we are close to the mesoscopic regime in the lowest temperature range [7]. This situation is responsible for the conductance fluctuations that we will discuss in the following paragraphs.

2D weak localization predicts that the resistance should be independent of magnetic field in the temperature range where it varies as lnT. This is not what is observed (Figure 32). The data in Figure 32 show that there is an additional contribution to the magnetoconductance of the carbon nanotube which is temperature independent up to the highest temperature investigated, including in the lnT variation range. This magnetoconductance was thus ascribed to the formation of Landau states which we discussed above for the case of the nanotube microbundle. Both 2D weak localization and "Landau level" contributions to the magnetoconductance can be separated as illustrated in Figure 32.

Typical magnetoconductance data are shown in Figure 33. At low temperature, reproducible aperiodic fluctuations appear in the magnetoconductance. The positions of the peaks and the valleys with respect to magnetic field are temperature independent. In Figure 34 we present the temperature dependence of the peak-to-peak amplitude of the conductance fluctuations for three selected peaks (see Figure 33) as well as the rms amplitude of the fluctuations, rms[ΔG]. It may be seen that the fluctuations have a constant amplitude at low temperature, which decreases slowly with increasing temperature with a weak power law at higher temperature. The turnover in the temperature dependence of the conductance fluctuations occurs at a critical temperature $T_c^* \approx 0.3°K$ [123] which, in contrast to the T_c [123] values discussed above, is independent of the magnetic field.

This behavior was found to be consistent with another quantum transport effect of universal character, the *universal conductance fluctuations (UCF)* [126,127]. Universal conductance fluctuations were previously observed in mesoscopic weakly disordered metals [128,129] and semiconductors [130,131] of various dimensionalities. In such systems, where the size of the sample, L, is smaller or comparable to both L_ϕ, the phase coherence length, and the thermal diffusion length $L_T = (\hbar D/k_B T)^{1/2}$, elastic scattering of electron wave functions generate an interference pattern which gives rise to a sample-specific, time-independent correction to the classical conductance [132]. The interference pattern, and hence correction to the conductance, can be modified by either applying a magnetic field or by changing the electron energy in order to tune the phase or the wavelength of the electrons, respectively [128–131]. The resulting phenomenon is called universal conductance fluctuations, because the amplitude of the fluctuations ΔG has a universal value: rms $[\Delta G] \approx q^2/h$ as long as the sample size $L < L_\phi, L_T$. When the relevant length scale, L_ϕ or L_T, becomes smaller than L, the amplitude of the observed fluctuations decreases due to self-averaging of the UCF in phase-coherent subunits. When the relevant length scale decreases with temperature, the amplitude of the fluctuations decreases as a weak power law: $T^{-\alpha}$ where α depends on dimensionality and limiting diffusion length, L_ϕ or L_T [132]. $\alpha = 1/2$ for a 2D system with $L_\phi \ll L, L_T$

The observed fluctuations in the field dependence of the conductance have therefore been interpreted in terms of universal conductance fluctuations for mesoscopic 2D systems.

So, in spite of the extremely small diameter of the nanotube compared to the de Broglie wavelengths of the charge carriers, the cylindrical structure of the honeycomb lattice gives rise to a 2D electron system for both weak localization and universal conductance fluctuation effects. Both the amplitude and the temperature dependence of the conductance fluctuations were found to be consistent with the universal conductance fluctuations models applied to the particular cylindrical structure of nanotubes [7].

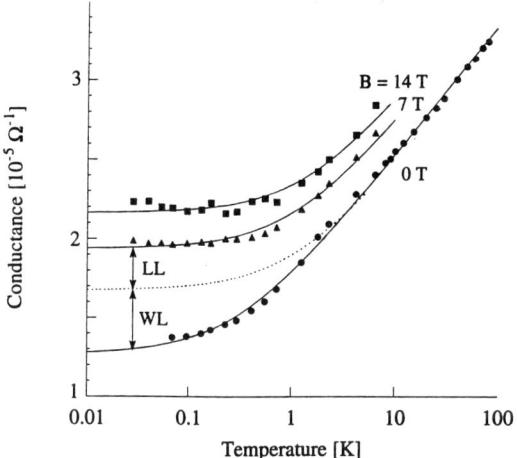

Figure 32 Electrical conductance as a function of temperature at the indicated magnetic fields. The dashed line separates the contributions to the magnetoconductance due to both the of Landau levels (LL) and weak localization (WL). The solid lines are theoretical fits.

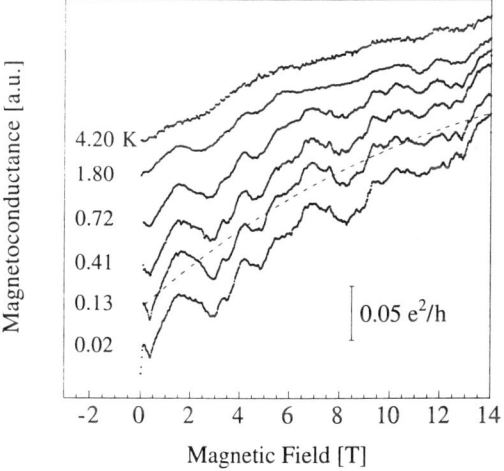

Figure 33 Magnetic field dependence of the conductance at different temperatures showing the aperiodic universal conductance fluctuations.

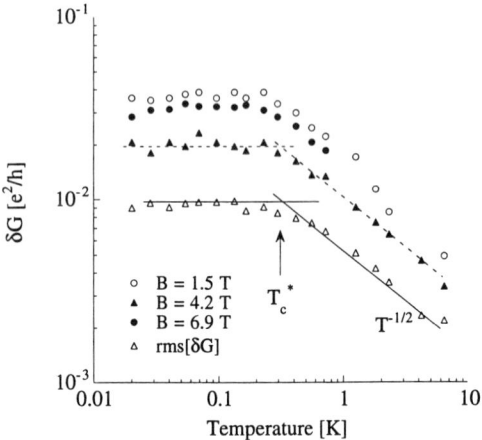

Figure 34 Temperature dependence of the amplitude of the periodic conductance fluctuations (δG) for the first three peaks selected in Figure 33 as well as rms [δG].

6.7 SAMPLE CHARACTERIZATION

6.7.1 Introduction

It was shown how the low temperature electrical resistivity, the so-called residual resistivity, reveals information about the lattice defects which are the dominant charge carriers scatterers at low temperature, while the interpretation of the temperature variation of the ideal electrical resistivity yields information about electron-phonon and electron-electron interactions. We have seen also that thermal conductivity measurements yield information about lattice defects and, in addition, about other phonon scattering events.

In this section we will show how by using the results of measurements of the electrical resistivity, the magnetoresistance, the thermal conductivity and the thermoelectric power, one may characterize and determine the defect structure of various types of carbon fibers, whether pristine or intercalated. This approach usefully complements the information obtained by means of more powerful techniques which probe the material at the microscopic level, such as SEM, high resolution TEM, x-ray diffraction, STM, and AFM. Contrary to these microscopic techniques, which are very localized and thus only probe a tiny portion of the sample, electrical and thermal transport data give an overall view of the entire macroscopic sample. In some cases transport data are also sensitive to defects which could not be detected by other techniques.

Though some data obtained on other types of carbon fibers may be used to illustrate a particular situation, we shall mainly concentrate here on pitch-derived carbon fibers. We have seen how thermal conductivity measurements performed in the liquid nitrogen temperature range allow to determine the in-plane coherence length in pristine fibers. Electrical resistivity measurements, though revealing very large differences for samples with different defect structures, are more delicate to analyze in detail in the case of pristine fibers. In principle, magnetoresistance measurements probe the mobilities, thus are essentially sensitive to the scattering mechanism. The thermoelectric power is very sensitive to the carrier densities, but though it depends on the nature of the scattering mechanism, it is not affected by its intensity. As it is the case for the electrical resistivity, the thermoelectric power data are delicate to interpret in pristine carbons and graphites.

Usually, the more complicated or delicate the method used, the most direct information it provides. Needless to say that, because of the particular shape of the cross sections and their lack of homogeneity, the methods that do not require the determination of the sample cross section should be preferred. This is the case for the magnetoresistance and the thermoelectric power.

6.7.2 Electrical Resistivity

Graphite, which is a typical semimetal, is among the materials which electrical resistivity is the most difficult to interpret in a quantitative manner. This is due to the fact that there are two types of carriers, electrons and holes, and that, because of their small densities, they are very sensitive to temperature. This is contrary to the case of metals where the carrier densities are temperature insensitive. It is thus easy for the case of metals to determine the temperature variation of the mobility by measuring that of the resistivity. If we are able to determine the carrier density by means of another measurement, i.e. the Hall effect, then we may determine the value of this mobility at any temperature. So, in the case of graphites a first complication stems for the fact that the carrier density varies with temperature. Also, since there are two types of carriers, we need to measure four parameters to have access to the mobility values.

On the other hand, in *pristine carbons and graphites* not only the mobilities, but also the carrier densities are very sensitive to defects. A difference in carrier concentrations for instance may be induced by a relatively small amount of defects or impurities and may cause dramatic changes in the electronic properties. This carrier concentration is also expected to be very sensitive to the 3D order. For example, crystal defects associated with layer disorder are known to operate as electron traps. Thus, their presence will cause a decrease of the electron density with respect to that of the holes. So, one should take into consideration not only the scattering mechanisms introduced by these defects but also their effect on the carrier concentrations and other electronic parameters, which have a direct impact on the electrical properties.

As a result, for a given situation, it is difficult to distinguish to what extent the mobilities or the carrier densities are influenced by the presence of defects in pristine carbons and graphites. So, when we compare two samples, the results of electrical resistivity measurements will only indicate whether there are differences in crystalline perfection between them, with no indication as to whether it should be attributed to the mobilies or the carrier densities.

Since the residual mobility is limited by scattering by static lattice defects, its value should provide in principle information about these defects, if the other electronic parameters are known. The static defects usually considered are point defects, such as impurities and vacancies, and linear defects such as dislocations. In 3D materials, surface defects such as sample boundaries for single crystals and grain boundaries for a polycrystalline material should also be considered. These are large-scale defects with respect to the carrier wavelengths. One case of practical importance is when large-scale defect scattering dominates, which is believed to be the case for graphites and its intercalation compounds in the liquid helium range. In that case the residual mean free path, l_r, is temperature independent and one may use the residual resistivity as a tool to determine the size of these defects, provided that the other electronic parameters are known. We have shown above (Section 6.3.4) that this is also the case for phonons, for which the mean free paths are of the order of the dimensions of the crystallites at low temperatures.

In pristine fibers of poor quality, the size of the crystallites is equal to a few tens of nm, whereas for high quality vapor deposited fibers, they are of the order of a few micrometers, almost like in HOPG.

Even in the simplest case of a dominant boundary scattering with a single type of carrier, the situation is not easy to analyze the electrical resistivity, when applied to pristine graphites. Inspecting Equation (5), we may see that we need to know about three electronic parameters, in addition to the mean free path, l:

- the density of charge carriers
- the effective mass of the carriers
- the velocity of the carriers

So, even for poorly graphitized fibers, where it is assumed that one type of carrier, (i.e., holes, and dominates), all three parameters will be different for different precursors or heat treatment. This is contrary to the case of phonons where, as we have seen above (Section 6.3), one needs to know the specific heat

Electrical and Thermal Transport in Carbon Fibers 443

and velocity of sound: two parameters that are not very different from one graphite to another.

For these reasons, we should consider for the electrical resistivity two temperature regions:

- the regions where the charge carrier densities are temperature independent. The limits of this temperature range vary from one type of graphite to another. The upper temperature limit should be lower for highly graphitized fibers, since they are more semimetallic, (i.e., they charge carrier densities are lower). Usually, below ~50°K we may assume that the charge carrier densities are constant
- the region where the charge carrier densities are temperature dependent (above ~50°K), which is a very delicate region to analyze since, in addition to l, v, m*, and N may all vary with temperature.

High resolution measurements in the liquid helium range may provide additional useful information too through localization effects. However, this effect which was recently discovered in pristine graphites needs to be further investigated first.

Thus, though the electrical resistivity is probably the parameter which is the easiest to measure and the most sensitive to defects in pristine fibers, it is difficult to state how these defects modify quantitatively the electrical resistivity leading to the very large differences which are observed on different samples.

For *GICs*, the static defects usually considered are also point defects and dislocations. Moreover, in a 2D material, a linear defect in the plane of the carrier propagation such as Daumas-Hérold domain boundaries should be considered as large-scale linear defects for charge carriers and are equivalent to a surface defect in 3D solids. This is also the case for phonons. When large-scale defect scattering dominates, which is most probably the case for all acceptor GICs of stage n ≥ 2 in the liquid helium range [14], the mean free path, l_r, is temperature and energy independent.

The residual resistivity was used as a tool to determine the size of large scale defects in GICs. It was shown [14,16,43] that for a stage-1 compound when $\rho_r \gg \rho_i$ the hole mean free path may be expressed:

$$l_r = \frac{3\pi\hbar b \gamma_0 I_c}{2q^2 \rho_r \varepsilon_F} \tag{50}$$

where b is the in-plane nearest neighbor distance, γ_0 the overlap energy, ε_F the Fermi energy, I_c the c-axis repeat distance and ρ_r the measured residual resistivity.

For a stage-2 compound in the residual range we have [43]:

$$l_r = \frac{I_c}{\rho_r} \cdot \frac{\pi \hbar}{q^2} \cdot \frac{1}{(k_{F1} + k_{F2})} \tag{51}$$

where k_{F1} and k_{F2} are the Fermi wave vectors in the hole bands 1 and 2 respectively [43,85].

In HOPG-acceptor GICs, ρ_r is of the order of 10^{-6} Ωcm, $I_c \sim 10^{-9}$m and $\varepsilon_F \sim$ 1eV. In that case, if we compute by means of relation (6.50) the mean free path for defect scattering, we find that it is of the order of a few 10^2 nm for a stage-1 compound, a value much smaller than in the pristine material. For intercalated BDFs heat treated at high temperature, which have a structure close to that of HOPG, the value of l_r should be smaller, but close to that obtained for intercalated HOPG. For most pitch-derived carbon fibers, whose structure is generally more disordered than that for HOPG and BDF, ρ_r is higher and l_r smaller. In the pristine material the mean free path is of the order of the crystallite size, which is smaller in PDF than in BDF or HOPG. The relative decrease of l_r after intercalation is less pronounced for PDFs. Thus, in addition to the defect structure of the pristine host material, one should take into account the defects introduced during the intercalation process. The relative effect of the defects introduced by intercalation is more important when the host material is of higher crystalline perfection. This results from the combination of the mean free paths for various scattering events as may be seen from Matthiessen's rule (Equation 1).

Equations (50) and (51) show that the measurement of the residual resistivity ρ_r and a knowledge of the Fermi wave-vector, or energy, allow a direct estimation of the size of the large-scale defects. So, in the case of GICs the situation is simpler than in the pristine material.

6.7.3 Magnetoresistance

The *"classica"* (Boltzmann) *magnetoresistance* essentially probes the carrier mobility. For the case of disordered systems, the *negative magnetoresistance* of quantum origin due to localization effects, which have been discussed in detail above, depends also on defects. So, for the purpose of comparative characterization at low temperature, we do not have to consider of the origin of the magnetoresistance, since both positive and negative magnetoresistances have a common origin: scattering of charge carriers by defects.

For the *"classica"* magnetoresistance, measurements could be performed at a fixed temperature, (i.e., 77°K, the liquid nitrogen boiling point at atmospheric pressure). 77°K is the ideal choice, since mobilities, and thus magnetoresistances, are higher than at room temperature. Also, liquid nitrogen can be easily handled.

Electrical and Thermal Transport in Carbon Fibers 445

One of the great advantages of using magnetoresistance as a characterization tool is that there is no need to determine the sample dimensions. In the simplest cases, mobilities may be directly estimated.

Concerning the "quantum" magnetoresistance, we have seen in Section 6.2.6 that the analysis, in the frame of the 2D weak localization theory, of the negative magnetoresistance of pristine fibers as a function of magnetic field at 4.2 and 77°K allows the determination of two essential parameters: the sheet resistance, R, and the characteristic field for scattering by magnetic impurities, H_s. We were able to show that the negative magnetoresistance resulting from the 2D weak localization was due to the contribution of the turbostratic layers. The strongest effects were observed for d_{002} around 0.340 nm. We concluded that, since with decreasing d_{002} the effects tend to disappear, the 3D graphitic structure was growing at the expense of the quasi 2D turbostratic structure. On the other hand, with increasing d_{002}, the disappearance of the negative magnetoresistance accompagnying the weak localization effects was attributed to the drastic decrease of the homogeneity factor λ and thus to the increase of disorder in the fibers. Thus, we have shown that the analysis of the negative magnetoresistance enables to determine the fraction of turbostratic layers in the carbon fibers and the degree of disorder.

Also, the decrease of H_s with decreasing d_{002} confirmed a well-known fact, (i.e., that the concentration of magnetic impurities decreases with increasing graphitization).

The good agreement between the results obtained on various pitch-derived fibers confirms that magnetoresistance measurements may be used as a "quantum too" to characterize carbon fibers at the level of the microstructure. Indeed, the fact that weak localization effects are used for this purpose demonstrates the direct relation between the quantum aspect of scattering and the material defect structure.

6.7.4 Thermal Conductivity

As was discussed above (Section 6.3.4), the measurement of the thermal conductivity at low temperature allows us to determine the in-plane coherence lengths. Thermal conductivity is among the transport coefficients described the one which, at the present time, is the most reliable to analyze and which provides a straightforward information.

We have shown that he temperature dependence of the lattice thermal conductivity from 3 to 300°K may be fitted using Kelly's model for the low temperature thermal conductivity of graphite with three adjustable parameters l_B, the phonon mean free path for boundary scattering, A, the constant for point defects scattering, and C_{44}, the shear modulus. The most important result was that the phonon mean free path for boundary scattering was found almost equal to the in-plane coherence length as determined by x-ray diffraction. One may thus use thermal conductivity measurements as a tool to determine this parameter, especially for high L_a values where x-ray are inadequate. So thermal conductivity measurements may be a very useful substitute for x-ray analysis for large in-plane coherence length values.

We have also observed that the value of A decreases with decreasing d_{002}. This showed, in agreement with results obtained by other techniques, that the concentration of point defects, (i.e., of impurities or vacancies), decreases with graphitization. Finally, the decrease of C_{44} with decreasing d_{002} was found to coincide with the expected dependence of the shear modulus on defects due to a decrease of the interlayer bonding with the increase of d_{002}. Contrary to sonic methods, the determination of C_{44} by thermal conductivity measurements presents the advantage to give the dislocation-free modulus or local modulus.

For the purpose of characterization, measurements could be performed in the liquid nitrogen temperature range, but they should be extended up to room temperature in order to be able to perform a thorough analysis of the results.

The main drawback of the method is that the experimental set up is very delicate to realize.

6.7.5 Thermoelectric Power

Contrary to the magnetoresistance, the diffusion thermoelectric power is very sensitive to the carrier densities, since it depends mainly on the Fermi energy, and to a lesser extent on the nature of the scattering mechanism, but not on the magnitude of the relaxation times of the carriers. Thus, for a single carrier system it allows the determination of the carrier density. However, as it is the case for the electrical resistivity, the fact that there are two types of carriers in graphite complicates the analysis of the results. The presence of phonon drag is also an additional problem to take into account.

In conjunction with the other measurements, the thermoelectric power may be used as a check that the carrier densities vary or not, for example when there is a phase transition in the sample.

Though some information might be obtained from measurements above 77°K, the liquid helium range is important too.

As for the case of the magnetoresistance, the measurement of the cross section is not needed. The main drawbacks for the case of graphite is that due to the presence of two types of carriers and of phonon drag, which may persist up to relatively high temperatures for the case of intercalated compounds.

Electrical and Thermal Transport in Carbon Fibers 447

6.7.6 Comparison of The Methods

In conclusion, we may state that, using the four methods described, a lot of information may be obtained concerning the electronic and lattice properties of graphite fibers, including various crystalline defects. These methods complement the information obtained by means of the more powerful techniques which probe the material at the microscopic level. There is still some progress to be made to understand the conduction mechanisms in such complicated systems like pristine carbons and graphites, but for the purpose of characterization a fairly good amount of information on the defects could be obtained by using more than one of the methods described.

The lattice thermal conductivity is the most direct and accurate tool to compare the in-plane coherence lengths in a pristine carbon fiber. It allows the estimation their average crystallite size. A more thorough analysis of the temperature dependence of the lattice thermal conductivity may give an indication of the density of point defects.

The quantitative influence of defects on the electrical resistivity is less obvious. However, an electrical resistivity measurement performed on different samples is a very sensitive way of detecting minute differences in their defect structure.

Magnetoresistance essentially probes the carrier mobility in a classical system, while the thermoelectric power is very sensitive to the carrier density (Fermi level) and depends, but to a lesser extent on the nature of the scattering mechanism. Both effects have the great advantage that they do not require the measurement of the cross sectional area of the samples and that they provide almost independent complementary data.

6.8 COMPOSITES

6.8.1 Introduction

Because of their geometry and mechanical properties, if we want to use the exceptional transport properties of fibers and their intercalation compounds to realize a practical conducting system, we need to embed them in a matrix. Alternatively, one may think of improving the electrical or thermal properties of polymeric materials, which are generally electrical insulators and poor thermal conductors, by realizing a fiber-polymer matrix composite. We will briefly discuss in this section the case of electrically and thermally conductive carbon fiber composites. An extensive analysis of these particular class of composites may be found elsewhere [133].

There is an increasing demand today for materials with high thermal conductivities in order to improve heat exchanges in practical devices or in manufacturing processes. This is particularly true for the case of space and airborne systems. Among the numerous advantages of using polymeric matrix composites as heat exchangers are their very low specific gravity as compared to metals and their alloys, their relatively high chemical resistance, their ease of processing and energy saving. Also, it is important to evacuate heat in electronic devices in order to realize efficient heat conductors which are nevertheless electrically insulating.

Recent studies have shown that we are now in a position to tailor materials at the atomic, molecular, and microstructural levels to obtain the desired properties. This is particularily true for the case of the thermal conductivity [3].

In a previous section, the temperature variation of the thermal conductivity below room temperature for different types of carbon fibers have been compared. By measuring single fibers, it was shown that by choosing adequately the precusor and by increasing the heat treatment temperature higher thermal conductivities are obtained. This is due to the modification of the microstructure and the improvement of the lattice perfection of the carbon fibers. In some way we are tailoring the conductivity at the *microscopic level*. It was also demonstrated that the thermal conductivity is directly correlated with the tensile modulus and electrical conductivity [5].

Composites allow tailoring of the properties of a given material at a *macroscopic level*. Though R&D in this area is mainly oriented towards the improvement of mechanical properties, they are also the choice candidates to realize efficient heat transfer devices and in some cases heat hyperconductors, with the highest thermal conductivities which could be attained in a practical material.

6.8.2 Electrical Conductors

Polymeric materials are generally electrical insulators. By adding to polymeric materials electrically conductive entities, such as carbon blacks or fibers, a transition from insulating to electrically conductive behavior occurs for an amount of filler exceeding the percolation threshold. These composites find many practical applications such as for emi shielding.

In Figure 35 we present a schematic diagram illustrating the typical dependence of the electrical resistivity on the fiber volume fraction [133]. The arrangement of the fibers for four typical situations are shown in the insets. Figure 35 shows that we may classify the composite samples into three categories which are related to the size of the aggregates of fibers in the electrically insulating matrix. For low concentrations, the fibers are randomly distributed in the matrix forming small aggregates or single inclusions which are separated by the polymer matrix. Since the latter has a very high electrical resistivity, the electrical resistivity of the composite is very high. This dilute concentration regime is presented in inset (a) of Figure 35. An increase in the fiber concentration leads to an increase of the number and size of the aggregates. Eventually some growing aggregates get in contact with their nearest neighbors and merge into larger clusters (inset (b) to Figure 35). For yet larger concentrations, an "infinite" cluster is formed (inset (c) to Figure 35). Around this concentration, which is called the percolation threshold, the electrical resistivity of the composite dramatically drops by 10 to 15 orders of magnitude. This results from the creation of a continuous path of electrically conductive fibers across the entire sample.

If we further increase the fiber concentration, the additional fibers generate new electrically conductive paths inside the infinite cluster or link to this cluster aggregates which were still isolated. This leads to a progressive decrease of the composite electrical resistivity (inset (d) to Figure 35).

It is worth noting that the percolation threshold is very sensitive to the filler geometry. For the case of fibers, it decreases when their length to diameter ratio increases.

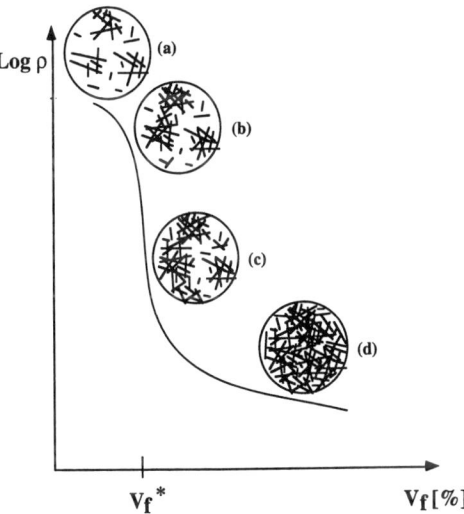

Figure 35 Schematic representation of the room temperature electrical resistivity of a chopped fiber-polymer matrix composite versus fiber volume fraction. The insets represent schematically the fiber arrangement in the composite for four concentration regimes (from reference [133]): (a) the dilute concentration range. The fibers are dispersed in the matrix, forming small aggregates or single inclusions separated by the polymer. (b) some aggregates reach their neighbours and merge into larger clusters, (c) an "infinite" cluster is created at the V_f^* concentration, (d) new paths inside the infinite cluster are created.

6.8.3 Thermal Conductors

For unidirectional polymer matrix composites containing highly conducting continuous carbon fibers the result on the thermal conductivity is rather trivial. It was shown experimentally in that case that the thermal conductivity of the composite along the fiber axis direction is, as predicted, equal to the product of the fiber volume fraction and thermal conductivity [2,3]. In order to check that, thermal conductivity measurements were performed on a single fiber, or on a small bundle of fibers, and the results were compared with those obtained for the composite as a function of fiber volume fraction [2]. This is shown in Figure 36. With the composites investigated, thermal conductivities higher than that of pure copper could be attained. This is not unexpected, since, as was shown above, continuous pitch-derived carbon fibers with thermal conductivities around 1000 $Wm^{-1}K^{-1}$ are now available.

In Figure 37 we present the room temperature thermal conductivity of chopped fiber-polycarbonate matrix composites versus the thermal conductivity of the fibers [134], all other parameters being almost the same. We may see that for carbon fibers with a thermal conductivity of 550 $Wm^{-1}K^{-1}$, one may elaborate a composite with a thermal conductivity of ~10 $Wm^{-1}K^{-1}$, which compares well with that of metallic alloys.

Figure 38 shows the relation at 300°K between the thermal conductivity of a chopped fiber-polycarbonate matrix composite versus fiber volume fraction for the same fibers. The heat flow is parallel to the plane of the composite which is in the form of a plate.

We have seen above (Section 6.8.2) that when the fiber concentration is large enough, an infinite cluster of fibers is formed and the whole composite becomes electrically conductive The transition from electrically insulating to electrically conductive state is explained in the frame of the percolation theory. However, such an abrupt transition does not occur in the case of thermal conductivity. This is due to the fact that the mechanisms responsible for electrical and thermal conduction are different. Electrical conductivity is always due to charge transport, while the thermal conductivity in the fibers and the polymeric matrix is due to phonons. While the polymeric matrix should be considered as an electrical insulator, it is only a poor thermal conductor and thus may still transfer some heat from one fiber to the other.

The case of composites with chopped fibers is far more complicated to analyze than for unidirectionnal composites with continuous fibers. For composites with chopped fibers, the relation between the thermal conductivity of the fiber and that of the composite is not staightforward as for the case of unidirectional composites with continuous fibers. In addition to the volume fraction, there is a large variety of parameters to control in order to improve the thermal conductivity.

One of these parameters is the fiber length for which results have been reported elsewhere [133,134] and which effect can be appreciated in Figure 38. Factors other than fiber length, conductivity, and concentration that control the thermal conductivities of the composite with short fibers are the length to diameter ratio, the average orientation of the fibers in the composite, and the fiber-matrix interface. Also, the electrical and thermal conductivities of the composite will be quite different according to the direction of the axis of the fiber with respect to the electrical current or the heat flow.

In some cases, the presence of the fiber may have a direct effect on the properties of a relatively important portion of the matrix in its vicinity (nucleation, transcrystallinity). For composites which are relatively poor conductors, this may have an impact on their thermal conductivity. So, one should investigate this third phase, the interphase. So, while one starts with a system of two constituents, he is left now with at least three different phases (the matrix itself may contain amorphous and crystalline phases): the matrix (excluding the interphase), the fibers and the interphases (modified matrices).

As may be seen from Figure 36, a composite with a thermal conductivity higher than 200 $Wm^{-1}K^{-1}$ has already been prepared with commercially available fibers and measured. If we extrapolate these results, we see that thermal conductivities of the order of 600 $Wm^{-1}K^{-1}$ may be easily realized, provided there is a sufficient quantity of fibers with thermal conductivities of 900 $Wm^{-1}K^{-1}$ to fabricate the composite.

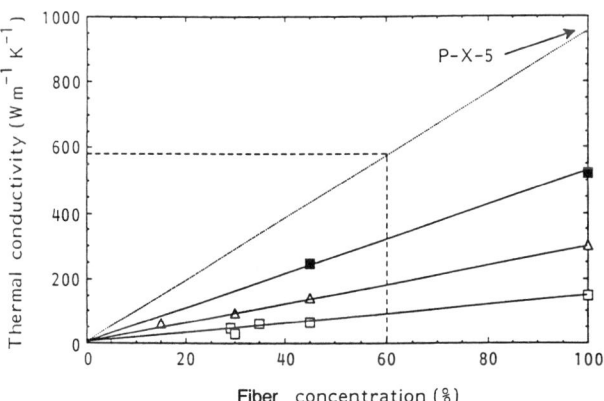

Figure 36 Thermal conductivity in the fiber direction of a polymer/carbon fiber unidirectional composite as a function of fiber content for different types of pitch-derived carbon fibers. The ordinates at 100% concentration are the thermal conductivities of the corresponding fibers [2]. The white squares are relative to P55 fibers from Amoco, the triangles to P100 and the black squares to P120.

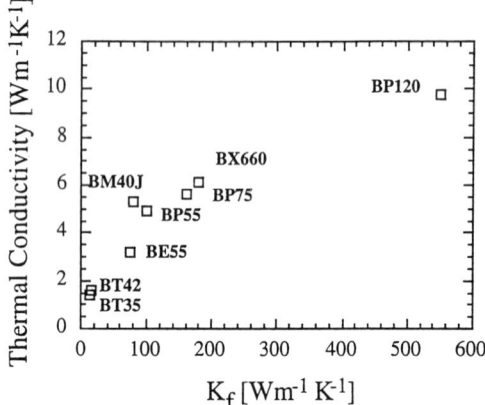

Figure 37 Room temperature thermal conductivity of chopped fiber-polycarbonate matrix composites versus the thermal conductivity of the fibers [133]. For each experimental point the corresponding type of fiber is indicated. The first letter (here B for Brabender kneading) indicates the fabrication process for the composite.

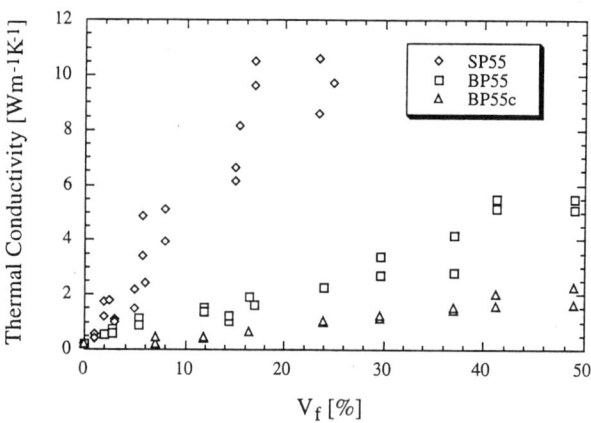

Figure 38 Room temperature dependence of the thermal conductivity of a chopped fiber (P55)-polycarbonate matrix composite versus fiber volume fraction [133]. The heat flow is parallel to the plane of the composite in the form of a plate. The different sets of experimental points are relative to composites with fibers of different average lengths: roughly 40 mm for BP55c, 100 mm for BP55, and 360 mm for SP55. From this figure one may also appreciate the effect of fiber length on the conductivity for a given volume fraction.

6.9 CONCLUSIONS

In conclusion, we may state that in many aspects of measuring the electrical and thermal transport on carbon fibers, whether pristine or intercalated or embedded in a polymer matrix, are quite rewarding. This is due to their variety of structures, their geometry and they ability to be intercalated. Some observations regarding quantum transport or the separation of electronic and lattice thermal conductivities could hardly be made on bulk carbons and graphites. Regarding practical applications, their exceptionally high thermal conductivities offer unique possibilities for heat transfer devices.

Though we have reported some results pertaining to PAN-based and vapor grown carbon fibers, we have mainly concentrated here on pitch-based fibers. We have also briefly reviewed some of the fascinating aspects of carbon nanotubes. We have mainly emphasized the fact that the transport properties of carbon fibers, as many other properties depend dramatically on the degree of disorder, which is well described by the main structural parameters. Conversely, the analysis of transport properties allow access to these parameters.

We have shown that the analysis of the field dependence of the negative *magnetoresistance* at 4.2 and 77°K using the 2D weak localization theory allows the determination *for pristine fibers* of the sheet resistance, R, and the characteristic field for scattering by magnetic impurities, H_s. These results confirmed that the negative magnetoresistance and the 2D weak localization in pristine carbon fibers should be attributed to the turbostratic layers. When the interlayer spacing decreases, these effects tend to disappear since the 3D graphitic structure is growing at the expense of the quasi 2D turbostratic structure. On the other hand, when the interlayer spacing increases, the disappearance of weak localization effects and of the resulting negative magnetoresistance has been related to the dramatic increase of inhomogeneities in the samples and thus to the increase of disorder. The analysis of the negative magnetoresistance thus enables to characterize the portion of turbostratic layers in the carbon fibers and the degree of disorder in their microstructure. These results were found to be consistent with those of the *zero-field electrical resistivity*.

The decrease of H_s with decreasing interlayer spacing suggested, in agreement with observations made by means of other techniques, that the concentration of magnetic impurities decreases during the graphitization process.

The good agreement between the results obtained on various pitch-derived fibers and pyrocarbons confirms that, on a general basis, magnetoresistance measurements may be used as a "quantum tool" to characterize carbon fibers at the level of the microstructure. Indeed, the fact that weak localization effects are used for this purpose demonstrates the direct relation between the quantum aspect of scattering and the material defect structure.

We have also discussed the temperature variation of the *thermal conductivity* of *pristine fibers*. It was shown that the temperature dependence of the lattice thermal conductivity may be fitted using Kelly's theory for the low temperature thermal conductivity of graphite with three adjustable parameters: l_B, the phonon mean free path for boundary scattering, A, the constant for point defects scattering, which is proportional to the concentration of defects, and C_{44}, the interlayer shear modulus. The most important result was the demonstration that the phonon mean free path for boundary scattering was almost equal to the in-plane coherence length as determined by x-ray diffraction. This enables to use thermal conductivity measurements as a tool for the determination of this parameter, especially for high L_a values where x-rays are inadequate. So thermal conductivity measurements may prove to be a very useful substitute for x-ray analysis for large in-plane coherence length values.

We have also shown that the value of A decreases with decreasing d_{002}. This shows that the concentration of point defects, (i.e., of impurities or vacancies), decreases with graphitization. This is in agreement with results obtained from magnetoresistance measurements and from other techniques.

Finally the decrease of C_{44} with decreasing d_{002} was found to be in agreement with the expected dependence of the shear modulus on defects due to a decrease of the interlayer bonding with the increase of d_{002}. Contrary to sonic methods, the determination of C_{44} by thermal conductivity measurements yields the local dislocation-free modulus.

These *global characterization techniques* present the advantage of giving an overall view over the entire sample, contrary to the powerful microscopic techniques which only probe a tiny portion of the sample.

Though the electrical and thermal conductivities of pristine carbon fibers are generated by different entities, charge carriers for the electrical resistivity and phonons for the thermal conductivity, these transport properties depend both dramatically on the structure of the fibers. They both increase with the in-plane coherence length. As a corollary, there is a direct relation between these transport coefficients for the same precursor and, once the electrical resistivity is measured, one can determine the thermal conductivity.

We have also reviewed the available experimental works which were performed on the temperature variation of the zero-field transport properties of *fibrous acceptor GICs*: electrical resistivity, thermal conductivity, and thermoelectric power.

We have first analyzed the ideal and residual *electrical resistivity* data and the effects of 2D weak localization and of carrier-carrier interaction. It was shown how the residual resistivity allows the estimation of the size of the large-scale lattice defects, which were found to be much smaller than in the pristine material. The low temperature behavior of the ideal electrical resistivity was attributed to hole-hole interaction in the presence of weak disorder, while electron-phonon scattering was predominant in the higher temperature range. As for the case of bulk GICs, the model of a 2D hole gas introduced by the intercalate and interacting with the phonons and defects of the host layers gave a good picture of electronic transport in the Boltzmann conductivity range. We have also discussed the limits of conductivity which could be attained in these compounds.

The *thermal conductivity* of three fibrous intercalation compounds was also analyzed. As is the case for bulk compounds, the general trend observed was a decrease of the total thermal conductivity at high temperature and an increase at low temperature with respect to that of the pristine material. It was shown that phonons may also probe the lattice defects. The extra contribution due to the intercalate phonons was clearly evidenced.

The experimental observations concerning the *thermoelectric power* confirmed the general trend observed for bulk compounds, which thus seems to be universal. At high temperature all compounds exhibited similar temperature variation, while in the low temperature region a linear function of the temperature was observed for the diffusion thermoelectric power.

We have also briefly reviewed the experimental data pertaining to the electronic properties of *carbon nanotubes*. From the temperature dependence of the electrical resistance of a microbundle it was shown that the measured nanotubes were semimetallic and behaved like turbostratic graphite. Magnetoresistance data obtained for a *single microbundle* were consistent with the formation of Landau levels proposed by Ajiki and Ando. Further results obtained on *single multiwalled nanotubes* revealed two additional features of quantum transport in low-dimensionality mesoscopic systems: 2D weak localization and universal conductance fluctuations.

In a different context, we have also considered electrically and thermally conducting *polymer-matrix composites filled with carbon fibers*. We have shown that in unidirectional composites with highly conductive continuous carbon fibers, room temperature thermal conductivities comparable to that of pure copper are now readily attained, and that hyperconductors of heat with much higher thermal conductivities are within the reach. With chopped fibers, thermal conductivities superior to that of metallic alloys may be realized. However, for composites with short fibers, there is still systematic work to be done in order to identify and control the large variety of parameters responsible for the improvement of the thermal conductivity. In some way, one may tailor the thermal conductivity of composites to the desired values for practical applications.

We have started our introduction by qualifying carbon fibers as "dirty physical systems." In our conclusion we may state that working on such real systems are surely rewarding in shedding some light on the most sophisticated aspects of quantum transport phenomena which were discovered the last decade on other "dirty systems," such as thin metallic films or wires, and that were ingeniously analyzed at that time by solid state theorists.

REFERENCES

1. Allen, G. and Issi, J-P., *Proceedings of ECCM-1*, p. 447 (1985).
2. Nysten, B. and Issi, J-P., *Composites*, 21, 339 (1990).
3. Issi, J-P., Nysten, B., Jonas, A., Demain, A., Piraux, L. and Poulaert, B.,in *Thermal Conductivity* 21, C.J. Cremers and H.A. Fine, editors, Plenum Press, New York (1990) p. 629.
4. Bertram, A., Proceedings of the conference on carbon and carbonaceous composite materials, Malenovice, October 1995, World Scientific Publishing, NJ (in press).
5. Nysten, B., Piraux, L. and Issi, J.-P., in *Thermal conductivity* 19, D. W. Yarbrough editor, Plenum Press, New York, 341 (1987).
6. Langer, L., Stockman, L., Heremans, J.P., Bayot, V., Olk, C.H., Van, C., Haesendonck, Bruynseraede, Y., and Issi, J.-P., *J. Mater. Res.* 9, 927-932 (1994).
7. Langer, L., Bayot, V., Grivei, E., Issi, J.-P., Heremans, J.P., Olk, C.H., Stockman, L., Van Haesendonck, C. and Bruynseraede, Y., *Phys. Rev. Letters* 76, 479 (1996).
8. Ebbesen, T. W., *Physics Today*, 49, 26 (1996).
9. Piraux, L., Issi, J.P., Michenaud, J.P., McRae, E., and J.F. Marêché, *Solid State. Commun.* 56, 567 (1985).
10. Bayot, V., Piraux, L., Michenaud, J-P. and Issi, J-P., *Phys Rev B* 40, 3514 (1989).
11. Nysten, B., Issi, J-P., Barton Jr., R., Boyington, D. R., and Lavin, J. G., *J. Phy. D: Applied Physics*, 24, 714 (1991).
12. Nysten, B., Issi, J-P., Barton Jr., R., Boyington, D. R., and Lavin, J. G., *Phys. Rev.* B44, 2142 (1991).
13. Lavin, J.G., Boyington, D.R., Lahijani, J., Nysten, B., and Issi, J.-P., *Carbon* 31, 1001 (1993).
14. Issi, J-P., in *Graphite Intercalation compounds*, Springer Series in Materials Science, volume II, editors H. Zabel and S.A Solin, Springer-Verlag, Berlin, (1992).
15. Issi, J-P., *Materials Science Forum* 91-93, 471 (1992).
16. Issi, J.-P., and Piraux, L., *Annales de Physique* 11(2), 165 (1986)
17. Nysten, B., Piraux, L., and Issi, J.P., *Synthetic Metals* 12, 505 (1985).
18. Dresselhaus, M.S., Dresselhaus, G., Sugihara, K., Spain, I.L., and Goldberg H.A., *Graphite Fibers and Filaments*, Springer Series in Materials Science 5, Springer-Verlag (1988).
19. *Graphite Intercalation compounds*, Springer Series in Materials Science, volumes I &II, H. Zabel and S.A Solin editors, Springer-Verlag (Berlin, 1992).

20. Abrahams, E., Anderson, P.W., Licciardello, D.C., and Ramakrishnan, T.V., *Phys. Rev. Lett.* 42, 613 (1979).
21. Bergmann, G., *Phys. Rev.* B28, 2914 (1983).
22. Hikami, S., Larkin, A.I., and Nagaoka, *Prog. Theor. Phys.* 63, 707 (1980).
23. Altshuler, B.L., Aronov, A.G., and Lee, P.A., *Phys. Rev. Letters.* 44, 1288 (1980).
24. Fukuyama, H., *J. Phys. Soc. Jpn* 48, 2169 (1980).
25 Piraux, L., *J. Mater. Res.* 5, 1285 (1990).
26. Piraux, L., Bayot, V., Michenaud, J.-P., Issi, J.-P., Marêché, J.F., and McRae, E., *Solid State Commun.* 59, 711 (1986).
27. Piraux, L., Bayot, V., Gonze, X., Michenaud, J.-P., and Issi, J.-P., *Phys. Rev.* B36, 9045 (1987).
28. Chieu, T.C., Dresselhaus, M.S., and Endo, M., *Phys. Rev.* B26, 5869 (1982).
29. Piraux, L., Issi, J.-P., Salamanca-Riba, L., and Dresselhaus, M.S., *Synthetic Metals* 16, 93 (1986).
30. *Localization, Interaction and transport Phenomena*, Kramer, B., Bergmann, G., Y. Bruynseraede editors, Springer Series in solid-state Sciences, vol. 61, Springer-Verlag, Berlin (1985).
31. Bergmann, G., *Phys. Rep.* 107, 1 (1984).
32. Lee, P.A., and Ramakrishnan, T.V., *Rev. Mod. Phys.* 57, 287 (1985).
33. Fukuyama, H., in *Percolation, Localization and Superconductivity*, A.M. Goldman and S.A. Wolf editors (Plenum Press, New York) p. 161 (1984).
34. Bright, A.A., and Singer, L.S., *Carbon* 17, 59 (1979).
35. Bayot, V., Piraux, L., Michenaud, J-P., Issi, J-P., Lelaurain, M. and Moore, A., *Phys. Rev. B* 41, 11770 (1990).
36. Klein, C.A., *J. Appl. Phys.* 35, 2947 (1964).
37. Bright, A.A., *Phys. Rev. B* 20, 5142 (1979).
38. Hishiyama, Y., *Carbon* 8, 259 (1970).
39. Hishiyama, Y., Ono, A. and Hashimoto, M., *Japan. J. Appl. Phys.* 10, 416 (1971).
40. Endo, M., Hishiyama, Y., and Koyama, T., *J. Phys. D: Appl. Phys.* 15, 353 (1982).
41. Endo, M., *J. Mater. Sci.* 23, 598 (1988).
42. Nysten, B., Ph. D Thesis, Université Catholique de Louvain, Louvain-la-Neuve (1991) (in French).
43. Issi, J.-P., Proceeding of the 10th course of the Erice Summer School, International School of Materials Science and Technology, Erice Sicily July 5-15, Nato ASI series; Series B, Physics (Plenum Press, New York) p 347 (1986).
44. Mrozowski, S., and Chaberski, A., *Phys. Rev.* 104, 74 (1956).

45. Robson, D., Assabghy, F.Y.I., and Ingram, D.J.E., *J. Phys. D: Appl. Phys.* 5, 169 (1972).
46. Robson, D., Assabghy, F.Y.I., Cooper, E.G. and Ingram, D.J.E., *J. Phys. D : Appl. Phys.* 6, 1822 (1973).
47. Gijs, M., Van Haesendonck, C., and Bruynseraede, Y., *J. Phys. F. : Met. Phys.* 16, 1227(1986).
48. Franklin, R.E., *Acta Cryst.* 4, 253 (1951).
49. Bourrat, X., Roche, E.J., and Lavin, J.G., *Carbon* 28, 236 (1990).
50. Bourrat, X., Roche, E.J., and Lavin, J.G., *Carbon* 28, 435 (1990).
51. Elman, B.S., Braunstein, G., Dresselhaus, M.S., and Venkatesam, T., *Nucl. Instrum. Methods Phys. Res.* B 7/8, 493 (1984).
52. Nysten, B., Issi, J-P., Barton, R., Boyington, D.R., and Lavin, J.G., *Ext. Abstr. Carbone 90 Conf., Paris, France* 630, (1990).
53. Issi, J-P., Boxus, J., Poulaert, B., and Heremans, J., *Thermal Conductivity* 17, 537, edited by J. G. Hust, Plenum Press, New York (1982).
54. Piraux, L., Issi, J.P., and Coopmans, P., *Measurement* 5, 2 (1987).
55. Piraux, L., Nysten, B., Haquenne, A., Issi, J-P., Dresselhaus, M.S., and Endo M., *Solid State Commun.* 50 697, (1984).
56. Ziman, J.M., *Electrons and Phonons*, Clarendon Press, Oxford (1960).
57. Berman, R., *Thermal Conduction in Solids* (Clarendon Press, Oxford, 1976).
58. Issi, J.-P., Heremans, J., and Dresselhaus, M.S., *Phys. Rev.* B27, 1333 (1983).
59. Klemens, P. G., *Encyclopedia of Physics*, Flügge, S., ed. (Springer-Verlag, Berlin, 1956) vol. XIV, p.198.
60. Kelly, B.T., *Physics of Graphite*, (Applied Science Publishers, London,1981).
61. di Vittorio, S.L., Dresselhaus, M.S., Endo, M., Issi, J-P., Piraux, L., and Bayot, V., *J. Mater. Res.* 6, 778 (1991).
62. Kelly, B.T., *Carbon* 5, 247 (1967); *Ibid.* 6, 71 (1968),; *Ibid.* 6 ,485 (1968).
63. Kearney, M.J., and Butcher, P.N., *J. Phys. C : Solid State Phys.* 21, L265 (1988).
64. Bayot, V., Piraux, L., Michenaud, J-P., and Issi, J-P., *Phys. Rev. Letters* 65, 2579 (1990).
65. Kelly, B.T., in *'Chemistry and Physics of Carbon 5'*, P.L. Walker Jr, ed., Marcel Dekker, New York (1969), 119.
66. Dreyfus, B., and Maynard, R., *J. Physique* 28, 955 (1968).
67. Issi, J-P., Heremans, J., and Dresselhaus, M.S., *Phys. Rev. B* 27, 1333 (1983).
68. Elman, B.S., Braunstein, G., Dresselhaus, M.S., and Venkatesam, T., *Nucl. Instrum. Methods Phys. Res. B* 7/8, 493 (1984).
69. Komatsu, K., *J. Phys. Soc. Japan* 10, 346 (1955).

70. Dolling, G., and Brockhouse, B.N., *Phys. Rev.* 128, 1120 (1962).
71. Blakslee, O.L., Proctor, D.G., Seldin, E.J., Spence, G.B., and Weng T., *J. Appl. Phys.* 41, 3373 (1970).
72. Seldin, E.J., and Nezbeda, C.W., *J. Appl. Phys.* 41, 3389 (1970)
73. Issi, J-P., *Aust. J. Phys.* 32, 585 (1979)..
74. Blatt, F. J., *Physics of Electronic Conduction in Solids*, McGraw-Hill, New York (1968).
75. Heremans, J., and Beetz, C.P., *Phys. Rev. B* 32, 1981 (1985).
76. Endo, M., Tamagawa, I., and Koyama, T., *Jpn. J. Appl. Phys.* 16, 1771 (1977).
77. Bayot, V. *Conduction et désordre dans le graphite*, Ph. D Thesis, UCL, Louvain-la-Neuve (*in french*), (1991).
78. Kaburagi, Y., Hishiyama, Y., Baker, D.F., and Bragg, R.H., *Phil. Mag. B* 54, 381 (1986)
79. Baker, D.F., and Bragg, R.H., *J. Non-Crystalline Solids* 58, 57 (1983)
80. Mott, N.F., and Davis, E.A., *Electronic Processes in Non-Crystalline Materials* (Oxford University Press, Oxford) (1979)
81. Zvyagin, I.P., *Phys. Status Solidi B* 58, 443 (1973)
82. MacDonald, D.K.C., *Principles of Thermoelectricity* (John Wiley & Sons Inc, New York) (1964).
83. Dresselhaus, M.S., and Dresselhaus, G., *Advances in Physics* 30, 139 (1981).
84. Chieu, T.C., Dresselhaus, M.S., Endo, M., Moore, A.W., *Phys. Rev. B* 27 3686 (1983).
85. Blinowski, J., Nguyen Hy Hau, Rigaux, C., Vieren, J.-P., Le Toullec, R., Furdin, G., Hérold, A., and Mélin, J., *J. Phys.* 41, 47 (1980).
86. Issi, J.-P., Poulaert, B., Heremans, J., and Dresselhaus, M.S., *Solid State Commun.* 44, 449 (1982).
87. Ansart, A., Meschi, C., and Flandrois, S., *Synthetic Metals*, 23, 455 (1988), Proc. 4th Int. Symp. on GICs, Jerusalem.
88. Messchi, C., PhD thesis, Université de Bordeaux I (unpublished) (1988), *in french;* C. Meschi, J. P. Manceau, S. Flandrois, P. Delhaes, A. Ansart and L. Deschamps, Ann. de Physique, 11, 199 (1986), colloq. 2, suppl. 2.
89. Manini, C., Marêché, J-F., and McRae, E., *Synth. Metals*, 8, 261 (1983).
90. Hambourger, P. D., Jaworske, D. A., and Gaier, J. R., *Extended Abstracts of the Symposium on Graphite Intercalation Compounds*, edited by P. C. Eklund, M. S. Dresselhaus and G. Dresselhaus, Materials Research Society, p.208 (1984).
91. Daumas, N., and Hérold, A., Hebd. Séance Acad, C.R.. *Sci. Paris* C 268, 373 (1969).
92. Piraux, L., Bayot, V., Issi, J-P., Dresselhaus, M. S., Endo, M., and Nakajima, T., *Phys. Rev.* B41, 4961 (1990).

93. Lee, P.A., and Ramakrishnan, T.V., *Phys. Rev.* B26, 4009 (1982).
94. Altshuler, B.L., Aronov, A.G., and Khmelnitzkii, D.E., *J. Phys.* C 15, 7367 (1982).
95. Takayama, H., *Z. Phys.* 263, 329 (1973).
96. Pietronero, L., and Strässler, S., *Synthetic. Metals* 3, 213 (1981).
97. Inoshita, T., and Kamimura, H., *Synthetic. Metals* 3, 223 (1981).
98. Issi, J-P., Heremans, J., and Dresselhaus, M. S., in *Physics of Intercalation Compounds*, editors L. Pietronero and E. Tosatti, Springer, Berlin (1981), p 310.
99. Issi, J.-P., *Mat. Res. Soc. Symp. Proc.* 20, 147 (1983) Elsevier Science Publishing Co., Inc.
100. Piraux, L., Kinany-Alaoui, M., Issi, J-P, Pérignon, A., Pernot, P., and Vangelisti, R., *Phys. Rev.* B38, 4329 (1988).
101. Boxus, J., Poulaert, B., Issi, J.-P., Mazurek, H., and Dresselhaus, M.S., *Solid State Commun.* 38, 1117 (1981).
102. Poulaert, B., Heremans, J., Issi, J.-P., Zabala Martinez, I., Mazurek, H., and Dresselhaus, M.S., *Extended Abstracts 15th Biennial Carbon Conf.*, 92 (1981).
103. Elzinga, M., Morelli, D.T., and Uher, C., *Phys Rev.* B26, 3312 (1982).
104. Blatt, F.J., Zabala Martinez, I., Issi, J.-P., Shayegan, M., and Dresselhaus, M.S., *Phys. Rev.* B27, 2558 (1983).
105. Kinany-Alaoui, M., Piraux, L., Issi, J-P, Pernot, P., and Vangelisti, R., *Solid State Commun.* 68, 1065 (1988).
106. Heremans, J., Issi, J.-P., Zabala Martinez, I., Shayegan, M., and Dresselhaus, M.S., *Phys Lett.* 84A, 387 (1981).
107. Piraux, L., Nysten, B., Issi, J.-P., Marêché, J.F., and McRae, E., *Solid. State Commun.* 55, 517 (1985).
108. Piraux, L., Nysten, B., Issi, J.P., Salamanca-Riba, L., and Dresselhaus, M.S., *Solid State Commun.* 58, 265 (1986).
109. Dreyfus, B., and Maynard, R., *J. Phys.* 28, 955 (1967).
110. Callaway, J., *Phys. Rev.* 113, 1046 (1959).
111. Piraux, L., Issi, J-P., Marêché, J-F., and McRae, E., *Synthetic Metals*, 30, 245 (1989).
112. Iijima, S., *Nature* 354, 56 (1991).
113. Charlier, J.C., and Michenaud, J.P., *Phys. Rev. Lett.* 70, 1858-1861 (1993), Charlier, J.C., *Carbon Nanotubes and Fullerenes*. PhD thesis, Catholic University of Louvain, May 1994.
114. White, C.T., Roberston, D.H., and Mintmire, J.W., *Phys. Rev.* B 47, 5485 (1993).
115. Saito, R., Dresselhaus, G., and Dresselhaus, M.S., *J. Appl. Phys.* 73, 494 (1993); Dresselhaus, M. S., Dresselhaus, G., and Saito R., *Solid State Commun.* 84, 201-205 (1992).

116. Iijima, S., *Mater. Sci. and Eng.* B19, 172 (1993).
117. Iijima, S., Ichihashi, T., and Ando, Y., *Nature (London)* 356, 776 (1992).
118. Ebbesen, T.W., and Ajayan, P.M., *Nature (London)* 358, 220 (1992).
119. Ebbesen, T.W., Hiura, H., Fujita, J., Ochiai, Y., Matsui, S., and Tanigaki, K., *Chem. Phys. Lett.* 209, 83-90 (1993).
120. Dresselhaus, M.S., Dresselhaus, G., and Saito, R., *Carbon* 33, 883 (1995).
121. Heyd, R., Charlier, A., Marêché, J.F., McRae, E., Zharikov, O.V., *Solid State Communications*, 89 (12), 989 (1994).
122. Seshardi, R., Aiyer, H.N., Govindaraj, A., Rao, C.N., *Solid State Communications* 91 (3), 195 (1994).
123. Tc* is taken from the intersect of the T-1/2 dependence and the saturation.
124. Whitesides, G.M., Weisbecker, C.S.(private communication).
125. Ajiki, H., and Ando, T., *J. Phys. Soc. Jpn.* 62, 1255 (1993).
126. Al'tshuler, B.L., *JETP Lett.* 41, 648 (1985).
127. Lee, P.A., and Stone, A.D., *Phys. Rev. Lett.* 55, 1622 (1985).
128. Umbach, C.P., Washburn, S., Laibowitz, R.B., Webb, R.A., *Phys. Rev.* B 30, 4048 (1984).
129. Washburn, S., Umbach, C.P., Laibowitz, R.B., and Webb, R.A., *Phys. Rev.* B 32, 4789 (1985).
130. Licini, J.C., Bishop, D.J., Kastner, M.A., and Melngailis, J., *Phys. Rev. Lett.* 55, 2987 (1985).
131. Skocpol, W.J., Mankiewich, P.M., Howard, R.E., Jackel, L.D., Tennant, D.M., and Stone, A.D., *Phys. Rev. Lett.* 56, 2865 (1986).
132. Lee, P.A., Stone, A.D., and Fukuyama, H., *Phys. Rev.* B 35, 1039 (1987).
133. Demain, A., *Thermal Conductivity of polymer-chopped carbon fiber composites,* Ph. D Thesis, Université Catholique de Louvain, Louvain-la-Neuve (1994).
134. Nysten, B., Jonas, A., and Issi, J-P., in *Thermal Conductivity* 21, Edited by C.J. Cremers and H.A. Fine, Plenum Press, New York, 647 (1990).

7
CARBON FIBER APPLICATIONS

Serge Rebouillat
DuPont de Nemours International S. A., Geneva, Switzerland
Jimmy C. M. Peng, and Jean-Baptiste Donnet
Ecole Nationale Supérieure de Chimie and Université de Haute-Alsace, Mulhouse, France
Seung-Kon Ryu
Chungnam National University, Taejon, Korea

7.1 INTRODUCTION

High strength, high toughness, and low weight are the most important characteristics of an ideal engineering material. Conventional engineering materials, the metals and their alloys, are strong and tough, but not light. Some conventional materials are strong but not very tough, whereas certain plastic materials are light but lack strength and toughness. Several types of fibrous materials have been developed during the last decade and have been successfully used as reinforcing components for certain applications. Although composites can be 10 times more expensive than typical aerospace-grade aluminum, the flexibility they offer in design and consolidation of parts allows large complex structures to be fabricated to exacting specifications. The properties of some of these materials are compared in Table 1. High-performance carbon fibers with improved strength and stiffness and very low density are one of the more recent developments. These carbon fibers are elastic to failure at normal temperature, which renders them creep-resistant and nonsusceptible to fatigue. They are chemically inert except in strongly oxidizing environments or in contact with certain molten metals. Carbon fibers have exceptional thermophysical properties and excellent damping characteristics. The physical properties of carbon fibers which make them versatile materials for certain applications are given in Table 2. It may,

Table 1 Comparison of Mechanical Properties of Various Fiber Composites (Carbon Fiber Volume Fraction, $V_f = 0.6$ Unidirectional/Epoxy) and Metals

	Tensile strength (GPa)	Young's modulus (GPa)	Specific tensile strength (MPa)	Specific tensile modulus (GPa)	Density (g/cm^3)
Fiber composites					
UHM carbon fiber[a]	1.38	290	820	173	1.68
HM carbon fiber	1.52	207	974	133	1.56
UHT carbon fiber[b]	1.90	124	1242	81	1.53
HT carbon fiber	1.52	103	993	67	1.53
Aramid fiber	1.38	76	1000	55	1.38
S-glass fiber	1.82	53	875	26	2.08
Boron fiber	2.76	248	1484	133	1.86
Metals					
Steel (150 type c35)	0.42	206	54	26.4	7.8
Magnesium alloy (DTD 88c)	2.80	42	156	23	1.80
Aluminum alloy (6061-T6)	0.26	69	101	27	2.56
Titanium alloy (6A1-4V)	0.98	112	220	25	4.45

[a] UHM=ultra-high modulus.
[b] UHT=ultra-high strength.

however, be mentioned that specific strength and stiffness are the basic characteristics which are useful in almost all applications. Carbon fibers, however, have some shortcomings as well. They are brittle, and have a low impact resistance and low break extension and very small coefficient of linear expansion. They have a high degree of anisotropy both in the direction of the fiber axis and perpendicular to it. They remain as expensive materials and can, therefore, be used only where cost is not a major factor. However, the price of these fibers is likely to decrease with increased production.

The three main sectors of carbon fiber applications are: 1) the high technology sector, which includes aerospace and nuclear engineering ; 2) the general engineering and transportation sector, which includes engineering components such as bearings, gears, cams, fan blades, and automobile bodies ; 3) sporting goods, such as golf clubs, bicycles, etc. In 1993 U.S. consumption of carbon fibers was 2.8 million kg. Nearly half of that went to aerospace applications, and sporting goods accounted for about 23% of consumption. The

Table 2 Applications and Characteristics of Carbon Fiber Composites in Industry

Applications	Characteristics
Aerospace, road, train, and marine transport, sporting goods	High strength, specific toughness, and light weight
Missiles, aircraft brakes, and aerospace antenna	High dimensional stability, low coefficient of thermal expansion, and low abrasion
Audio equipment, speakers of Hi-fi equipment, robot arms	Good vibrational damping, strength, and toughness
Automobile hoods, casting and bases for electronic equipment, brushes	Electrical conductivity
Surgery and x-ray equipment, implants, medical applications in prostheses	Biological inertness and x-ray permeability
Textile machinery, general engineering	Fatigue resistance, self-lubrication, high damping
Chemical industry, nuclear field, valves, seals	Chemical inertness, high corrosion resistance
Large generator retaining rings, radiological equipment	Electromagnetic properties

requirements of the three sectors are fundamentally different. For example, the large-scale use of carbon fibers in aircraft and aerospace is driven by maximum performance and fuel efficiency, while the cost factor and the production requirements are not critical. The use of carbon fibers in general engineering and surface transportation is dominated by cost constraints, high production rate requirements, and generally less critical performance needs. This leads to very different consideration regarding material forms, acceptable matrices, the significance of hybridization, potential manufacturing methods, and on-line quality control requirements. The availability of carbon fibers has thus made it possible to make entirely different kinds of high performance articles. Carbon fibers, however, are seldom used alone. They are used for reinforcing certain matrix materials to form composites.

Carbon fibers are unidirectional reinforcements and can be arranged in such a way in the composite that it is stronger in the direction which must bear loads. Another arrangement of fibers can be used to prevent deflection on loading in one direction and not in the other. But in general, the processed bodies are expected to bear loads, resist deflection, and give high performance in more than one direction. Consequently, the carbon fibers have to be arrayed in more than one direction in the composite to attain better efficiency. This selective reinforcement by carbon fibers allows designers to fashion articles suited to their needs. The designer may arrange more fibers in one direction than in the other, or use two types of carbon fibers, one known for its high strength to bear loads in one direction and the other for its high stiffness to resist deflection in the other direction. In addition, the structure can be constructed by using two different kinds of fibers materials such as Kevlar®* or glass and carbon fibers. The structures constructed using Kevlar® and carbon fibers simultaneously in the same composite have impact strengths better than those reinforced with Kevlar® fibers alone [1]. These composites in which more than one kind of reinforcement is used are called hybrids.

Since a composite material needs several different orientations of the reinforcing fiber for most applications, it is better to use the fiber in fabric form. Carbon fiber, being brittle, is difficult to obtain in fabric form, but due to advances in sizing and weaving technology; it is now possible to obtain carbon fiber cloth of superior quality in good quantity. There are several weave styles applicable, such as: (1) Unidirectional, the warp tows are held parallel by very light polyester or glass warp threads; (2) twill, each end floats more than one and under a number of crossing threads; and (3) satin, each end passes under

---

*Kevlar® is a registered trademark of DuPont.

Carbon Fiber Applications

one crossing thread and over N threads. For example, the 5HS shown in Figure 1 has the ends and picks passing under one and over four ends. There is also (4) leno, wrap ends in the fabric are twisted to cross one another between picks ; (5) basket, each end and pick passes over more than one end and under an equal number of crossing threads. For example, a 4 × 4 weave represents that each end and picks passes over four and under four crossing threads. Carbon fiber cloth has certain distinct advantages in making composites. The fabric is easy to handle and can be laid up faster in the desired shapes. It is bi-directional and can, therefore, be woven with two different types of fibers in the warp and in the fill. The labor cost in making composites in complex designs is much less and can compensate for the high cost of the woven material. Some representative companies that currently weave carbon fiber fabrics are given in Table 3.

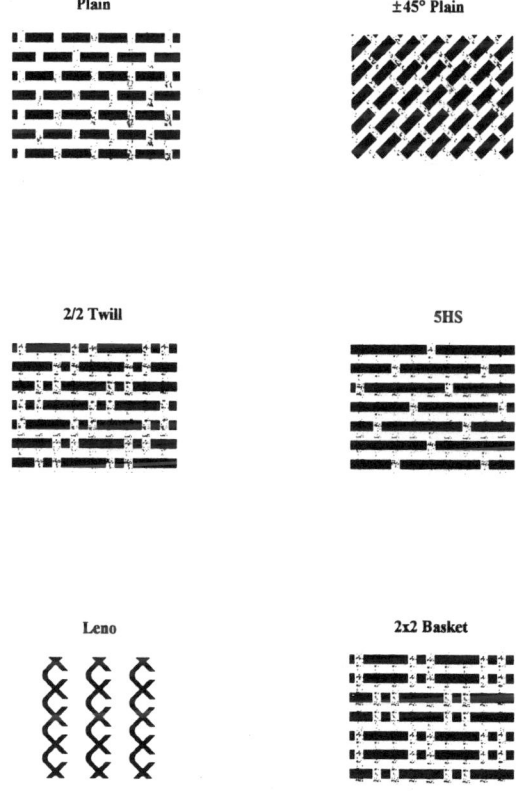

Figure 1   Different carbon fiber weave styles.

Table 3 Representative Manufacturers of Woven Fabric Carbon Fibers

Manufacturer	Trademark	Country
Brochier SA	Lyvertex	France
Carr Reinforcements Ltd.	Carrfiber	UK
CS-Interglas	Interglas	Germany
Fiber Materials Inc.	FMI	USA
Hercules Inc.	Magnamite	USA
Mitsubishi Rayon Co. Ltd.	Pyrofil	Japan
North American Textiles Inc.	Nortex	USA
Plastic Developments Ltd.	PDL	UK
Sigri Great Lakes Carbon Gmbh	Sigratex	Germany
Swiss Silk Zurich	Teletex	Switzerland
Textile Technologies Inc.	TTI	USA
Toho Rayon Co Ltd.	Besfight	Japan

Source: from ref. 2.

The physical properties of carbon fiber-reinforced composite materials depend considerably on the nature of the matrix, the fiber alignment, the volume fraction of the fiber and matrix, and on the molding conditions. Therefore, by choosing suitable conditions, it is possible to make composites suited to a particular need. Several types of matrix materials such as glass and ceramics [2–5], metals [2,6–12], and plastics [2,13–22] have been used as matrices for reinforcement by carbon fibers. However, from the point of view of mechanical properties, density, and fiber-matrix cohesion, epoxy resins are frequently the best choice, although these matrices are not suitable for high-temperature applications. Several high-temperature plastic matrices have now been developed which may improve the temperature range of application of these composites. For very high-temperature applications, carbon-carbon composites consisting of a fibrous carbon substrate in a carbonaceous matrix have been developed [23–36].

7.2 CARBON FIBER COMPOSITES

High-performance composites are materials that have outstanding strength-to-weight properties which other materials lack. They generally consist of a matrix material and a reinforcing agent. Several types of matrices, such as plastic

Carbon Fiber Applications

(which may be thermosets, polyesters, or phenolics), metals, and carbon have been used for reinforcement by different types of fibrous fillers, including glass fibers, silicon carbide fibers, polyester and nylon fibers, and high-performance carbon and aramid fibers. Glass fibers have been used as reinforcing fibers for the last 25 years because they provide a good balance of properties, but they have poor modulus and the composites lack impact resistance and stiffness. Thermoplastic fibers are lightweight, less brittle, but they are not heat resistant. High-performance carbon and aramid fibers are now widely used to reinforce plastics and carbon matrices for high-performance composites because of their outstanding specific strength, stiffness, and high resistance to heat and chemical attack. For specific properties composites are also being prepared by using more than one reinforcing fiber.

Just as there are different types of reinforcements and matrices, there are different types of composite compounds, including prepregs (short for products made by preimpregnating woven or continuous reinforcement), sheet molding compound (SMC), and bulk molding compound (BMC). These different composite compounds can be prepared using different techniques, such as hand lay-up, compression molding, or injection molding, depending on the type of the fibrous materials and the matrix material. Three main types of composite materials which have been investigated and generally prepared using carbon fibers are: (1) polymer matrix composites (CFRP); (2) carbon-metal composites (CFRM); and (3) carbon-carbon composites (CFRC).

7.2.1 Carbon Fiber-Polymer Matrix Composites

Carbon fibers are considered excellent reinforcement for plastic materials because : (1) they possess the desirable properties of low density (1.44–2.7 g/cm^3) and high strength (3–4.5 GPa); (2) their high modulus makes the reinforced structure stiff; and (3) plastics have good adhesion characteristics toward carbon fibers and can make sound structures. When combined with a resin binder to support the applied load, these fibers provide mechanical capability equal to or exceeding those of most metals. The structural and mechanical performance of a composite is determined by a number of factors including volume fraction, length distribution, orientation of the fiber, stacking sequence and number of angle plies in laminates. There are several methods by which carbon fiber-reinforced polymer (CFRP) structures are prepared.

7.2.1.1 Carbon fiber-polymer matrix composite fabrication methods

There are a number of manufacturing processes been developed for the carbon fiber-polymer matrix composite with a common ground in selecting most efficient and economic process. They are: (1) selection of the cheapest and most reliable raw materials; (2) use of fast automated equipment to avoid high labor cost; and (3) use of iterative on-line control to assure quality and minimize post-fabrication inspection. The fabrication processes are classified as: (a) molding processes which use fibers already converted into web or sheet forms, and (b) direct roving process where the dry reinforcements in the form of a continuos tow and the resins are sprayed directly to the mold surface.

7.2.1.1.1 Molding processes

The common feature of the molding processes is that the reinforcement is first converted into a sheet or web form, such as a woven cloth, a prepreg or sheet molding compound, a continuous or discontinuous random mat. Then, the dry web of fibers is impregnated with resin either before or during the final molding operation. A sufficient molding process requires a low viscosity of resin to fully impregnate the fibers, which can be achieved by diluting with a solvent and impregnation may be assisted by pressure or vacuum. The molding processes have a number of variants, depending on the viscosity of flow, use of vacuum, pressure and heat, the design, and size of mold cavity.

Sheet and Bulk Moldings. Sheet and bulk moldings offer the automotive industry the capability for high volume production. The feed stock of sheet molding is a sheet of about 6 mm thick, while bulk molding is normally in the form of a rope 30–50 mm diameter. The sheet molding compound is prepared by chopping continuous strand rovings onto a plastic film that has been previously coated with a resin paste. The paste and fibers are gently mixed together and a sheet product is formed. The curing time depends on the temperature, resin and parts thickness. Rolls of sheet or bulk molding compounds are stored until the viscosity has increased to a predetermined level. Unsaturated polyester or vinyl ester resins are the major polymers used in the sheet and bulk moldings. In the automobile industry, calcium carbonate is used extensively as a filler in these polymers to control flow during molding due to its low cost and ability to provide smooth molded surfaces.

Leaky Mold Technique. In this method, weighed quantities of aligned carbon fiber tow are introduced into an excess of a low-viscosity and low-temperature curing resin contained in a mold. The resin is allowed to wet the carbon fibers. The excess of the resin is then forced out of the leaky mold by applying slight pressure. The gap between the top force and the walls of the mold is kept between 2 and 4 thousandths of an inch [16], allowing rapid draining of the resin but preventing extrusion of the carbon fibers. A typical CFRP contains equal parts by weight of the resin and fiber, corresponding to a fiber volume fraction of about 40–43% [16].

Injection Molding. In the case of carbon fiber-reinforced thermoplastics the fabrication is usually carried out by injection molding [16] which requires a very high pressure, typically 100–200 MPa. The feed stock is usually pre-compounded molding pallets containing 20 to 40% by weight of very short fibers (Figure 2). The compound is hopper-fed and plasticized by shear deformation under heat, producing a viscous homogeneous mix. The injection molding process is capable of producing very complicated parts to very accurately controlled dimensions. The shortage of this method is the cost very high when compared with other fabrication processes.

Hot Compression Molding. This method is similar to the sheet molding process, generally used for mass production of structures using prepregs. This technique is often applied to phenolic resins which require much higher consolidation pressure. The layers of fibers are impregnated with heat-setting resin in dilute solution and dried to yield prepreg tapes. The prepregs are then cut into the required size and stacked at the required orientation in steel molds and consolidated under heat and pressure. The curing temperature may vary from 90 to 320°C, depending on the resin matrix [16]. The hot compression molding process is widely applied for the mass production of relatively small complicated parts using multi-cavity tools.

Vacuum-Bag Molding. In this technique the prepreg layer is consolidated by applying a pressure of about 1 atmosphere. This is carried out by evacuating the air held between a mold tool, the prepreg, and the impervious plastic film, which is firmly attached to the mold. The prepreg assembly is generally heated in an oven to cure it. The technique is very useful when expensive autoclaves and presses are not available [41].

Resin Injection Molding. In this technique, the pre-formed fibers are placed in the mold and the resin is subsequently injected to impregnate the fiber and fill the mold cavity. The mold design (Figure 3) is an important factor for the process, which needs to be constructed so that the resin reaches all the areas in a short time before the onset of gelation without causing movement of the pre-formed fibers. Resins with low viscosity are commonly used in this process to allow good penetration of the resin into the fiber. In application, large and complex products, such as a car body shell, can be made inexpensively and efficiently using this technique.

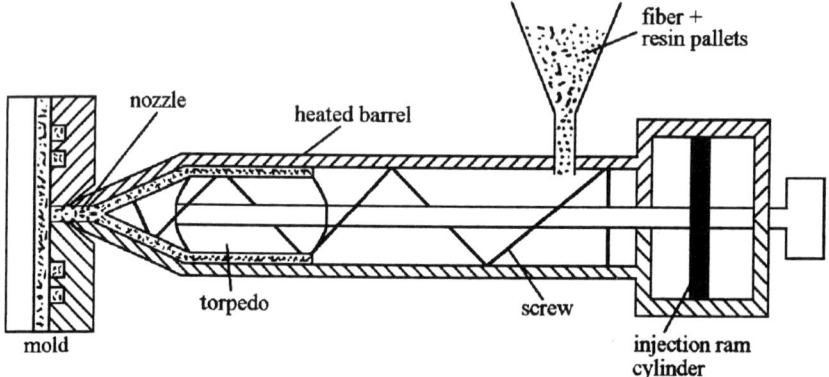

Figure 2 Injection molding process of carbon fiber/polymer composite fabrication.

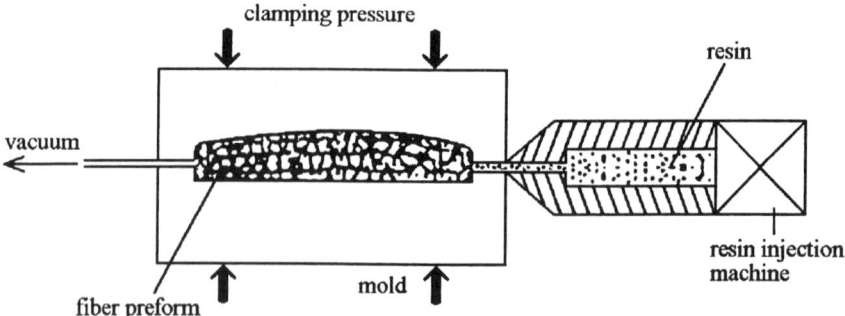

Figure 3 Resin injection molding method of manufacturing carbon fiber/polymer matrix composites.

Carbon Fiber Applications

Autoclave Molding. This technique is similar to the vacuum-bag technique except that pressure is applied from outside on the plastic membrane. This helps in removing the air from the bag faster and ensures better consolidation. This process requires an expensive autoclave and is generally used to fabricate aerospace structures, such as high quality flat or curved panels [41,42]. A single layer of a fine peel-ply is stacked on both sides of the laminate which is built on a metal mold plate. A perforated release film is covered on the surface of the laminate and bleeder or breather clothes will be added if necessary. As shown in Figure 4, the whole assembly is put into a vacuum bag which is sealed to the mold plate. Then, the mold is transferred into the autoclave in a vacuum system where the temperature and pressure are controlled and all the volatile materials are removed. The consistent process which allows a very uniform pressure over the entire surface of the molding makes this technique very attractive in the high technique areas.

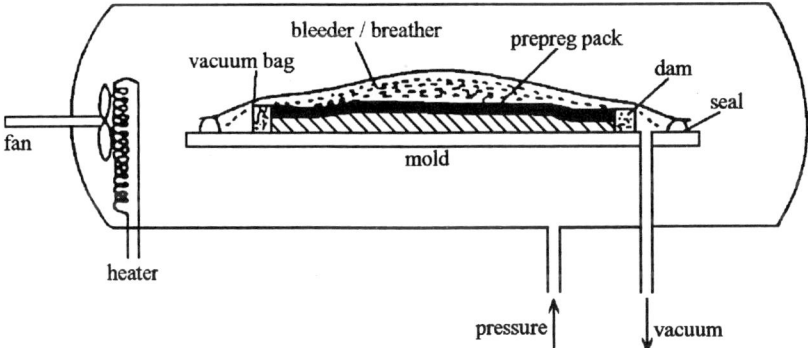

Figure 4 Autoclave molding method of manufacturing carbon fiber/polymer composites.

7.2.1.1.2 Direct roving process

Filament Winding Technique. This technique involves the building up of layers of fibers impregnated with the resin by winding them on a rotating frame. Generally, continuous lengths of carbon fibers are used in this technique. Carbon fibers are passed through a bath containing the thermosetting resin and wound around a mandrel in the conventional manner according to a programmed lay-up pattern, and then cured to yield a plastic structure which is strengthened and stiffened in the desired directions. There are two types of patterns used in this technique: helical winding in which a fixed angle is chosen; and biaxial winding where two winding angles (0° and 90°) are selected. Helical winding is more common because of its simplicity, but biaxial winding enables the structure with special design and property. Filament winding has been used to make aerospace hardware, sporting goods, and components for the automotive industry [17].

Pultrusion Technique. The molding and filament winding techniques alone could not meet every design requirement of CFRP, and consequently, an alternative process has been developed [13,37–40] which is capable of preparing elongated structural forms of CFRP with predominantly unidirectional reinforcement. This method involves drawing the fibers through a bath of liquid resin and then directly and continuously through a heated die to produce a continuous section.

The method, known as pultrusion, is capable of realizing the high-potential strength and stiffness of the carbon fiber by accurate placement of the reinforcement. The resin-impregnated tows of carbon fiber are drawn through an orifice under tension into a die. Hydrostatic pressure is applied in the orifice construction to drain the excess resin and to expel the entrapped air. This produces a unidirectionally aligned wet-stage composite with a controlled volume fraction of the fiber and resin, depending on the cross-sectional areas of the fiber and orifice. The drawing force and drawing rate are determined by the shape, size, volume fraction, and resin viscosity, The wet-stage composite is allowed to stay in the elongated die until gelation, partial curing, and solidification of the composite into the required shape has taken place. There are several critical factors controlling the success of this operation which include heat transfer, flow of liquids through porous media and cure kinetics. Spencer [13] has described pilotscale pultrusion equipment (Figure 5), which is capable of handling up to 200 tows (1 tow = 10,000 filaments) and giving a range of CFRP profiles. The production rate can range between 70 and 150 mm/min [13].

The versatility of the pultrusion method can possibly be exploited to produce curved and tapered products by incorporating multidirectional reinforcement by integrating tape or filament-winding operations as a continuous process [12]. It is suggested [13] that the pultrusion technique may also be used to process reinforced ceramics and cements or high-temperature polymers such as polyimides, as well as thermoplastics such as polysulfones. The pultruded CFRP has been found to have better mechanical properties than the CFRP prepared by the molding techniques (Table 4).

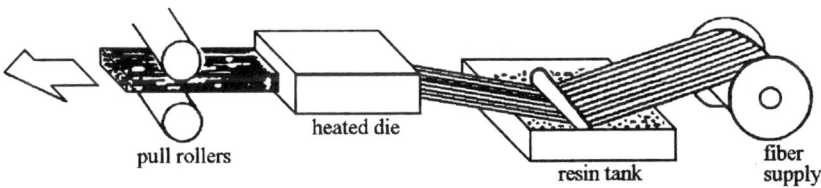

Figure 5 Design of a commercial pultrusion machine of fabricating carbon fiber/polymer composites [13].

Table 4 Comparison of Mechanical Properties Between Pultruded and Molded CFRP

	Flexural strength (GN/m^2)	Flexural modulus (GN/m^2)	Interlaminar shear strength (MN/m^2)	Izod impact strength (J)
Pultruded composite	1.44	148	83.4	27
Molded composite[a]	1.26	135	71.2	26.6

[a]data obtained by leaky mold technique.
Source: from ref. 13.

7.2.1.2 Carbon fiber-polymer matrix composite properties

7.2.1.2.1 Effect of carbon fiber form on CFRP properties

Carbon fiber reinforcements are now available in a wide range of forms, such as long and continuous tows, yarns and cloth prepregs, woven fabrics, mats, braids, pellets, and chopped fibers (Figure 6). Each form can be used to prepare CFRP structures, depending on the need of the properties of the composite and product. They may be dispersed in a thermosetting or thermoplastic matrix with random or directional orientation. However, the properties of the composites will be different, depending on the size of the fibers and their orientation. Milled fibers, which have an average length of about 300 μm, have also been found to improve the properties considerably [20], although improvement is better when longer fibers are used. Typical properties of reinforced nylon 6/6 prepared by injection molding techniques are given in Table 5. The longer carbon fiber reinforcement significantly increases both the strength and the modulus while maintaining or increasing good electrical conductivity.

Continuous carbon fibers are used as reinforcements when an ultimate in performance and weight reduction is needed and when thermal expansion is to be kept at very low levels. These are available in several forms, such as yarns or tows, containing from 400 to 160,000 individual filaments [20], unidirectional impregnated tapes, plies at selected fiber orientation, and fabrics of many weaves and weights. The choice of the form of the continuous fiber depends on the fabrication process. For example, filament winding processes generally use lower filament count yarns to minimize looseness of winding, whereas the compression molding processes make use of unidirectional tape.

Carbon fibers can also be used in the form of fabrics when complex part shapes and processes require careful orientation of the reinforcement. Carbon fabrics are available in several designs and can have a plain or a satin weave. Commonly used 8 harness satin-weave fabrics retain most of the fiber characteristics in the composite and can be easily draped over complex mold shapes. Plain weave fabrics are less flexible and are suitable for flat or simple contoured parts and have less translation of fiber properties into the composite. PAN carbon fabrics are usually prepared by weaving carbonized yarn, whereas pitch-based carbon fibers are usually woven at an intermediate processing stage and then converted into a high-modulus product while in the fabric form. The properties of carbon fabric-epoxy composites for some fabric materials are given in Table 6. Although the strength properties of pitch-based carbon fabrics are not as good as those for PAN-based products, the inherent lower cost potential for the pitch materials and their continually improving properties make this product attractive.

Figure 6 Different carbon fiber forms. (From Torayca Bulletin, courtesy of Toray Industries, Inc.)

Table 5 Effect of Carbon Fiber Forms on the Mechanical Properties of Nylon 6/6 Reinforced Composites by the Injection Molding Techniques

	30% chopped carbon fiber	30% milled carbon fiber	30% long carbon fiber
Tensile strength (MN/m^2)	245	111	280
Compressive strength (MN/m^2)	—	98	—
Flexural strength (MN/m^2)	357	177	372
Tensile modulus (GN/m^2)	10.2	8.9	11.6
Flexural modulus (GN/m^2)	20.2	10.4	21.6
Coefficient of thermal expansion (10^{-4} in/in-°C)	0.19	0.27	—
Electrical resistivity (Ω-cm)	10^2	10^3	—

Source: from ref. 20.

Table 6 Effect of Different Weaving Technique on the Mechanical Properties of PAN- and Pitch-Based Carbon Fiber/Epoxy Composites

	Tensile strength (MN/m^2)	Flexural strength (MN/m^2)	Shear strength (MN/m^2)	Tensile modulus (GN/m^2)	Flexural modulus (GN/m^2)
Satin weave					
PAN (Thornel 300)	628	835	63	70	66
Pitch (Thornel PVCB-20)	383	489	31	42	41
Plain weave					
PAN (Thornel 300)	559	767	59	66	66
Pitch (Thornel PVCB-20)	349	420	24	98	93

Source: from ref. 20.

7.2.1.2.2 Effect of resin matrix on CFRP properties

In an advance composite material, the matrix resin is the continuous phase which binds the fibers and transfers the stress to the fiber across the interface. The matrix resin has a significant influence on some of the mechanical properties of the reinforced carbon fiber composites, such as the interfacial strength, interlaminar fracture toughness, etc. The matrix resins are generally divided into two groups: *thermosetting* and *thermoplastics*. Thermosetting resin possesses a lower molecular weight and be able to be irreversibly cross-linked with carbon fiber. On the other hand, thermoplastics are normally in the solid form with a high molecular weight, can be compressed mixing with carbon fiber upon heating. Some of the characteristics of these two types of polymer matrixes are listed in Table 7. Most common thermoset resins are epoxy, vinylester, and polyesters, etc. Epoxy resins are the most commonly used resin in the carbon fiber composites because of their good mechanical properties and ease of processing and handling. Mechanical properties of some representative thermoset resins are compared in Table 8.

Several high-temperature plastic matrices have now been developed which improve the temperature range of applications of these composites. Polyimide resins are one such class which can produce carbon fiber-reinforced plastic laminates for use up to 400°C [47]. These composites have good thermal resistance, which extends their long-term use potential to higher temperatures (Table 9). They can be used profitably as compressor blades and for rotating parts and are particularly useful in internal engine components because of their good stability in mineral oil.

The importance of the matrix in determining the performance properties of high-strength carbon fiber composites should not be underestimated. It has been examined using three types of epoxy resins: a conventional bisphenol-A epichlorchydrin resin (Epikote DX 231) cured with a dicyanodiamine-based curing agent, a precondensate resin (Epikote 210) cured with a boron trifluoride amine complex, and a conventional cycloaliphatic resin (Epikote DX 284) cured with 4,4-diaminodiphenyl sulfone, by Reader and Morley [46]. Some of the mechanical properties of these composites, at room temperature, before and after immersion in water for 4 weeks, are given in Table 10. The composites based on the conventional resin showed good mechanical properties up to 70°C and maintained these properties during prolonged aging at this temperature. The composites prepared with the precondensate resin, however, gave good performance up to 150°C and withstood aging at this temperature. The unconventional resin gave composites with rather higher interlaminar shear strength (ILSS) and good retention properties at temperatures between 170 and 180°C, which showed indifference to prolonged aging at 150°C. The composites based on all three resins showed good moisture tolerance.

Table 7 Characteristics of Thermoset and Thermoplastics Resins

Characteristics	Thermoset resin	Thermoplastic resin
Properties		
Formulation	complex	simple
Viscosity	low	high
Solvent resistance	excellent	poor to good
Fracture toughness	poor to good	good to excellent
Mechanical property (−50 to 90°C, hot/wet)	fair to good	fair to good
Composite fabrication		
Fiber impregnation	easy	difficult
Processing time	long	short to long
Processing temperature/pressure	low to moderate	high
Prepreg stability	poor	excellent

Table 8 Typical Mechanical Properties of Some Common Thermoset Resins used in the Carbon Fiber Composites

	Epoxy	Vinyl ester	Polyester	Polyimide (PMR15)
Tensile strength (MPa)	50–100	60–90	40–90	56
Compressive strength	100–200	110–120	90–250	187
Tensile modulus (GPa)	3–6	3–3.5	2–4.5	3.2
Density (g/cm^3)	1.1–1.4	1.1	1.2–1.5	1.43
Elongation (%)	1–6	2–5	2	—
Thermal expansion coefficient (10^{-6}/K)	60	80	100–200	50

Table 9 Mechanical Properties of Some Polyimide Reinforced Carbon Fiber Composites

	Young's modulus (GPa)	Tensile strength (MPa)	Elongation (%)
HM carbon fiber	350	1700–2100	0.5
HT carbon fiber	210	2300–3000	0.9–1.1
Kevlar 49 fiber	130	2700	2.1
S-Glass fiber	86	4400–4800	5.4

Source: from ref. 60 and 61.

Carbon Fiber Applications

Table 10 Mechanical properties of carbon fiber composites prepared from different resins [46]

Mechanical properties	Conventional resin (Epikote D × 231)	Precondensate resin (Epikote 210)	Nonconventional resin (Epikote D × 284)
(A) Initial properties at 23°C			
Flexural modulus (GN/m^2)	131	151	124
Flexural strength (MN/m^2)	1480	1130	1430
ILSS (MN/m^2)	72	79	100
(B) Properties after immersion in water for 4 weeks at 23°C			
Flexural modulus (GN/m^2)	122	153	113
Flexural strength (MN/m^2)	1450	1130	1300
ILSS (MN/m^2)	66	74	100

7.2.1.2.3 Carbon fiber hybrid composites

Fiber-reinforced composites are one means of providing materials that exhibit specific strength and stiffness. However, property requirements for composite materials are becoming increasingly complex because the materials have to operate under varying physical, mechanical, and chemical environments. Consequently, the level as well as the combination of composite properties has to be more severe and complex. Thus there has been a spurt in the development of a variety of reinforcement fibers with different mechanical, physical, and electrical properties. It is now realized that the use of more than one reinforcement material in composites is important since no one fiber can tackle all the technical and economic requirements of aerospace, transport, sports, and other industries. Thus it is important to consider a more general composite behavior rather than a single property such as stiffness. For example, in the case of structural applications, response to impact loading or damage tolerance is an important area of concern. Unlike metals, which deform by plastic flow, most organic matrix composites can only absorb energy elastically or by various fracture processes. Energy absorption during loading is related to the area under

7) show that high-modulus carbon and boron fibers require relatively little energy to reach failure at strains on the order of 1–1.5%, while Kevlar® and glass fiber, although they fail at similar stresses, do so at higher strains, signifying higher energy absorption. Thus it is reasonable to believe that combining different fibers in a matrix will permit another degree of freedom in engineering a balance in composite properties. Such combinations of fibers are called "hybrids" and can constitute carbon and glass fibers, carbon and boron fibers, carbon and Kevlar® fibers, glass and aramid fibers, and several other combinations.

There are several ways in which different fibers or materials can be combined to reinforce the same matrix. The three basic levels that the material combination can be achieved are shown in Figure 8. The advantage of using yarns in producing hybrids is the handling of some materials in the form of reinforcing fabrics and in facilitating the processing of brittle fibers. Hybrid fabrics, usually in woven or unidirectional tape versions, are probably the most widely recognized form of multimaterial reinforcement. Hybridization within the matrix can be achieved (Figure 8) by using two or more fibers in the same fabric (intraply hybrid) or by mixing single-fiber plies through the composite cross section (interply hybridization).

Generally, the preferable hybrids consist of Kevlar® and carbon fibers because these hybrids are cost-effective and show marked improvement in impact resistance and fracture strength. However, it should be mentioned that although impact resistance is enhanced by the addition of Kevlar® to a carbon fiber composite, properties such as modulus, flexural, and compressive strength are attenuated. Better damage tolerance, in terms of residual strength after impact, was reported in these interply composites when the stacking sequence involved Kevlar® outer plies, indicating a shielding of the ''core'' carbon fiber plies [43].

Thus high-modulus composites from hybrid materials offer attractive alternatives to metal structures. Hybrids of high-strength, high-modulus carbon fibers with glass and aramid fibers are used in tubular structures for making golf clubs and bicycle frames, to reduce cost. Silicon-glass-carbon fiber hybrids have been tried on helicopter blades. It is possible to create complex hybrids, for example, by using a hybrid filament to produce a hybrid yarn and then a hybrid intraply fabric. A typical case is a metallized glass fiber-Kevlar® fiber hybrid which can be used to fabricate the conductive lightweight, stiff composite needed in some microwave and lightning strike-protection applications. Thus the hybrids, while providing quite a new generation of materials, may afford cost benefits or penalties, depending on whether cheaper or more expensive fiber reinforcements are being replaced in the original reinforcement.

Carbon Fiber Applications

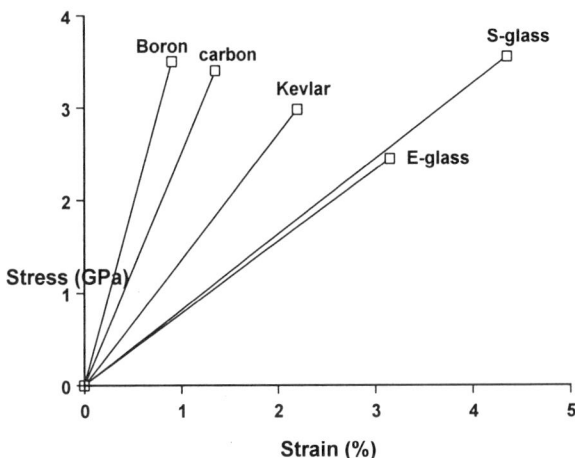

Figure 7 Stress-strain curves of common reinforcement fibers [43].

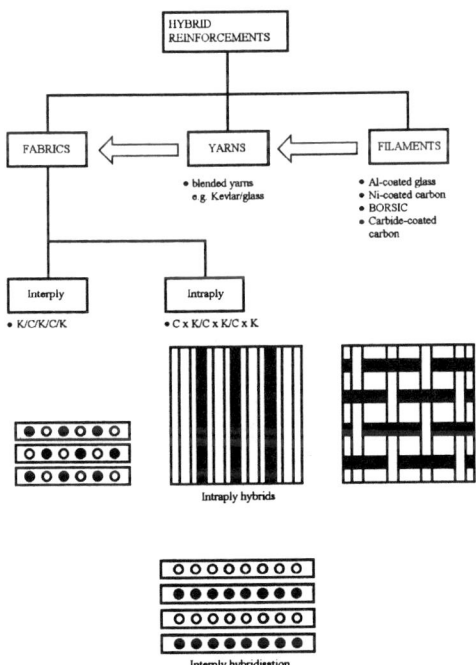

Figure 8 Hybrid fabric reinforcement [43].

A new type of super hybrid material has recently been developed by Haga et al. [22], which is composed of an aluminum alloy sheet with an unidirectional Kevlar® epoxy prepreg sheet and an unidirectional carbon epoxy prepreg sheet, abbreviated as A, K, and C, respectively. They found that the AKC laminate exhibited higher tensile strength and modulus by about 30% and 10% than AK laminate hybrid. The fracture strain of AKC laminate is increased by about 15% than the value of KC laminate and this enhancement is contributed from the incorporation of aluminum in the hybrid composite. Some of the mechanical testing results are listed in Table 11. The hybrids were tested have the following volume fractions: AKC, 3:1:1, AK, 3:2, KC, 1:1. As indicated, the compressive properties is enhanced by 12% for the new hybrid compared with AK laminate. The compressive strength of AKC laminate has been improved to about one-half of its tensile strength. The authors concluded that: (1) the AKC new super hybrid composite has both high strength character of aluminum/carbon fiber/epoxy (ACE) hybrid, and the high ductility of AK materials and (2) significant improvement of compressive strength than AK hybrid.

Another type of hybrid composite, which contains two different matrices rather than two reinforcements has been developed to make conventional carbon fiber-reinforced composites tougher, that is, more tolerant to damage and stress concentration. These are based on the interleaf concept [45], in which a thin, high-elongation resin leaf is placed between each fiber ply to provide

Table 11 Some Mechanical properties of the Tested Hybrid Composites

Hybrid composites	AKC	KC	AK
Tensile strength (MPa)	1080	1970	820
Compressive strength (MPa)	530	520	330
Tensile modulus (GPa)	79.7	94.6	71.8
Compressive modulus (GPa)	79.1	86.0	70.3
Shear strength (MPa)	127	30	117
Shear modulus (GPa)	17.2	2.9	15.3
Tensile failure strain (%)	2.25	1.95	2.26
Compressive failure strain (%)	1.44	0.84	1.59

Source: from ref. 22.

toughness (Figure 9). Each ply has its fibers in an epoxy matrix for high hot-wet performance. The two resins remain discrete during cure, so interleaf toughness does not compromise matrix hot-wet compression strength. One such composite, CYCOM-HST-7, developed by Kreiger [45], is twice as tough as the conventional CFRP and maintains a 50% weight advantage over aluminum in the critical allowance of hot-wet compression.

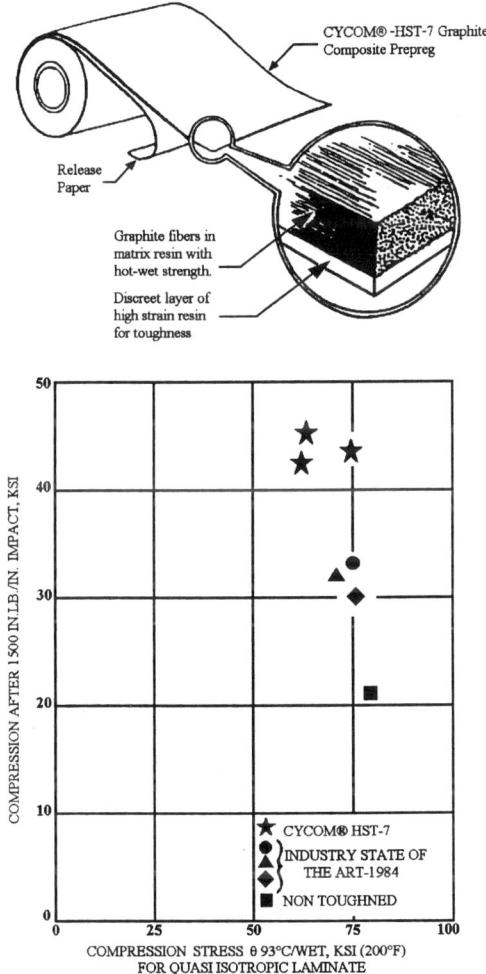

Figure 9 Description of Cycom-HST-7 graphite prepreg (top) and its comparison with other composites [45].

7.2.2 Carbon-Carbon Composites (CCC)

Carbon-carbon composites are considered as one of the most attractive materials in the space shuttle applications, particularly used as components withstand high temperature (up to 1800°C), such as nozzles, leading edges, nose cone etc.) due to a combination of good strength retention and low density [51]. Today, the US and France are leading in the development and production of CCCs with a multi-million dollar business (about 150 million US dollars). The major application of CCCs is the aircraft brake discs, and Schmidt [25] indicated that in 1996 structural 2-D carbon-carbon composites cost $440–2870/kg, and large 3-D billets cost $1100–3300/kg. Coated carbon-carbon composites for antioxidation cost three to ten times more than regular grades.

The carbon matrix is usually obtained from chemical vapor deposition or thermal decomposition of a carbon source such as phenolic resin. The main drawback of carbon-carbon composite is the formation of gaseous oxides of carbon at temperature exceeding 450°C, leading to a significant loss of strength. At a higher temperature application in the range of 1000~2000°C, it is necessary to protect carbon-carbon composites against oxidation. There are two approaches for that purpose, one is to deposit diffusion barriers to prevent oxygen reaching to the carbon surface, and the other method is to deposit a layer of inhibitors to slow down the oxidation process.

The selected coating materials such as silicon-based ceramics (SiC or Si_3N_4) need to have the following characteristics: a low oxygen permeability, good oxidation resistance, good adhesion to carbon, and a thermal expansion match with carbon. As seen in Figure 10 [52], the carbon fiber reinforced silicon carbide matrix has improved the tensile strength property compared with other types of fibers. Tensile strength reduces at approximately 1200°C and the limit of use is reached at 1500°C. The low density of carbon-carbon composites result in a higher specific strength than the other types of fiber composites or metal alloys at high temperature (Figure 11). The weight saving is critical in the aircraft application, for example, conventional steel brakes weigh about 1100 kg, but instead carbon-carbon composite brakes only weigh 700 kg. Carbon-carbon composites can reach much higher strength at the same density.

Carbon-carbon composites consist of a fibrous carbon substrate in a carbonaceous matrix. Although both constituents consist of the same element, carbon, the composite behavior is very complex and depends on the type of the reinforcing fiber, the fiber volume fraction, and the fiber arrangement, on the one hand, and on carbonaceous matrix parameters such as matrix structure (which may range from carbon to graphite), the processing conditions, and temperature and treatment of the composite, on the other hand. The selection of the precursor matrix and the carbonization conditions are generally those that produce the maximum carbon yield.

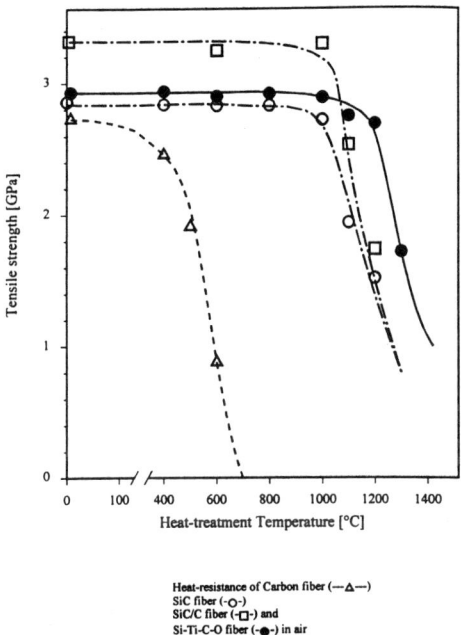

Figure 10 Tensile strength versus temperature for different types of fibers [52].

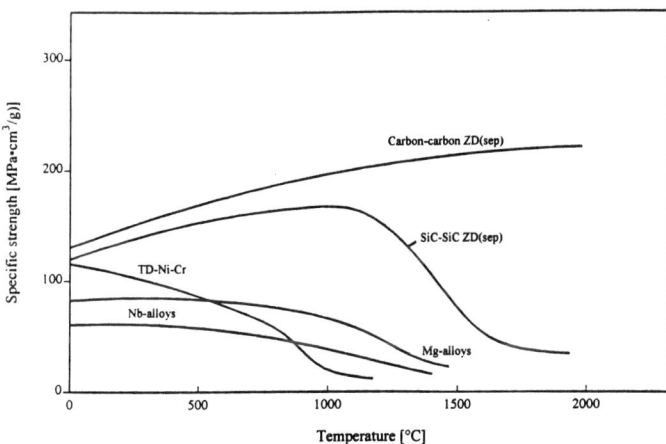

Figure 11 Specific strength of carbon-carbon composites at different temperatures in comparison with other materials [52].

7.2.2.1 Preparation of carbon-carbon composites

Preparation of carbon-carbon composites essentially involves the impregnation of continuous carbon filaments with an organic matrix precursor, which may be a pitch or a resin or other organic precursors [24–28, 62–66] by a wet winding technique. The impregnated fibers are heated to remove the solvent and increase the molecular weight of the resin. The fibers are then oriented in the desired arrangements and in the required volume fraction and molded into prepregs, which are further pressed and cured into laminates [27]. The laminates are carbonized in an inert atmosphere at 800–1100°C, at rates varying between 5 and 20°C per hour depending on the thickness of the material [28]. The weight loss and the volume shrinkage of the organic carbon matrix produce a porous carbon-carbon composite that must be densified. The densification of the carbonized composite is carried out either by chemical vapour deposition (CVD) of carbon from a hydrocarbon gas (usually methane or benzene), or by multiple impregnation and carbonizations with a liquid organic resin or pitch.

Chemical Vapor Deposition. The chemical vapor deposition of pyrolytic carbon on carbon substrates is an important process for fabricating carbon composites which are used in many aerospace applications. The CVD process is simpler and cheaper than other methods used for carbon-carbon composite fabrication, such as pitch impregnation. It also has the advantage of obtaining a wide range of composite properties by setting the experimental conditions. The CVD technique is based on the diffusion of a carbonaceous gas into the porous substrate and the diffusion-controlled reaction of the gas with the heated substrate, resulting in the deposition of carbon and elimination of gaseous by-products. The quality and structure of the resulting deposited carbon depends on the density and surface area of the substrate and on experimental conditions, such as temperature, pressure, and rate of flow of the gas. Several CVD techniques have been developed for the infiltration of the porous substrate, but the isothermal and temperature gradient techniques are more commonly used [27,29–31,64,66]

In the isothermal technique (Figure 12) the substrate is heated by an induction-heated susceptor made of graphite so that the substrate and gas are maintained at an uniform temperature. The hydrocarbon gas decomposes at about 1100°C and at a reduced pressure (50 torr) in contact with the hot substrate, depositing carbon. This technique, however, results in overcrusting of the outer surface by deposited carbon, which has to be machined away to achieve maximal density. In general, infiltration requires about 2000 hours [27].

The infiltration by the temperature gradient technique (Figure 13) involves shorter times. A nonconducting sleeve, which serves as an insulator and channels the reactants and the by-products, is placed between a graphite mandrel, which acts as a susceptor, and the induction coil. The porous substrate is not induction heated, owing to its low density, and therefore has poor coupling. The inner side of the substrate is in direct contact with the mandrel and is therefore at a higher temperature. The outer surface of the substrate is in contact with cool gases flowing over its surface. Consequently, the deposition starts at the mandrel and progresses radically through the substrate as the compacted parts of the substrate become inductively heated [27].

Multiple Impregnation Technique. The densification of the carbonized composite is carried out in a combined vacuum-pressure impregnation step followed by recarbonization of the impregnated material (Figure 14). The porous composite is first evacuated at 400 Pa and then dipped into the impregnating liquid for about 20 hours under a pressure between 1 and 200 Pa, depending on the precursor material [27]. The recarbonization is then carried out between 1000 and 2880°C. This densification cycle has to be repeated 4 – 12 times, depending on the thickness and the required final density of the composite [27,32–34].

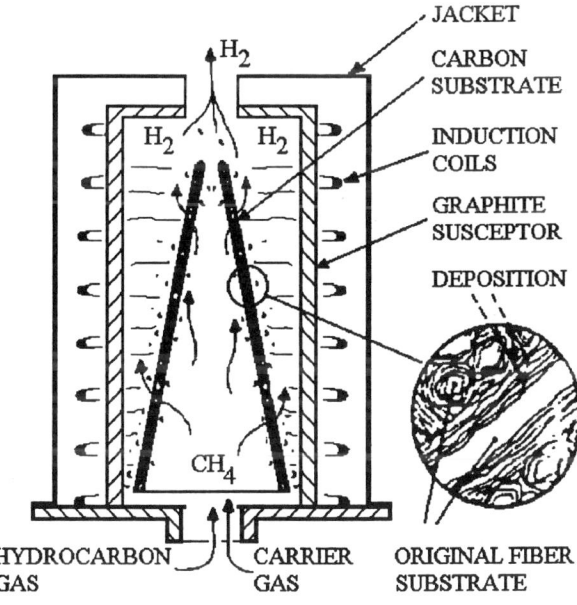

Figure 12 Isothermal technique to deposit pyrolytic carbon on carbon substrates [30].

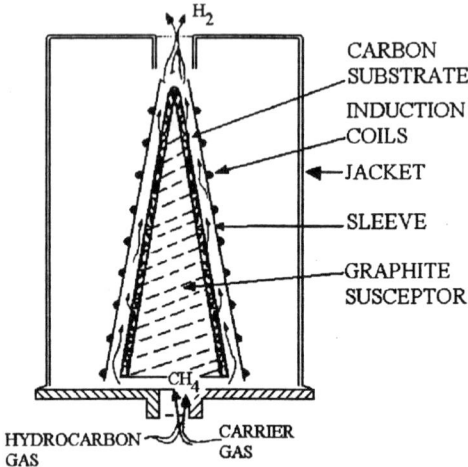

Figure 13 Temperature gradient technique of chemical vapor deposition [30].

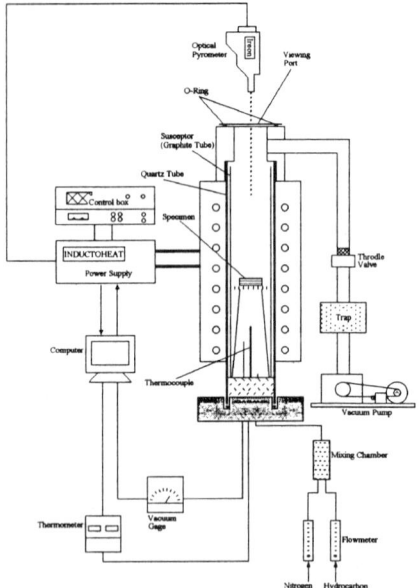

Figure 14 Schematic diagram of multiple impregnation technique [34].

7.2.2.2 Effect of matrix precursor and carbon fiber type on carbon-carbon composite properties

Two types of precursor materials have been used in the fabrication of carbon-carbon composites: thermosetting resins such as phenolics and polyimides, and thermoplastic precursors such as pitches. Phenolic resins and polyimides are easy to handle and give considerable carbon yield (60–65%), but they undergo a high isotropic shrinkage. Consequently, pitches, which under certain conditions [67–70] can yield high carbon contents, are preferred because of their lower shrinkage. Synthetic pitches such as isomeric polyphylenes also give high carbon yields, but they are difficult to handle [71]. The shrinkage of the carbon-carbon composite also depends on the nature of the carbon fiber surface and decreases with decreasing concentration of active surface groups on the carbon fiber surface. The presence of these active surface groups increases adhesion between the fiber and the matrix, which along with high cross-sectional shrinkage, causes pyrolysis cracks in the matrix perpendicular to the fiber axis and leads to shear stress concentration, which may result in fiber damage.

Interactions at the matrix fiber interface during the processing of carbon-carbon composites mainly involve the nature and type of carbon fiber, the amount of precursor, and the composite shrinkage. Fitzer and co-workers [35,55] and Huettner [36] have shown that to achieve high values of flexural strength of the composite and translation of fiber strength, high-modulus PAN-based or MP pitch-based carbon fibers should be preferred, and if possible, without surface treatment. Diefendorf [57] has recently reached a similar conclusion and listed experimental results in Table 12. As shown, the high modulus surface-treated fibers still fractured at the matrix failure strain.

Table 12 The Effect of Carbon Fiber Type on the Mechanical Properties of Unidirectional Carbon-Carbon Composites ($V_f = 55\%$)

	High modulus surface-treated	High strength surface-untreated
Tensile strength (MPa)	575	600
Compressive strength (MPa)	380	285
Bending strength (MPa)	850–1000	1250–1600
Tensile modulus (GPa)	220	125
Compressive modulus (GPa)	250	10
Fracture toughness (kJ/m^2)	20	70

Source: from ref. 57.

Table 13 Effect of Matrix Precursor on Carbon-Carbon Composite Modulus Increase

Matrix	Heat treatment at 1000°C	Heat treatment at 2600°C
Pitch	130%	210%
Phenolic resin	110%	140%

Source: from ref. 57.

On the other hand, the high strength surface-untreated fibers exhibited higher strengths and fracture toughness values. In carbon-carbon composites the matrix contributes appreciably to the stiffness of the composite. Table 13 shows modulus of composites with different matrix precursors, such as phenolic, and CT-pitch after final heat treatment temperatures of 1000 and 2600°C. In all cases, the experimental modulus exceeds the theoretical expected ones even for glass-like carbon matrices and for nongraphitized composites [36]. This has been attributed to a preferred orientation of the matrix carbon layers, which are aligned "epitactical" around the carbon fiber surface [36,57,58].

7.2.2.3 Effect of final HTT and number of impregnation on bulk properties of carbon-carbon composites

Carbonization of all carbon-carbon composites is usually carried out at 1000°C when the residual carbon still contains appreciable quantities of hydrogen. Therefore, the bulk properties of carbon-carbon composites are influenced considerably by the final graphitizing heat treatment temperature and the number of impregnation (densification cycles) (Figure 15). The graphitizing treatment causes shrinkage of the secondary carbon and also alters the microstructure of the binder as well as of the impregnation coke. In the case of pitch precursors, the preferred orientation of the binder in impregnation cokes is initiated through a mesophase intermediate in a direction parallel to the fiber, due to the thin slit faces enhancing the Young's modulus of the composite. Thus the quality of the binder coke is one of the more important parameters that determine the behavior of carbon-carbon composites. Diefendorf [57] reported that the final heat treatment temperature had an influence on the mechanical properties of carbon-carbon composites. For example, the 2000°C heat treatment on a high modulus (400 GPa) carbon fibers produced modulus decreasing from 180 GPa at room temperature to 170 GPa. However, the tensile strength remained almost constant, varying from 0.95 GPa at room temperature to 1.1 GPa at 2000°C.

Carbon Fiber Applications

Carbon-carbon composites translate the outstanding physical and mechanical properties of carbon fiber into a bulk carbon. The applications of these composites are well established in certain high-performance areas such as space vehicles, rockers, and ultrasonic planes. Being high-temperature materials, carbon-carbon composites are also being introduced in general engineering applications where high strength, high stiffness, low thermal shock, low weight, high fracture toughness, and chemical resistance are required. However, the susceptibility of these materials to oxidation and dusting out of the matrix under extreme stress conditions limit their applications. However, these problems can be partly solved by fabricating carbon-carbon composites with hybrid carbon-polymer matrices. Such composites are fabricated by using the conventional carbon precursor impregnation process, followed by a final impregnation by a cross-linking resin which is not subsequently carbonized [35]. Such carbon-carbon composites have strength, modulus, and interlaminar shear strength comparable to those of carbon fiber-reinforced polymers (CFRPs) (Figure 16).

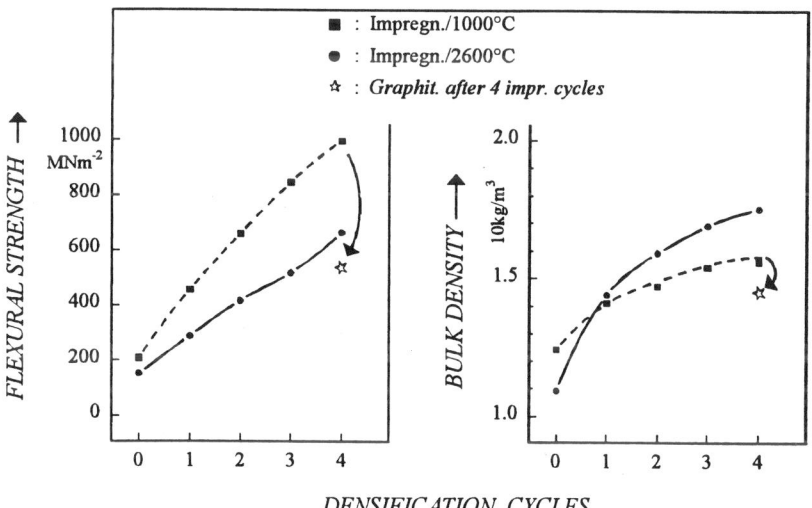

Figure 15 Flexural strength and bulk density of unidirectional carbon-carbon composites using Sigrafil HM PAN-based carbon fiber and coal tar binder pitch containing 10 wt% sulfur [35].

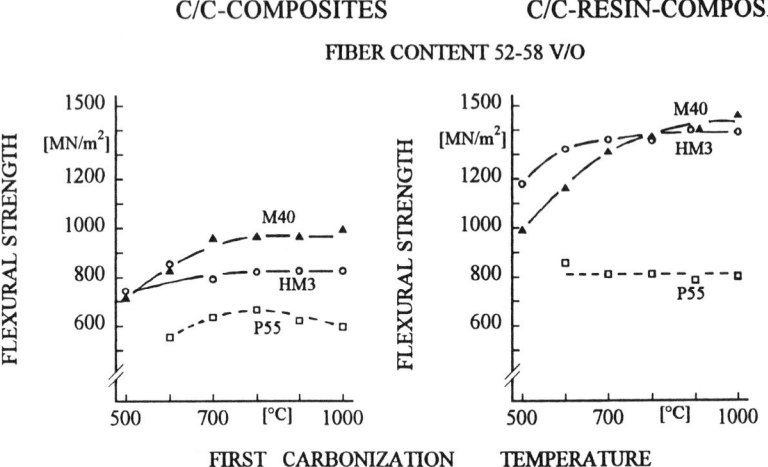

Figure 16 Comparison of flexural strength of unidirectional carbon-carbon composites with hybrid composites in which the final impregnation is made with an epoxy resin [35].

7.2.3 Carbon Fiber Reinforced Metal Composites (CFRM)

CFRM are a new class of advanced materials which are consisted of thin metallic sheets bonded together with fiber reinforced adhesive matrixes. The characteristic feature of CFRM is their exceptional fatigue resistance and as well as some other advantages they offer, such as higher strength and lower density than conventional aluminum and better machinability and impact resistance than thermoset composites. Most metals are fairly strong, ductile, and isotropic. In the nineteenth century a considerable amount of effort was devoted to optimizing these properties by refining, alloying, and heat treatment, and considerable success was achieved. However, several limitations still exist. One of these is a lack of high-temperature strength, that restricts their applications. Consequently, attempts are being made to reinforce metals with carbon fibers to make composites useful for aerospace and general engineering applications. Several metals have been reinforced by high-performance carbon fibers, but the three that have shown the most economic promise are aluminum for aerospace applications, high-temperature oxidation-resistant nickel, and lead-tin alloys as bearing metals. However, there are still problems to be overcome before any of these can reach the stage of commercial exploitation.

7.2.3.1 Fabrication of carbon-metal composites
The fabrication of metal matrix composites is more difficult than that of the organic matrix systems because higher temperatures and pressures are needed. The use of high temperatures may result in degradation of the carbon fiber. Several well-known methods for preparing metal-metal composites have been used to prepare carbon-metal composites. These methods are broadly classified into: (1) diffusion coating [2,74]; (2) electroless/electroplating deposition [7,8,76]; (3) liquid metal infiltration [9,10]; and (4) casting [115].

Diffusion coating involves the flow of metal around the fiber to provide metal-fiber contact at temperatures below that of the matrix melting-point under pressure. The diffusion coating technique deforms the precursor wires into close contact and to knit the wires into a composite by diffusion through the control of time, pressure, and temperature. This technique has been applied to fabricate carbon fiber reinforced aluminum, copper, magnesium, lead, and tin metal matrix composites.

Electroless deposition is a process where the metal ions are reduced to metals in aqueous solution by a suitable reducing agent. Electroplating comprises electrodeposition of matrix material in successive layers to build up the desired thickness of composite material. To prevent metal hydroxide precipitation in alkaline solutions, complex agents are commonly used in the acidic range acting as a buffer. The surface of the fibers is carefully catalyzed so that plating occurs only on the fiber and not on the container. However, this process needs subsequent hot processing of the fibers after electrodeposition to obtain dense composites. This may again involve fiber damage.

Liquid metal infiltration technique, which is more commonly used, involves the least fiber damage. However, the process depends largely on the wetting of the fiber by the liquid metal and on surface tension between the fiber and the liquid to achieve complete penetration of the fiber. However, in certain cases, the liquid metal may react with the carbon fiber, resulting in fiber degradation. In such a case the fibers are first coated with another metal, which prevents interaction with the metal matrix and promotes wetting.

The *casting technique* of continuous carbon fiber CFRM is widely used when the materials are used in complex-shaped parts [115], such as the joints or components of an internal combustion engine. It is also a favored method to produce thick materials or composites involved complex fiber orientations. However, the casting is not proper to make thin or flat sheet type of composites because it is difficult to supply molten metal through thin cross sections over long distances. In this technique, fibers are first put into a casting mode of desired configuration, and then the molten metal is added to form the composite part. Woven fibers are used in the epoxy reinforcement but not in the case of metals, because of the high modulus of metals resist the fiber-straightening process and lead to fibers breaking. For this reason, fiber-reinforced castings often use fibers oriented in a variety of directions in successive layers.

7.2.3.2 Carbon fiber-reinforced aluminum
Carbon-aluminum composites have been prepared by electroplating [76], hot processing of the coated fibers [74], and liquid metal infiltration of bare and coated fibers [9]. Composites with good tensile strengths at low and high temperatures have been obtained, but very few data are available for more complex mechanical properties, such as stress rupture, impact, and fatigue [79]. Paprocki et al. [10] have produced carbon fiber-reinforced aluminum by liquid metal infiltration. Tows of multifilament fibers were coated with a fine layer of titanium and boron by the reduction of titanium tetrachloride and boron trichloride with zinc vapor. This coating activates the fiber surface, promotes wetting and infiltration by molten aluminum, and results in close bonding of the fiber to the metal. The composite so obtained was found to have a high specific strength and specific modulus, in accord with the rule-of-mixture value determination. As shown in Figure 17, the molten aluminum reacts with carbon to form aluminum carbide above ~500°C and structural studies of the composite material showed a relative uniform infiltration of the metal around the carbon fibers. However, defects, such as voids, are present in the casting due to poor wettability.

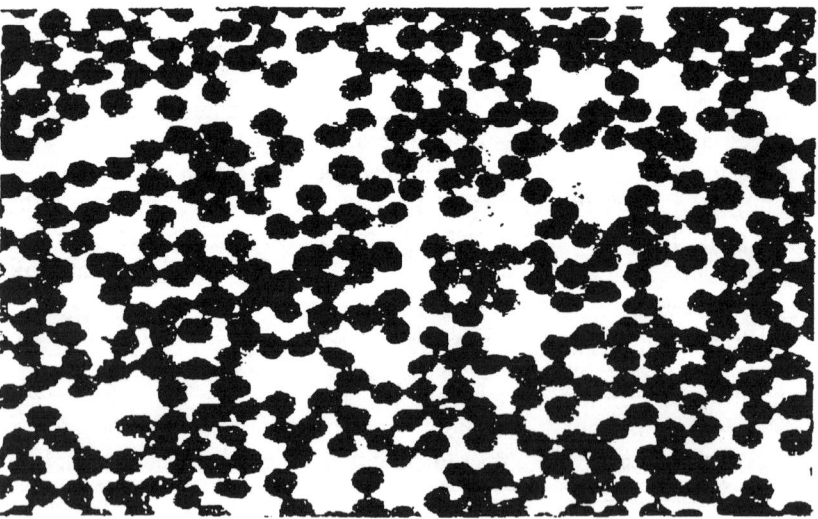

Figure 17 Micrograph of the carbon fiber-reinforced aluminum composites.

Carbon Fiber Applications

The mechanical properties of carbon fibers and their resulting aluminum composites [10] (Table 14) show a good translation of the fiber properties into the composites. The high-modulus carbon fibers did not show any damage due to the formation of carbides and were very stable in aluminum, indicating that these composites should retain their properties at relatively high temperatures. These workers also prepared carbon-magnesium composites which possessed even better mechanical properties [10].

7.2.3.3 Carbon fiber-reinforced nickel

Carbon fiber-nickel composites are prepared by a electroless deposition process; hot-cleaned carbon fibers are first subjected to an ultrasonic vibration to activate the surface, and then are dipped in nickel chloride solution with the reducing agent, sodium hypophosphite, and a buffer system, mixture of ammonium chloride and sodium citrate to maintain the pH between 9 and 10. The reaction temperature is kept between 85 and 90°C.

$$Ni^{2+} + H_3PO_2 + H_2O \rightarrow Ni + H_3PO_3 + 2H^+$$

Table 14 Mechanical Properties of Carbon Fibers and their Aluminum Reinforced Composites

	Tensile strength (MN/m^2)	Tensile modulus (GN/m^2)
Carbon fiber		
Union Carbide VS-00-54	1750–2270	690
Celanese GY-70-SE	1920	485–550
Carbon fiber/aluminum		
VS-00-54 / Al-201 (Fiber volume 48–52%)	1045–1080	348
GY-70-SE / Al-201 (Fiber volume 37–38%)	800–835	209

Source: from ref. 10.

Carbon-nickel composites have been prepared with good initial mechanical properties [6–8,78]. These properties, however, deteriorate on exposure of the composites to high temperature or to oxidizing environments. These limitations of carbon-nickel composites have been overcome to a certain extent by a protective coating of the fiber. This, however, involves difficult procedures and is, therefore, not attractive.

7.2.3.4 Carbon fiber-reinforced magnesium

The carbon fiber-magnesium composite is usually produced by a liquid metal infiltration process. In this process, the multifilament carbon fiber tow is first chemically activated by depositing a thin layer of titanium boride; then, the molten magnesium is wetted on this surface-activated yarn tow to produce a carbon fiber-reinforced magnesium wire. Another technique, *casting,* is also commonly applied in the magnesium reinforced fibers. This technology is much further developed for carbon fiber reinforced magnesium than other metals because the magnesium coating is air-stable. Some typical mechanical properties of the casting magnesium reinforced fibers are listed in Table 15. All the fibers used in the table are different grades of pitch-based carbon fibers. The magnesium infiltration of the fibers is generally excellent evidenced by the microstructure in Figure 18.

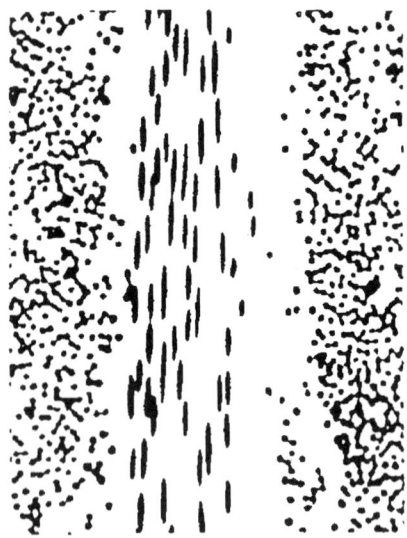

Figure 18 Microstructure (100 ×) of the carbon fiber reinforced magnesium composite by the casting process [115].

Table 15 Mechanical Properties of Carbon Fiber/Magnesium Reinforced Composites Fabricated by the Casting Process

Pitch-based carbon fiber	P55	P55	P55	P75	P100
Casting	rod	plate	plate	hollow cylinder	hollow cylinder
Fiber content/ orientation	40%/0°	40%/0°	20%/0° + 20%/90°	40%/ ± 16° + 9%/90°	40%/ ± 16°
Tensile strength (MPa)	720	480	—	450	560
Compressive strength (MPa)	—	20	240	61	38
Tensile modulus (GPa)	172	159	90	179	228
Compressive modulus (GPa)	—	21	90	86	30

Source: from ref. 115.

7.3 APPLICATIONS

The basic characteristics important to all applications of carbon fiber composites are high specific strength and stiffness together with lightweight. Composites with properties superior to those of steel and lighter in weight by 40–70% have been realized. There is a greater degree of flexibility in carbon fiber-reinforced materials, in that the strength and stiffness can be varied significantly in different areas of a composite part by selecting the type, form, and suitable orientation of the fiber in the composite and by controlling the local concentration of the fiber. In addition, several other physical properties of carbon fiber composites, such as their thermal stability, electrical conductivity, and corrosion resistance, can be varied by a suitable choice of the matrix material and by varying the processing conditions for preparation of the composite. We have divided the carbon fiber composite applications on the basis of their most significant physical properties in a particular application, but by no means does this imply that the other physical properties do not play their role.

7.3.1 High Strength and Stiffness Materials

7.3.1.1 Aerospace applications
Aerospace has been a primary influence in the progress of CFRP and still constitutes the largest market sector [75,77]. It is estimated to account for 55–65% of the current world output of carbon fibers and is likely to be the source of sustained growth of at least 10% per year during the next decade. This expansion is likely to take place in using CFRP in helicopters and in civilian aircraft. In addition, consumption in military aircraft and other space and defense applications is also likely to increase. Carbon fiber-epoxy composites have accounted for the greatest proportion of applications, although there have been important developments in carbon-carbon composites, which have applications in aircraft brakes, nozzles, and other high-temperature equipment.

Figures 19 to 22 illustrate some of the areas in which advanced composites are considered to be competitive with metals. Carbon fibers are preferred in those stiffness-critical components where design specifications on deflection limits, buckling, and dynamic response cannot be met without them. Important advances include major use of CFRP for control surfaces (e. g., spoilers and ailerons) and tail plane assemblies for the Airbus 320 and Boeing 777 aircraft. These highly successful passenger planes not only use CFRP extensively for secondary structures such as fairings, brakes, spoilers, ailerons, landing gear doors and rudders, but also employ CFRP in the vertical and horizontal fins in the A310 and A320 series, making them the first passenger planes to use a composite in their primary structure.

The new generation of military combat aircraft (Figure 21) fabricated by French company, Dassault Aviation, has also utilized composites in a large portion of structure design [53]. There are a considerable utilization of new materials (Figure 22) in this aircraft: (1) Titanium structures formed by Super Plastic Diffusion Bonding technique, (2) Aluminum structures by Super Plastic Forming technique, and (3) Composites materials. Two main groups of composites, PEEK/carbon fiber, and epoxy/carbon fiber composites, are applied in the structure The composites materials represent approximately 24% of the airframe mass and 70% of the wet surface.

The most significant area of application of CFRP has been in the main and tail rotor blades of helicopters [83] shown in Figure 23, where the prospects for improved fatigue resistance over all metal blades and the greater freedom in design and fabrication have been recognized and exploited rapidly. The use of composite unidirectional prepreg permits better optimization of rotor blade mechanical and dynamic characteristics, and they have demonstrated better static and fatigue strength than those of metal blades.

Carbon Fiber Applications

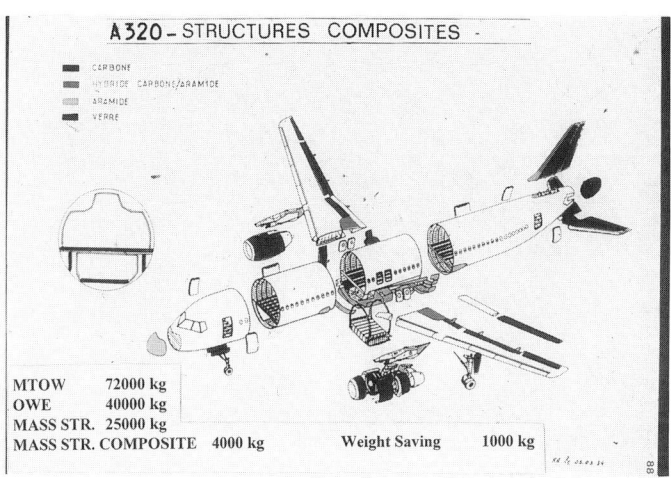

Figure 19 Carbon-carbon and carbon fiber-aramid hybrid composites used as structure components on the A320 airbus (Original photo from AEROSPATIALE, with permission).

Figure 20 Boeing 777 commercial airplane (Original photo from Toray Industries Inc., with permission).

Figure 21 French Rafale military aircraft (Original photo from Dassault Aviation, with permission).

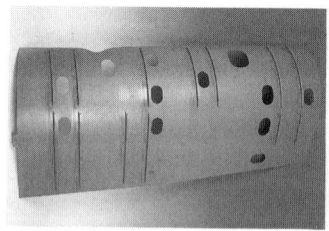

Figure 22 Carbon fiber/PEEK or epoxy reinforced composites used as the frame components such as wing structures and exhaust pipe on the Dassault Rafale combat aircraft (Original photo from Dassault Aviation, with permission).

Carbon Fiber Applications

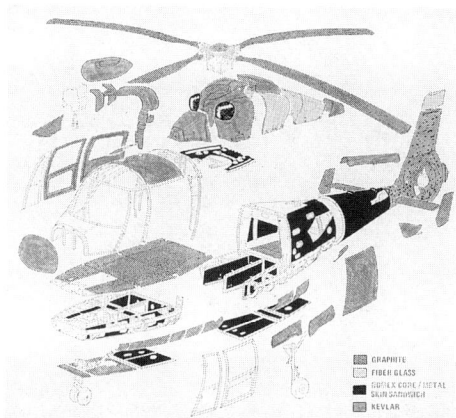

Figure 23 Application of carbon fiber composites in helicopter construction (Original photo from AEROSPATIALE, with permission).

The largest pieces of CFRP structure in service to date are to be found on a space vehicle, the NASA Space Shuttle. The payload bay doors of the Shuttle (Figure 24), which are 18.3 meter long and 139.4 square meter in area, are fabricated of CFRP [82]. This has resulted in an estimated weight saving of 23% over the same structure in aluminum alloy. The carbon-carbon composites (RCC) were used as components of the nose cap and panels due to their excellent thermal and oxidation protection at >1500°C, and low coefficient of thermal expansion. AEROSPATIALE has largely used carbon fiber composites as the frame structures on the launchers and satellites (Figure 25) due to their characteristic high stiffness, light weight and good thermal stability at the extreme environment.

Generally, laminated CFRP structures are used in these aerospace applications. Laminated parts now flying successfully on several aircraft include land gear doors on the F-14 fighter and wing spoilers on more than 100 Boeing 737s in the NASA fleet [17]. The floor decking of the Boeing 747 is also made from laminated carbon fiber prepreg and Dupont's Nomex® aramid [17]. The weapon bay door alone saves about 19% in weight and about 15% in cost over the metal door [84]. The high specific strength of carbon fiber-reinforced materials is also made use of in pressure vessels and rocket motor cases. The use of CFRP in the bypass duct of the Rolls-Royce Turbofan engine enhances fuel consumption efficiency and noise reduction. The range of composite items, such as (1) vertical and horizontal stabilizers; (2) control panels; (3) doors and access panels ; and (4) air brakes are commonly known in the use on production aircraft.

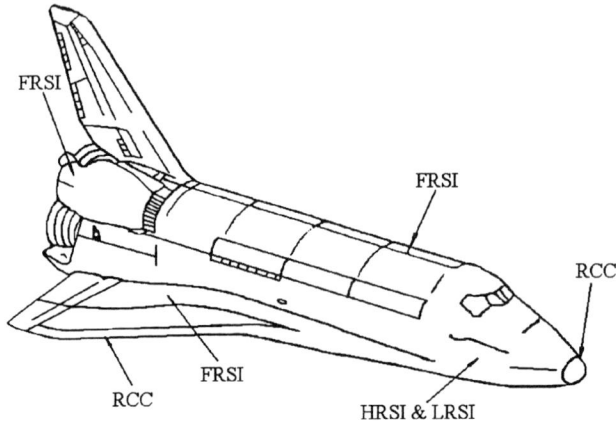

Figure 24 A schematic drawing of space shuttle.

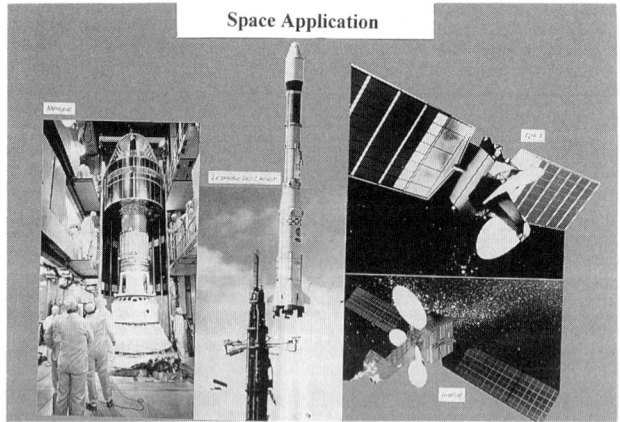

Figure 25 Carbon fiber composites applied in the aerospace industry, such as the launchers, communication, and weather satellites (Original photo from AEROSPATIALE, with permission).

Carbon Fiber Applications

7.3.1.2 Application in automobiles industry

Unlike the aerospace sector, in which CFRP design, manufacture, and applications developed very rapidly, the use of CFRP in transport vehicles has been slow and more uncertain. This is because cost is an important factor in this sector. Even though there is a drive toward producing lightweight, fuel-efficient vehicles, it has been difficult to justify the high cost of CFRP on cost-performance benefits. Nevertheless, CFRP has created an awareness within the vehicle industry of the broad scope for CFRP composites use in load-bearing parts.

7.3.1.2.1 Passenger vehicles

Two major load-bearing components which are being examined closely are the drive shaft [85] and the body spring [86]. A CFRP drive shaft fabricated by a wet filament winding technique has a weight about half that of its steel counterpart, with comparable flexural and torsional strength and stiffness. The longitudinal and circumferential stiffness of the shaft can be varied almost independently by changing the yarn winding pattern so that the drive shaft transmits torque but damps out noise vibrations. CFRP leaf spring development began in the 1970s and showed considerable promise for low weight and fatigue tolerance. The weight of the body spring made from CFRP is almost 25% that of the comparable metal part. The lifetime of these components is also much better because of the high fatigue resistance properties of carbon fibers. But these CFRP body springs were soon replaced by hybrid CFRP/Graphite Reinforced Polymer (GRP) designs. CFRP composites can also be used to make helical springs, which require a high degree of stiffness and fatigue strength in torsional shear.

Lovins et al. have designed a new model of *supercar* for General Motors concept car-*Ultralite* [88]. The body structure of this supercar (Figure 26) was made of carbon fiber composite sandwich structure with polyurethane foam between layers, performing an exceptional high strength with a weight only 190 Kg. *Ultralite* concept car is only composed of six components-rear panel, two doors, right and left halves, and floor tub. The cost of this highly composite composed concept car is quite high, about $13,000 for the materials alone. The authors suggested the cost can be reduced by a full-scale production through a judicious combination of aramid, glass, carbon fiber composites.

In their light vehicle project, Ford Motor Company has demonstrated the possibility of using CFRP by replacing steel parts such as doors, bonnets, drive shafts, structural brackets, wheels, and frames, resulting in a weight reduction of about 33%. Numerous CFRP constructions in racing car bodies have indicated that CFRP can withstand stressed applications in very demanding vehicle service environments. It is apparent that although CFRP will be a preferred choice for many applications in road transport, the choice narrows down to some critical parts when the cost of CFRP is considered. This indicates that CFRP in these applications will be used only on a highly selective basis where stiffness or other properties of the CFRP are essential. However, as applications expand, there is going to be a substantial market potential for CFRP. It is estimated that for every 100 g of carbon used on a typical medium volume passenger vehicle, there will be an increase in demand of 1000 tons in CFRP. This appears to be a powerful incentive [83].

7.3.1.2.2 Racing car

The history of present day all-carbon fiber racing cars dates from 1980. Formula 1 race cars are generally recognized to represent the state of the art in the high performance automobile industry. Much of its reputation comes from the use of the advanced materials, especially fiber reinforced composites [117]. The use of composites in Formula 1 racing car has increased dramatically over the past ten years. More and more metal components, such as wings etc., have been replaced by the carbon fiber composite materials. Figure 27 shows a schematic drawing of the composite parts of the 1991 McLaren MP4/6 racing car [116]. This Formula 1 racing car consists of approximately 75% by weight of carbon fiber composites in the structure, resulting in a higher performance and safety for its occupant. Crash safety is a very important aspect of racing car design. The enormous energy absorbing capability of fiber reinforced composites has largely reduced the instance of serious injury in the race. The structural components of racing cars are mainly composed of continuous fiber reinforcement, thus, the stiffness and strength of the composite is dominated by the fiber properties. PAN-based carbon fibers are dominantly used in the manufacture of Formula 1 due to an excellent balance of mechanical properties and good handling. New products with higher strength and modulus have been continuously developed, such as pitch-based fiber reinforced materials. Ford GT-40 sports car whose body is made from bundles of short carbon fibers applied as a grid pattern over the hood, fender, and roof. The use of carbon fiber reduced the weight of the body by about 40% (from 31.5 kg to 19.1 kg), increased the stiffness so that the aerodynamic shape was maintained under wind pressure at high speeds, eliminated drumming inside the car caused by roof vibration, and substantially increased the life of the GRP body [60]. This car has won the Le Mans race several times. The light weight and extra rigidity of the CFRP cowling helps ensure racing performance.

Carbon Fiber Applications

Figure 26 A schematic presentation of General Motors concept car-*Ultralite* [after ref. 88].

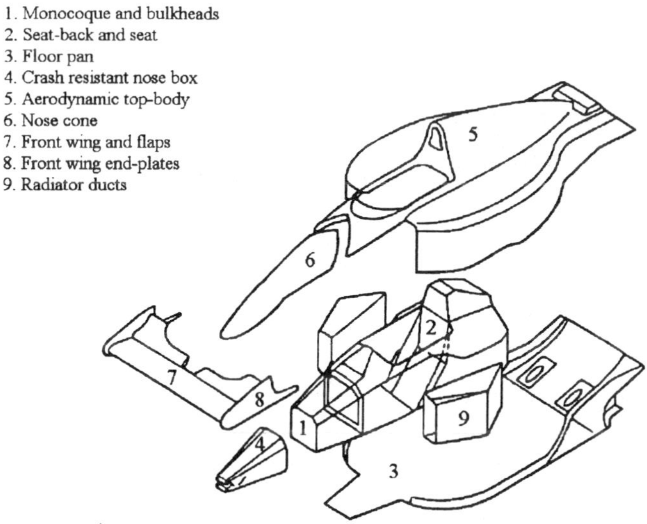

1. Monocoque and bulkheads
2. Seat-back and seat
3. Floor pan
4. Crash resistant nose box
5. Aerodynamic top-body
6. Nose cone
7. Front wing and flaps
8. Front wing end-plates
9. Radiator ducts

Figure 27 Simplified modification of structural components of 1991 Mclaren MP4/6 racing car made from carbon fiber composites [after ref. 116].

7.3.1.3 Marine applications

In the past five years, due to an increased demand of light-weight and corrosion-reduced structure, there is an increased interest in the development of the application of composites in the structure components of ships and submarines (i.e., load-bearing structures and auxiliary items, such as gratings, ventilation ducts). A significant technical limitation of the composites used on the ships or submarines is the combustible nature (fire, smoke, and toxicity) of the organic matrixes based composites. Because the safety requirement is more stringent, the composite materials used on the US Navy vessels is requested to be sufficiently fire resistant not to be a source of spontaneous combustion. Among the many possible thermoset resins, phenolic resins have the inherent characteristics of low flammability, produce little smoke on burning, and have good thermal stability.

The stiffness, rapid recovery, and fatigue resistance of CFRP laminates can be exploited to construct backup beams for bumpers and collision beams in doors, all of which represent weight savings as well [16]. CFRP has also been investigated to effect weight reductions in surface vessels and hovercraft [18] and in making transportable bridges. The *Asean Lady* motor cruiser, the world's largest CFRP boat, has a 48-m-long hull and deck made up of Torayca CFRP to reduce weight and thus enable higher speeds (Figure 28).

Figure 28 *Asean Lady* motor cruiser. (Original photo from Toray Industries Inc., with permission.)

Carbon Fiber Applications

7.3.1.4 Applications in the railway industry

The application of composites in today's rail vehicles is mainly adopted on the three-dimensional molded nose cap at the front part of trains and internals of passenger train [93]. The carbon fiber reinforced composite incorporated train structure saves about 30–40% weight than a conventional steel cab, and exhibits a strong impact resistance to prevent penetration by a 0.9 kg steel cube when traveling at 350 Km/hour. A recent French TGV high speed train (Figure 29) was fabricated with different types of composites for the nose section, to provide extra rigidity and resistance to impact when runs as high as 300 km/hour. There are some applications in the train construction composite materials can find advantages where aluminum and steel have a traditional stronghold, for example bodyshells. The body shell of a train accounts for a major portion of its mass and price. Recently, a Swiss company Schindler Waggon has successfully developed an entirely composite-structured train bodyshell. This new system involved glass and carbon fiber composites performs an excellent corrosion resistance and better thermal insulation. On the other hand, rather than designing the bodyshell by a whole composite structure, the Japanese Railway Co. has developed some aluminum-CFRP structures for the train roof shells, resulting in a 30% reduction in weight than an equivalent aluminum structure.

Figure 29 Application of carbon fiber composites in the structure of a modern high speed train-*French TGV*.

7.3.1.5 Applications in sports and leisure-time activities

The high specific strength and stiffness coupled with rapid damping and good fatigue resistance (Figure 30) of carbon fibers make them versatile materials for certain sport and leisure-time products (Figure 31), such as golf club shafts, fishing rods, telescoping casting rods, vaulting poles, ski poles, tennis and badminton racquets, bows and arrows, hockey sticks, wind-surfing masts, racing car shafts, paddles and their shafts, racing-bicycle frames, and race cars.

Filament-wound CFRP golf clubs are very popular in the United States and have consumed a good proportion of the carbon fibers. Golfers are ready to pay the high cost of such clubs if they can add more distance to their drives. Reinforcement by carbon fibers results in a lighter shaft, permitting proportionally more weight than usual in the head of the golf club. Fiber-reinforced laminated bows are stiffer, recover faster, and throw arrows farther. Arrows are made from carbon fibers by a combined filament winding-pultrusion technique [87].

The most important qualities of a ski pole are minimum weight and adequate strength combined with maximum stiffness. The performance also depends on impact resistance and the location of the center of gravity. Skis fabricated from CFRP laminates are superior for expert skiers competing for seconds or even fractions of a second. It is interesting to observe that a decrease of 60 g per pair of skis (first obtained by Exel Company using a high-strain carbon fiber) could decrease average times by about 16 seconds per 10 kilometers [87]. Kevlar® fiber composites are close competitors due to their light weight and excellent vibration damping properties. But Kevlar® can be used only on the bottom side of skis because of its moderate compression strength. CFRP laminates can also be used in ice hockey sticks especially in the shafts. The stick shaft construction resembles that of skis-sandwich construction with wood in the middle [87].

A wind surfing mast is simply a fiber-reinforced epoxy tube. As the mast has to bend according to the position of the sail and has to withstand very hard blows, glass fiber is the preferred material in mast fabrication. However, carbon fiber, which is stiffer than glass fibers, can be used on the mast sides so that the mast bends in line with the sail section but remains vertically as stiff as possible against this line. The use of CFRP provides the mast with sufficient stiffness in the direction where stiffness is needed and suitable flexibility in the direction where flexibility is needed. Another article with a large market in the sports and leisure field is the racing bicycle. CFRP racing bicycles are well known and have been produced for several years, but a real breakthrough will occur when the bicycle frame tubes can be produced at prices competitive with metal tubes.

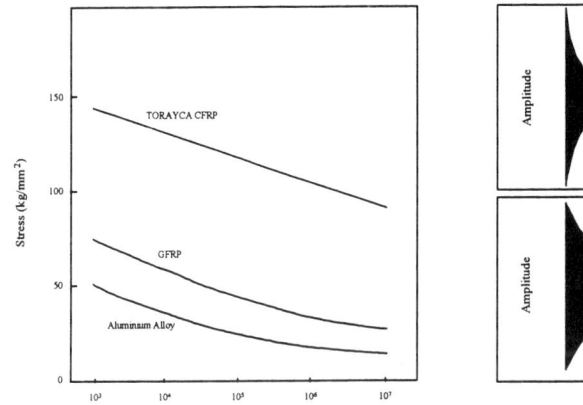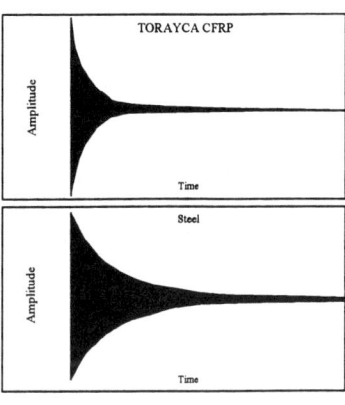

Figure 30 Fatigue strength (left) and damping characteristics (right) of carbon fibers. (From Torayca Bulletin, courtesy of Toray Industries Inc.).

Figure 31 Carbon fiber composites used in the sports products application (Original photo from Toray Industries Inc. with permission of reprint).

7.3.2 Thermoresistant Materials Application

Carbon fibers have exceptional thermophysical properties, which combined with their stiffness, make them extremely useful in composites for higher-temperature applications and distinguish them from other reinforcing agents. They have good thermal conductivity which helps heat dissipation in components such as gears, bearings, brake disks, and other friction-related products, and in cryogenic processing equipment. They also have high thermal stability and are therefore the only fibrous material that can be used above 500°C. Vapor grown carbon fiber (VGCF) was fabricated with aluminum matrix composites for analysis of thermophysical properties [86]. The resulting composites exhibit the highest thermal conductivity than the other types of carbon fibers. A thermal conductivity of 910 W/m-K was observed by the authors [86] in the aluminum reinforced VGCF composites, a value which is more than twice that of copper. Therefore, the VGCFs are attractive materials in the applications of electronic devices. On the other hand, low thermal conductivity perpendicular to the fiber axis makes them useful materials for heat shields in missiles [1,19,89].

7.3.2.1 Aircraft and vehicle brakes
Brakes convert the kinetic energy of a moving vehicle into heat which is stored in the brake material until it can be dissipated. Consequently, aircraft brakes experience very high temperatures. The conventional metal brake materials undergo considerable changes in their strength and frictional properties at these high temperatures and may sometimes even melt. When made from carbon-carbon composites, aircraft brakes are found to have remarkably stable frictional characteristics to temperatures well above 2000°C. Carbon is stronger at 2000°C than at room temperature, whereas steel melts at 1600°C. The carbon fibers also provide the resistance to thermal shock that is so important in this application. Carbon-carbon composite brakes are lighter because the density of this material is one-fourth the density of steel.

About twenty years ago, Goodyear developed CFRC air brakes for F-15 fighter planes which weigh about 30 kg (as opposed to 100 kg), and observed that CFRC brakes perform about three to five times better than their steel counterparts in actual number of landings before burning out [90]. Thus the value of carbon-carbon composite brakes is not just in weight saving but also in performance. Societé Européenne de Propulsion (SEP) has also developed carbon-carbon composite disk brakes for the Mirage 2000 and supersonic Concorde. The typical braking effort on landing and the performance characteristics of brake disk materials are compared in Figure 32 on an arbitrary scale [91]. The carbon-carbon composite brakes are far superior to metal brakes due to their excellent thermal shock resistance and abrasion resistance. The weight saving could be considerable using carbon-carbon composite brakes on a large aircraft. The weight could be reduced from about 1000 kg of conventional steel brakes to 700 kg, resulting in a weight saving of 300 kg. Brake disks have been fabricated from both rayon and PAN precursor filaments, using resin coke or pyrolytic carbon or a combination as the binder material. The PAN filaments give a density increase of about 12% over rayon filaments in similar matrices. This has the added advantage that a smaller physical package can be used for a given kinetic energy absorption [90].

CFRP composites have also been used for automotive brake linings, in combination with conventional cast iron disks for use in automobiles. Studies of frictional values and wear have shown that the linings are very promising, although the effective use of these materials will depend on their economic feasibility. Carbon fiber composite discs and pads were first used in the racing car by Brabham in 1982, with a significant weight saving and a higher friction coefficient advantages over the cast iron components it replaced. Figure 33 is a schematic drawing of the fiber reinforced components used on the brakes of Formula 1 race car.

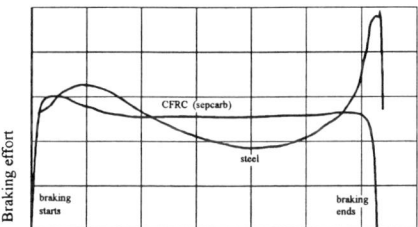

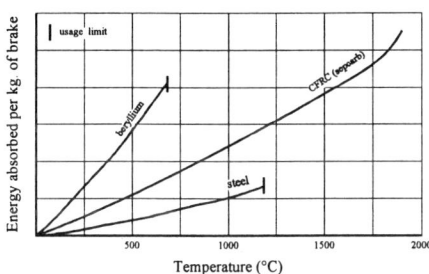

Figure 32 Comparison of performance of a carbon-carbon composite brake and a conventional steel brake (From SEP Bulletin SEPCARB, courtesy of Société Européenne de Propulsion).

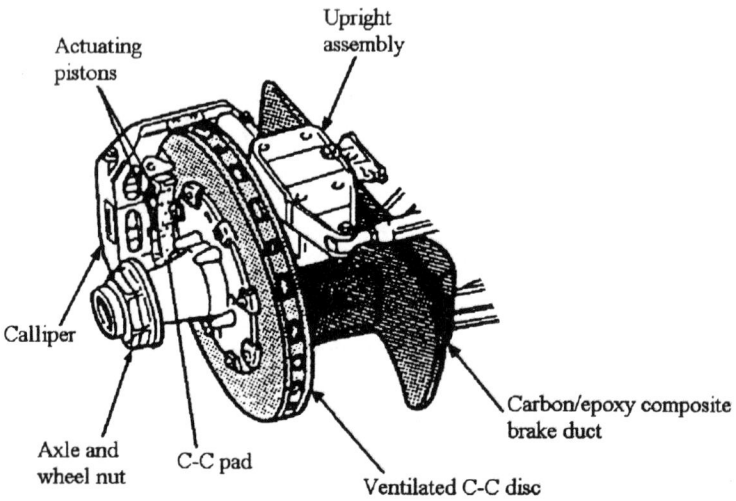

Figure 33 Brakes and suspension structure on Formula 1 race car [after ref. 116].

7.3.2.2 Aerospace antenna
The low coefficient of linear expansion of carbon fibers makes it possible to design structures with almost zero or very low linear or planar thermal expansion. This is a valuable characteristic for precision instrument components such as telescopes and the alignment of aerospace antenna. British Aircraft Corporation (BAC) has developed radar antennas based on CFRP aluminum honeycomb construction capable of retaining dimensional stability in widely different environmental conditions. A 183-cm-high carbon sandwich antenna was made from an assembly of CFRP laminates bonded to an aluminum honeycomb core built by the Hughes Aircraft Company for the Canadian Anik communication satellites. The CFRP components can be used only in a limited temperature range. However, investigations are now under way to develop resin materials that would enhance the temperature range for CFRP applications. Back in the 1970s, Fitzer and Heym [19] suggested the use of polyimide matrices for reinforcement by carbon fibers for high-temperature applications. However, these resin matrices are expensive and therefore have limited applications. BAC has used carbon fiber-reinforced polyimide type Keremid

Carbon Fiber Applications

601 for making missile fins [19]. These fins have been successfully flown on test missiles. There also appears to be an excellent prospect of using carbon fiber-reinforcing polyimide composites in engine service where the temperature may exceed 450°C [17]. A large-diameter (ca. 45 m), high-precision telescope using CFRP has been made by Mitsubishi Electric Company.

7.3.2.3 Heat shields in missiles
Heat shields protect missile payload from the high temperatures and severe aerodynamic conditions experienced during the reentry of the missile into the atmosphere. They are generally composed of a relatively brittle phenolic matrix and a low-modulus carbon fabric [1]. The phenolic matrix pyrolyzes during reentry of the missile, forming a carbon-carbon composite which is eroded by the heat and abrasion. Carbon fiber reinforcement prevents the shield from cracking, while its high longitudinal thermal conductivity and low specific heat minimize temperature gradient over the surface, preventing nonuniform erosion of the shield, which would unbalance the missile. The carbonized heat shield also insulates the payload, as the thermal conductivity of the shield is very near normal to the fabric.

7.3.3 Chemical and Corrosion Resistance Materials

7.3.3.1 Chemical industry
Carbon fibers show chemical inertness to a wide range of reagents and hostile environments with exceptional stiffness and lightness. Consequently, carbon fibers in the form of felt, mat, or cloth impregnated by a resin matrix can be used as protective lining for chemical plants and in anticorrosion equipment, including bulk liquid transport. A significant application is the construction of acid distribution troughs made from carbon fiber felt cladded on GRP and a foam polyurethane core [86]. Carbon fiber-reinforced troughs are stronger and lighter than traditional troughs made from asbestos-reinforced phenolic resins for the same cost and do not deform under their own weight. Another important application is the construction of CFRP blades for rotary compressors. This would allow them to run without oil-mist lubrication, to produce oil-free compression [16]. The possibility of fabricating aircraft fuel tanks has already been evaluated [92].

7.3.3.2 Nuclear field

Another important area in which high strength and high stiffness coupled with chemical inertness of carbon fibers outweigh the cost disadvantage is in the construction of a gas centrifuge for uranium enrichment. In 1975, Tewary reported that GRP and metals reacted with fluorine in uranium hexafluoride involving in the uranium enrichment and therefore required a protective coating [18]. Carbon fiber composites need no extra coating and thus result in weight saving without losing strength.

It should, however, be pointed out that although the advantages of improved resistance to acids, alkalis, and reducing agents are real, they cannot, at present, be fully utilized, as the resin matrices' are usually more vulnerable chemically than are the carbon fibers. It will, therefore, be necessary to develop greatly improved resins before carbon fiber composites can be used in the chemical industry. This problem is being solved by the development of CFRC, as carbon is one of the most corrosion-resistant materials available. Graphite tubes whose outer shells have been reinforced by carbon fibers can be used as heat exchanger tubes (Figure 34) and for the treatment of wastewater and the washing out of pollution gases.

Figure 34 Graphite heat exchanger tubes reinforced by a carbon fiber shell [35].

7.3.4 Electrical Conductivity Materials Application

Carbon fibers have good electrical conductivity and are therefore attractive fillers in polymers for making conducting composites used for electromagnetic interference shielding (Figure 35), lightning-strike tolerant aircraft surfaces, and large conducting space structures [94]. A carbon fiber loading of 10% can produce electrostatic dissipating materials which can be used to protect electronic circuits and avoid spark generation. The conductivity continues to increase almost linearly with loading, permitting broader control of composite conductivity to meet different applications. The conductivity of the composite can be controlled by proper selection of both carbon fiber and matrix material, by special additives or mixtures, by optimizing the material molding conditions and participation in design and tooling [95].

Some of end-use requirements for composites conductivities are :

For static dissipating 10^4–10^6 Ωcm
Functional elements in high impedance circuits 10^2–10^3 Ωcm
Shielding from radio-frequency interferences 1–10^2 Ωcm

Radio-frequency (RF) screening requirements are specific to individual applications and may vary from 10 dB (90% screening) to 80 dB (99.99% screening). Table 16 gives some conductivity data for composite materials prepared by using glass and carbon fibers with various thermoplastic as matrix materials. The data show clearly that carbon fiber-reinforced composites are available which can be used successfully for electrostatic dissipation and for radio frequency screening. Carbon fiber-reinforced composites can also be used as heater elements, both as space heaters at low temperatures and as furnace heaters at high temperatures [41].

Nickel-coated carbon fibers (NCG) have been described [49] which have an unique combination of electrical conductivity and remarkable reinforcing ability. NCG fiber composites can be used in the computer field for electromagnetic interference (EMI) shielding and electrostatic dissipation for the protection of microcircuitry. The EMI shielding effect varies with fiber loading in the composite (Table 17) and can therefore be controlled depending on the end-use requirement of the material. It has been shown that composite panels containing NCG fiber fabric as the top successfully withstand simulated zone I lightening. This is because a 3 K plain weave provides about 74 m² of conductive surface area per square meter of fabric. Thus NCG fabric can be substituted for the topmost ply of carbon fiber in a laminate, thus providing heightened strike protection.

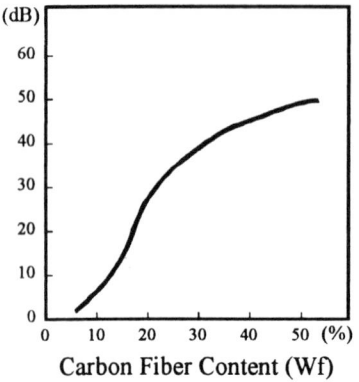 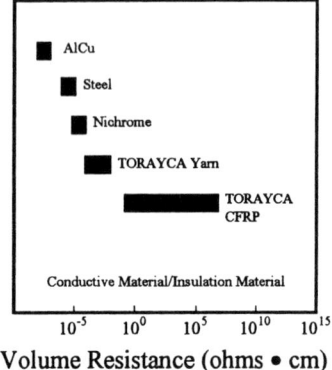

Figure 35 Electronic properties of carbon fibers. Left figure: EMI shielding (CFRTP mm²). Right figure: Electrical conductivity. (From Torayaca Bulletin, courtesy of Toray Industries Inc.)

Table 16 Comparison of Conductivity Data (Ωcm) of Glass Fiber and Carbon Fiber Reinforced with Various Thermoplastic Composites

Reinforced thermoplastic materials	Carbon fiber (30% by weight)	Glass fiber (20% by weight)
Polyamide 6.6	10^3	10^{15}
Polyamide 6	10^5	10^{15}
Polycarbonate	$10^4 - 10^7$	10^{15}
Polyphenylene oxide	10^4	10^{15}

Source: from ref. 95.

Table 17 Fiber loading in Relation to EMI Shielding

NCG fiber loading (wt %)	Shielding effectiveness[a] (dB)
10	25–45
15	35–55
20	45–65

[a] ASTM ES7-83 dual-chamber method at frequencies from 30 to 1000 Mhz.

7.3.6 Biological Inertness and X-Ray Permeability Materials

7.3.6.1 Medical applications
Carbon is well known to have the best biocompatibility of all known materials; it is compatible with blood, soft tissues, and bones. This biocompatibility, coupled with the specific strength and pliability of carbon fibers, make them useful as replacements for damaged or absent tendons and as surgical implants for regenerating tissues, which can grow around, along, and through the fiber bundle [35,36,81,97–101] (Figures 36 and 37). Low temperature isotropic carbon (LTIC) was demonstrated as an excellent materials in the construction of cardiovascular devices, such as artificial heart valves [97]. Carbon fiber implants are largely used in the clinical devices, such as in ligamentous reconstruction of the knee and in Achilles tendon ruptures. In addition, carbon fibers can be used in x-ray equipment because they are more permeable to x-rays than are the conventional metallic components.

Implant materials for endoprostheses were first developed in the 1960s [102], but interest in carbon as an implant material was not raised until 10 years later by the pioneering work of Bokros et al. [103]. Today, carbon in its many forms, is widely used and is considered an interesting implant material for many endoprosthetic devices. Hüttinger gave a very good review of its uses [105,106]; a complete review of biomedical applications of carbon and graphite through 1980 has been summarized by Jenkins [107]. For example, in recent years more than 400,000 carbon components for heart valves have been implanted.

The first successful attempts to replace tendon and ligament were reported by Jenkins [98]. In a series of experiments the Achilles tendon of a number of sheep were excised and replaced with a double-plaited strand of flexible carbon fiber tow. In a controlled series of sheep the Achilles tendon was also excised and either replaced with nylon or not replaced at all. The control sheep never walked satisfactorily, whereas the sheep with the inserted carbon fiber walked normally within 2 to 3 weeks. The histological examination of sections following implantations showed no more than the inflammatory reaction expected in a wound. The strength of the implant increased with time and corresponded with the stimulation of an in growth of new tissues into the carbon fiber scaffold [98]. The carbon fiber was also found to plant from where the site of the implantation.

Fitzer et al. [35,109] indicated the possibility of using carbon fiber-reinforced composites as plates for internal fixation of bones in osteo synthesis and endo prosthesis (Figure 36). Carbon fiber-glass hybrids have been used successfully for making prosthetics (Figures 37) [35,86,99]. Although more expensive, these prosthetics are light and strong compared to the conventional metal prosthetics. Furthermore, in the case of carbon fiber-reinforced prosthetics, the pattern of the limb to be fitted can be molded directly, which greatly facilitates the fitting process.

The nature of interaction between carbon fiber reinforced composites and biological environment is strongly influenced by the properties of carbon matrix in which carbon fibers are placed [104]. Blazewicz et al. [97] evaluated the carbon fiber composites performance in the biological environment. The matrix of four types of composites is phenol-formaldehyde resin. After the carbonization stage composites A and B were impregnated with pitch, composites C and D with phenol-formaldehyde resin. Composites B and D were further deposited with a layer of pyrocarbon by CVD technique. The four composites were immersed in the isotonic solution in which the natural biological environment was simulated, to investigate the interaction of composites with the tissue environment. Experimental results from ultrasonic measurements and infrared studies indicated that biodegradation of a composite occurred, a rebuilding of surface layers taking place. It is evident that a strong biological interaction between carbon phase and tissue environment happened by distinct changes in the chemical composition shown in Table 18, especially true for the phenol-formaldehyde reinforced composites C and D.

Figure 36 Carbon fiber prosthetics for joints. (From SEP Bulletin, courtesy of Société Européenne de Propulsion.)

Figure 37 Carbon fiber composite hydraulic motor for artificial heart implant. (From SEP Bulletin, courtesy of Société Européenne de Propulsion.)

Table 18 Chemical Composition of Carbon Fiber Composites Before and 10 Weeks after Immersed in Isotonic Solution

Composite	Element content percentage			
	C	O	H	N
Initial				
A	98.6	1.02	0.28	0
B	99.36	0.36	0.27	0
C	96.03	1.75	0.53	1.70
D	96.48	2.47	0.35	0.69
After 10 weeks immersed in isotonic solution				
A	97.1	2.53	0.28	0
B	98.2	1.51	0.29	0
C	87.0	10.7	1.19	1.11
D	91.6	6.81	0.52	1.1

Source: from ref. 97.

7.3.5.2 X-ray equipment

CFRP x-ray pressure plates have an x-ray absorbency of about one-sixth that of similar aluminum plates (Figure 38) and can therefore give better resolution in x-ray pictures [86]. The high permeability to x-rays, the high stiffness to resist deflection, and high oscillation damping characteristics of carbon fiber composites are utilized for constructing the patient's coach in computerized tomography equipment and for cantilever tables in therapeutic x-ray equipment [59]. Filament-wound CFRP rings are also used in the tomography scanner as a cover for the x-ray tubes. The possibility of using carbon fiber composites as collimators and filters for elastic waves, useful to construct ultrasonic focusing devices for medical and surgical applications, has also been indicated [18].

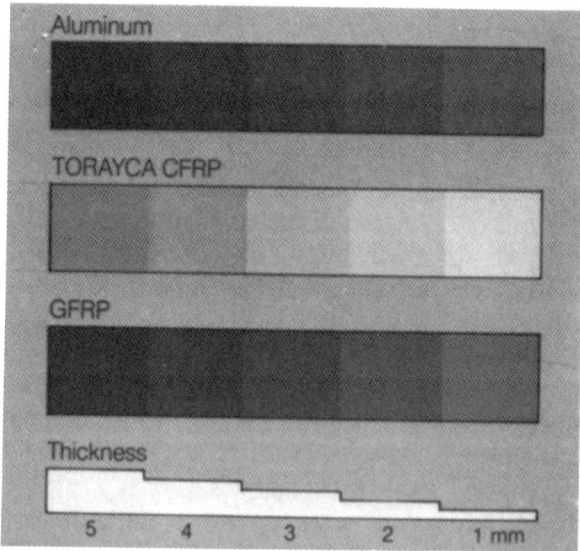

Figure 38 X-ray transmission in carbon fibers. (From Torayaca Bulletin, courtesy of Toray Industries Inc.)

7.3.6 Fatigue Resistance Materials

7.3.6.1 Applications in textile machinery
Textile machines involve a combination of high-speed reciprocation, oscillation, and rotatory motion, where inertia, strength, fatigue, and natural frequency of metal components limit machine operating speeds [111,112]. Since CFRPs provide advantages in these properties, they are being used extensively to make textile machine components. The common components being made from CFRP are heeled frames, picking sticks, slays in weaving looms, reciprocating guide bars, and push rods in knitting machines. These components are generally made from protruded carbon fiber-epoxy strip and from injection-molded carbon fiber-reinforced nylon [86]. The high thermal stability and the high specific stiffness requirement of a looper bar in a lockstitch machine can be met only by CFRP. The reciprocating guide bar reciprocates longitudinally at the rate of 750 cycles per minute. The hollow steel bar breaks in less than 4×10^8 cycles, probably due to fatigue failure caused by the vibration wave transmitted through them from the driving cam. The use of CFRP has increased the speed by 10% or more and has increased the life of the bar to at least 5×10^8 cycles [113]. This is due to the 25–30% lower stress and better vibration damping characteristics of CFRP (Table 19). The life of the composite bar is expected to be at least double at the standard speed. This provides an advantage to the user even at double the price of the conventional steel bar as the losses incurred due to shutdown of the machine to change the guide bar are avoided.

7.3.6.2 Applications in general engineering
Several general engineering components, such as bearings, gears, cams, and fan blades, and certain components for domestic electrical equipment are also made from carbon fiber composites. These components are generally made from injection-molded carbon fiber-reinforced thermoplastics such as nylon. Compressor blades for oil-free compressors are made using thermosetting matrices [86]. Carbon-carbon composites can also be used in the field of general engineering where high temperature, high strength, and corrosion resistance are required. Bolts, screws, nuts, and washers made from CFRC are used in semiconductor industry and in the fabrication of furnaces where high temperature and severe chemical conditions are present [36]. In addition, they can be used in the field of apparatus construction in the form of tubes, crucibles, sleeves, and other auxiliary aids, such as load-bearing plates, rods, and fastening elements.

A composite prepared by incorporating exfoliated graphite in carbon fiber-epoxy matrix is enhanced in ductility by about 80% [114]. This is attributed to the ability of the exfoliated graphite to deform by shear among the graphite layers and to act effectively as a soft matrix. Although the strength of the composite is decreased by the addition of the exfoliated graphite, the lubricity is greatly enhanced so that the composite material could be valuable for seals, gaskets, reflectors, and erosion barriers [114].

7.4   ACTIVATED CARBON FIBERS

The production of highly effective fibrous carbon adsorbents with smaller diameter, excluding or minimizing external and intra-diffusional resistance to mass transfer, and therefore, exhibiting high sorption rates, is a challenging task for investigators in the science and technology of carbon fibers. With low hydrodynamic resistance, to be used in thin layers for the treatment of high gas flows, these materials will increase efficiency and permit a far greater flexibility and simplification in the design of sorption processes for environmental pollution control. The application of activated carbon fiber is extended not only adsorbents but also catalyst supports, electrical and electronic materials for its unique pore structures and useful level of electrical conductivity. Thus efforts have been under way to prepare cheaper activated carbon fibers with controlled pore size distribution and high surface area.

Table 19 Stress in Bars of Different Materials

Materials and conditions	Stress (MPa)
Cam speed 780 cycles/min	
Throw reversible angle 17	
UHM/epoxy	25.5
VHS/epoxy	31.4
HM/epoxy	25.1
steel	88.2
Cam speed 950cycles/min	
Throw reversible angle 27	
UHM/epoxy	22.8
VHS/epoxy	23.6
HM/epoxy	21.0
steel	74.4

7.4.1 ACF Preparation

Activated carbon fiber (ACF) has been obtained from an appropriate fibrous precursor by an adequate carbonization and activation process. Among various combinations of precursor and its procedure, it is desirable to obtain activated carbon fiber having a high adsorption capacity with a high conversion ratio to activated carbon fiber and a high mechanical durability as well. ACFs are now manufactured from regenerated cellulose (viscose rayon), phenolic resin (Kynol), polyacrylonitrile (PAN), and coal tar pitch-based fibers. The carbonization process of the organic polymers is itself accompanied by the formation of porous fiber structure [118]. At the same time, an increase in the specific surface area, due to the development of pores, occurs only during low temperature stages (400–600°C) of carbonization when the strength of the material is low. The adsorption capacity at these stages remains low, which is due to the deposition of some tarry substance in the entrances of sorbing pores which prevents adsorption at these sites. Further increase of the carbonization temperature, as a rule, does not remove the tar from the pores but rather results in its carbonization, which finally seals the pores against any external sorbate. To restore sorption properties to these materials it is necessary to carry out a process of activation.

The activation process of carbonized fibers is basically analogous to those for producing nonfibrous forms of activated carbon, and can be separated into three main groups: (1) vapor/gas activation [119,120]; (2) chemical activation [121–123]; and (3) a mixed approach including vapor/gas activation of chemically pre-activated samples [124].

Activation of carbonaceous materials with vapor or gaseous agents involves, along with physical processes (evaporation from the surface of the activated material, removal of the tarry products condensed in the pore by the flow of activating gas, increase in the volume of the material as a result of diffusion and evaporation of the tarry products), also some chemical transformations (interactions with the surface carbon and the carbon of the flow structures, chemical reactions with the tarry products in the pore). The preferred activating gases are atmospheric oxygen, carbon dioxide, steam, and their combinations [119,120].

Chemical activation is brought about by the action of various chemical agents producing a dehydration effect. Among the most commonly used chemical activations are orthophosphoric acid [121], zinc chloride, aluminum, magnesium, or iron halides followed by activation in CO_2 or in steam in the temperature range 600–1000°C [122,123]. Chemical activation is primarily directed at freeing the sealed micropores from the tar, thus forming channels for the evolution of volatile products during carbonization. Contrary to vapor/gas activation, chemical activation is not characterized by the burn-off of certain structures in carbonaceous material. Because of this the adsorption capacity of such adsorbents is somewhat lower than that produced during vapor/gas activation. Among the advantages of chemical activation are high yields of the activated material, a short duration of activation and good sorption characteristics. But chemical activation has some drawbacks. It is accompanied by the formation of high ash content carbon adsorbents which narrows the scope of their application in medicine and in the refinement of industrial solutions.

Activated carbon fiber was first patented by Abbot [125]. Viscose textiles were pyrolized and activated to form activated carbon textiles [123,126,127]. Economy [129] and McNair [128] showed the preparation of activated carbon fibers in 60% yield from phenolic fiber. The carbonization was carried out in nitrogen atmosphere to about 800°C. The activation was conducted in steam and nitrogen mixture atmosphere between 750–1000°C. The degree of activation, and consequently, the surface area of the final product, depended on the temperature and the time of activation. Toho Beslon [130] developed activated carbon fiber from polyacrylonitrile resin. By preoxidation of the raw fiber at 225–255°C for 0.5 to 2 hrs by air, oxygen is introduced to the fiber, then activation by steam is carried out. Toyobo Co. [131] first commercialized activated carbon fiber from viscose rayon in felt or paper form. Kuraray Chemicals [132] studied in detail the production conditions of activated carbon fibers from phenolic resin fiber. Osaka Gas Co. started development of carbon fiber production in 1980, and from 1985, development research for manufacturing process of activated carbon fiber from coal tar pitch by a joint development with Unitika. Suzuki [133] introduced the Japanese manufacturers, their products, and several reference data. In USA, Ashland Petroleum Co. developed pitch based activated carbon fiber. In Russia, research, manufacturing and problems of consumption of activated carbon fibers are dealt with at the Leningrad Institute of Chemical Fibers and Composites and the Institute of General and Inorganic Chemistry (Academy of Sciences of Russia).

7.4.2 ACF Characterization

Exact characterization of less-crystalline microporous solids having large micropore volume such as activated carbon is strongly desired to develop both adsorption science and advanced technology. Activated carbon is representative of less-crystalline microporous solids [134]. However, it is not so easy to characterize activated carbon. Structural key factors for elucidation of activated carbon structures are, microstructures, micrographitic structures, atomic compositional structures and electronic structures [135,136]; these structures are associated with different characteristics of activated carbons. Activated carbon fibers have more uniform microporous structures and slightly oriented micrographitic structures [137]. Activated carbon fiber is the best model of the less-crystalline microporous solids [138].

Since activated carbon fibers are considered to have rather defined pore structures and surface characteristics, fundamental studies on adsorption have been tried various aspects. Ryu [139] determined the BET specific surface area of isotropic pitch based activated carbon fibers from the adsorption isotherm of N_2 at 77K. The microporous volume was determined in two different manners; (1) from the adsorption isotherm of N_2 at 77K using the α_s-method [140] and (2) from the CO_2 uptake at 273K according to the Dubinin-Radushkevich (D-R) equation [141]. Specific surface area, S_{BET}, and micropore volume, V_{DR} and $V_{\alpha s}$, corresponding to carbon fibers activated at different levels of weight loss are listed in Table 20.

External surface area, S_{ext}, estimated from the linear part of the α_s-plot [140] are also indicated. The unactivated carbon fiber possesses an external surface measured by N_2 adsorption close to its geometric surface area. There is however a slight uptake of CO_2 indicating that some microporosity is already present in the unactivated fiber. These micropores are however inaccessible to N_2 at 77K but can only be reached by CO_2 at 273°K. The micropore volumes of the carbon fibers increase significantly upon activation. However, the increase of V_{DR} and $V_{\alpha s}$ do not follow the same trend. After 60% weight loss the value of the micropore volume measured by D-R equation levels off whereas the one determined by the α_s-method steadily increased with the extent of activation.

Kaneko [142] determined the He adsorption isotherm and compared to N_2 adsorption isotherm. Figure 39 shows the adsorption isotherms of He at 4.2K and N_2 at 77K on PAN-based activated carbon fiber. The amount of adsorbed He and N_2 are expressed by their volumes using the observed He density (0.205 gml^{-1}) and the liquid N_2 density(0.807 gml^{-1}), respectively. Both isotherms are of Type I. The He adsorption isotherm has a sharper uptake below P/Po = 0.02, but the amount of He adsorption approaches to that of N_2 adsorption with the increase of P/Po. The micropore size distributions of PAN and cellulose based activated carbon fibers by He and N_2 adsorption were calculated and compared in Figure 40. Small angle x-ray scattering and adsorption isotherms of He at 4.2K and N_2 at 77K on pitch based activated carbon fiber and activated carbon fiber treated at 1473K (ACF-1473) in Ar were determined by Setoyama [143]. The N_2 adsorption provided completely different peak positions of the micropore size distribution of ACF and ACF-1473, while He adsorption led to slight difference in the mean micropore width. The presence of necked micropores in ACF-1473 was evidenced by the distinct adsorption behaviors of He and N_2.

Table 20 Gas Adsorption Characteristics of Activated Carbon Fiber

Weight loss (%)	S_{BET} (m^2/g)	V_{DR} (cm^3/g)	$V_{\alpha s}$ (cm^3/g)	S_{ext} (m^2/g)
0	0.2	0.16	—	—
22	335	0.23	0.19	6
38	500	0.26	0.28	5
61	900	0.39	0.50	7
71	1360	0.39	0.73	17
78	1580	0.33	0.85	27

Source: from ref. 139.

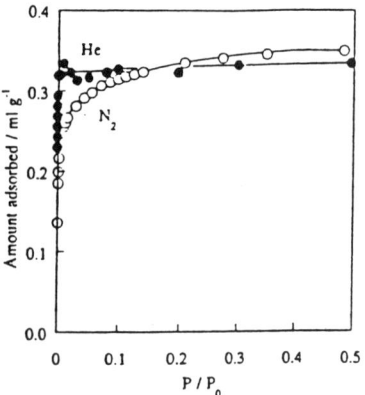

Figure 39 He and N_2 adsorption isotherms by PAN-based ACF at 4.2 and 77K, respectively [142].

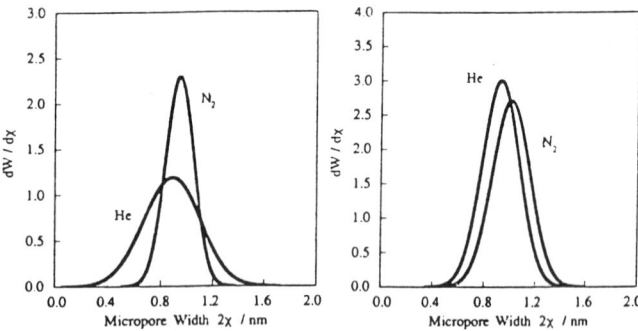

Figure 40 The micropore size distributions of PAN and cellulose based ACF from He and N_2 [142].

Economy [144] showed that the activated carbon fibers have pore radii between 5-14Å. Jaroniec [145] analyzed nitrogen adsorption isotherms in detail and both cellulose based and PAN-based activated carbon fibers have 85% micropores of uniform size and 15% of fine mesopores. Ehrburger [146] compared the change in microporosity of pitch based ACF activated in steam and in CO_2 respectively. During activation with steam, enlargement of the micropores takes place more effectively than with CO_2 and the shift from micropores to meso and macropores during steam gasification is more pronounced. Kaneko [147] studied adsorption of H_2O, NH_3, and SO_2 on activated carbon fibers and examined the effect of oxidation on adsorption characteristics. Fu [148] studied the reduction characteristics of different activated carbon fibers reacting with different metal ions such as Au(III), Fe(III), Ag(I), and Hg(II) by use of x-ray diffraction, x-ray photoelectron spectroscopy (XPS), and other analytic methods.

The direct observation of surface pores of activated carbon fiber has not been possible with conventional techniques such as electron microscopy, because of the resolution limit. Ishizaki [149] tried a scanning electron microscopy (SEM) study of surface pores of activated carbon fiber. Oshida [150] investigated the microstructure of the pores of activated carbon fibers using transmission electron microscopy (TEM) combined with computer image analysis. Donnet [151] studied the scanning tunneling microscopy (STM) of activated carbon fibers. At atomic resolution, the STM technique allows us to observe directly the changes in surface structure and the micropores before and after activation. Figure 41 shows the STM images of fibers before and after steam activation.

Kuriyama [152] measured the x-ray diffraction on activated carbon fibers which shows a broad diffraction line at the position of the (002) line in graphite. This implies that activated carbon fibers are a graphitic material with a high degree of disorder. Kuriyama measured also the photoconductivity activated carbon fibers [153]. From the Raman measurements the ACFs remain graphitic as the activation process proceeds to the greatest achievable specific surface area. Temperature dependence of the electrical conductivity and magnetic field dependence of the magnetoresistance of activated carbon fibers were measured by Vittorio [154].

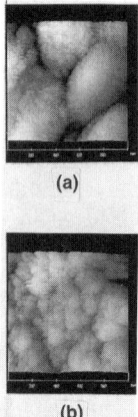

Figure 41 Large scale STM image of (a) pitch-based carbon fiber and (b) activated carbon fiber [151].

7.4.3 ACF Applications

Thin fiber clearly assure fast intraparticle adsorption kinetics compared with palletized or granular activated carbons commonly employed in gas phase and aqueous phase adsorption. This becomes more important in designing adsorption units where intraparticle diffusion resistance is the most significant factor, resulting in considerable decrease in the size of adsorption units. Activated carbon fibers in diverse textile forms are increasingly used in the form of various air filters [155,156], air conditioners [155], gas masks [156], and in appliances for water treatment [157,158].

Activated carbon fiber may be immediately used for purification of nitrogen oxide containing fumes [159]. The introduction of iron compounds into the composition of activated carbon fibers increases the sorption capacity of the material toward NO_x [160]. Effective conversion of NO_x to N_2 can be achieved by passing NO_x mixed with NH_3 through stacked packing of activated carbon fiber pretreated with sulfuric acid and annealed at 350°C [161,162]. Copper doping of the dispersed α-FeOOH particles enhanced the NO adsorptivity of pre-oxidized ACF remarkably [163]. Mochida [164] reported that PAN and pitch-based activated carbon fibers can adsorb, oxidize, and hydrate SO_2 into aq.H_2SO_4 by additional amount of H_2O onto the activated carbon fiber. Activated carbon fibers pretreated with salts of Mg, Al, Ca, Fe, and Zn are twice as effective than granular activated carbon in the adsorption of ammonia [165]. In recent years activated carbon fiber have found wide application in the recovery of some solvent vapors. Recovery of such vapors is important in excluding toxic substances from air and as a safety precaution in production. The adsorption/desorption rates obtained with activated carbon fibers are 10-100 times higher than those for granular activated carbon which makes it possible to curtail the duration of the adsorption-desorption cycle by 80% [166]. Figure 42 shows the adsorption/desorption curves of activated carbon fiber and granular activated carbon for toluene.

Adsorption of ethylmercaptan is observed corresponding to the content of nitrogen, which is the unique character of activated carbon fiber prepared from PAN [133]. Activated carbon fibers have high reduction-adsorption capacity for Pt(IV) from solutions containing 600–1000 ppm H_2PtCl_6 [167]. The adsorption capacity of a cigarette filter [168] from activated carbon fibers is reported to be more than double that of the filters made from cellulose acetate. Activated carbon fibers are also used for the production of protective clothing used in various chemical industries for work in extremely hostile environments [155,156]. A good separation of C_{60} and C_{70} was observed on preparative separation of fullerene mixture on activated carbon fibers/silica gel stationary phase [169].

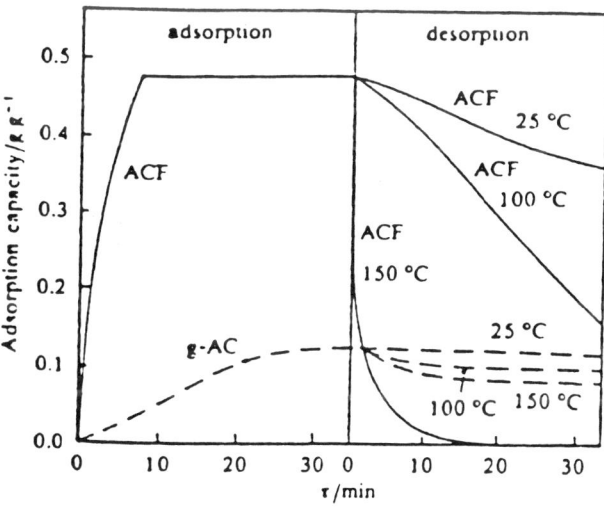

Figure 42 Adsorption/desorption curves of ACF and g-AC for toluene, adsorption capacity as a function of time [166].

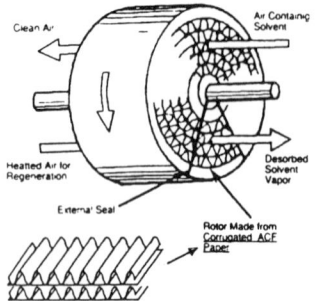

Figure 43 Rotor-type contactor made of corrugated ACF paper [170].

The adsorbed fiber form demands a special approach to the design of recovery apparatus different to that where granulated or powdered carbons are employed. Matsuo [170] developed a rotor type adsorber as shown in Figure 43, which employs the rotating adsorbent unit made of corrugated paper (honeycomb rotor) made from ACF and wood pulp for the continuous operation of adsorption and regeneration of adsorbent.

Activated carbon fibers are recognized as efficient catalysts or supports for the catalytically active phase in heterogeneous catalysis [171,172]. The use of carbon materials as supports for the active phase is favored because of their developed specific surface area, the possibility of producing the materials with homogeneous surface properties, sufficient strength, heat and chemical stability and their resistance towards the active metal phase.

Among other advantageous properties of carbon supports in comparison to commonly used mineral ones are heat conductance which facilitate homogeneity of the temperature regime of the catalytic processes and electrical conductance capable of promoting electrocatalytic processes. Thus, the use of the fibrous forms of carbon supports has extra advantages. The catalytic activity of the element-carbon fibers depends on the relevant processing route: whether the carbon fibers have been first soaked in the solution of a metal compound the initial raw material have been pyrolyzed in, with subsequent heat-treatment or the presence of metal compound.

The copper-carbon fibers produced by carbonization of copper 6-carboxyl cellulose salt demonstrate high catalytic activity towards dehydration of cyclohexanol and isopropanol [173]. The efficiency of this catalyst is due to the presence of finely dispersed reduced copper in the structure of the element-carbon fiber. High catalytic activity and selectivity is achieved with platinum-carbon fibers, produced with platinum below 3 wt.% on carbonization of H_2PtCl_6 soaked rayon, in the processes of configurational isomerization of cis-

1,2-dimethylcyclopentane, dialkylcyclobutanes, and benzene dehydrogenation [174].

Activated carbon fibers have been successfully used as an electrode of the electric double layered capacitor and developed to practical application [175]. The ACF capacitor has an extremely large specific capacity, and is expected to be in great demand for a back-up power source for computer memory because of their properties. The experimental ACF capacitor is composed of a pair of ACF felt electrode, and electrolyte is the same as for the secondary battery [176]. Filter paper was used as the separator of two ACF electrodes. Before and after charging of ACF capacitor, no changes are observed in the x-ray diffraction pattern of the ACF electrode. This indicates that charged ion species are only stored as electric double layers on the large activated carbon fiber surface by d.c. electric field, and are not doped in the fibers under this charging voltage as in secondary battery, where electrochemical interaction occurs.

A large number of studies so far carried out for obtaining more improved carbon materials having a larger Li doping/undoping capacity appear to have been concentrated in the modification of internal structure of carbon and little attention has been paid on the improvement of the surface configuration [177]. The doping/undoping reactions, however, occur at the carbon surface and hence the improvement of the surface condition is considered to be also important. Takamura [178] and Awano [179] have investigated on the surface to improve the characteristics of the doping/undoping reactions of activated carbon fiber by finding out an effective treatment. Activated carbon fibers were heat-treated and the cyclic voltammograms (cv) was compared before and after the heat-treatment. Remarkable improvements were found by the heat-treatment but the behavior indicated very sensitive to the voltage scanning rate, showing that two types of electrochemical Li reserve mechanism are involved for the sample: ordinary type doping and Faradaic adsorption of Li on the carbon surface.

REFERENCES

1. Towne, M. K., and Dowell, M. B. Papers of the American Association for Textile Technology, Inc., Mod. Tex. 51 (May 1976) .
2. Carbon and High Performance Fibers—Directory and Databook, edition 6, Chapman & Hall, London, UK, 1995.
3. Sambell, R . A. J., Bowen, D. H ., and Phillips, D. C . J. Mater. Sci. 7, 663 (1972).
4. Phillips, D. C., Sambell, R. A. J., and Bowen, D. H. J. Mater. Sci. 7, 1454 (1972).

4. Phillips, D. C., Sambell, R. A. J., and Bowen, D. H. J. Mater. Sci. 7, 1454 (1972).
5. Brazel, J. P., in 17th Conference on Metal, Carbon/Graphite, and Ceramic Matrix Composites(1993), NASA Conf. Publ. 1994, p. 933.
6. Abraham, S., Pai, B. C., Satyanarayana, K. G., and Vaidyan, V. K., J. Mater. Sci. 25, 2839 (1990).
7. Jackson, P. W., and Marjoram, J. R. J. Mater. Sci. 5, 9 (1970).
8. Essock, D. M., in 17th Conference on Metal, Carbon/Graphite, and Ceramic Matrix Composites(1993), NASA Conf. Publ. 1994, p. 878.
9. Morris, A. W. H. Carbon Fibers—Their Composites and Applications, Proc.1st Intern. Carbon Fiber Conf., The Plastics Institute, London (1971), p. 119.
10. Paprocki, S., Kizer, D., and Meyerer, W. Enigma Eighties Econ. Energy Book 2, 24(2), 1451 (1979).
11. Rashid, M. S., and Wirkus, C. D. Bull. Am. Ceram. Soc. 51, 836 (1973).
12. Kitamura, A. Personal communication, 1987.
13. Spencer, R. A. P. Carbon Fibers—Thetr Place in Modern Technology, Proc. 2nd Intern. Carbon Fiber Conf., The Plastic Institute, London (1974), p. 140.
14. Wrzesien, A. P. Carbon Fibers—Their Place in Modern Technology, Proc. 2nd Intern. Carbon Fiber Conf., The Plastic Institute, London (1974), p. 161.
15. Morgan, P. E. Carbon Fibers—Their Place in Modern Technology, Proe. 2nd Intern. Carbon Fiber Conf., The Plastics Institute, London (1974), p. 154.
16. Phillips, L. N. Proc. Inst. Mech. Eng. 185, 52 71, 793(1970–1971).
17. Dowell, M. B. Plastics Eng. 31 (April1977).
18. Tewary, V. K. Proc. Nucl. Phys. Solid State Phys. Symp.18A, 15 (1975).
19. Fitzer, E., and Heym, M. Chem. Ind. (London) 663(1976).
20. Shepler, R. E., Towne, M. K., and Saylor, D. K. Machine Design 3 (May 1979).
21. Jeffries, R. Nature 232, 304 (1971).
22. Haga, O., Koyama, H., Kawada, K., Adv. Composite Mater. 5, 139 (1996).
23. Jamet, J. F., and Hordonneau, A., Spécial Composites Avancés, Nov–Dec, 1994, p.37.
24. Fitzer, E., Hiittner, W., and Manocha, L. 14th Bienn. Conf. Carbon, Pennsylvania State University, University Park, Pa., 1979, extended abstracts, p . 240.

25. Highlights from SAMPE Isse '96, Composites & Adhesives Newsletter, p. 15, July–September 1996.
26. Markovic, V., Marinkovic, S., and Marsh, H. 15th Bienn. Conf. Carbon, Philadelphia, 1981, extended abstracts, p. 272.
27. Fritz, W., Huttner, W., and Hartwig, G. Proc. ICMC Symposium 1978, Met. Mater. Compos. Low Temp., 1979, p. 245.
28. Hill, J., Thomas, C. R., and Walker, E. J. Carbon Fibers—Their Place in Modern Technology, Proc. 2nd Intern. Carbon Fiber Conf., The Plastics Institute, London (1974), p.122.
29. Fitzer, E., Fritz, W., and Schoch, G., High Tem.-High Pres. 24, 343 (1992).
30. Buckley, J. D., Ceramic Bulletin 67, 364 (1988).
31. Sheehan, J. E., in 17th Conference on Metal, Carbon/Graphite, and Ceramic Matrix Composites(1993), NASA Conf. Publ. 1994, p. 920.
32. Fitzer, E., and Terwiesch, B. Carbon 11, 570 (1973).
33. Evangelidis, J. S. 13th Bienn. Conf. Carbon, Irvine, Calif., 1977, extended abstracts, p. 16.
34. Park, H. S., Choi, W. C., Kim, K. S., J. of Adv. Mater. p. 34, July 1995..
35. Fitzer, E. Carbon 25, 163 (1987).
36. Huttner, W. Toyobashi Carbon Conf., Toyohashi University, Toyohashi, Japan, 1984.
37. Paton, W., Lockhart, A. H., and Montgomery, I. W. Carbon Fibers—Their Place in Modern Technology, Proc. 2nd Intern. Carbon Fibers Conf., The Plastics Institute, London (1974),p. 148.
38. Patton, W. Composite 3(3), 119 (1972).
39. Hancox, N. L., Garnet, G., and Spencer, R. A. P. 3rd Conf. Res. Proj., Plastic Institute, London, 1971.
40. Hardesty, E. E. 28th Reinforced Plastics Tech. Conf., Soc. Plastics Ind., 1973.
41. Fitzer, E. (ed.) Carbon Fibers and Their Composites, Springer Verlag, New York (1985).
42. Brunsch, K. Autoclave and compression molding, in Carbon Fibers and Their Composites (edited by E. Fitzer), SpringerVerlag, New York (1985), p. 143.
43. Weuver, J. V., and Taylor, E. 3rd Intern. Conf. Carbon Fiber CF III, London, PRI, 1985, paper 30.
44. Dorey, G., Sidey, G. R., and Hutchings, J. Composites 9, 25(1978).
45. Kreiger, R. B. 3rd Intern. Conf. Carbon Fiber CF III, London, PRI, 1985, paper 23.
46. Reader, C. E. L., and Morley, J. M. Proc. 4th Intern. Conf. Carbon Graphite, Soc. Chem. Ind., London (1976), p. 238.

47. Rogers, R. F., and Kingston Lee, D. M. Proc. 3rd Conf. Ind. Carbon Graphite, Soc. Chem. Ind., London (1971), p. 503.
48. Dorey, G. Phys Technol. 11, 56 (1980).
49. Luxen, B. A. 3rd Intern. Conf. Carbon Fiber CF III, London, PRI, 1985, paper 8.
50. Subramanian, R . V ., Sundaran, V., and Patel, A. K . 33rd Ann. Tech. Conf., Soc. Plastics Ind., Reinforced Plastics Composites Inst., Section 20, 1978, pp . 1–8.
51. Tenney, D. R., Lisagor, W. B., and Dixon, S. C., Materials and structures for hypersonic vehicles, IACS-88-2.3.1., (1988).
52. Rieck, U., Bolz, J., and Müller-Wiesner, D., Int. J. of Materials and Product Technology 10, 303 (1995).
53. Paté, F., Nouvelle Revue D´aéronautique et D´Astronautique, No. 3, p.31 (1995).
54. Thomas, C. R., and Walker, E. J. High Temp. High Pressures 10, 79 (1978).
55. Fitzer, E., Geigl, K. N., and Hûttner, W. Carbon 18, 383 (1980) .
56. Thomas, C. R., and Walker, E. J. Proc. 5th Intern. Conf. Carbon Graphite, vol. 1, Soc. Chem. Ind., London (1978), p. 520.
57. Diefendorf, R. J., in 17th Conference on Metal, Carbon/Graphite, and Ceramic Matrix Composites(1993), NASA Conf. Publ. 1994, p.911.
58. Zimmer, J. E., and White, J. L. Carbon 21, 323 (1983).
59. Bacos, M. P., J. de Physique III, 3, 1895 (1993).
60. Chand, N., and Soni, R. P., Research and Industry 40, 5 (1995).
61. Vu-Khanh, T., Poly. Composites 8, 363 (1987).
62. Bradshaw, W. G., and Pinoli, P. C. 12th Bienn. Conf. Carbon, Pittsburgh, Pa., 1975, extended abstracts, p. 297.
63. Brunnel, L. R. 12th Bienn. Conf. Carbon, Pittsburgh, Pa., 1975, extended abstracts, p. 333.
64. Pacault , A ., Loll, P ., Delhaes , P ., and Pierre , A . 1 3th Bienn. Conf. Carbon, Irvine, Calif ., 1977, extended abstracts , p . 164.
65. Dacie, B., and Marinkovic, S. Carbon '80 3rd Intern. Carbon Conf., Baden Baden, 1980, preprints, p. 613.
66. Ismail, I. M. K., Mat. Res. Soc. Symp. Proc. 250, 71 (1992).
67. Fitzer, E., and Bûrger, A. Chem. Ing. Tech. 42, 1203 (1970).
68. Fitzer, E., and Terweisch, B. Carbon 10, 383 (1972).
69. Fitzer, E., Heym, M., and Rhee, B. Carbon '76 2nd Intern. Carbon Conf., Baden Baden, 1976, preprints , p . 508.
70. Fitzer, E., and Tillmanns, H. 12th Bienn. Conf. Carbon, Pittsburgh, Pa., 1975, extended abstracts, p. 217.
71. Fitzer, E., and Greiser, F. Carbon '76 2nd Intern. Carbon Conf., Baden Baden, 1976, preprints, p. 513.

72. Fitzer, E., Geigl, K. H., and Huttner, W. Proc. 5th Intern. Conf. Carbon Graphite, 1978, Soc. Chem. Ind., London (1978),p. 493.
73. Fitzer, E., Geigl, K. H., and Huttner, W. 14th Bienn. Conf. Carbon, Pennsylvania State University, University Park, Pa., 1979, extended abstracts, p. 236.
74. Lalacona, F. P. N71 34502 (1971), quoted in C. F. Old and M. G. Nicolas paper in Carbon Fibers—Their Place in Modern Technology, The Plastics Institute, London (1971), p. 89.
75. Dauphin, J., High Temp. Chem. Processes 3, 139 (1994).
76. Baker, A. A., et al. Composite 2, 154 (1971).
77. Reinforced Plastics, September 1996, p.44.
78. Wang, W. L., Zhang, G. D., Jin, C., and Wu, R. J. Proc. Intern. Symp.on Carbon, Toyohashi, Japun, 1982, p. 447.
79. Old, C. F., and Nicholas, M. G. Carbon Fibers—Their Place in Modern Technology, Proc. 2nd Inter. Carbon Fiber Conf., The Plastics Institute, London (1974), p. 89.
80. Howlett, B. A., Minty, D. C., and Old, C. F. Carbon Fibers—Their Composites and Applications, Proc. Intern. Carbon Fiber Conf., The Plastics Institute, London (1971), p. 99.
81. Old, C. F., and Nicholas, M. G. Carbon Fibers—Their Place in Modern Technology, Proc. 2nd Intern. Carbon Fiber Conf., The Plastics Institute, London (1974), p. 93.
82. Thomas, D. K. Plastics Rubber Intern. 8, 53 (1983).
83. Paton, W . 3rd Intern. Conf. Carbon Fiber CF III, London, PRI, 1985, paper 1.
84. Forsyth, R. B . 20th Natl. Symp. Exhibition, SAMPE, San Diego, Calif., April 29—May 1975.
85. Thompson, A. W. New Fibers and their composites, Philos. Trans. R. Soc. London, A294, 577 (1980).
86. Ting, J-M., Lake, M. L., Duffy, D. R., J. Mater. Res. 10, 1478 (1995).
87. Lapinleimu, R. 3rd Intern. Conf. Carbon Fiber CF III, London, PRI, 1985, paper 20.
88. Langdon, R., Automotive Engineer, p. 68, June/July 1996.
89. Laramee, R ., Lamere, G., Prescott, B ., Mitchell, R., Scottosanti, P., and Dahle, D. 13th Bienn. Conf. Carbon, Irvine, Calif., 1977, extended abstracts , p . 74.
90. Kirkhard, F. P. 1 2th Bienn. Conf. Carbon, Pittsburgh, Pa., 1975, extended abstracts, p. 279.
91. Sepcarb Carbon/Carbon Composites Materials Activity Bulletin, Société Européanne de Propulsion, Bouteaux, France (1981).
92. Van Auken, R. L. 27th SPI Reinforced Plastics/Composites Conf., Washington, D.C., 1972, p. 13 D.

93. Robinson, M., Carruthers, J., "Composites make tracks in railway engineering," Reinforced Plastics, November 1995, p. 20
94. Jaworske, D. D., Gaier, J. R., Hung, C. G., and Banks, B .A. SAMPE Q. 18, 9 (1986).
95. Ballace, B. 3rd Intern. Conf. Carbon Fiber CF III, London, PRI, 1985, paper 10.
96. Naitove, M. H. FRP '78: All eyes on automotive plastics, Technology 24, 67 (1978).
97. Blazewicz, M., Chlopek, J., Wajler, C., Kus, W. M., Corecki, A., Materials Engineering 7, 339 (1996).
98. Jenkins, D. H. R. 13th Bienn. Conf. Carbon, Irvine, Calif., 1977, extended abstracts, p.178.
99. Johnson, G. R. Conf. Design. Fiber Reinforced Mater., Inst. Mech. Eng., London, 1977, paper C2333/77.
100. McKenna, G. B., and Statton, W. O. Appl. Polymer Symp. 31, 335 (1977).
101. Fitzer, E., Huttner, W., and Wolter, D. 13th Bienn. Conf. Carbon, Irvine, Calif., 1977, extended abstracts, p. 180.
102. Charuley, J. Triangle 4, 175 (1960).
103. Bokros, J. C., et al. Chem. Phys. Carbon 9, 103 (1973).
104. Blazewicz, M., Wajler, C., Gorecki, A., Kus, W., "Chemical testing of carbon implant in vitro", Ceramics in Substitutive and Reconstructive Surgery, Elsevier Science Publisher, B.V. 239, 1991..
105. Huttinger, K. J. Intern. Symp. Carbon, 1982, extended abstracts, p. 29.
106. Huttinger, K. J., et al. Biomaterials 1, 67, 73 (1980); Proc.6th Intern. Conf. Carbon Graphite, 1982.
107. Jenkins, G. M. Chem. Phys. Physiol. Meas.1, 171 (1980).
108. Burri, C., and Neugebauer, R. Replacement of Lipaments by Carbon Fibers, Springer, New York (1985).
109. Fitzer, E., Huttner, W., Claes, L., and Kinzl, L. Carbon 18, 383 (1980).
110. Shorter, J. J. 3rd Intern. Conf. Carbon Fiber CF III, London, PRI, 1985, paper 24.
111. Trowin, E. M. Reinforced Plastics Cong., British Plastics Federation, Brighton, U.K., 1976, paper 23. Lucas, R. Carbon Fibers—Thetr Place in Modern Technology, Proc. 2nd Intern. Carbon Fiber, London (1974), p. 266.
112. Lucas, R. Carbon Fibers - Their Place in Modern Technology, Proc. 2nd Intern. Carbon Fiber Conf., The Plastics Institute, London (1974), p. 266.
113. Lovell, D. R. 3rd Intern. Conf. PRI, 1985, paper 27.
114. Li, Pontong, Li, Xiaoming, and Chung, D. D. L. 18th Bienn Conf. Carbon, 1987, extended abstracts and program, p. 239.

115. Goddard, D. M., Burke, P. D., Kizer, D. E., Bacon, R., and Harrigan, W. C., in 17th Conference on Metal, Carbon/Graphite, and Ceramic Matrix Composites(1993), NASA Conf. Publ. 1994, p. 867.
116. Savage, G., Polymers & Composites, p. 617, October 1991.
117. European Plastics News, p. 34, February 1996.
118. Bohra, J. N., Awasthy, B. R., and Chari, S. S., Fiber Sci. Technol. 14: 221 (1981).
119. Biederman, D. L., Miles, A. J., Vatola, F. J., and Walker, P. L., Carbon, 14: 351 (1976).
120. Olander, D. R., and Balock, M. J. Catal. 60: 41 (1979).
121. Mortka, S., and Tomassi, W., Biul. Wojsk. Akad. Tech. 26(9): 51 (1977)
122. Babicki, R., and Perzynski, B., PL Patent 96334, (1975), Chem. Abstr. 90, 56690 (1979).
123. Ailey, A., and Maggs, A., GB Patent 1,301,101 (1969). 1,310,011 (1973).
124. Gavrilov, M. Z., Ermolenko, I. N., and Efimova, T. A., Dokl. Akad. Nauk BSSR, 22 : 53 (1978).
125. Abbot, W. F., US Patent 3,053,775 (1962).
126. Doying, E. G., US Patent 3,256,206, Union Carbide Co. (1966)
127. Freeman, J. J., Gimblett, F. G. R., and Sing, K. S. W., Carbon, 27 : 85 (1989).
128. McNair, R. N., and Arons, G. N., Adsorptive Textile Systems Containing Activated Carbon Fibers, Carbon Adsorption Handbook, Chapter 22, 819-859, Ann Arbor Science Pub. (1977).
129. Economy, J., and Lin, R. Y., J. Mater. Sci. 6, 1151 (1971).
130. Japanese Patent 132,193 (1976).
131. Japanese Patent 30,810 (1978).
132. Japanese Patent 7,583 (1980).
133. Suzuki, M., Carbon, 32 : 577 (1994).
134. Rodriguez-Reinoso, F., Molina-Sabio, F. M., Munecas, M. A. J., Phys. Chem. 96, 2707 (1992).
135. Kuriyama, K., and Dresselhaus, M. S., Phys. Rev. B. 44, 8256 (1991).
136. Imai, J., and Kaneko, K., Langumuir, 8, 1695 (1992).
137. Imai, J., Souma, M., Ozeki, S., Suzuki, T., and Kaneko, K., J. Phys. Chem. 95, 9955 (1991).
138. Kaneko, K., Shimizu, K., and Suzuki, T., J. Phys. Chem. 97, 8705 (1992).
139. Ryu, S. K., Rhee, B. S., Lee, J. K., Pusset, N., and Ehrburger, P., Proceedings of Carbon'90, Paris, 1990, pp 96-97.
140. Rodriguez-Reinoso, F., Martin-Martinez, J. M., Prado-Burguete, C., and McEnaney, B., J. Phy. Chem. 91, 515 (1987).

141. Dubinin, M. M., Progress in Surface and Membrane Science, vol. 9, Academic Press, New York, 1975, p. 1–70.
142. Kaneko, K., Setoyama, N., and Suzuki, T., Characterization of Porous Solids III, Studies in Surface and Catalysis, vol. 87, Elsevier Science B. V., 1994, p 593.
143. Setoyama, N., Ruike, M., Kasu, T., Suzuki, T., and Kaneko, K., Langmuir, 9, 2613 (1993).
144. Economy, J., and Lin, R. Y., Appl. Polymer Symp. 29, 199 (1976).
145. Jaroniec, M., Gilpin, R. K., Kaneko, K. and Choma, J. Langmuir, 7, 2719 (1991).
146. S. K. Ryu, H. Jin, D. Gondy, N. Pusset, and P. Ehrburger, Carbon, 31: 841 (1993).
147. Kaneko, K., Katori, T., Shimizu, K., Shindo, N., and Maeda, T. J., Chem. Soc. Faraday Trans. 88, 1305 (1992).
148. Fu, R. W., Zeng, H. M., and Lu, Y., Carbon, 31, 1089 (1993).
149. Ishizaki, N., Science and Industry (Jap.), 59(5), 171, (1985).
150. Oshida, K., Kosiso, K., Matsubayashi, K., Takeuchi, K., Kobayashi, S., Endo, M., Dresselhaus, M. S., and Dresselhaus, G., J. Mater. Res., 10, 2507 (1995).
151. Donnet, J. B., Qin, R. Y., Park, S. J., Ryu, S. K., and Rhee, B. S., J. Mater. Sci., 28, 2950 (1993).
152. Kuriyama, K., and Dresselhaus, M. S., J. Mater. Res., 6, 1040 (1991).
153. Fung, A. W. P., Rao, A. M., Kuriyama, K., Dresselhaus, M. S., and Dresselhaus, G., In Defects in Materials, Proceedings MRS Symposia, Boston, 209, 1991, p335. Materials Research Society Press. Pittsburgh, PA.
154. Vittorio, S. L. di., Dresselhaus, M. S., Endo, M., Issi, J. P., Piraux, L., and Bayot, V., J. Mater. Res. 6, 778 (1991).
155. Arons, J. N., MacNair, R. N., and Erikson, R. L., Text. Res. J. 43, 539 (1973).
156. Swiss Patent 514,500 (1970).
157. Jung, C. H., Jung, H. H., Oh, W. Z., Ryu, S. K., and Rhee, B. S., Proceedings the 22th Biennial Conf. on Carbon, San Diego, 1995, pp 474-475.
158. Kawazoe, K., and Sakoda, A., Kagaku Kogaku, 50(3), 182 (1986).
159. Richter, E., Knoblauch, K., Jontgen, H., DE Offen Patent 3423761, 1984; Chem. Abstr. 1986, 104, 135350; Knoblauch,K., Richter, E., Jontgen, H., Chem.-Ing. Techn. 57, 239 (1985).
160. Kaneko, K., Ozeki, S., and Inouye, K., Nippon Kagaku Kaishi, 7, 1351 (1985).
161. Komatsubara, Y. Ida, S., and Fujitsu, H., Fuel, 63, 1738 (1984).

162. Mochida, I., Kawano, S., Kisamori, S., Fujitsu, H., and Maeda. T., Carbon, 32, 175 (1994).
163. Wang, Z. M., Shindo, N., Otake, Y., and Kaneko, K., Carbon, 32, 515 (1994).
164. Mochida, I., Kuroda, K., Kawabuchi, Y., Kawano, S., Matsumura, Y., and Yoshikawa, M., Proceedings the 22nd Biennial Conf. on Carbon, San Diego, 1995, pp 576-577.
165. Brun'ko, T. G., Ermolenko, I. N., Laserko, G. A., and Morozova, A. A., Vestsi Akad. Navuk BSSR, Ser. Khim. Navuk (6), 45 (1979).
166. Ohmori, S., SEN-I GAKKAISHI, 41(6), 167, (1985).
167. Fu, R. W., Zeng, H. M., and Lu, Y., Carbon, 33, 657 (1995).
168. Br. Patent 1,256, 048 (1970)
169. Manolova, N., Rashkov, I., Legras, D., Delpuex, S., and Beguin, F., Carbon, 33, 209 (1995).
170. Matsuo, T., Komagata, H., and Suzuki, Y., Kagaku Kogaku, 43, 232 (1979).
171. Rikatskaya, T. L., Litvinskaya, V. V., Abramova, N. N., Red'ko, T. D., and Popova, N. A., Zh. Prikl. Khim. 60, 1415 (1987).
172. Juntgen, H., Fuel, 65, 1436 (1986).
173. Ermolenko, I. N., Safonova, A. M., Bel'skaya, R. I., Malashevich, Z. V., and Berezovik, G. K., Vestsi Akad. Navuk BSSR, Ser. Khim. Navuk (5), 17 (1976).
174. Ermolenko, I. N., and Safonova, A. M., Vestsi Akad. Navuk BSSR, Ser. Khim. Navuk (6), 33 (1985).
175. Endo, M., Okada, Y., and Nakamura, H., Synthetic Metals, 34, 739 (1989).
176. Yoshida, A., Tanahashi, I., Takeuchi, Y., and Nishino, A., IEEE Transections on Components, Hybrids, & Manufacturing Technology, Vol. CHMT-10, No.1, Mar. (1987).
177. Dahn, J. R., et al., Phy. Rev. B. 42, 6424 (1994).
178. Takamura, T., Awano, H., Ura, T., and Ikezawa, Y., Proceedings of the 36th Battery Symp. in Japan, Kyoto, 1995, pp 91–92.
179. Awano, H., Aida, T., Ikezawa, Y., and Takamura, T., Proceedings. of the 36th Battery Symp. in Japan, Kyoto, 1995, pp 129–130.

Author Index

Abbot, W. F. 526, 539
Abhiraman, A. S. 354, 366, 9, 13, 74–75
Abraham, S. 468, 498, 534
Abrahams, E. 376, 379, 416, 457
Abramova, N. N. 532, 541
Adams, D. F. 345, 365
Adams, W. W. 313, 316–317, 343–344, 346, 348, 366
Agrawal J. P. 6, 73
Ahearn, C. 164–165, 225
Ahmed, M. 82
Aida, T. 533, 541
Ailey, A. 525, 539
Aiyer, H. N. 434, 463
Ajayan, P. M. 433–434, 463
Ajiki, H. 435, 463
Akovali, G. 121, 155, 188–189, 226, 252, 305
Alcaniz Monge, J. 168–169, 224, 271, 307

Alig, R. 84
Allen, G. 372, 456
Allen, S. R. 29, 78 , 314, 317, 342, 365
Allred, R. E. 213, 228
Altshuler, B. L. 376, 379, 419, 438, 457, 460–463
Anderson, C. W. 211, 228
Anderson, D. P. 313–317, 346, 352, 366
Anderson, P. W. 376, 379, 416, 457
Anderson, R. 95, 153
Anderson, S. H. 137, 158
Ando, T. 435, 463
Ando, Y. 433, 463
Annis, B. K. 148, 159, 281, 308
Ansart, A. 415, 459
Arai, Y. 315, 318, 320, 346, 366
Arefieva, E. F. 66–67, 83
Armstrong–Carroll, E. 188, 226
Aronov, A. G. 376, 379, 419, 457, 460

Assabghy, F. Y. I. 386–389, 389, 407, 458
Atkinson, K. E. 184–185, 187, 226
Atsushi, T. 12–13, 74
Audier, M. 67, 84
Auguié, D. 126, 130, 133, 156–157
Auslender, V. L. 361, 368
Awano, H. 533, 541
Awasthy, B. R. 525, 539
Ayache, J. 132, 157

Babicki, R. 525, 539
Bacon, G. E. 86, 152
Bacon, R. 3, 32, 34, 36–40, 72, 78, 79,117, 155,237, 303, 332–333, 335–336, 338–339, 349, 364,495, 498, 538
Bacos, M. P. 522, 536
Badami, D. V. 237, 303
Bader, M. G. 178–179, 226,256, 306, 355, 367
Bahl, O. P. 2, 10–17, 20–30, 72, 74–79, 172, 100–101, 104, 153–154, 181–182, 225–226, 252, 305
Bai, S. J. 313–314, 316–317, 342, 365
Bailey, J. E. 23–24, 77
Baillie, C. A. 178–179, 226, 257, 306
Baired, T. 66–67, 83
Bajaj, P. 5–6, 72–73
Baker, A. A. 495–496, 537
Baker, D. F. 408–409, 459
Baker, R. T. K. 66, 68, 83
Balaev, G. A. 357, 367
Balduhn, R. 128, 156
Balik, K. 163, 225
Ballace, B. 517–518, 538
Balock, M. 525, 539
Ban, K. 51, 248, 304
Ban, L. L. 114, 154
Banks, B. A. 517, 538
Bansal, R. C. 12–13, 17, 26, 74, 80, 85, 152, 162, 172, 176, 224–225, 233, 253, 255–256, 303, 305, 306
Barber, M. A. 66, 68, 83

Barbier, B. 121, 155,161, 224
Barnet, F. R. 112, 154
Barnett, B. C. 10, 74
Barr, J. B. 51, 54, 82, 126, 132, 156
Barton, R., Jr. 373, 375, 390, 395–396, 400, 402, 456
Barton, R. 253, 306–390, 458
Barton, S. S. 253, 305
Basch, R. 46, 81
Basche, M. 163, 168, 224
Bascom, W. D. 277, 308, 340, 349, 364
Bastianelli, V. 7, 73
Bastick, M. 110, 116, 119, 154–164, 168, 224
Bayor, A. G. 8–9, 73
Bayot, V. 87, 152,372, 376, 381, 389–389, 399–400, 407, 417–420, 423, 430, 434, 435, 437–438, 456–459, 529, 540
Bayramli, E. 121, 155, 188–189, 226, 252, 305
Beamson, G. 276, 307
Beck, S. 197, 227
Beetz, C. P. 407, 459
Beetz, C. P., Jr. 321–322, 324, 330, 363
Beguin, F. 531, 541
Behar, F. 126, 156
Bel'skaya, R. I. 532, 541
Bell, J. P. 195–196, 226
Bellamy, M. 210, 228
Benardin, J. 253, 305
Benedict, J. P. 95, 153
Bennett, S. C. 108, 116, 119–120, 154–155, 327–328, 363
Beny-Bassez, C. 85, 93–94, 151
Berezovik, G. K. 532, 541
Bergmann, G. 376, 379–380, 457
Berlin, À. À. 17, 76, 355–367
Berman, R. 393–395, 423–424, 458
Bernard, C. 192, 227
Bernardo, M. D. 249, 251, 305
Bert, M. 6, 73
Bertram, A. 372, 382, 388–389, 391–401, 456
Bertrand, P. 292–294, 309

Beslon, T. 14, 75
Beziers, D. 259, 306
Bhardwaj, A. 276, 308
Bhardwaj, I. S. 276, 308
Biederman, D. L. 525, 539
Biensan, P. 99–100, 153
Billmayer, F. W. 23, 77
Binnig, G. 98, 153, 281, 308
Bishop, D. J. 438, 463
Blackketter, D. M. 213–214, 228
Blackman, B. R. K. 181, 226
Blackslee, O. L. 124, 126, 156, 403, 459
Blatt, F. 405–406, 414, 459
Blatt, F. J. 423, 460
Blazeuiex, S. 27, 78
Blazewicz, M. 519–521, 538
Blinowski, J. 413–414, 422, 430, 444, 459
Bobka, R. J. 232–233, 303
Bodenstein, P. 88, 152
Boehm, H. P. 121, 155, 253, 255, 257, 305, 306
Bogdanov, V. V. 362, 371
Bogdanov, W. W. 362, 368
Bogoeva-Gaceva, G. 269–270, 307
Bohra, J. N. 525, 539
Bokros, J. C. 519, 538
Boll, D. J. 340, 349, 364
Bolz, J. 486–489, 536
Bonnamy, S. 87, 92, 95, 103, 114, 126–130, 132–140, 142, 144–147, 152, 156–158
Bouix, J. 203, 205, 228
Boulmier, J. L. 87, 103, 107, 139, 141, 147, 154, 159
Bouraoui, A. 85, 88, 90, 151
Bourrat, X. 87, 103, 114, 139–140, 147, 152, 158, 283, 309, 389, 458
Boutevin, B. 259, 306
Bowden, M. 345, 365
Bowen, D. H. 468, 533
Bower, W. N. 355, 367
Bowie, S. H. V. 85, 151

Bowles, K. J. 237, 304
Boxus, J. 392, 423, 428, 458
Boyer, A. 99, 103, 125, 153
Boyington, D. R. 149, 159, 373, 375, 390, 395–396, 400, 402–403, 456, 458
Bradley, R. H. 248, 251–252, 276–278, 304–305, 307
Bradley, W. L. 267, 306
Bradna, P. 292, 309
Bradshaw, W. G. 194, 227, 488, 536
Bragg, R. H. 408–409, 459
Brandrup, J. 18, 23, 76,, 77
Braun, M. 50, 82
Braunstein, G. 390, 402, 458
Brazel, J. P. 468, 534
Brendle, M. 181–184, 186, 226, 252, 305
Brezovska, S. 269, 307
Bright, A. A. 140–141, 158, 381, 386–389, 389, 457
Brindley, G. W. 87, 152
Brockhouse, B. N. 403, 459
Broido, A. 36, 79
Brombley, J. 26, 78
Brooks, C. S. 232, 264–265, 303
Brooks, J. D. 41, 46, 80, 128, 156
Brown, N. M. D. 281, 283, 308
Brun'ko, T. G. 530, 541
Bruneton, E. 120, 155
Brunnel, L. R. 488, 536
Brunsch, K. 473, 535
Bruynseraede, Y. 372, 376, 388, 434–435, 437–438, 456, 458
Buchanan, F. J. 209, 228
Buckley, J. D. 197, 227, 468, 488–490, 535
Budnizky, G. M. 367
Budtov, V. P. 360, 361, 368
Buechler, M. 128, 135, 156, 339, 364
Buesking, K. W. 206–207, 227
Bunsell, A. R. 328, 363
Burevski, D. 269–270, 307
Bûrger, A. 491, 536
Burke, P. D. 495, 498, 538

Burlant, W. J. 18, 76
Burri, C. 538
Butcher, P. N. 400, 458
Buttry, D. 173–175, 215–216, 229, 259–260, 262–263, 274–275, 306, 307

Callaway, J. 424–425, 460
Cano, R. J. 185–186, 226
Carlier, V. 292–294, 309
Carruthers, J. V. 537
Cartwright, J. M. 346, 365
Casella, G. I. 161, 173, 224, 272, 276, 278, 290, 307–309
Caseneuve, C. 252, 305
Castle, J. E. 257, 289, 306, 309
Cataldi, T. R. I. 161, 224–276, 278, 308
Cazeneuve, C. 289, 309
Cazorla-Amoros, D. 168–169, 224, 271, 307
Celanese Corporation 12–13, 74
Cermia, E. 13, 75
Cha, Q. 51, 82
Chaberski, A. 386, 457
Chabra, P. 172, 176, 225
Chakraborty S. 203, 228
Chand, N. 480, 506, 536
Chang, G. 15, 76
Chang, W. C. 172, 225
Chao, H. T. 172, 225
Chari, S. S. 12–13, 75, 100, 102, 153, 525, 539
Charlier, A. 433–434, 460, 463
Charuley, J. 519, 538
Chataignier, E. 259, 306
Chatzi, E. 271, 307
Chavan R. B. 5, 72
Chen, C. I. 211, 228
Chen, K. J. 335, 337, 364
Cheng, T. H. 211, 228
Chhabra, P. 233, 255–256, 303, 306
Chi, W. 57–58, 82–83
Chiche, P. 128, 156

Chien, T. C. 66, 83, 143, 159, 365, 376, 412–413, 457, 459
Chin, J. W. 346, 365
Chitrapa, P. 67, 84
Chiu, H. T. 196, 227
Chlopek, J. 27, 78, 519–521, 538
Chlyabich, P. P., 361, 368
Cho, Y. I. 136, 157
Choe, I. W. 14, 75
Choi, W. C. 468, 489–490, 535
Choma, J. 529, 540
Chou, S. T. 211, 228
Chouliotis, A. 274, 307
Christin, F. 192, 227
Christoforov, E. J. 368
Christopher, D. J. 345, 365
Chun, B. W. 257, 306
Chung, D. D. L. 137, 158, 524, 538
Chwastiak, S. 48, 81, 129, 131, 157
Claes, L. 519, 538
Clark D. 164, 224
Clarke, A. J. 23–24, 77
Clinard, C. 129–130, 133–134, 157
Cochran, R. C. 213, 228
Cochrane, W. 85, 151
Cofer, C. G. 203–205, 228
Cole, K. C. 272, 274, 307
Coleclough, W. F. 176, 225
Coleman, M. M. 18–19, 22, 76–77
Colthup, N. B. 275, 307
Combaz, A. 107, 110, 139, 147, 154
Commercon, P. 226
Conder J. R. 244–245, 304
Connolly, M. P. 239, 303
Cooper, E. G. 386–389, 407, 458
Cooper, G. A. 27, 29, 78
Cooper, W. T. 246, 304
Coopmans, P. 392, 458
Cordner, L. 340, 349, 364
Corecki, A. 519–521, 538
Cory, M. T. 33, 78
Costello, J. A. 203, 227
Couzi, M. 93, 152
Cranch, G. E. 36, 79
Crane, L. W. 161, 224

Author Index

Crasto, A. S. 314–317, 343, 346, 365, 366
Cray, R. D. 32, 37–78
Criscione, J. M. 203, 227, 228
Cuesta, A. 93, 152, 248, 251, 304
Cupp, K. H. 16, 76
Curtis, C. J. 330, 364
Curtis, P. T. 345, 365

Dacie, B. 488, 536
Dahle, D. 512, 537
Dahn, J. R. 533, 541
Dai, S. R. 324, 363
Daley, M. 240, 304
Damiano, D. 173, 225, 290, 309
Damodaran, S. 354, 366
Danghan, W. 14, 75
Daozhi, L. 161, 224
Dasch, C. J. 68, 84
Daukeys, R. J. 161, 224
Dauksch, H. 119, 120, 155, 173, 225
Daulan, C. 99, 153
Daumas, N. 417, 459
Daumit, G. P. 295, 309
Dauphin, J. 500, 537
David, A. 43, 80
Davis, C. Q. 161, 193, 224
Davis, E. A. 408, 459
Davis, L. E. 279, 308
Davis, R. 257, 306
Deduit, J. 128, 156
Dekanski, A. 269–270, 307
Delacroix, A. 215, 229
Delgass, W. N. 188, 226, 276, 279–279, 299–302, 308
Delhaes, P. 488, 536
Delmonte, J. 3, 72, 85, 151
Delpuex, S. 531, 541
Demain, A. 372, 374, 395, 447–452, 456, 463
Desai, P. 9, 74, 354, 366
Désarmot, G. 106, 108, 110–112, 114, 116, 118, 121, 154–155, 259,

[Désarmot, G.]
306, 327–328, 330, 332–333, 363
Desimoni, E. 161, 173, 224–225, 276, 278, 290, 308, 309
Desprès, J. F. 98, 120, 155
DeTeresa, S. J. 313–316, 340, 342, 344–347, 365
Deurbergue, A. 99, 101–102, 104–108, 110, 113–116, 119, 141, 153–155
Devilbiss, T. A. 277, 308
Devour, M. G. 65, 67, 83–84
Dhamelincourt, P. 93, 152
Dhami, T. L. 15–16, 25, 76, 181–185, 186, 226, 252, 305
di Vittorio, S. L. 399, 458
Didchenko, R. 238, 303
DiEdwardo, A. H. 14, 75
Diefendorf, R. J. 41, 43, 46, 48, 79, 81, 85, 97, 116, 118, 131, 152–153, 157, 335, 337, 364, 491–492, 536
Diehl, E. 121, 155, 253, 255, 257, 306
Dietz, V. R. 29, 78, 234, 303
Diffu, J. V. 36, 79
Dillard, D. A. 346, 365
Dilsiz, N. 121, 155, 188–189, 226, 252, 305
Dix, R. 172, 176, 225
Dixon, S. C. 486, 536
Dobb, M. G. 140–141, 158, 313, 316–317, 349–350, 366
Dolling, G. 403, 459
Dong, S 181, 185–184, 186, 226, 252, 305
Donnellan, T. M. 213, 228
Donnet, J. B. 2, 12–13, 15, 17, 26, 29, 72, 74, 80, 85, 119, 120–121, 123–124, 148, 152, 156, 162, 173, 177, 181–184, 186, 221–222, 224–226, 228, 237, 244–247, 249–250, 252–254, 281–287,

[Donnet, J. B.]
303–306, 308–309, 324–325, 363, 529, 540
Dorey, G. 116, 155, 535–536
Dowell, M. B. 466, 468, 474, 503, 512, 515, 533–535
Doying, E. G. 526, 539
Draper, N. R. 363
Drbohlav, J. 139, 158
Dresselhaus, G. 68, 84, 85, 87, 143, 151–152, 159, 283, 309, 375, 378, 381–383, 389, 400, 402, 412–414, 433, 456, 459–463, 529, 540
Dresselhaus, M. S. 64, 68, 83–84, 85, 87, 143, 151–152, 159, 283, 309, 365, 375–376, 378, 381–383, 389, 390, 392, 396, 399–400, 402, 412–414, 418–420, 423, 425–428, 430–431, 433, 456–463, 527, 529, 539, 540
Dreyfus, B. 402, 424, 458, 460
Druin, M. L. 168, 172, 176, 225
Drushel, H. B. 253, 306
Drzal, L. T. 120, 155, 210, 216, 220–221, 228, 233, 244, 252, 303, 305, 324, 326, 340, 353, 363, 366
Dubinin, M. M. 527, 539
Dubinskaya, A. M. 17, 76
Dubois, J. 103, 154
Duffy, D. R. 505, 512, 515, 519, 522–523, 537
Duffy, J. V. 33, 78, 194, 227
Duflos, J. L. 31, 78
Duncan, R. G. 75, 335, 364
Dunham, M. G. 50, 53, 82, 136, 157
Dupont 49
Duvis, T. 196, 227
Dynes, P. J. 161, 224, 251, 305

East, G. C. 9, 10, 74
Ebbesen, T. W. 372, 433–434, 456, 463
Ebert, E. 189, 226
Eby, R. K. 316–317, 343–344, 365

Economy, J. 203–205, 228, 240, 304, 526, 529, 539–540
Edie, D. D. 10, 50, 53, 74, 82, 85, 136–137, 149, 152, 157–159, 185–186, 226, 316–317, 346, 366
Edison, T. A. 1, 32, 72
Effler, L. J. 148, 159, 281, 308
Efimova, T. A. 525, 539
Ehrburger, P. 119, 155, 177, 221–222, 225, 228, 237, 239, 254, 303, 304, 306, 527– 529, 539–540
Eiji, K. 57, 82
Elings, V. B. 122, 148, 159
Elkind, J. L 95, 153
Elman, B. S. 390, 402, 458
Elzinga, M. 423, 426, 460
Emig, G. 191–194, 227
Endo, M. 64–68, 83, 87, 114, 137, 140, 143, 144, 150, 152, 155, 157–159, 197–199, 227, 328, 363, 365, 376, 382, 386–389, 389, 392, 396, 399, 407, 412–413, 418–420, 423, 427, 430, 457–459, 529, 533, 538–541
Endo, T. 131, 157
Enikolopov, N. S. 355–367
Erdman, J. G. 126, 156
Ergun, S. 85, 91, 151
Erickson, P. W. 161, 194, 224, 227
Erikson, R. L. 531, 540
Erinc, N. K. 121, 155, 188–189, 226, 252, 305
Ermolenko, I. N. 525, 530, 533, 539, 541
Espinat, D. 126, 156
Essock, D. M. 468, 495, 498, 534
Esteve, R. N. 39, 79
Evangelidis, J. S. 468, 489, 535
Everall, N. J. 345, 365
Everett, D. H. 247, 304
Evrtors, I. L. 20, 77
Ewins, P. D. 349, 366
Exxon Research Engineering Co. 48, 80

Author Index

Fain, C. C. 10, 74 137, 149, 158, 159
Fang, C. K. 172, 225
Farrow, G. J. 184–187, 226
Faserforsch, J. U. 6, 73
Fateley, W. G. 275, 307
Favry, Y. 328, 363
Fawaz, S. A. 343, 365
Feetes, F. S. 66, 68, 83
Fellers, J. F. 148, 159, 281, 308
Ferguson, J. 12–13, 74
Ferment, G. R. 168
Ferris, R. J. 316, 342, 344, 365
Fester, W. 17, 76
Fiedler, A. K. 23, 77, 101, 153
Figueiredo, J. L. 44–45, 80
Fischbach, D. B. 85, 88, 93, 151–152, 330, 349, 364
Fitchmun, D. R. 360, 368
Fitzer, E. 17, 20–26, 45, 76–77, 80, 85, 101, 112, 120, 128, 152–153, 155–177, 197, 202, 217, 219, 222–223, 225–229, 468, 471, 473, 488–489, 491, 493–494, 512, 514, 516–517, 519, 534–538
Fitzgerald, J. D. 133–134, 140, 142, 157–159
Flandrois, S. 99–100, 153, 415, 459
Fleurot, O. 137, 149, 158–159
Fluck, N. 184–185, 187, 226
Foch, P. 99, 103, 125, 153
Forsyth, R. B . 503, 537
Fortin, F. 147, 159
Foster, P. J. 274, 307
Fourdeux, A. 97, 108, 116–117, 119, 153, 155, 237, 303
Fourneaux, R. 121, 155, 336, 364
Fowkes, F. M. 252, 305
Fox, N. K. 10, 74, 137, 158
Franklin O. 177, 225, 276, 278, 308
Franklin, R. E. 87, 152, 388, 458
Franquinet, P. 292, 294, 309
Fratzl, P. 237, 303
Freeman, J. J. 526, 539
Friedel, G. 101, 153

Friedlander, H. N. 18, 23, 76
Frimpong, S. 237, 304
Fritz, W. 468, 489, 535
Frohs, W. 20–21, 26, 77
Frushour, B. 9, 74, 100, 153
Fryer, J. R. 66–67, 83
Fu, R. W. 529, 531, 540, 541
Fujikata, Y. 200–202, 227
Fujikawa, H. 200, 227
Fujino, H. 349, 366
Fujita, J. 433–434, 463
Fujita, K. 176, 225
Fujitsu, H. 530, 540
Fujiyama, S. 132, 157
Fukuyama, H. 376, 379–380, 416, 438, 457, 463
Fuller, M. P. 272, 307
Fung, A. W. P. 529, 540
Furdin, G. 413–414, 422, 430, 444, 459
Furuyama, M. 137, 140, 144, 150, 158, 314, 318, 346, 366
Futura, T. 128, 157

Gaier, J. R. 415, 459, 517, 538
Galiotis, C. 313–346, 365–366
Gallis, H. M. 39, 79
Gardiner, D. J. 345, 365
Gardner, S. D. 171–172, 225
Garnet, G. 474, 535
Garton, A. 274, 304
Gates, J. 345, 365
Gavrilov, M. Z. 525, 539
Ge, K. 15, 77
Geige, K. H. 120, 155
Geigl, K. H. 536
Geigl, K. N. 491, 536
Geller, A. A. 8, 74
Geller, B. E. 8, 74
Gerber, C. 281, 308
Gerrard, D. L. 345, 365
Ghiorse, S. R. 237, 304
Gibson, J. 65, 83
Gijs, M. 388, 458
Gilbert, A. H. 341–342, 365

Gillespie, D. 253, 305
Gilpin, R. K. 529, 540
Giltrow, J. D. 359, 368
Gimblett, F. G. R. 526, 539
Giorgetta, S. 299, 301, 302, 309
Gkogkidis, A. 202, 227
Glogar, P. 163, 225
Goan, J. C. 232, 303
Goddard, D. M. 495, 498, 538
Gold, G. S. 232–265, 303
Goldberg, H. A. 85, 151, 283, 309, 375, 378, 381–383, 389, 400, 402, 412, 456
Goldstein, E. L. 260–261, 306
Goma, J. 96, 98, 114, 153
Gondy, D. 239, 304, 529, 540
Gonze, X. 376, 417, 419–420, 457
Goodhew, P. J. 23–24, 77
Gorbatkina, Y. A. 357, 367
Gorecki, A. 520, 539
Govindaraj, A. 434, 463
Grabandt, O. 178, 225
Grachev, V. I. 5
Grant, B. 66–67, 83
Grant, T. S. 267, 307
Grasselli, J. G. 275, 307
Grassie, N. 5–7, 18, 73, 76, 100, 153
Gregg, S. J. 355, 367
Greinke, R. A. 129–131, 157
Greiser, F. 491, 536
Griffiths, P. R. 272, 307
Grivei, E. 372, 376, 434–435, 437–438, 456
Grove, D. 9, 74
Grüber, P. E. 163, 170, 224, 232, 303
Guescis, R. 39, 79
Guigon, M. 85, 99, 101–102, 105–106, 108, 110–112, 114, 116–119, 120–121, 123, 141–143, 147, 151, 153–155, 159, 283, 308, 327–328, 330, 332–333, 363
Guilpain, G. 324–325, 363
Guinier, A. 85, 87, 150
Guniaev, G. M. 355, 357, 367

Guo, H. 51, 56, 57, 59–63, 82, 140, 141, 158, 349, 366
Guoxiang, L. 161, 224
Gupta, A. K. 5, 73
Gupta, D. 6, 15–16, 73, 75
Gupta, N. 67, 84
Gupta, S. K. 23, 67, 78, 85
Gurley, J. A. 122, 148, 159
Gustafson, R. R. 161, 193, 224, 257, 306
Gutzeit, C. L. 36, 79
Guyot, A. 6, 73

Haga, O. 468, 484, 534
Hahn, H. T. 313–316, 346, 349, 365, 366
Hall, D. L. 233, 244, 303
Hallum, J. V. 253, 306
Hamada, T. 137, 140, 144, 150, 158
Hambourger, P. D. 415, 459
Hammer, G. E. 252, 305
Hamoudi, A. 6, 73
Hancox, N. L. 474, 535
Hansma, P. K. 281, 308
Haquenne, A. 392, 396, 458
Harada, T. 66, 83, 200–202, 227
Hardesty, E. E. 474, 535
Harding, D. R. 208, 228
Harper, A. M. 137, 158
Harrigan, W. C. 495, 498, 538
Harris, P. S. 66, 83
Harrison, B. H. 253, 305
Harrison, M. G. 137, 158
Hart, V. E. 34, 36, 79
Hartland, H. G. 16, 76
Hartwig, G. 468, 489, 535
Haruta, K. 83
Harvey, J. 177, 225
Hasegaqwa, I. 66, 83
Hashimoto, K. 65, 83, 200–202, 227
Hashimoto, M. 382, 386, 457
Hashimoto, T. 68, 85
Hautz, R. C. 3–5, 23, 72
Hawkins, H. T. 197, 227

Author Index

Hawthorne, H. M. 57, 80, 83, 339, 348, 349, 364
Hay, J. N. 18, 76
Hayashi, R. 209, 228
Hayes G. J. 316–317, 346, 366
Hayes, J. M. 246, 304
He, Q. 161, 193, 224, 257, 306
Hechler, J. 272, 274, 307
Heck, W. 121, 155, 253, 255, 257, 306
Heidenreich, R. D. 114, 154
Heins, M. 20–21, 26, 77
Helminiak, T. E. 316, 346, 348–349, 351, 365–366
Henderson, G. W. 339, 364
Henrich, G. 253, 306
Henrichsen, R. E 330, 364
Herai, T. 209, 228
Heremans, J. 372, 376, 392, 402, 407, 414, 423, 425, 431, 434–435, 437–438, 456–460
Hergue, J. J. 221–222, 228
Hérold, A. 413–414, 417, 422, 430, 444, 459
Herqué, J. J. 176–177, 221, 225, 254, 306
Herrick, J. W. 163, 170, 194, 224, 227, 232, 303
Herris, P. S. 66, 68, 83
Hess, N. M. 114, 154
Heyd, R. 434, 463
Heym, M. 20, 77, 468, 491, 512, 514, 534, 536
Hideyuki, N. 82
Higuchi, M. 314–315, 317–318, 346, 366
Hiittner, W. 468, 488, 534
Hikami, S. 376, 379–380, 457
Hill, J. 468, 488, 535
Hippo, E. J. 240, 304
Hiramatsu, T. 349, 366
Hiroshi, N. 50, 82
Hirsch, A. 215, 229
Hirsch, P. B. 85, 97, 151
Hishiyama, Y. 65, 82, 137–138, 140, 142, 144, 147, 158, 382, 386, 408–409, 457, 459

Hiura, H. 433, 434, 463
Hobson, J. P. 246, 304
Hodge, D. J. 252, 305
Hoffman, W. P. 122, 148, 159, 170, 224, 234, 236, 253, 257–258, 281, 283–286, 288–301, 304, 306, 308–309
Hofmann, A. 291, 309
Höhmann-Wien, S. 257, 306
Hohnson, F. C. 18, 76
Honda, H. 41, 43, 47, 49, 51, 55, 79, 81–82, 128, 132–134, 157
Hono, T. 132, 157
Hopfgarten, F. 289, 309
Hordonneau, A. 468, 534
Hoses, T. N. 3, 72
Houwelingen, G. D. B. van 178, 225
Howard, R. E. 438, 463
Howarth, C. R. 274, 307
Howie, A. 85, 97, 151
Howlett, B. A. 537
Hsu, F. M. 266, 306
Hsu, S. E. 211, 228
Hsueh, C. H. 361, 368
Hu, Z. J. 50, 82
Huafu, W. 161, 224
Huang, L. 15, 76
Huang, Y. 237–239, 304
Hueber, H. 253, 305
Hung, C. G. 517, 538
Hurley, W. C. 281, 283, 299, 308
Hutchings, J. 535
Hüttinger, K. J. 253, 257–258, 300–301, 306, 309, 519, 538
Huttner, W. 120, 155, 197, 226, 468, 489, 491–492, 519, 523, 535–538
Hyvernat, P. 130, 133, 157

Ibarra, L. 259, 306
Ichihashi, T. 433, 463
Ida, S. 530, 540
Ihnatowicz, M. 128, 156
Iijima, S. 433, 460, 463
Ikeda, M. 252, 305

Ikezawa, Y. 533, 541
Ilvoas, A. M. 177, 225
Imai, J. 527, 539
Imamura, T. 41, 43, 47, 49, 51, 55, 80, 81–82
Inaba, T. 41, 43, 80
Inagaki, M. 87, 89, 132, 137–138, 140, 142, 144, 147, 152, 157–158, 328, 363
Ingram, D. J. E. 107, 154, 386–389, 389, 407, 458
Inoshita, T. 422, 460
Inoue, T. 41, 43, 49, 79, 81
Inouye, K. 530, 540
Iroh, J. O. 195, 226
Irvine, T. F. 136, 157
Ishida, H. 271, 274–276, 307
Ishioka, M. 65, 83, 197–199, 227
Ishitani, A. 143, 159, 161, 224, 269, 276, 307
Ishizaki, N. 529, 540
Ismail, I. M. K. 234, 236, 238, 303, 488, 536
Issi, J. P. 149, 159, 372–376, 381, 385, 389–389, 392–396, 399–400, 402–403, 405–406, 412, 414, 417–428, 430–431, 434–435, 437–438, 443, 448, 450–451, 456–460, 529, 540
Itoi, M. 217–218, 228
Iund, C. R. F. 67, 85
Iwashita, N. 87, 89, 137–138, 140, 142, 144, 147, 152, 158

Jackel, L. D. 438, 463
Jackson, P. W. 468, 495, 498, 534
Jaeger, H. 222–223, 229
Jaeger, J. 253, 305
Jain, M. K. 13, 75
Jain, R. K. 172, 225
Jakubowski, J. 194, 226
James, R. W. 85, 150
Jamet, J. F. 468, 534
Janevski, A. 269–270, 307
Jang, J. 271, 307

Jansen, H. 112, 154, 332, 334, 364
Japan Exlan Co. Ltd. 5, 6, 8, 13–14, 72–73, 75
Jaroniec, M. 529, 540
Jaworske, D. A. 415, 459
Jaworske, D. D. 517, 538
Jeffries, R. 468, 534
Jenkins, D. H. R. 519, 538
Jenkins, G. M. 97, 153, 519, 538
Jenneskens, L. W. 178, 225
Jensen, R. M. 340, 349, 364
Jeziorowski, H. 267, 307
Ji, Y. 137, 158
Jiang, H. 314–315, 317–318, 320, 346, 354, 366
Jillin Chemical Industries Corp. 6, 73
Jin, C. 498, 535
Jin, H. 239, 304, 529, 540
Johnson W. 3, 72
Johnson, D. J. 85, 87, 97, 103, 107, 108, 112, 114, 116–117, 119, 120, 140–141, 151–152, 154–155, 158, 237, 281, 303, 308, 313–317, 327–328, 346, 349–350, 363, 366
Johnson, G. R. 519, 538
Johnson, J. W. 20, 27, 76, 78, 120, 155, 330, 364
Johnson, W. 27, 78, 327–328, 363
Joiner, J. C. 237, 303
Jonas, A. 372, 374, 395, 448, 450–451, 456, 463
Jones, B. F. 75
Jones, B. H. 335, 364
Jones, C. 179, 184–185, 187, 226, 269, 307
Jones, F. R. 210–211, 228
Jones, G. A. 237, 303
Jones, S. P. 149, 159
Jones, W. R. 330, 364
Jonson, D. J. 237, 303
Jortner, J. 304
Joseph, D. 147, 159
Ju, C. P. 241–243, 304
Jung, C. H. 530, 540

Author Index

Jung, H. H. 530, 540
Juntgen, H. 532, 541

Kabadi, V. N. 289–290, 309
Kaburagi, Y. 137–138, 140, 142, 144, 147, 158, 408–409, 459
Kaelble, D. H. 161, 224, 251, 305
Kagaku, G. 1, 60, 72
Kagan, D. F. 360, 368
Kalvin, I. L. 14, 75
Kamimura, H. 422, 460
Kanamaju, K. 34, 78
Kaneko, K. 527–529, 530, 539–541
Karandikar, A. V. 8, 74
Kargin, V. A. 27, 78
Kasotochkin, V. I. 27, 78
Kastner, M. A. 438, 463
Kasu, T. 527, 540
Katagiri, G. 349, 366
Kato, T. 65, 83
Katori, T. 529, 540
Katsuki, H. 65, 83
Kawabata, S. 319, 347, 366
Kawabuchi, Y. 530, 541
Kawada, K. 468, 484, 534
Kawade, M. 314–315, 340–341, 348, 364
Kawamura, K. 97, 153
Kawano, S. 530, 540
Kawazoe, K. 530, 540
Kaznelson, M. D. 357, 367
Kearney, M. J. 400, 458
Keizo, O. 12, 13, 74
Kelly, A. 324, 363
Kelly, B. T. 396, 399–400, 402, 427, 458
Kenichi, M. 57, 82
Kennedy, B. J. 209, 228
Kennedy, J. M. 316–317, 346, 366
Kenneth J. 237, 304
Khmelnitzkii, D. E. 419, 460
Kihn, Y. 121, 155
Kilzer, F. J. 36, 79
Kim, D. P. 203, 228
Kim, J. K. 209, 228

Kim, K. S. 468, 489–490, 535
Kimura, H. 128, 134, 157
Kimura, K. 58, 83
Kinany Alaoui, M. 423, 430–431, 460
King, M. W. 16, 76
Kingston Lee, D. M. 479, 535
Kinloch, A. J. 181, 226
Kinua, M. A. 194, 227
Kinzl, L. 519, 538
Kirby, J. R. 18, 23, 76–77
Kirkhard, F. P. 512, 537
Kisamori, S. 530, 540
Kiselev, G. A. 5
Kitamura, A. 468, 475, 534
Kitano, A. 314–315, 346, 366
Kizer, D. 468, 495–498, 534, 538
Klein, C. A. 381, 435, 457
Klemens, P. G. 393, 458
Klepeis, S. J. 95, 153
Klimenko, I. B. 5
Knoblauch, K. 530, 540
Knovich, M. M. 20, 77
Ko, T. H. 12–15, 74–76, 164, 224
Kobayashi, K. 203, 228
Kobayashi, S. 529, 540
Kobets, L. P. 194, 227, 357, 367
Kobori, K. 335, 337, 364
Koch, B. 299, 301–302, 309
Koenig, J. L. 93, 121, 152, 271, 274–276, 307, 345, 365
Koenig, M. F. 210, 228
Kogure, K. 97, 140, 142, 158, 338, 364
Koitabashi, T. 41, 42, 45, 79
Kolloid Z. Z. 253, 306
Komagata, H. 532, 541
Komatsu, K. 403, 458
Komatsubara, Y. 530, 540
Konkin, À. À. 356–357, 367
Korai, Y. 47, 49, 82, 132, 137, 139, 147, 157–159
Kosiso, K. 529, 540
Kovac, C. A. 47, 81
Kowalski, I. M. 330, 331, 364
Koyama, H. 468, 484, 534

Koyama, T. 64–68, 83, 114, 155, 382, 386, 407, 457, 459
Koyamo, K. 34, 78
Kozey, V. V. 320, 366
Kozlowski, C. 177, 225
Krassowski, W. N. 362, 371
Kraus, I. G. 355, 367
Kreiger, R. B. 484, 485, 535
Krekel, G. 253, 257–258, 300, 306, 309
Kromp, K. 237, 303
Krucinska, I. 336, 364
Kubomura, K. 314, 315, 317–318, 346, 366
Kucera, M. 163, 225
Kuleznev, V. N. 361, 368
Kumar, S. 313–317, 320, 343, 346, 348, 349, 351–354, 365, 366
Kundra, K. D. 22, 75, 77
Kunzru, D. 67, 84
Kurauchi, T. 211–213, 228
Kureha Chemical Ind. Co. 41–42, 79, 162, 224
Kuriyama, K. 527, 529, 539–540
Kuroda, K. 530, 541
Kus, W. M. 519–521, 538

Labes, M. M. 195, 226
Lafdi, K. 92, 95, 101, 114, 121, 126–127, 129–140, 142, 144–147, 152–153, 155–158
Lahaye, J. 253, 305
Lahijani, J. 149, 159, 373, 403, 456
Laibowitz, R. B. 438, 463
Lake, M. L. 84, 505, 512, 515, 519, 522–523, 537
Lalacona, F. P. 495–496, 537
Lamere, G. 512, 537
Langdon, R. 505, 507, 537
Langer, L. 372, 376, 434–435, 437–438, 456
Lankaster, J. K. 359, 368
Lapinleimu, R. 510, 537
Laramee, R. 512, 537
Larkin, A. I. 376, 379, 380, 457
Larsen J. V. 161, 224

Laserko, G. A. 530, 541
Laureyns, J. 93, 152
Lavielle, L. 244, 304
Lavin, J. G. 97, 101, 140, 142, 144, 149, 153, 158–159, 283, 309, 338, 364, 373, 375, 389–390, 395–396, 400, 402–403, 456, 458
Layden, G. K. 20, 77
Le Berre, A. 215, 229
Le Coustumer, P. 121, 155
Le Toullec, R. 413–414, 422, 430, 444, 459
Lee, J. K. 527–528, 539
Lee, M. C. 267, 307
Lee, P. A. 376, 379, 419, 438, 457, 460–463
Legras, D. 531, 541
Legras, R. 292–294, 309
Lehmann, S. 232, 303
Lehurreau, P. 177, 225
Lei, J. 14, 75
Lelaurain, M. 87, 152, 381, 389–389, 457
Leonhardt, G. 291, 309
Lesko, J. J. 346, 365
Lespade, P. 93, 152
Lewis, I. C. 46–47, 50, 81, 126, 129, 131–133, 156–157
Lewis, R. T. 133, 157
Li, A. 45, 63, 81
Li, C. H. 164, 224
Li, C. T. 335, 364
Li, M. 190, 226
Li, P. 524, 538
Li, Q. 215, 229
Li, R. 42, 43, 80
Li, X. 524, 538
Liang, J. L. 195, 196, 226
Licciardello, D. C. 376, 379, 416, 457
Licini, J. C. 438, 463
Lin C. H. 74
Lin, J. S. 196, 227
Lin, L. 14, 15, 75, 137, 158
Lin, R. Y. 526, 529, 539–540

Lin, S. H. 265, 306
Lin, S. S. 289, 309
Linares–Solano, A. 168, 169, 224, 271, 307
Ling, C. H. 12, 13, 74
Ling, L. C. 50, 82
Ling, X. 252, 276–278, 305, 307
Lipatov, Y. S. 358, 359, 367, 368
Lisagor, W. B. 486, 536
Lisicki, Z. 44, 80
Little, J. A. 209, 228
Little, R. W. 36, 79
Litvinskaya, V. V. 532, 541
Liu, G. 45, 80
Liu, M. 190, 226
Liu, S. 14, 75
Lloyd, D. R. 244, 304
Lloyd, P. 216, 228
Lloyd, P. F. 120, 155
Lloyd, P. M. 210, 228
Lockhart, A. H. 474, 535
Loison, R. 99, 103, 125, 153
Loll, P. 488, 537
Longuet–Escard, J. 88, 152
Lovell, D. R. 311, 313, 363, 523, 538
Lowell, L. P. 232–233, 303
Lu, Y. 273, 307, 529, 531, 540–541
Lucas, J. P. 267–269, 307
Lucas, R. 523, 538
Lumsdon, J. 345, 365
Lustan, V. G. 190–191, 226
Luthra, K. L., 206, 227
Luxen, B. A. 517, 536
Lyubchev, L. 14–15, 75
Lyubcheva, H. 14–15, 75

Ma, C. C. M. 172, 225
MacDonald, D. K. C. 408, 459
MacNair, R. N. 531, 540
Macturk, K. S. 316–317, 343–344, 365
Madhukar, M. 220–221, 228, 353, 366
Madronero, A. 295, 296, 309
Madrosky, S. L. 34, 36, 79
Maeda, K. 129–131, 157, 203, 228

Maeda, T. 529, 530, 538–540
Maggs, A. 525, 539
Magnan, C. 85, 151
Mahapatro, B. 12–13, 74
Mahy, J. 178, 225
Mai, Y. W. 209, 228
Maire, J. 85, 88, 150–151
Malashevich, Zh. V. 532, 541
Malaskonov, I. 14, 75
Malda, K. 41, 43, 80
Malinsky, Y. M. 360, 368
Maniette, Y. 96, 153
Manini, C. 415, 459
Manjeet, B. 5, 72
Mankiewich, P. M. 438, 463
Manocha, L. M. 13, 20–23, 28–29, 75, 77, 101, 153, 172, 208, 225, 228, 468, 488, 534
Manolova, N. 531, 541
Manson, J. A., 361, 368
Mansur, F. T. 163, 170, 224, 232, 303
Marchand, A. 93, 152
Marêché, J. F. 372, 376, 389, 417, 423, 426–427, 430, 434, 456–457, 460–463
Marêché, J. F. 415, 426, 428, 459–460
Marinkovic, S. 22, 77, 468, 488, 535, 536
Marjoram, J. R. 468, 495, 498, 534
Mark, H. B. 253, 305–306
Markovic, V. 468, 488, 535
Marks, B. S. 194, 227
Marom, G. 14–15, 75, 341–342, 365
Marsh, H. 29, 78, 468, 488, 535
Marsh, P. A. 119, 155
Marshall, P. 121, 155
Martin-Martinez, J. M. 527, 539
Martinez-Alonso, A. 93, 152, 249, 251, 305
Masayoshi, W. 12–13, 75
Masuda, T. 65, 83, 200–202, 227
Mathur, R. B. 10–17, 20, 22, 24–27, 29–30, 74–78, 100, 102, 104, 153–154
Matkowsky, R. D. 23, 77
Matsbara, I. 335, 337, 364

Matsubara, K. 197–199, 227
Matsubayashi, K. 529, 540
Matsuhisa, Y. 349, 366
Matsui, J. 121, 155, 244, 304
Matsui, S. 433–434, 463
Matsumoto, M. 68, 84, 137, 140, 144, 150, 158
Matsumoto, T. 44, 51, 60–62, 80, 82, 328–329, 363
Matsumura, Y. 45, 80, 530, 541
Matsuo, T. 532, 541
Matta, V. K. 10–11, 13–14, 16, 74, 76
Mattson, J. S. 253, 305, 306
Mattsumoto, T. 65, 83
Matyushova, V. G. 358, 367
Mauri, R. E. 194, 227
Maus, L. 161, 224, 251, 305
Mayer, J. 299, 301–302, 309
Mayer, R. M. 29, 78
Maynard, R. 402, 424, 458, 460
Mayor, R. M. 27, 29, 78
Mazurek, H. 423, 428, 460
Mc Guchan, R. 100, 153
Mc Hugh, J. J. 137, 158
McCabe Michael, 14, 16, 75
McEnaney, B. 527, 539
McGarry, F. J. 342, 365
McGuchan, R. 5, 7, 18, 74, 76
McHenry, E. R. 48, 81, 129, 157
McIntyre, J. E. 9–10, 74
Mckee, D. W. 163, 170, 197, 209, 224, 228, 234, 303
McKenna, G. B. 519, 538
McNair, R. N. 526, 539
McNeil, I. 6, 73
McRae, E. 372, 376–389, 415, 417, 423, 426–428, 430, 434, 456–457, 459–463
Mehta, V. R. 320, 348, 353, 366
Melanitis, N. 313, 345–346, 365–366
Mélin, J. 413–414, 422, 430, 444, 459
Melngailis. 438, 463
Mencik, Z. 360, 368

Mercher, J. A. 233, 244, 303
Mercuri, R. A. 203, 227
Mérienne, M. C. 121, 155
Mering, J. 85, 87–88, 150, 152
Meschi, C. 415, 459
Metelkin, V. I. 362, 371
Metjolkin, W. J. 362, 371
Mewis, J. 132, 157
Meyerer, W. 468, 496, 497, 534
Michenaud, J. P. 372, 376, 381, 389–389, 400, 417, 419–420, 433, 456–458, 460
Middlemiss, B. A. 252, 305
Migliaresi, C. 341–342, 365
Miles, A. J. 525, 539
Milne, J. M. 330, 364
Mimeault, V. J. 163, 170, 224, 303
Mintmire, J. W. 433, 460
Minty, D. C. 537
Mitchell, R. 512, 537
Mitsubishi Chemical Co. and Parent Co. 5, 6, 8–10, 12–14, 72–75
Mitsuru, U. 50, 82
Mittal, J. 14–17, 24–27, 29–30, 76–78, 104, 154
Miwa, M. 314–315, 340–341, 348, 364
Miyachi, H. 335, 337, 364
Mladenov, I. 14–15, 75
Moalli, J. E. 342, 365
Mochida, I. 29, 41, 43, 45, 47, 49–50, 52, 78, 80–82, 129–132, 137, 139, 147, 157, 158–159, 530, 540
Molchanov, B. I. 359, 368
Moldenaers, P. 132, 157
Molina-Sabio, F. M. 526, 539
Molleyre, F. 110, 116, 119, 154, 164, 168, 224
Molyneux, C G. 345, 365
Monin, J. C. 126, 156
Montes-Moran, M. A. 249, 251, 305
Montgomery, I. W. 474, 535
Monthioux, M. 87, 103, 114, 139, 147, 152

Moore, A. 381, 389–389, 412–413, 457, 459
Moore, D. R. 36, 79
Morawski, J. C. 330, 364
Morelli, D. T. 423, 426, 460
Moreton, R. 106, 154
Morgan, P. E. 468, 535
Morida, K. 1, 60, 72
Morii, T. 267, 307
Morita, K. 143, 159, 335, 337, 364
Morita, T. 314–315, 346, 366
Morley, J. M. 479, 481, 535
Morone, A. 161, 224, 276, 278, 308
Morozova, A. A. 530, 541
Morris, A. W. H. 468, 495, 496, 534
Mortka, S. 525, 539
Morton, R. 10, 12, 27, 74, 78
Moshkovskii, Y. S. 17, 76
Mosseri, R. 101, 153
Motojima, S. 66, 83
Mott, N. F. 408, 459
Moulder, J. F. 279, 308
Moutand, G. M. 31, 78
Moyer, R. O. 36, 79
Mrozowski, S. 103, 154, 386, 457
Muilenberg, G. E. 279, 308
Mukai, S. R. 200–202, 227
Mukai, S. R. 65, 84
Muller, D. J. 17, 20, 22–23, 25, 76–77, 101, 153
Muller, T. 5, 6, 72
Müller-Wiesner, D. 486–489, 536
Munecas, M. A. J. 526, 539
Murata, K. 68, 84
Murata, Y. 143, 159
Murayama, K. 143, 159
Murdie, N. 241–243, 304

Nagaoka, 376, 379–380, 457
Nagpal, K. C. 10, 11, 16, 74
Naitove, M. H. 538
Nakajima, A. 143, 159
Nakajima, T. 418–420, 423, 427, 430, 459
Nakamizo, M. 47, 81
Nakamura, E. 41, 43, 80
Nakamura, H. 533, 541
Nakanishi, Y. 33, 36, 39, 78, 176, 225, 357–358, 367
Nakao, F. 178, 225, 233, 303
Nakayama Y. 161, 224, 270, 276, 307
Narcy, B. 120, 155
Nardin, M. 324, 325, 363
Naslain, R. 192, 227, 336, 364
Nay, J. C. 188, 226
Nazem, F. F. 136, 157
Neill, W. K. 36, 79
Netravali, A. N. 189, 226
Neugebauer, R. 538
Newman, J. W. 42, 80
Nezbeda, C. W. 403, 459
Ng, C. B. 339, 364
Nguyen, H. H. 413–414, 422, 430, 444, 459
Nicaise, G. 107, 154
Nicholas, M. G. 496, 519, 537–538
Nicholson, R. B. 85, 97, 151
Nikkoso Co. Ltd. 5–6, 14, 72–73, 75
Nimtz, F. G. 8, 74
Nippon carbon Co. Ltd., 14, 75
Nippon Steel Corp. 57, 83
Nishida, T. 137, 140, 144, 150, 158
Nishino, A. 533, 541
Nitto Baseki Co. Ltd. 14, 75
Noël, D. 272, 307
Noguchi, K., 314, 315, 346, 366
Nongaillard, B. 297–309
Norr, M. K. 112, 154
Northold, M. G. 112, 154, 332, 334, 364
Novak, R. C. 163, 224
Nukui, A. 45, 80
Nysten, B. 99, 149, 153, 159, 372–375, 382, 384, 390, 392, 395–396, 400, 402–403, 407, 409–411, 423–424, 426–428, 430, 448, 450–451, 456–458, 460–463

Oberlin, A. 64, 66, 67–68, 83, 85–87, 92–99, 101–121, 126–147,

[Oberlin, A.]
  151–159, 283, 308, 327–328, 330, 332–333, 363
Oberlin, M. 96, 98, 110, 114, 116, 119, 130, 133, 153–154, 157
Ochiai, Y. 433, 434, 463
Odom, E. M. 345, 365
Ogawa, T. 252, 305
Oh, W. Z. 530, 540
Ohmori, S. 531, 541
Ohsawa, T. 314–315, 340, 341, 348, 364
Ohtsuba, R. 29, 78
Oi, S. 47, 81
Oka, H. 50, 82
Okada, T. 197–199, 227
Okada, Y. 533, 541
Okhuysen, B. 213, 228
Okuda, K. 57, 82
Olander, D. R. 525, 539
Old, C. F. 496, 519, 537–538
Olk, C. H. 372, 376, 434–438, 456
Olsson, D. M. 323, 363
Ono, A. 382, 386, 457
Ono, T. 143, 159
Oshida, K. 87, 152, 529, 540
Oshmyn, V.G. 355–367
Ota, E. 131, 157
Otake, Y. 532, 541
Otani, S. 1, 41–45, 49–50, 57–58, 72, 79–84, 131–132, 157
Ouaftouh, M. 297–309
Ourak, M. 297–309
Overbury 274, 307
Owens, T. W. 281, 283, 299, 308
Own, S. H. 322, 323, 324, 325, 363
Oya, A. 131, 157, 168–169, 224, 271, 307
Ozeki, S. 527, 530, 539–540

Pacault, A 488, 536
Padhya, M. R. 8, 74
Padmanabam, M. 6, 73
Pai, B. C. 468, 498, 532
Paiva, M. C. 249, 251, 305

Palazotto, A. N. 343, 365
Paliwal D. K. 5, 73
Pallozzi, A. A. 232, 303, 355, 367
Palma, E. 259, 306
Pankratov, V. A. 358, 367
Papadopoulos, D. S. 208, 228
Papaspyrides, C. D. 196, 227
Papirer, E. 120, 155, 173, 225, 248, 305
Paprocki, S. 468, 496–497, 534
Parisi, J. P. 259, 306
Park, C. R. 313, 315–317, 346, 349–350, 365–366
Park, H. S. 468, 489–490, 536
Park, N. A. 136, 157
Park, S. J. 247, 304, 529, 540
Park, Y. D. 48, 81
Parker, E. 103, 107, 112, 154
Parks, W. G. 39, 79
Parratt, N. J. 362, 368
Parrish, R. G. 140, 144, 158
Parsons, J. L. 18, 76
Pashley, D. W. 85, 97, 151
Paté, F. 500, 536
Patel, A. K. 536
Patel, G. C. 9, 10, 74
Paton, W. 474, 500, 506, 535, 537
Patron, L. 7, 73
Patscheider, J. 299, 301–302, 309
Patton, W. 474, 534
Pavlov, V.V. 357, 367
Peacock, J. A. 252, 305
Peebles, L. H. 18, 20, 23, 47, 76, 77, 82, 311, 334, 342, 348, 363
Peeper, R. T. 16, 76
Pelet, R. 126, 156
Peng, J. C. 173–175, 215–216, 229, 259–260, 262–263, 274–275, 306–307
Pennock, G. M. 133–134, 140, 142, 157–159
Peppas, N. A. 267, 307
Pérignon, A. 423, 430–431, 460
Pernot, P. 423, 430–431, 460

Perol, N. 253, 305
Perov, B. V. 190, 191, 226
Perret, R. 87, 97, 108, 116–117, 119, 152–155, 237, 281, 303
Peryshkin, N. G. 359, 368
Perzynski, B. 525, 539
Petcavich, R. J. 22, 77
Petearich, R. R. 18, 76
Peterlik, H. 237, 303
Peterlin, A. 360, 368
Peters, E. M. 36, 79
Petrarce, A. 39, 79
Pfeiffer, J. P. 126, 156
Phan, H. T. 281–283, 299, 308
Philips, L. N. 16, 20, 76, 468, 470, 508, 515, 534
Philips, D. C. 468, 533
Pierre, A. 488, 536
Pietronero, L. 422, 460
Piggott, M. R. 324, 363
Pilon, A. 274, 307
Pinoli, P. C. 488, 536
Pinson, J. 161, 224
Pipes, R. B. 217–218, 228
Piquero, T. 203, 205, 228
Piraux, L. 66, 83, 372–374, 376, 381, 385, 389–389, 392, 395–396, 399–400, 403, 414, 417–421, 423–424, 426–428, 430–431, 443, 448, 450, 456–460, 529, 540
Pitt, W. G. 188, 226
Pittman, C. U. 171, 172, 225
Pizana, C. C. 244, 304
Platonava, N. V. 5
Pluedmann, E. P. 271, 307
Poceva, J. 269, 307
Poleunis, C. 292–294, 309
Pollack, S. S. 126, 156
Popova, L. A. 360, 368
Popova, N. A. 532, 541
Popovska, N. 14, 75, 191–194, 227
Porter, R. S. 316, 342, 344, 365
Potter, R. T. 349, 366

Poulaert, B. 372, 374, 392, 395, 414, 423, 428, 448, 450, 456, 458–460
Prado-Burguete, C. 527, 539
Prandy, J. M. 313–316, 346, 365
Pregermain, S. 128, 156
Prescott, B. 512, 537
Price, J. 121, 155
Proctor, D. G. 124, 126, 156, 403, 459
Prosen, S. P. 194, 227, 232, 303
Puri, B. R. 253, 255–256, 305–306
Pusset, N. 239, 304, 527–529, 539–540

Qian, S. A. 47, 81, 137, 158
Qin, R. Y. 57, 59, 82, 121, 123–124, 148, 156, 244–247, 249–250, 281–285, 287, 304, 308, 529, 540
Quate, C. F. 281, 308

Rabotnov, J. N. 190–191, 226
Racke, P. B. 22, 77
Rad, V. N. P. 168
Rafalkes, I. S. 66, 67, 83
Ramakrishnan, T. V. 376, 379, 416, 419, 457, 460
Ramer, J. K. 50, 82
Rand, B. 85, 138, 152, 158, 164, 165, 225, 264, 306
Rannou, I. 87, 152
Rao, A. M. 529, 540
Rao, C. N. 434, 463
Rappeneau, J. 87, 152
Rashid, M. S. 468, 534
Rashkov, I. 531, 541
Raskovic, V. 22, 77
Ravey, J. C. 126, 156
Reader, C. E. L. 479, 481, 535
Red'ko, T. D. 532, 541
Reger, E. O. 362, 371
Reich, L. 18, 76
Reid, J. D. 33–34, 78–79
Reifsnider, K. L. 346, 365
Reynolds, W. N. 85, 106, 151, 154, 327, 330, 364
Rhee, B. S. 48, 81, 491, 527–528, 530–529, 536, 539–540

Rhodes, J. M. 14, 75
Rice, B. P. 313–314, 316–317, 342, 365
Rich, M. J. 120, 155, 210, 216, 228
Richter, E. 530, 540
Rieck, U. 486–489, 536
Rieux, J. P. 177, 225
Rigaux, C. 413–414, 422, 430, 444, 459
Riggs, D. M. 41, 43, 46, 48, 79, 81, 131, 157
Riggs, R. P. 14, 76
Riggs, W. M. 279, 308
Rikatskaya, T. L. 532, 541
Roberston, D. H. 433, 460
Robin, P. L. 107, 154
Robin, R. P. 14, 76
Robinovich, E. Y. 66–67, 84
Robinson, K. E. 137, 149, 158–159
Robinson, M. 509, 536
Robinson, R. 232, 264, 303, 306
Robson, D. 386–389, 389, 407, 458
Roche, E. J. 140, 144, 158, 283, 309, 389, 458
Rodda, E.T. 67, 84
Rodriguez-Reinoso, F. 526, 539
Rogers, K. 137, 158
Rogers, R. F. 479, 535
Rohrer, H. 98, 153, 281, 308
Rosembanm, S. J. 20, 77
Rosenwaig, A. 272, 307
Rosovitsky, V. F. 358, 367
Ross, R. A. 185–186, 226
Ross, S. E. 36, 79
Rousseaux, F. 95, 152
Rouvaen, J. M. 297–309
Roux, J. C. 99–100, 153
Rouxhet, P. G. 107, 128,154, 156
Rouzaud, J. N. 85, 87, 93–94, 103, 114, 126, 139, 147, 151–152, 156
Rudzinski, W. 247, 304
Ruike, M. 527, 540
Ruland, W. 39, 79, 85–97, 108, 116–117, 119, 150, 152–155, 237, 281, 303, 332, 364

Ryu, S. K. 239, 304, 527–529, 539–540
Ryuichi, H. 50, 82

Saacco, A., Jr. 67, 69, 84
Saadaoui, H. 99, 100, 153
Saal, R. N. J. 126, 156
Sach, R. S. 163, 168, 224
Sack, H. 126, 156
Sadler, R. 289–290, 309
Sadoc, J. F. 101, 153
Safonova, A. M. 533, 541
Saito, F. 315, 318, 320, 346, 366
Saito, R. 433, 460, 463
Sajiki, Y. 137, 140, 144, 150, 158
Sakaguchi, Y. 57, 82
Sakai, M. 132, 157, 328, 363
Sakai, Y. 132, 157
Sakanishi, A. 139, 158
Sakata, H. 143, 159
Sakoda, A. 530, 540
Salamanca Riba, L. 376, 423, 426–427, 430, 457, 460
Salvi, A. M. 161, 173, 224–225, 276, 278, 290, 308–309
Sambell, R. A. J. 468, 533
Sanada, Y. 128, 134, 157
Sanchez, M. 121, 155, 161, 224
Sandle, N. K. 15, 77
Sannda, Y. 50, 82
Sano, H. 203, 228
Sappock, R. 253, 255, 257, 306
Sasaki, T. 139, 158
Sato, N. 211–213, 228
Sato, T. 315, 318, 320, 346, 366
Satyanarayana, K. G. 468, 498, 534
Saunders, S. C. 322–325, 363
Savage, G. 506–507, 514, 539
Savvateev, S. G. 362, 371
Saylor, D. K. 468, 476–477, 535
Schalamon, W. A. 39, 79, 332–333, 364
Schikner, R. C. 137, 158
Schmidt, D. L. 32, 37–78, 197, 227
Schoch, G. 191–194, 227, 468, 488, 535

Schram, E. P. 203, 227
Schreiber, H. P. 244, 304
Schul, Z. 43, 81
Schultz, J. 244, 252, 304–305, 324–325, 363
Schurmus, H. 22, 77
Schutzenberger, L. 65, 83
Schuyten, H. A. 33, 78
Schwenker, F. 34, 79
Sclavons, M. 292–294, 309
Scola, D. A. 195, 196, 226, 232, 264–265, 303
Scottosanti, P. 512, 537
Seiferling, M. 257, 306
Seldin, E. J. 124, 126, 156, 403, 459
Sellitti, C. 274–276, 307
Sema, I. 33, 36, 39, 78
Semenova, G. P. 357, 367
Serin, V. 121, 155
Seshardi, R. 434, 463
Setoyama, N. 527–529, 540
Sevely, J. 121, 155, 336, 364
Shanahan, M. E. R. 252, 305
Shang, Y. 48, 50, 53–54, 81
Sharma, R. C. 6, 73
Sharp, J. A. 46, 80
Sharp, J. V. 327, 363
Shayegan, M. 423, 460
Shchukina, L. A. 357, 367
Sheehan, J. E. 206–207, 209, 227–228, 468, 488, 535
Shen, C. 71, 84
Shen, Z. M. 1, 45, 48, 50–51, 53–54, 56–63, 72, 80–83, 281, 283, 286, 309
Shepler, R. E. 468, 476–477, 535
Sherwood, P. M. A. 177, 210, 225, 228, 276, 278, 308
Shi, J. L. 137, 158
Shieldlin, A. 14–15, 75
Shimizu, K. 41, 43, 50, 52, 80, 82, 132, 157, 527, 529, 539–540
Shindo, A. 3, 27, 33, 34, 36, 39, 72, 78–79
Shindo, N. 529, 530, 538–541

Shinohara, A. H. 315, 318, 320, 346, 366
Shoji, Y. 12–13, 75
Shorter, J. J. 538
Shul, G. S. 357, 367
Sidey, G. R. 535
Silvaggi, A. F. 237, 303
Silver, D. S. 253, 300, 306, 309
Simitzis, J. 15, 76
Simms, J. R. 271, 307
Simon, H. 244, 304
Sinclair, D. 338, 364
Sines, G. 97, 140, 142, 158, 338, 364
Sing, K. S. W. 526, 539
Singer, L. S. 42, 48, 80–81, 128–131, 140–141, 156–158, 381, 386–389, 389, 457
Sirota, A. G. 359–361, 368
Sivaram, P. 13, 74
Sivy, G. T. 18–19, 76
Skocpol, W. J. 438, 463
Skourlis, T. 196, 227
Smiley, R. J. 188, 226, 276, 279–279, 299–302, 308
Smith R. A. 253, 305
Smith, A. W. 203, 227
Smith, H. 363
Smith, R. N. 253, 305
Smith, S. B. 313, 346, 366
Smith, T. G. 161, 224
Soeda, F. 161, 224, 269, 276, 307
Sohi, M. M. 349, 366
Sone, Y. 50, 52, 81
Soni, R. P. 480, 506, 536
Souillart, C. 103, 154
Souma, M. 527, 539
Southall, J. M. 345, 365
Spain, I. L. 85, 151, 283, 309, 375, 378, 381–383, 389, 400, 402–412, 456
Speight, J. G. 126–127, 156
Spence, G. B. 403, 459
Spencer, R. A. P. 468, 474, 533, 535
Sperling, L. H. 361, 368
Spescha, G. 299, 301–302, 309

Sposili, R. 213, 228
Srinivasagopalan, S. 330, 349, 364
Stami Carbon B.V. 8, 73
Standage, A. E. 23, 77
Starr, T. 311–312, 314–318, 363
Statton, W. O. 519, 538
Steffens, D. A. 338–339, 349, 364
Stepanitsov, I. E. 190–191, 226
Stevenson, W. T. K. 139, 158
Stille, J. K. 23, 77
Stockman, L. 372, 376, 434–435, 437–438, 456
Stone, A. D. 438, 463
Strässler, S. 422, 460
Strauss, S. 34, 36, 79
Strife, J. R. 206, 227
Strong, S. L. 34, 36, 79
Stuez, D. E. 16, 76
Stumm, W. 262, 306
Stypka, T. 336, 364
Subramanian, R. V. 194, 226, 322–325, 363, 536
Sugawara, L. 128, 157
Sugihara, K. 85, 151, 283, 309, 375, 378, 381–383, 389, 400, 402, 412, 456
Suhng, Y. 195, 226
Sullivan, B. 206–207, 227
Sun, S. 190, 226
Sundaran, V. 536
Sutherland, I. 252, 276–278, 305, 307
Sutter, J. K. 208, 228
Suzuki, M. 526, 531, 539
Suzuki, T. 527–529, 539
Suzuki, Y. 532, 541
Swain, R. E. 346, 365
Swan, J. W. 1, 32, 72

Tai, N. H. 172, 225
Takahashi, K. 87, 152
Takamine, M. 44, 81
Takamura, T. 533, 541
Takano, N. 147, 159
Takayaki, I. 51, 55, 82
Takayama, H. 419, 460

Takemoto Oil Fat Co. Ltd. 14, 75
Takenaka, Y. 178, 225, 233, 303
Takeshita, K. 29, 41, 43, 78, 80, 129–131, 157
Takeuchi, K. 87, 152, 529, 541
Takeuchi, Y. 533, 540
Talyor, G. H. 41, 46, 80
Tamagawa, I. 407, 459
Tamaru, K. 33, 78
Tanabe, Y. 328, 363
Tanahashi, I. 533, 541
Tandon, D. 240, 304
Tang, M. M. 34, 78
Tang, W. K. 36, 80
Tanigaki, K. 433–434, 463
Tanimoto, T. 267, 307
Tascon, J. M. D. 93, 152, 249, 251, 305
Tate, K. 139, 158
Taylor, E. 481–483, 535
Taylor, G. H. 128, 133–134, 140, 142, 156–159
Teghtsoonian, E. 339, 348–349, 364
Tennant, D. M. 438, 463
Tenney, D. R. 486, 536
Teranishi, H. 31, 78
Terriere, G. 141, 159
Tersoff, J. 281, 308
Terweisch, B. 491, 536
Terwiesch, B. 468, 489, 535
Tesner, P. A. 66–67, 83
Tesoro, G. C. 36, 79
Tetlow, P. L. 313, 346, 366
Tewary, V. K. 468, 508, 516, 522, 534
Than, E. 291, 309
Thomas, C. R. 468, 488, 535–537
Thomas, D. K. 503, 538
Thomas, R. B. 66, 83
Thomas, Y. 259, 306
Thompson, A. W. 505, 537
Thomson, G. P. 85, 151
Tibbetts, G. G. 65–67, 83–84
Tietz, J. V. V. 335, 364
Tillmanns, H. 491, 536
Ting, J. M. 84, 537
Ting, H. Y. 12–13, 74

Ting, J. M. 505, 512, 515, 519, 522–523, 538
Tissot, B. 126, 156
Toho Beslon Co. Ltd. 6, 22, 73, 77
Toka rsky, E. W. 97, 116, 118, 153
Tokai Electrode Manufg. Co. Ltd, 75
Tomanova, A. 163, 225
Tomassi, W. 525, 539
Tombrel, F. 87, 152
Tomio, A. 41, 43, 79
Tomioka, T. 137, 140, 144, 150, 158, 315, 318, 320, 346, 366
Toray Industries, Inc. 5–6, 8–9, 13–14, 17, 27, 72–73, 75–76, 78
Torner, R, V. 362, 371
Toshima, H. 132, 157
Tournant, R. 128, 156
Towne, M. K. 466, 468, 476–477, 512, 515, 533–535
Toyobo Co. Ltd. 8–9, 73
Toyokazu, M. 12, 13, 74
Tressler R. E. 203, 227
Trostyanskaya, E. B. 194, 227
Trowin, E. M. 523, 538
Tsamopoulous, J. A. 67, 85
Tse, M. K. 232, 303
Tsushima, E. 314–315, 340–341, 348, 364
Tsutomu, N. 82
Tsutsumi, K. 234–235, 248, 252, 265–266, 303–306
Tuinstra, F. 345, 365
Tuinstra, F. 93, 121, 152
Turner, N. H. 234, 303
Turner, W. N. 18, 76
Tyson, C. N. 87, 97, 108, 117, 119, 152, 155, 237, 303
Tyson, W. R. 324, 363

Uchiyama, Y. 203, 228
Ueda, K. 44, 80
Uher, C. 423, 426, 460
Uhlmann, D. R. 20, 77
Umbach, C. P. 438, 463
Union Carbide Corp. 29, 78

Upadhyaya, D. 213–214, 228
Ura, T. 533, 541
Utevsky, L. E. 361, 368

Vaidyan, V. K. 468, 498, 534
Vaidyanathan, N. P. 289–290, 309
Vaidyanathan, R. 289–290, 309
Van Auken, R. L. 515, 537
Van De Mark, M. R. 260–261, 306
Van Haesendonck, C. 372, 376, 388, 434–435, 437–438, 456, 458
Van Krevelen, D. W. 99, 103, 125, 126, 128, 153
Vandygriflt, W. G. 95, 153
Vangelisti, R. 423, 430–431, 460
Varshavsky, V.Y. 357, 367
Vastola, F. J. 525, 539
Veldhuizen, L. H. 112, 154, 332, 334, 364
Venema, A. 178, 225
Venkatesam, T. 390, 402, 458
Venner, I. G. 48, 82
Verdu, M. 295–296, 309
Verhoest, J. 22, 77
Verhovetz, A. P. 361, 368
Vezie, D. L. 313, 346, 366
Vidano, R. 93, 152
Vieren, J. P. 413–414, 422, 430, 444, 459
Villeneuve, J. F. 336, 364
Villey, M. 87, 103, 107, 110, 139, 147, 154
Vincent, C. 203, 205, 228
Vincent, H. 203, 205, 228
Vittorio, S. L. di. 529, 540
Voet, A. 119, 155, 253, 305, 330, 364
Volfson, S. A. 355–367
Volk, H. F. 203, 227
Vorpagel, E. R. 101, 153
Vu-Khanh, T. 480, 536
Vukov, A. J. 166–168, 224

Wadsworth, N. J. 164, 224
Wagner, C. D. 279, 308
Wagner, H. D. 341–342, 365

Wagoner, G. 335–336, 364
Wajler, C. 519–521, 538
Waldman, D. A. 189, 226
Walker, E. J. 468, 488, 535–537
Walker, P. L. 525, 539
Wang, C. S. 313–314, 316–317, 342–343, 365
Wang, C. Y. 50, 82
Wang, D. 210, 228
Wang, H. H. 74
Wang, K. L. 211, 228
Wang, M. J. 244–246, 304
Wang, S. 274, 307
Wang, T. 177, 225, 276, 278, 308
Wang, T. K. 281, 283, 286, 309
Wang, W. L. 498, 537
Wang, X. 273, 307
Wang, Y. 48, 50, 53–54, 82, 206, 228
Wang, Z. M. 206, 228, 530, 541
Ward, T. C. 244, 304
Warner, S. B. 20, 77
Warren, B. E. 87–88, 152
Washburn, S. 438, 463
Washiyama, M. 132, 157, 349, 366
Waterbury, M. C. 324, 326, 340, 363
Watt 10, 12, 27, 74
Watt, W. 3, 20, 23, 27–28, 72, 76–78, 103, 107, 112, 120, 154–155, 164, 224
Watts, J. F. 181, 226 257, 289, 306, 309
Watts, W. H. 36, 80
Weaver, J. W. 33–34, 78–79
Webb, R. A. 438, 463
Weber, W. 253, 306
Wei, W. C. 207–208, 228
Wei, Y. 190, 226
Weibel, E. 281, 308
Weibull, W. 16, 76
Weisbecker, C. S. 434, 463
Weiss, R. 120, 155, 177, 217, 219, 223, 225, 228–229
Weisweiler, W. 189, 226
Weitzsacker, C. L. 210, 228
Well, H. 176, 225

Welte, D. H. 126, 156
Weng, L. T. 292–294, 309
Weng, T. 124, 126, 156, 403, 459
Weuver, J. V. 481–483, 535
Whelan, M. J. 85, 97, 151
White, C. T. 433, 460
White, J. L. 42, 47, 80–81, 101, 103, 128, 135, 153–154, 156, 283, 308, 339, 364, 492, 536
Whitesides, G. M. 434, 463
Wightman, J. P. 226, 346, 365
Williams, W. S. 338–339, 349, 364
Winter, L. L. 3, 72
Wintermantel, E. 299, 301–302, 309
Wirkus, C. D. 468, 534
Wolff, S. 244–246, 304
Wolter, D. 197, 226, 519, 538
Wrightman, J. P. 277, 308
Wrzesien, A. P. 468, 534
Wu, F. 273, 307
Wu, H. D. 211, 228
Wu, R. J. 498, 537
Wu, T. M. 207–208, 228
Wu, Z. 171–172, 225
Wudl, F. 215, 229

Xie, Y. 177, 225, 276, 278, 308
Xue, G. 273, 307

Yakovlev, V. M. 357, 367
Yamada, K. 41–42, 45, 80, 209, 228
Yamada, T. 209, 228
Yamada, Y. 41, 43, 47, 49, 51, 55, 79, 81–82, 132, 137–138, 140, 142, 144, 147, 157–158, 217–218, 228
Yanagida, K. 139, 158
Yang, C. Q. 271–272, 307
Yasuda, E. 328, 363
Yen, T. F. 126, 156
Yip, P. W. 289, 309
Yoji, M. 12–13, 74
Yokogawa, K. 147, 159
Yokoyama, A. 41–42, 45, 79–80
Yoneshiga, I. 31, 78

Author Index

Yoon, S. 50, 82, 137, 147, 158–159
Yoshida, A. 137–138, 140, 142, 144, 147, 158, 533, 541
Yoshida, S. 168–169, 224
Yoshikawa, M. 530, 541
Yoshiro, K. 57, 82
You, H. X. 281–283, 308
Young, C. L. 244, 245, 304
Young, D. A. 253, 305
Young, R. J. 120–121, 143, 159, 237–239, 304
Yua, G. R. 74
Yuang, Y. 120–121, 143, 159
Yumitori, S. 210–211, 228

Zabala Martinez, I. 423, 460
Zang, H. 15, 76
Zekina, I. G. 359, 368

Zeng, H. M. 529, 531, 540–541
Zeng, S. M. 132, 157
Zha, Q. F. 137, 158
Zhang, F. 50, 53–54, 82
Zhang, G. D. 498, 537
Zhang, J. 211, 228
Zharikov, O. V. 434, 463
Zheng, J. M. 50, 82
Zhou, B. L. 206, 228
Zhou, J. 267–269, 307
Zillikha, A. 14–15, 75
Zima, J. 292, 309
Ziman, J.M. 393–395, 404, 458
Zimmer, J. E. 101, 128, 135, 153, 156, 283, 308, 492, 536
Zou, Y. L. 189, 226
Zvyagin, I. P. 408, 459

SUBJECT INDEX

Acoustic emission, 211–212
Active surface area (ASA), 232, 234–236
Activated carbon fiber, 70, 240, 263–264, 524–530
   application, 530
   characterization, 526
   preparation, 525
Additives, 101, 198
Adhesion, 163, 176, 181, 189, 197, 210
   amine immobilization, 217
   composite properties, 353, 464, 491
   surface acidity, 257
   surface area, 232
Adsorption, 263–270, 527, 530
   activated fiber, 527, 530
   microstructure, 232, 234
   moisture, 267
   organic compounds, 264–265
   surface energy, 244
   surface treatment, 171

Anisotropic content (AC), 48
Anisotropic pitches, 50, 125, 128, 133
Anodic oxidation, 177, 278
Applications in
   aerospace, 500
   aerospace antenna, 514
   aircraft brakes, 512
   chemical industry, 515
   electrical, 517
   general engineering, 523
   heat shields, 515
   leisure time activities, 510
   marine transport, 508
   medical, 519
   missiles, 515
   nuclear field, 516
   railway, 509
   road transport, 505
   sports, 510
   textile machinery, 523
Atomic force spectroscopy (AFM), 98, 299

Subject Index

Attenuated total reflection (ATR) spectroscopy, 274
Auger electron spectroscopy (AES), 288

Basic structural units (BSU), 87, 92, 99, 101, 103
BET analysis, 219, 232, 265
Bimodification, 17
Black fibers, 23
Boltzmann Zero-Field Resistivity, 375
Bonding, 36, 209, 269, 402, 494
    electropolymerization, 194
    oxidative treatment in gases, 163
oxidative treatment in liquids, 170
plasma treatment, 184
whiskerization, 190
Bragg law, 86
Brooks and Taylor spheres, 128, 132
Buckling, 342, 349, 353, 500

Callaway formalism, 425
Calorimetry, 244
Carbonization,
    acrylic fibers, 39
    mesophase pitch fibers, 59
    PAN fibers, 26
Carbon-carbon composites (CCC), 239, 484
    effect of fiber types, 491
    effect of final HTT, 492
    effect of matrixes, 491
    preparation of, 488
Carbon-metal composites (CFRM), 494
    aluminum reinforced, 496
    magnesium reinforced, 498
    nickel reinforced, 497
    preparation of, 495
Carbon nanotubes, 433
    electrical resistivity, 434
    magnetoresistance, 434
    quantum transport, 437

Carbon-polymer composites (CRFP), 469
    effect of carbon fiber form, 476
    effect of resin matrix, 478
    fabrication methods, 469
    hybrid composites, 481
Carbon vapor deposition, 234
Casting, metal matrix composites, 495
Ceramic matrix composites, 203
Chemical vapor deposition, 197, 206, 488
Chemisorption, 173, 235, 255
Coatings of carbon fibers, 202
    boron-based coatings, 203
    metal, 207
    silicon and alumina-based coatings, 206
    zircon based coatings, 207
Coefficient of thermal expansion, 335, 501
Compressive properties, 338
mechanisms of compressive failure, 348
    stresses, matrix shrinkage, 340
    test methods, 338
Contact angle, 181, 189, 257, 251, 270
Copolymers, 18, 358
Covalent bonding, 259, 361
Creep, temperature, 338
Cyclization, 17–18, 20, 22, 25, 57

Degradation of rayon, 34
    in flame retardants, 36
    in inert atmosphere, 33
    in reactive atmosphere, 34
Density, 66, 141, 241, 360
Diameter of fibers/filaments, 16, 169, 263, 339, 435,
Diffuse reflectance IR, 272
Disclinations, 99, 123, 134, 147, 389
Dispersive free energy, 244
Dormant mesophase, 132
Dry spinning, 9

Subject Index

Electrical resistivity, 376
   electron-phonon interaction, 385
   magnetoresistance, 386
Electrical properties, 441
Electrochemical oxidation, 177
Electrodeposition, 323, 495
Electro-initiated polymerization, 194
Electroless deposition, 495
Electron diffraction, 86
Electron energy loss spectroscopy
   (EELS), 123
Electron phonon interaction, 385
Electron spectroscopy for chemical
   analysis (ESCA), 123
Electron spin resonance (ESR)
   techniques, 141, 189
Electroplating, 495
Elemental analysis, 23, 189
Elongation-to-break (see also Tensile
   properties), 311
Etched fibers, 335
Etching, 95, 335
   electrochemical, 177
   plasma treatment and, 180
Exfoliation of microfilaments, 119

Fiber anisotropy, 348
Fiber diameter, (*see* Diameter of
   fibers/filaments)
Fiber formation
   rayon-based fibers, 31
   pitch-based, 41
   VGF, 67
Fiber-matrix shear strength, 213
   interfacial shear strength, 213
   test methods (*see also* Interlaminar
   shear strength), 213
Filament texture,
   PAN fibers, 114
   pitch-based, 137
Fine structure, 326, 352
   basic structural units, 103
   microfibrils, 118
   skin-core texture, 112
Flaws, 114, 132, 210, 323,

Fractionated pitch-based fibers, 147
Fragmentation test, 179, 184, 324
FTIR photoacoustic spectroscopy, 271
Functional groups, 121, 161, 173, 178,
   194, 221, 247, 253, 279, 360

Gauge length, 321, 324–326, 343
Glass transition temperature, 211, 340,
   349
Grain boundaries, 114, 144, 389, 442
Graphite intercalation compounds
   (GIC), 373, 385, 405
Graphitic oxide, 173, 177
Graphitic structure, 39, 389, 445, 453
Graphitization, 33, 39, 55, 87, 99, 139,
   233, 388
   PAN-based fiber, 29
   rayon-based fiber, 39
   pitch-based fiber, 61

Heat treatment,
   and coatings, 203
   electroless nickel-coated fibers, 497
   rayon precursor fibers, 31
Heteroatom, 59
High-modulus carbon fiber,
   PAN-based, 12
   pitch-based, 43
   rayon-based, 32, 39
High-temperature processing,
   acrylic-based fibers, 26
   carbon-carbon composite, 492
   pitch-based fibers, 59, 61
   rayon-based fibers, 39
Hot compressing, 471
Hot processing, 495
Hummer's reagent, 173, 176
Hybrids, 466, 482, 484, 519

IFSS (*see* Interfacial shear strength)
ILSS (*see* Interlaminar shear strength)
Image analysis, 97, 129, 134, 141
Impact strength, 195, 475
Infiltration, 203, 488, 495, 498
Infrared spectroscopy, 271

570 *Subject Index*

Interfacial shear strength, 184, 189, 213, 220, 249, 267
Interlaminar shear strength, 161, 166, 168, 173, 213, 219, 221, 259, 479, 493
Interphase, 196, 210–212, 346, 357, 451
Inverse gas chromatography, 244
Isotropic pitches, 45, 126, 128, 147

Kelly's theory, 400
Kink band formation, 344
Kinking, 348

Ladder polymer, 20, 22, 26, 101
Lamellar models, 110
Laser optical diffraction, 97
Linz-Donawitz converter gas (LDG), 197
Liquid pulse injection (LPI) technique, 65, 200
Local molecular orientation (LMO), 92
Low temperature heat treatment, rayon-based fibers, 33

Magnetoresistance, 386
Matrices,
 adhesion mechanism, 214
Matrix crack, 211
Matrix deformation, 210
Matrix interactions,
 carbon-carbon composites, 202, 491, 513
 metal composites, 495
 polymer composites, 469
Matrix morphology, carbon-carbon composites, 486
Matrix properties and shear strength, 218–221
Mechanical properties, 15, 20, 39, 61, 123, 140, 165, 182, 196, 213, 220, 262, 355, 448, 478, 496
Melt spinning, 10, 41, 51, 53, 57
Mesophase dormant, 43, 49
Mesophase pitch, 43, 48, 56, 60, 131, 136, 194, 237, 283, 328,

Metal matrix composites, 494
Microbuckling, 348–349, 353
Microetching, plasma treatment, 180
Microfibrils, 349
Microvoid content, 349, 352
Modulus, 38, 49, 140, 172, 330
 Young's, 328, 330
 torsional, 330, 348, 352
Mott formula, 405

Near-field scanning microscopies, 98
Neomesophase, 43
Neutron spectrometry, 402
Nuclear magnetic resonance (NMR), 125

Onion-skin structure, pitch-based fibers, 51
Onion-radial transverse structure, 52
Open-wedge radial texture, 137, 138
Optical microscopy, 85
Orientation of fibers, 356
Oriented core texture, pitch-based fibers, 141
Oxidation, 161–170,
 anodic, 177
 catholic, 177
 during stabilization step, 17, 57
 in gases, 161
 in nitric acid solution, 170
 in oxygen or air, 163
 in solutions, 176

PAN-based carbon fibers, 2
 carbonization of PAN fibers, 26
 HM PAN fibers, 114
 HTS PAN fibers, 110
 precursor of PAN fibers, 4, 12
 stabilization of PAN fibers, 17–25
PAN fibers adhesion in composites, 213
Photoelectron spectroscopy, 173, 257, 276, 529
Piezoresistance, 347
Pitches, 42–47, 129–136, 491
 anisotropic mesophase, 46

Subject Index

[Pitches]
    coal tar, 41
    isotropic, 45
    petroleum, 41
Pitch-based carbon fibers, 41
    anisotropic pitches, 46
    Brooks and Taylor mesophase spheres, 42
    fractionation process, 135
    gas-sparge pitches, 129
    heterogeneity, 135
    mesophase pitches, 128
    molecular weight spectra, 129
    precursors, 44–46
    spinning, 50, 136
    stabilization, 57
Plasma treatment, 181–188
    radio-frequency (RF) plasma, 188
    organic compounds plasma, 188
    oxygen-containing groups, 181
Poisson's ratio, 336, 340
Polar component, 181, 245, 360
Polymerization, 6, 34, 50, 101, 188, 261
Polymer coating, 189, 194
Pores, 26, 44, 115, 203, 237, 355
Precursors, 20, 31, 66, 355
    cellulosic, 34
    coal, 41
    heterocyclic polymers, 44
    phenolic, 2
    pitch, 44
    polyacrylonitrile (PAN), 4
    rayon, 34
Preferred orientation, 20, 37, 57, 87, 133, 237, 333, 492
Preoxidation, 525
Pristine carbon fibers, 372, 425, 440, 453
Processing (*see also* High-temperature processing),
    carbon-carbon composites, 486
    metal composites, 494
Pyrolytic coating, 33, 46, 65, 197

Quinolin insoluble (QI), 44, 46, 125, 125, 131

Radial orientation, 51
Radial-type MP-based fibers, 47
Raman spectroscopy, 87, 95, 143
Random structure, pitch-based fibers, 52
Rayon-based carbon fibers, 31
Resistivity-temperature relationship, 391
Resistivity, electrical, 376
    magnetoresistance, 386
    phonon, 385
    zero-field resistivity, 381

Scanning Auger microprobe (SAM), 289
Scanning electron microscopy (SEM), 51, 87, 96, 137, 183, 327
Scanning tunneling microscopy (STM), 99, 119, 1480, 240, 281–288
Secondary ion mass spectroscopy (SIMS), 292–296
    time-of-flight secondary ion mass spectroscopy (TOF SIMS), 294
Seebeck coefficient, 403
Selected area diffraction (SAD), 86, 97, 110, 144
Shear bands, 350
Shear modulus, 332, 348, 399, 454
Shear rate, 50, 53
Shear strength, 178–180, 217–222
Sheet structure, 110–112
Short-beam shear test, 214
Shrinkage, 20, 340, 491
    acrylic-based fiber carbonization, 20, 22
    carbon-carbon composites, 491–492
Simple two band (STB) model, 380, 435
SIMS (*see* Secondary ion mass spectroscopy)
Sinclair loop test, 338–339
Single-fiber fragmentation tests, 324
Single-filament compression tests, 347

Sizing, 210–213
Skin-core texture, 112
Softening point, 41, 44, 48, 57, 125, 136
Spinning, 8–9, 50
   acrylic-based fibers, 8
   dry–jet wet spinning, 8
   dry spinning, 9
   melt spinning, 9
   pitch-based fibers, 50
   wet spinning, 8
Squeeze casting, 495
Stabilization of fibers, 17, 57
Strength,
   compressive, 338
   flaws, 237
   gauge length, 324
   precursor properties, 12
   process conditions, 354
   torsional, 348
STM (*see* Scanning tunneling microscopy)
Strain, 32, 145, 311, 324, 330
Stress-strain curves, 330, 343
Stretching, and fiber orientation, 10
Structure,
   carbon-carbon composite matrix, 484
   electroless nickel-coated fibers, 495
   micro-, 114, 142
   PAN-based fibers, 99, 108
   pitch-based fibers, 125
   radial, 47
   random, 52
Structure-property relationships,
   compressive failure mechanisms, 348
   compressive property test methods, 338
   gauge length variations, 324
   high temperature creep, 338
   nonlinear elasticity, 330
   thermal and electrical properties, 393
Surface area, 164–171, 174–176, 232–236

Surface energy, 183, 244–248, 251
   dispersive component, 251
   polar component, 244–248
   contact angle measurment,
   two-liquid method (TLM),
Surface functional groups, 215, 253
   carbonyl, 176
   carboxyl, 179, 253, 258
Surface-enhanced Raman scattering, 93
Surface modification,
   anodic oxidation, 177–179
   coatings, 202
   plasma treatment, 180
   preferential adsorption, 262
   wet and dry oxidation, 163–
Surface properties,
   microroughness, 232–237
Surface treatment of carbon fibers, 161

Temperature (*see also* High-temperature heat treatment),
   anodic oxidation, 177
   carbon-carbon composites,
   coefficient of thermal metal composite fabrication, 495
   pitch-based fibers spinning, 50
   thermal and electrical properties, 371
Tensile properties (*see also* Mechanical properties),
   acrylic-based fibers heat
   coated fibers, 202
   gauge length, 324
   mechanisms of tensile
   test methods, 311
   variations in, 321
Tension, 17, 20, 26, 37, 46, 241, 340–343
Texture (*see also* Fine structure), 39, 56, 95, 112–120, 132–138 171, 286
Thermal conductivity, 392, 445
   electronic thermal
   lattice thermal conductivity,
Thermal expansion coefficient, 335–336, 340

Subject Index

Thermal shrinkage stress, 340–341
Thermoelectric power (TEP), 403–412
    diffusion thermoelectric power, 372, 404
    phonon drag thermoelectric power, 404
Thermoplastic resins, 476, 517
Thermoset plastics, 476–481
Torsional modulus, 348
Transmission electron microscopy (TEM), 95
Transverse compression modulus (TCM), 319
Transverse compression strength (TCM), 319
Turbostratic, 86–90

Ultrasonic spectroscopy, 297
Union Carbide fiber, 141
Universal conductance fluctuations (UCF), 438

Vapor grown carbon fibers (VGCFs), 65, 67–68, 71, 197–202, 369
    seeding catalysts method, 197
    fluidizing catalysts method, 197
Vermicular fibers, 66
Vickers microhardness, 128
Viscosity, 48–50, 128, 358, 470, 474
Void content, 239, 351

Water contact angles, 270
Weaving process, 466–468
Wedge disinclinations, 137
Wedge-type MP-based fibers, 138
Weibull parameters, 321
Wet oxidation, 217, 221
Wetting/wettability, 251
Whiskerization, 162, 192
Wide-angle X-ray diffraction spectroscopy, 238
Wide-angle X-ray scattering (WAXS), 86

X-ray diffraction, 86
X-ray-excited Auger electron spectroscopy (XAES), 289
X-ray photoelectron spectroscopy (XPS), 189, 278–280

Young's modulus, 9, 12, 29, 39, 70, 104, 142, 149, 237, 328, 332–334, 464